Springer Optimization and Its Applications

Volume 121

Aims and Scope

Optimization has been expanding in all directions at an astonishing rate during the last few decades. New algorithmic and theoretical techniques have been developed, the diffusion into other disciplines has proceeded at a rapid pace, and our knowledge of all aspects of the field has grown even more profound. At the same time, one of the most striking trends in optimization is the constantly increasing emphasis on the interdisciplinary nature of the field. Optimization has been a basic tool in all areas of applied mathematics, engineering, medicine, economics and other sciences.

The series *Springer Optimization and Its Applications* aims to publish state-of-the-art expository works (monographs, contributed volumes, textbooks) that focus on algorithms for solving optimization problems and also study applications involving such problems. Some of the topics covered include nonlinear optimization (convex and nonconvex), network flow problems, stochastic optimization, optimal control, discrete optimization, multi-objective programming, description of software packages, approximation techniques and heuristic approaches.

More information about this series at http://www.springer.com/series/7393

Neculai Andrei

Continuous Nonlinear Optimization for Engineering Applications in GAMS Technology

Springer

Neculai Andrei
Research Institute for Informatics
Center for Advanced Modeling & Optimization
Bucharest, Romania

Academy of Romanian Scientists
Bucharest, Romania

ISSN 1931-6828 ISSN 1931-6836 (electronic)
Springer Optimization and Its Applications
ISBN 978-3-319-58355-6 ISBN 978-3-319-58356-3 (eBook)
DOI 10.1007/978-3-319-58356-3

Library of Congress Control Number: 2017941512

Mathematics Subject Classification (2010): 65K05, 90C30, 90C51

Printed on acid-free paper

This Springer imprint is published by Springer Nature
The registered company is Springer International Publishing AG
The registered company address is: Gewerbestrasse 11, 6330 Cham, Switzerland

Preface

This book is for all those interested in using the GAMS technology for modeling and solving complex, large-scale, continuous nonlinear optimization problems or applications. Mainly, it is a continuation of the book *Nonlinear Optimization Applications Using the GAMS Technology*,[1] where a number of 82 nonlinear optimization applications are presented in the GAMS language. *The purpose of this book is to present the theoretical details and computational performances of the algorithms for solving continuous nonlinear optimization applications integrated in GAMS: MINOS, KNITRO, CONOPT, SNOPT, and IPOPT.*

The idea is to give a comprehensive description of the theoretical aspects and details on computational numerical linear algebra of the above algorithms working under GAMS specifications. The motivation and the significance of this enterprise are to stimulate the reader to understand and to use the most important capabilities and options of the optimization algorithms working in GAMS, for solving nonlinear complex optimization applications.

We consider only the local solutions of large-scale, complex, continuous nonlinear optimization applications. Other types of optimization problems like: mixed-integer programming, mixed complementarity, equilibrium constraints, network optimization, nonsmooth optimization, stochastic programming, and global solution of nonlinear optimization applications, operational in GAMS technology, are beyond the scope of this book.

The content of the book is organized in 21 chapters. Chapter 1 has an introductory character by presenting the variety of constrained nonlinear optimization methods, their theoretical support, and the structure of the book. In Chapter 2 we point out some key aspects of the mathematical modeling process in the context of mathematical modeling technologies based on algebraically oriented modeling

[1]N. Andrei, *Nonlinear Optimization Applications Using the GAMS Technology*, Springer Optimization and Its Applications 81, ISBN 978–1–4614-6796-0, doi:10.1007/978–1–4614-6797-7, © Springer Science+Business Media New York, 2013.

languages. Chapter 3 is a gentile introduction to the main aspects of the GAMS technology subject to modeling and solving continuous nonlinear optimization applications. GAMS is an algebraically oriented language that allows for a high-level algebraic representation of mathematical optimization models, a compiler that is responsible for user interactions by compiling and executing user commands given in GAMS source file which also has a set of solvers. In Chapter 4 a number of 18 real nonlinear optimization applications, both in algebraic and GAMS representations, are detailed. These optimization applications are a basis for seeing the performances of the algorithms described in this book. The rest of the chapters present on the one hand the relevant theoretical and computational results on some methods and techniques which proved to be the most effective and robust for solving nonlinear optimization problems and on the other hand detail on the algorithms MINOS, KNITRO, CONOPT, SNOPT, and IPOPT which work in the GAMS technology. Accordingly, in a professional and unique manner, the book emphasizes large-scale optimization techniques, such as penalty and augmented Lagrangian functions, linearly constrained augmented Lagrangian methods, sequential linear and sequential quadratic programming, and interior point methods, all equipped with line-search and trust-region globalization schemes. It is worth mentioning that not only the algorithms working under the GAMS specifications (MINOS, KNITRO, CONOPT, SNOPT, and IPOPT) are described in this book. To illustrate the performances of the methods integrated in the GAMS technology for solving nonlinear optimization problems and for making comparisons, we give some details for some other optimization algorithms based on penalty-barrier methods (SPENBAR), a SQP method using only equality-constrained sub-problems (DONLP), a sequential quadratic programming algorithm with successive error restoration (NLPQLP), and filter methods with sequential linear programming (filterSD) or with sequential quadratic programming (filterSQP), which are not integrated in GAMS. The purpose of including these algorithms is to emphasize the importance of combining or integrating different optimization techniques, globalization strategies and mechanisms, as well as advanced linear algebraic procedures in order to get efficient and robust nonlinear optimization algorithms.

A special attention is given to the Karush-Kuhn-Tucker (KKT) optimality conditions which characterize the local optimal solutions of nonlinear optimization problems. These necessary and sufficient optimality conditions are described in Chapter 5. Since the simple bound optimization problems have a very important role in the economy of nonlinear optimization, in Chapter 6, we present some of the main algorithms for solving this class of problems. In Chapter 7 the theory behind two very important concepts in nonlinear optimization, the penalty and the augmented Lagrangian, is introduced. These concepts are implemented in different forms and combinations in the frame of line search or trust region, leading thus to very efficient and robust algorithms.

Chapter 8 is dedicated to the SPENBAR algorithm based on a combination of the augmented Lagrangian with log-barrier function. The penalty parameters included in this augmented Lagrangian function are updated in such a way as to obtain a

KKT point for the optimization problem. A different and more sophisticated modification of the augmented Lagrangian is described in Chapter 9, where MINOS, one of the most respectable nonlinear programming algorithms, is described. This algorithm, integrated in GAMS, is based on the linearization of the constraints. In MINOS the modified augmented Lagrangian includes a penalty term based on the departure from the linearity of constraints. The modified augmented Lagrangian sub-problems are solved by using a reduced-gradient algorithm along with a quasi-Newton algorithm.

Chapter 10 is dedicated to quadratic programming. Both the equality-constrained quadratic programming and the inequality-constrained quadratic programming are shown, emphasizing the primal-dual active-set method by Goldfarb and Idnani. In Chapter 11 both the equality-constrained sequential quadratic programming and the inequality-constrained sequential quadratic programming with line search or trust region are introduced. Special attention is given to the sequential linear-quadratic programming method, which is one of the main ingredients in some nonlinear-constrained optimization algorithms.

A sequential quadratic programming method using only equality-constrained sub-problems – DONLP – is described in Chap. 12. This is an active-set sequential quadratic programming algorithm. The idea of this algorithm is to consider only the "near-active" or violated inequalities in the quadratic programming sub-problem and to treat them as equalities. Another sequential quadratic programming algorithm with successive error restoration – NLPQLP – is presented in Chapter 13. This is also an active-set method, as it provides an estimate of the active set at every iteration. It uses a quasi-Newton approximation to the Hessian of the Lagrangian, which is updated with the BFGS formula. To calculate the stepsize that minimizes an augmented Lagrangian merit function, a nonmonotone line search is used. Both the DONLP and the NLPQLP show the diversity of the active-set sequential quadratic programming methods. These are *pure* sequential quadratic programming methods.

In the rest of the chapters, we present only the algorithms integrated in the GAMS technology. In Chapter 14 the active-set sequential linear-quadratic programming – KNITRO/ACTIVE – is described. The interior point sequential linear-quadratic programming – KNITRO/INTERIOR – is presented in Chapter 19. Both these approaches are integrated in KNITRO, one of the most elaborated algorithms for solving general large-scale nonlinear optimization applications. KNITRO/ACTIVE is based on an active-set method based on sequential linear-quadratic programming using the projected conjugate gradient iteration. On the other hand, KNITRO/INTERIOR is based on the interior point methods in two implementations: KNITRO/INTERIOR-CG, in which the algorithmic step is computed by means of an iterative conjugate gradient method, and KNITRO/INTERIOR-DIRECT, where the step is computed by a direct factorization of the corresponding linear system associated to the interior point method. These two approaches communicate by the *crossover technique* used for the first time in linear programming by Megiddo.

Chapter 15 presents a pure sequential quadratic programming algorithm for large-scale nonlinear-constrained optimization – SNOPT. SNOPT implements a particular sequential quadratic programming that exploits sparsity in the constraint Jacobian and maintains a limited-memory quasi-Newton approximation to the Hessian of the Lagrangian function. Even if this algorithm contains a lot of very sophisticated ingredients, part of them taken from MINOS, its computational performances are modest, being superseded by other algorithms integrated in the GAMS technology, like KNITRO, CONOPT, and IPOPT.

A generalized reduced-gradient algorithm with sequential linearization – CONOPT – is explained in Chapter 16. This line search algorithm illustrates the importance of combining the generalized reduced-gradient method with sequential linear programming or with sequential quadratic programming. CONOPT integrates three active-set methods. The first one is a gradient projection method in the frame of generalized reduced-gradient method that projects the gradient of the objective function onto a linearization of the constraints. The second one is a sequential linear programming algorithm, while the third is a sequential quadratic programming algorithm. CONOPT includes algorithmic switches that automatically detect which method is the best.

Interior point and filter methods are the content of Chaps. 17 and 18, respectively. For the interior point method, a prototype is presented. Both the line-search and the trust-region interior point algorithms are discussed. At the same time, a variant of the line-search interior point algorithm is described in details, where a methodology emphasizing the development and the convergence of the interior point algorithms is given. The filter methods, which is a technique for the globalization of the nonlinear optimization algorithms, aim at avoiding the need to choose penalty parameters in penalty or augmented Lagrangian functions. Two filter methods are described: the sequential linear programming filter algorithm and the sequential quadratic programming one. These chapters represent the fundamentals for some algorithms for nonlinear optimization, which are much elaborated and which combine different strategies based on sequential linear or sequential quadratic programming or on filter line search, in the frame of interior point methods. All these are presented as KNITRO/INTERIOR in Chapter 19 and as IPOPT in Chapter 20. The last chapter contains some numerical studies and comparisons among the algorithms integrated in the GAMS technology.

The conclusion is that there is no algorithm able to solve any nonlinear-constrained optimization problem or application. The most powerful nonlinear optimization algorithms combine and integrate different optimization techniques (modified augmented Lagrangian, sequential linear or quadratic programming, interior point methods with filter line search or trust region) and include advanced computational linear algebra techniques.

The book is of great interest to all those using the GAMS technology for modeling and solving complex, large-scale, continuous nonlinear optimization problems or applications from different areas of activity. Mathematical programming researchers, theoreticians and practitioners in operations research, management consultants, practitioners in engineering, and basic science and industry researchers will find plenty of information and practical hints at using in an efficient and robust way the continuous nonlinear optimization software integrated in the GAMS technology. The book is also suitable for scientists working in the optimization theory, optimization algorithms design, and computational developments, as well as for college and graduate students in mathematics, science, operations research, and business, who are interested in advanced optimization algorithms and computer optimization software.

I am grateful to the *Alexander von Humboldt Foundation* for its appreciation and financial support during the 2+ years in different universities from Germany. My acknowledgments are also due to Elizabeth Loew and Razia Amzad and their colleagues at Springer, for their encouragement and superb assistance with the preparation of this book. Finally, fondest thanks go to my wife, Mihaela, who has constantly understood, helped, and supported me along the years.

Bucharest and Tohăniţa/Bran Resort
Bucharest, Romania
November 2016

Neculai Andrei

Contents

List of Figures

List of Tables

List of Algorithms

Chapter 1
Introduction

This book is on nonlinear optimization in the GAMS technology. Continuous nonlinear optimization problems have a simple mathematical model and always refer to a system its running we want to optimize. Firstly, it contains an *objective function* which measures the performances or requirements of the system. Often, this function represents a profit, a time interval, a level, a sort of energy, or combination of different quantities which have a physical significance for the modeler. The objective function depends on some characteristics of the system, called *variables* or *unknowns*. The purpose of any optimization problem is to find the values of these variables that minimize (or maximize) the objective function, subject to some *constraints* the variables must satisfy. Constraints of an optimization problem may have different algebraic expressions. There are static and dynamic constraints called *functional constraints*. The difference between these types of constraints comes from the structure of their Jacobian. Another very important type of constraints is the *simple bounds on variables*. Both the objective function and the constraints may depend on some *parameters* with known values which represent the constructive characteristics of the system under optimization. The process of identifying the variables, parameters, the objective functions, and constraints is known as *modeling*, one of the finest intellectual activities. It is worth saying that in this book, we assume that the variables can take real values and the objective function and the constraints are smooth enough (at least twice differentiable) with known first-order derivatives. When the number of variables and the number of constraints are large, the optimization problem is quite challenging.

If the objective function and the constraints are all linear, then the problem is known as *linear programming*, (see Dantzig, 1963; Vanderbei, 2001; Andrei 2011a) which represents a special chapter in the mathematical programming area, and we do not consider it in this book. If the objective function is a quadratic function and the constraints are all linear, we have a *quadratic programming* problem with plenty of theoretical and computational results and applications (see Chapter 10 of this book). There are optimization problems with only

N. Andrei, *Continuous Nonlinear Optimization for Engineering Applications in GAMS Technology*, Springer Optimization and Its Applications 121,
DOI 10.1007/978-3-319-58356-3_1

simple bounds on variables (see Chapter 6 of this book). The *general nonlinear optimization* problems include both functional constraints and simple bounds on variables.

Suppose one has an algorithm or a collection of algorithms able to solve general nonlinear optimization problems. For recognizing whether the algorithm found the optimal solution of the problem, one needs some mathematical tools. The most elegant and effective tool is the so-called *optimality conditions*. These are a nonlinear algebraic system of equations which may be used both to design nonlinear optimization algorithms and to check whether the current set of variables is indeed the solution of the problem. There are two types of optimality conditions. The *necessary optimality conditions* are conditions that must be satisfied by any solution point. The *sufficient optimality conditions* are those that, if satisfied at a certain point, guarantee that this point is in fact a solution. Observe the nuances of these two concepts (see Chapter 5 of this book). The necessary conditions are also called the *first-order conditions* because they involve the properties of both the gradients of the objective function and the constraints and also how these gradients are related to each other at a solution. In fact, these conditions are called KKT (Karush-Kuhn-Tucker) conditions and are the foundation of many algorithms for nonlinear optimization. The *second-order conditions*, both necessary and sufficient, use the second derivatives of the functions defining the problem and examine these derivatives to see whether this extra information can resolve the issue of increase or decrease in objective function values in undecided directions.

If the optimality conditions are not satisfied at a current point, then they must be used to see how the current solution estimation can be improved. This is the subject of *sensitivity analysis*, which shows the sensitivity of the current solution to changes in the structure and the data of the model of the optimization problem.

1.1 Nonlinear Optimization Modeling

A short glimpse at the optimization domain shows a substantial progress of the development of algorithms and computer codes to solve real large-scale nonlinear optimization applications. From the application viewpoint, one can see that a large part of time required for developing, improving, and maintaining an optimization model is spent with data preparation and transformation, as well as with report preparation and documentation. A nonlinear optimization model involves time and intellectual efforts to organize the data and write computer programs that transform the data into a form required by the optimization codes. A typical example is MPS representation of a linear programming problem. In nonlinear optimization the situation is more complex because the user must prepare the function defining the problem and their derivatives in a form admitted by a particular optimizer. Of course, the user must also transmit to the optimizer the simple bounds on variables, the initial point, and some other parameters associated to a particular optimizer. It is worth saying that such an approach of using an optimizer is only accessible to the specialist who wrote the optimizer and not to the wide audience of analysts in charge of some projects involving the optimizer. For example, it must be emphasized that the optimizers not imbedded in the GAMS technology discussed in this book (SPENBAR, DONLP, NLPQLP, filterSD or filterSQP) have very different (and difficult) communication facilities subject to transmitting the functions of the problem and their derivatives, simple bounds on variables, initial point, some other parameters, etc.

Currently, the advanced industrial usage of optimization algorithms involves the *algebraic oriented languages* that offer advanced capabilities for expressing an optimization model in a form immediately accepted by an optimizer. The most important modeling technologies with optimization facilities are GAMS and AMPL. In this book the GAMS technology was chosen to illustrate its possibilities and performances subject to both modeling and optimization facilities. This is the reason the GAMS technology is presented in Chapter 3.

1.2 Constrained Nonlinear Optimization Methods: A Critical Review

Consider a particular form of a nonlinear optimization problem:

$$\begin{aligned} &\min f(x) \\ \text{subject to}: \quad & \\ &c(x) = 0, \ \ x \geq 0, \end{aligned} \tag{1.1}$$

where $f: \mathbb{R}^n \to \mathbb{R}$ is the objective function and $c: \mathbb{R}^n \to \mathbb{R}^m$ is the constraint assumed to be twice continuously differentiable. In problem (1.1) we note the *multipliers* corresponding to the equality constraint by y and the multipliers of the inequality constraints by $z \geq 0$. Observe that (1.1) may have inequality constraints, simple bounds on variables, or even range constraints of the form $l_i \leq c_i(x) \leq u_i$, $i = 1, \ldots, m$, which are omitted here for simplifying the presentation (Leyffer & Mahajan, 2010). For solving (1.1), iterative methods are used in a reductionist approach. The problem (1.1) is approximated by a sequence of sub-problems. Their solving gives a sequence of solutions $\{x_k\}$, starting from an initial point x_0. The solution of a sub-problem is the initial guess for the next sub-problem in this sequence, and all this process has a number of the so-called *major iterations*. In this frame, every sub-problem is also solved by an iterative process through the so-called *minor iterations*. A simple prototype for solving (1.1) to generate the sequence $\{x_k\}$ is as follows:

Algorithm 1.1 Prototype for nonlinear optimization algorithm

1.	Choose an initial estimate x_0 and set $k=0$
2.	Evaluate a criterion for stopping the iterations
3.	At the current point x_k, elaborate a local model of (1.1)
4.	Approximately solve the local model (sub-problem) to determine an improved estimation x_{k+1} of the solution of (1.1)
5.	Set $k=k+1$ and continue with step 2 ♦

In the following, the main components of Algorithm 1.1 for solving the problem (1.1) are discussed in a *critical manner*. The real sense of "criticism" used here does not mean "destruction" but "creation." Consequently, some questions may arise: how the *approximate sub-problem* that determines an improved new iterate is computed, the *globalization strategy* that ensures convergence of the algorithm initialized in remote starting points by indicating whether a new estimate of the solution is better than the current estimate, the *refining mechanisms* that reduce the step computed by the local model to enforce the globalization strategy to work far away from the solution, and the *convergence tests* as well as *termination conditions* that check for the optimal solution or for the failure of the algorithm. *The nonlinear optimization algorithms are classified according to the implementation of each of these fundamental components of the Algorithm 1.1.*

1.2.1 Convergence Tests

Usually, the convergence tests are based on the Karush-Kuhn-Tucker conditions (see Theorem 5.15). For example, for problem (1.1) a suitable approximate convergence tests may be presented as

$$\|c(x_k)\| \leq \varepsilon_1, \tag{1.2a}$$

$$\|\nabla f(x_k) - \nabla c(x_k)y_k - z_k\| \leq \varepsilon_2, \tag{1.2b}$$

$$\|\min\{x_k, z_k\}\| \leq \varepsilon_3, \tag{1.2c}$$

where $\varepsilon_i > 0$, $i = 1, 2, 3$, are tolerances, and the min in (1.2c) corresponding to the complementarity condition is taken componentwise. (Here $\nabla f(x_k)$ is the gradient of f and $\nabla c(x_k)$ is the Jacobian of c computed at point x_k). It is quite possible that the convergence of the algorithm is not ensured to an approximate KKT point (see Theorem 5.15), for example, if the constraints do not satisfy a constraint qualification (see Remark 5.2). In these cases the test (1.2b) is replaced by

$$\|\nabla c(x_k)y_k + z_k\| \leq \varepsilon_2, \tag{1.3}$$

corresponding to a Fritz John point. It is also possible for the termination conditions to be modified to include the "relative error," as used in some algorithms.

The limits of these convergence tests are that the feasibility conditions and those of optimality are separated, so that we may obtain a feasible solution which is not optimal and vice versa. *However, the feasibility of the constraints is always the most important criterion for termination of iteration.*

1.2.2 Infeasible Points

As already seen in Algorithm 1.1, one needs an initial point x_0 to start the iterations. Computing of an initial point is in itself a problem. Unless the problem is convex or some restrictive assumptions are made, Algorithm 1.1 cannot guarantee convergence even to a feasible point. Besides, there are many cases in which the points of the sequence $\{x_k\}$ determined by Algorithm 1.1 must be feasible. In these cases, a procedure for certifying the feasibility is needed. An appropriate converge test and feasibility sub-problem are based on the following feasibility problem:

$$\min_{x \geq 0} \|c(x)\|, \tag{1.4}$$

which, as can be seen, may be formulated as a smooth optimization problem by introducing slack variables. For solving the problem (1.4), one can use Algorithm 1.1 because the feasibility problem can be reformulated as a smooth nonlinear optimization problem by introducing additional variables. The objective function of (1.4) can be replaced by a weighted norm. In this case suitable tests are.

$$\|\nabla c(x_k)y_k - z_k\| \leq \varepsilon \text{ and } \|\min(x_k, z_k)\| \leq \varepsilon,$$

where y_k are the multipliers or weights corresponding to the norm used in the objective of (1.4).

1.2.3 Approximate Sub-problem: Local Models and Their Solving

In step 3 of Algorithm 1.1, the elaboration of a local model of the problem (1.1) around the current point x_k is needed. *The key difference among the nonlinear optimization algorithms is how the local model is constructed around the current point.* The idea is to generate a local model of the problem for which a solution can be easily obtained and which can improve the current iterate. Three broad classes of local models are known: *sequential linear models, sequential quadratic models, and interior point models.*

Sequential linear programming (SLP) methods construct a linear approximation to (1.1). Usually, these linear programming programs are unbounded. Therefore, these methods require an additional trust-region constraint:

$$\min_d m_k(d) \triangleq \nabla f(x_k)^T d$$

subject to: (1.5)

$$c(x_k) + \nabla c(x_k)^T d = 0,$$
$$x_k + d \geq 0,$$
$$\|d\|_\infty \leq \Delta_k,$$

where $\Delta_k > 0$ is the *radius of the trust-region*. To ensure the convergence, in general, $\Delta_k \to 0$ must converge to zero. This method was used by Griffith and Stewart (1961) with some success, without introducing a trust-region constraint. The sequential linear programming method can be considered as steepest descent methods and typically converge only linearly. *If at the solution point there are exactly n active constraints and the normals of these constraints are linearly independent, then the sequential linear programming method reduces to the Newton method for solving a square system of nonlinear equations, which is superlinear convergent.*

Sequential quadratic programming (SQP) methods minimize a quadratic model $m_k(d)$ subject to a linearization of the constraints about the current point x_k:

$$\min_d m_k(d) \triangleq \nabla f(x_k)^T d + \frac{1}{2} d^T H_k d$$

subject to: (1.6)

$$c(x_k) + \nabla c(x_k)^T d = 0,$$
$$x_k + d \geq 0,$$

where $H_k \triangleq \nabla^2_{xx} L(x_k, \lambda_k)$ is the Hessian of the Lagrange function associated with problem (1.1) and λ_k are the Lagrange multipliers, to obtain the search direction $d \triangleq x - x_k$. The new iterate is given by $x_{k+1} = x_k + d$ together with the Lagrange multipliers λ_{k+1} associated with the linearized constraints of (1.6). *If the matrix H_k is not positive definite on the null space of the active constraint normals, then the quadratic sub-problem* (1.6) *is nonconvex and the SQP methods determine a local minimum of* (1.6). *In case of large-scale problems, the solution of the quadratic sub-problems can become computationally expensive because the null space method for solving the quadratic sub-problem requires the factorization of a dense reduced Hessian matrix.*

Sequential linear-quadratic programming (SLQP) methods combine the advantages of the SLP method (fast solution of the linear programming sub-problem) and the SQP methods (fast local convergence) by adding an equality constraint to the SLP method (Fletcher & Sainz de la Maza, 1989; Chin & Fletcher, 2003; Byrd, Gould, Nocedal, & Waltz 2004) (see Chapter 11). In these methods, two sub-problems are solved: the *first one* is a linear programming sub-problem which gives a step for the next iteration and also an estimate of the active set $A_k \triangleq \left\{ i : [x_k]_i + \hat{d}_i = 0 \right\}$ obtained from a solution $\hat{d}$ of (1.5). (Here $[x_k]_i$ is the component i of vector x_k). This estimation of the active set is used to construct and solve the *second problem* which is an equality-constrained quadratic programming one defined on the active constraints:

$$\min_d q_k(d) \triangleq \nabla f(x_k)^{\mathrm{T}} d + \frac{1}{2} d^{\mathrm{T}} H_k d$$

subject to: (1.7)

$$c(x_k) + \nabla c(x_k)^{\mathrm{T}} d = 0,$$
$$[x_k]_i + d_i = 0, \quad i \in A_k.$$

If H_k is positive definite on the null space of the constraint normals, then the solution of (1.7) *is equivalent to the following linear algebraic system:*

$$\begin{bmatrix} H_k & -\nabla c(x_k) & -I_k \\ \nabla c(x_k)^{\mathrm{T}} & 0 & 0 \\ I_k^{\mathrm{T}} & 0 & 0 \end{bmatrix} \begin{bmatrix} x \\ y_k \\ z_k \end{bmatrix} = \begin{bmatrix} -\nabla f(x_k) + H_k x_k \\ -c(x_k) + \nabla c(x_k)^{\mathrm{T}} x_k \\ 0 \end{bmatrix}, \tag{1.8}$$

obtained by applying the KKT conditions to the sub-problem (1.7). (Here $I_k = [e_i]_{i \in A_k}$ are the normals of the active inequality constraints, and y_k and z_k are the Lagrange multipliers.) It is always possible to choose a basis from the linear programming sub-problem (by simplex algorithm) such that the augmented matrix $[A_k \; I_k]$ has full rank. *If H_k is not positive definite on the null space of the active constraint normals, then the inertia of this matrix can be corrected by adding to H_k a multiple of the identity matrix, thus ensuring the descent step of the sub-problem* (1.7).

Sequential quadratic-quadratic programming (SQQP) methods have been recently introduced by Gould and Robinson (2008, 2010). From the very beginning, they construct and solve a quadratic programming sub-problem by using a positive definite Hessian of the Lagrangian. The solution of this convex sub-problem is used into an equality-constrained quadratic programming model with exact second derivative of the Lagrangian. More precisely, the SQQP procedure is a second derivative method that is globalized via the l_1 merit function. Given an estimate x_k

of the solution to the problem (1.1), a search direction at this point is generated from the combination of three steps: a *predictor step* defined as a solution to a strictly convex quadratic programming sub-problem, a *Cauchy step* (driving convergence of the algorithm) which is computed from a special univariate global minimization problem, and a *SQP step* computed from a local solution of a special nonconvex quadratic programming sub-problem. The algorithm is imbedded into the trust-region approach.

Remark 1.1 *As a general characterization of these sequential linear and quadratic programming methods, the following theoretical aspects can be mentioned. If H_k is the exact Hessian of the Lagrange function and the Jacobian of the active constraints has full rank, then the SQP methods are quadratic convergent near a minimizer that satisfies both a constraint qualification and a second-order sufficient condition (the Hessian of the Lagrangian is positive definite on the null space of the active constraint normals)* (Boggs & Tolle, 1995). *Moreover, under the additional assumption of strict complementarity, all four methods based on sequential linear or quadratic programming identify the optimal active set in a finite number of iterations. The above sequential methods are also known as active-set methods because the solution of each linear or quadratic sub-problem provides not only a suitable new iterate but also an estimate of the active set at the solution point* (Leyffer & Mahajan, 2010).

Additionally, two major concerns are associated to the SQP methods: incompatible linearized constraints and unbounded solutions of quadratic sub-problems. For handling unbounded solutions of (1.6), *two approaches are possible. The first one is to use a positive definite approximation to the Hessian in* (1.6). *Thus, a strictly convex-bounded quadratic program is obtained with a unique minimizer. The second approach allows for a nonconvex quadratic program by explicitly bounding the solution* via *a trust-region constraint. Both approaches are effective in practice. The issue of incompatible sub-problems is more challenging. Observe that the quadratic programming sub-problem* (1.6) *may be naturally incompatible. In this case different techniques have been suggested: "constraint shifting"* (Vardi, 1985), *the use of the "elastic mode"* (Gill, Murray, & Saunders, 2005) (*see* Chapter 15), *and a "feasibility restoration phase"* (Fletcher & Leyffer, 2002) (*see* Chapters 18 and 20), *to deal with incompatible sub-problems.* ♦

Interior point methods are an alternative approach to the active-set methods. Essentially, these are *perturbed Newton methods* which postpone the determination of active constraints set at the end of the iterative process for solving the problem. Plenty of interior point methods are known, but the most successful one is the primal-dual, which can be viewed as the Newton method applied to the perturbed first-order optimality conditions of problem (1.1). The first-order optimality conditions include *stationary condition* of the Lagrange function, *feasibility condition* (satisfying the constraints of the problem), and the *complementarity condition* (see Definition 5.14). In interior point methods, just these complementarity conditions are perturbed, as in the following nonlinear algebraic system:

$$F_\mu(x,y,z)=\begin{bmatrix}\nabla f(x)-\nabla c(x)^{\mathrm{T}}y-z\\ c(x)\\ Xz-\mu e\end{bmatrix}=0, \tag{1.9}$$

where $\mu>0$ is the barrier parameter, $X=\mathrm{diag}(x)$ is a diagonal matrix with components of the vector x on the main diagonal, and e is a vector with all components equal to one. Interior point methods start the computations from an interior point x_0, $z_0>0$ and generate a sequence of interior points x_k, $z_k>0$ as approximate solutions of the system (1.9), for a decreasing sequence of barrier parameters. Applying the Newton method to the nonlinear primal-dual system (1.9) around x_k, we get the *local model* (i.e., approximate sub-problem):

$$\begin{bmatrix}H_k & -\nabla c(x_k) & -I\\ \nabla c(x_k)^{\mathrm{T}} & 0 & 0\\ Z_k & 0 & X_k\end{bmatrix}\begin{bmatrix}\Delta x\\ \Delta y\\ \Delta z\end{bmatrix}=-F_\mu(x_k,y_k,z_k), \tag{1.10}$$

where H_k is an approximate to the Hessian of the Lagrange function and $Z_k=\mathrm{diag}(z_k)$ is a diagonal matrix with components of the vector z_k on the main diagonal. ($X_k=\mathrm{diag}(x_k)$). The next step is computed as

$$\left(x_{k+1},y_{k+1},z_{k+1}\right)=(x_k,y_k,z_k)+\left(\alpha_x\Delta x,\alpha_y\Delta y,\alpha_z\Delta z\right),$$

where the stepsizes α_x, α_y, and α_z are computed to ensure that $x_{k+1},z_{k+1}>0$ remain strictly positive. A simple variant of interior point algorithms is as follows:

Algorithm 1.2 Prototype of interior point algorithm

1.	Choose an initial point estimate (x_0,y_0,z_0) such that $x_0>0$ and $z_0>0$ Choose a value for the barrier parameter μ_0, the parameter $0<\sigma<1$, and a decreasing sequence ε_k. Set $k=0$
2.	Evaluate a criterion for stopping the iterations concerning the optimality of (x_k,y_k,z_k)
3.	Set $(x_{k,0},y_{k,0},z_{k,0})=(x_k,y_k,z_k)$ and $j=0$
4.	If $\left\|F_{\mu_k}\left(x_{k,j},y_{k,j},z_{k,j}\right)\right\|\le\varepsilon_k$, then go to step 6. Otherwise, continue with step 5
5.	Approximately solve the Newton system (1.10) for a new iterate $(x_{k,j+1},y_{k,j+1},z_{k,j+1})$. Set $j=j+1$ and go to step 4
6.	Reduce the barrier parameter $\mu_{k+1}=\sigma\mu_k$, set $k=k+1$, and go to step 2 ♦

Observe that this algorithm has two loops: one responsible for solving the problem and the other one for approximately solving the Newton system. Notice that this is a general prototype interior point algorithm and many other ingredients are necessary to be introduced in order to have an efficient and robust algorithm (see Chapters 17, 19, and 20).

In this point it is very instructive to see the relationship of interior point methods to the barrier methods developed by Fiacco and McCormick (1990). The importance of the barrier methods comes from the fact that they can provide polynomial-time algorithms for linear programming problems (Wright, 1991; Forsgren, Gill, & Wright, 2002; Nemirovskii & Todd, 2008). Barrier methods approximately solve a sequence of barrier sub-problems of the following form:

$$\min_{x\in\mathbb{R}^n} f(x) - \mu\sum_{i=1}^{n}\log(x_i)$$

subject to: (1.11)

$$c(x) = 0,$$

for a decreasing sequence of barrier parameters $\mu > 0$. The first-order optimality conditions for problem (1.11) are given by:

$$\nabla f(x) - \mu X^{-1}e - \nabla c(x)y = 0, \tag{1.12a}$$

$$c(x) = 0. \tag{1.12b}$$

Now, applying the Newton method to the nonlinear system (1.12), the following linear algebraic system is obtained:

$$\begin{bmatrix} H_k + \mu X_k^{-2} & -\nabla c(x_k) \\ \nabla c(x_k)^{\mathrm{T}} & 0 \end{bmatrix}\begin{bmatrix} \Delta x \\ \Delta y \end{bmatrix} = -\begin{bmatrix} \nabla f(x_k) - \mu X_k^{-1}e - \nabla c(x_k)y_k \\ c(x_k) \end{bmatrix}.$$

Introducing the first-order multiplier estimates $Z(x_k) \triangleq \mu X_k^{-1}$, which can also be written as $Z(x_k)X_k = \mu e$, the following linear system is obtained:

$$\begin{bmatrix} H_k + Z(x_k)X_k^{-1} & -\nabla c(x_k) \\ \nabla c(x_k)^{\mathrm{T}} & 0 \end{bmatrix}\begin{bmatrix} \Delta x \\ \Delta y \end{bmatrix} = -\begin{bmatrix} \nabla f(x_k) - \mu X_k^{-1}e - \nabla c(x_k)y_k \\ c(x_k) \end{bmatrix}.$$

This system is equivalent to the primal-dual Newton system (1.10), where

$$\Delta z = -X^{-1}Z\Delta x - Ze - \mu X^{-1}e$$

has been eliminated. *Therefore, the main difference between the classical barrier methods as they were elaborated by Fiacco and McCormick and the primal-dual interior point methods is that the matrix* Z_k *is not free for barrier methods but is chosen as the primal multiplier* $Z(x_k) \triangleq \mu X_k^{-1}$. *This freedom in the primal-dual interior point methods avoids some difficulties with ill-conditioning of the Hessian to the barrier function. This was the main reason for rejecting the barrier methods in solving nonlinear optimization problems.*

Remark 1.2 *As a characterization of the interior point methods, if there exists a compact set of isolated local minimizers of problem* (1.1) *with at least one point in the closure of the strictly feasible set, then it follows that the barrier methods converge to a local minimum of problem* (1.1) (Wright, 1991). *On the other hand, for the convex nonlinear optimization problems, the interior point methods, as in the case of linear programming, lead to polynomial-time algorithms* (Nesterov & Nemirovskii, 1994). ♦

1.2.4 Globalization Strategy: Convergence from Remote Starting Points

The local models discussed above guarantee convergence only in a small neighborhood of a regular point (see Definitions 5.13 and 5.16). But, initialization of the algorithms near the unknown optimal solution is a difficult task. Therefore, the globalization strategies are concerned with ensuring convergence from remote starting points to stationary points (see Theorem 5.3). Globalization strategy should not be confused with global optimization. To ensure convergence from the remote starting points, the progress of iterates generated by the approximate sub-problems must be monitored. For unconstrained optimization, monitoring is very simple because the progress of the algorithm can be measured by comparing the values of the objective function. For constrained optimization, it is necessary to take into account the constraint violation. Three broad classes of strategies exist: *augmented Lagrangian methods*, *penalty and merit function methods*, and *filter methods*.

Augmented Lagrangian methods for problem (1.1) use the augmented Lagrangian as

$$L(x,y,\rho) = f(x) - c(x)^{\mathrm{T}}y + \frac{\rho}{2}\|c(x)\|_2^2, \tag{1.13}$$

where $\rho > 0$ is a penalty parameter. In this context, there are two possibilities to develop algorithms based on the augmented Lagrangian:

(a) *Linearly constrained Lagrangian methods*. These methods minimize a modified Lagrangian subject to a linearization of the constraints. The modified Lagrangian is defined as

$$\bar{L}(x,y,\rho) = f(x) - p_k(x)^{\mathrm{T}}y + \frac{\rho}{2}\|p_k(x)\|_2^2, \tag{1.14}$$

where $p_k(x)$ collects the higher-order nonlinear terms from Taylor series at the current iterate x_k, i.e.:

$$p_k(x) = c(x) - c(x_k) - \nabla c(x_k)^{\mathrm{T}}(x - x_k). \tag{1.15}$$

Therefore, this approach consists in forming and solving the following local model:

$$\begin{aligned} &\min_x \bar{L}(x, y_k, \rho_k) \\ &\text{subject to:} \\ &c(x_k) + \nabla c(x_k)^{\mathrm{T}}(x - x_k) = 0, \\ &x \geq 0, \end{aligned} \tag{1.16}$$

for a fixed value of the penalty parameter ρ_k. The Lagrange multipliers are updated by using a first-order multiplier update rule:

$$y_{k+1} = y_k - \rho_k c(x_{k+1}), \tag{1.17}$$

where x_{k+1} is the solution of (1.16). Therefore, the local model (1.16) is used in an iterative scheme for a decreasing sequence of penalty parameters. This approach is implemented in MINOS (see Chapter 9). Another modification of the augmented Lagrangian as a combination of penalty and barrier terms is used in SPENBAR (see Chapter 8).

(b) *Bound-constrained Lagrangian methods*. These methods approximately minimize the augmented Lagrangian subject to simple bounds on variables, for problem (1.1):

$$\begin{aligned} &\min_x L(x, y_k, \rho_k) \\ &\text{subject to:} \\ &x \geq 0, \end{aligned} \tag{1.18}$$

for a sequence of increasing penalty parameters. The advantage of this approach is that efficient methods for simple bound on variables optimization can be applied like gradient projection conjugate gradient approach by Moré and Toraldo (1991), limited memory BFGS with bounds by Byrd, Lu, Nocedal, and Zhu (1995), spectral projected gradient method by Birgin, Martínez and Raydan (2000, 2001), truncated Newton with simple bounds by Nash (1984a, 1984b, 1985), etc. (see Chapter 6). The idea of an iteration which uses the local model given by (1.18) is as follows. For a fixed value of the penalty parameter ρ_k, solve the sub-problem (1.18). Thus an approximate solution x_k^c is obtained. If $\|c(x_k^c)\| \leq \eta_k$, then update the Lagrange multipliers as $y_{k+1} = y_k - \rho_k c(x_k^c)$, otherwise increase the value of the penalty parameter like, for example, $\rho_{k+1} = 10\rho_k$. Here $\{\eta_k\}$ is a forcing sequence which controls the progress to the feasibility of the nonlinear constraints. Each minimization of (1.18) can be started from the previous iterate. Representative for bound-constrained Lagrangian method is LANCELOT (Conn, Gould, & Toint, 1992b).

Remark 1.3 *As a general characterization of the augmented Lagrangian methods, the linearly constrained augmented Lagrangian method can be made globally convergent by adding slack variables in order to handle infeasible sub-problems* (Friedlander & Saunders, 2005). *On the other hand, a bound-constrained Lagrangian method converges globally if the sequence* $\{x_k\}$ *is bounded and if the Jacobian of the constraints at all limit points of* $\{x_k\}$ *has column rank no smaller than m (the number of constraints)* (Conn, Gould, & Toint, 1991). *The difficulty with the augmented Lagrangian methods is the procedure for selecting the penalty parameters. Introducing the filter methods by* Fletcher and Leyffer (2002) *avoids this difficulty.* ♦

Penalty and merit function methods combine the objective function and a measure of the constraint violation into a single function whose local minimizers correspond to the local minimizers of problem (1.1). Convergence from remote starting points can be ensured by forcing the descent of the penalty or the merit function using one of the mechanisms discussed in the next section. The exact penalty functions are alternative to augmented Lagrangian and are defined as

$$P_\rho(x) = f(x) + \rho\|c(x)\|, \tag{1.19}$$

where $\rho > 0$ is the penalty parameter. In (1.19) the most used norm is l_1. If $\rho > \|y^*\|_D$, where y^* is the Lagrange multiplier corresponding to the nonlinear constraints and $\|.\|_D$ is the dual norm of $\|.\|$ (i.e., the l_∞ norm in the case of the l_1 exact penalty function), then the local minimum x^* of the exact penalty function $P_\rho(x)$ is a local minimum of problem (1.1) (Fletcher, 1987). Therefore, the classical approach minimizes a sequence of penalty sub-problems $P_\rho(x)$ for an increasing sequence of penalty parameters. Some other merit functions are known, for example, the oldest one being the quadratic penalty function $f(x) + \rho\|c(x)\|_2^2$. The corresponding algorithm based on this penalty function is convergent to a local solution only if the sequence of the penalty parameters is divergent to infinity. The global convergence is also ensured by using the augmented Lagrange and the penalty functions of the type $f(x) + c(x)^T y + \rho\|c(x)\|$. A major difficulty with these penalty functions is the possibility that near a strictly isolated minimum, the step given by the Newton method should not be accepted. This phenomenon is known as the Maratos effect (Maratos, 1978). The remedy for the Maratos effect is the introduction of second-order corrections or the use of the nonmonotone line search. But the greatest difficulty with penalty methods is the selection and the updating of the penalty parameters.

Filter methods try to harmonize two objectives always present in any optimization problem: minimization of the objective function and minimization of the violation of the constraints (see Chapter 18). The filter methods keep a record of the objective function value $f_l = f(x_l)$ and of the constraint violation $h_l = \|c(x_l)\|$, for a number of previous iterates x_l, $l \in F_k$, where F_k is the filter at iteration k (Fletcher & Leyffer, 2002). A new iterate is acceptable in the filter if it improves either the objective function or the constraint violation, compared to all previous iterates recorded in the filter. In other words, the point $\widehat{x}$ is acceptable in the filter if $f(\widehat{x}) \le f(x_l) - \gamma h(x_l)$ or $h(\widehat{x}) \le \beta h(x_l)$, for any $l \in F_k$, where $\gamma > 0$ and $0 < \beta < 1$ are constants which ensure that the iterates cannot accumulate at infeasible limit points. To ensure the convergence to a local minimizer, the filter algorithm uses a standard sufficient reduction condition taken from the unconstrained optimization:

$$f(x_k) - f(x_k + d) \geq -\sigma m_k(d), \tag{1.20}$$

where $\sigma > 0$ is the fraction of predicted decrease and $m_k(d)$ is the model reduction from the approximate sub-problem. This condition is used only if the model predicts a decrease in the objective function. Therefore, the filter method uses a switching condition $m_k(d) \geq \gamma h(x_k)^2$ to decide when (1.20) should be enforced. A new iterate that satisfies both these conditions is called an iteration of type f, while an iterate for which the switching condition fails is called an iteration of type h.

Remark 1.4 *The general characteristic of the filter method is that it avoids the complications with selecting and updating the penalty parameter from the augmented Lagrangian or the penalty function methods. However, the difficulty with this method is the selection of the parameters which determine the convergence as well as the specification of the heuristics for ensuring the feasibility of the sub-problems and the convergence to a local solution.* ♦

1.2.5 Refining the Local Model

The idea of refining the local model is to reduce the stepsize that is computed by the approximate sub-problem. Two mechanisms for refining the local model are known: *line-search methods* and *trust-region methods*. Both these mechanisms can be used in the frame of any of the approximate sub-problems and of any globalization strategies for convergence from remote starting points. Therefore, a great variety of nonlinear optimization algorithms result with different capabilities for solving complex and large-scale optimization problems.

Line-search methods enforce convergence by using a backtracking line search along the direction s. For sequential quadratic programming methods, the searching direction is given by the quadratic programming problem (1.6), $s = d$. For the interior point methods, the searching direction $s = (\Delta x, \Delta y, \Delta z)$ is the solution of the primal-dual linear algebraic system (1.10). To ensure that the model produces a descent direction, one must have $\nabla P(x_k)^T s < 0$, where $P(x)$ is a merit or a penalty function. A very popular line search is the *Armijo rule* (Armijo, 1966) (see Andrei 2009, Chapter 4). A prototype algorithm with line search is as follows:

Algorithm 1.3 Prototype for line-search method for nonlinear optimization

1.	Choose an initial point $x_0 \in \mathbb{R}^n$, as well as $0 < \sigma < 1$. Set $k = 0$
2.	Evaluate a criterion for stopping the iterations. If x_k is the optimal solution of the problem, stop; otherwise, go to step 3

(continued)

3.	Approximately solve an approximate sub-problem of (1.1) around x_k to find a search direction s
4.	Verify whether s is a descent direction, if $\nabla P(x_k)^T s < 0$, for a given merit function $P(x)$
5.	Set $\alpha_0 = 1$ and $m = 0$
6.	If $P(x_k + \alpha_m s) \geq f(x_k) + \alpha_m \sigma s^T \nabla P(x_k)$, then continue with step 7; otherwise, go to step 8
7.	Set $\alpha_{m+1} = \alpha_m/2$ and evaluate $P(x_k + \alpha_{m+1} s)$. Set $m = m + 1$. Continue with step 6
8.	Set $k = k + 1$ and go to step 2 ♦

Observe that the algorithm has two loops. The inner loop determines the stepsize by backtracking. The outer loop is responsible for the search direction determination, i.e., the solution of the problem.

The line-search methods with filters can be defined in a similar way, but instead of checking the descent in the merit function, a filter method is used to check acceptance to a filter (Leyffer & Mahajan, 2010).

Trust-region methods restrict the step in an explicit way by adding a trust-region constraint of the form $\|d\| \leq \Delta_k$, where the used norm is l_1, l_2, or l_∞. Most methods use l_∞ norm in the trust-region constraints, which can be represented by simple bounds on variables. The trust-region radius $\Delta_k > 0$ is adjusted at every iteration, depending on how well the approximate sub-problem agrees with the nonlinear optimization problem (1.1). A prototype of the optimization algorithm with trust-region is as follows:

Algorithm 1.4 Prototype for trust-region method for nonlinear optimization

1.	Choose an initial point x_0 and a value for the trust-region radius $\Delta_0 > 0$. Set $k = 0$
2.	Evaluate a criterion for stopping the iterations. If x_k is the optimal solution of the problem, stop; otherwise, go to step 3
3.	Approximately solve the local model of the problem (1.1), i.e., the sub-problem (1.6) in which a trust-region constraint $\|d\| \leq \Delta_k$ is introduced
4.	If $x_k + d$ is sufficiently better than x_k, then increase Δ_k and continue with step 5; otherwise, reduce Δ_k, for example, $\Delta_k = \Delta_k/2$ and go to step 3
5.	Set $k = k + 1$ and go to step 2 ♦

The trust-region methods are related to *regularization techniques*, which refer to the positive definiteness of the Hessian matrix H_k. These techniques are implemented by adding a multiple of the identity matrix to the Hessian. Locally, the solution of the regularized problem is equivalent to the solution of a trust-region problem with a l_2 norm trust-region. Clearly, the disadvantage of the trust-region methods is that the sub-problems may become inconsistent as $\Delta_k \to 0$. This situation can be dealt with in three different ways: a *penalty function* approach (Nocedal & Wright, 2006) (see Chapter 7), a *restoration phase* in which the algorithm minimizes the constraint violation (Fletcher & Leyffer, 2003) (see Chapter 18), or a *composite step* approach (Omojokun, 1989) (see Chapter 11). All these refining mechanisms of the local model are used in nonlinear optimization algorithms. The difficulties are related to the computation of the stepsize, to the selection of the merit function, and to the techniques for regularization.

In conclusion, a large variety of constrained nonlinear optimization algorithms are known. All of them use four major ingredients: a *local model* of the problem around the current point, a *globalization strategy* which ensures the convergence of the algorithm from remote initial points, a *refining mechanism* which reduces the stepsize computed by the local model to enforce the convergence, and *some convergence tests* which certify the type of the limit point obtained by the algorithm. The diversity of optimization algorithms is given by the way in which all these ingredients are implemented. Besides, the optimization algorithms implement a multitude of other ingredients which are crucial in the efficiency and robustness for solving complex large-scale nonlinear optimization problems. Among these ingredients there are sparse or dense linear algebra techniques, factorization of indefinite matrices, inertia correction of matrices, Schur complement computation, identification of the active constraints, initialization of the algorithms with respect to the primal and dual variables, updating procedures for barrier parameter or a trust-region radius, updating the parameter of the central trajectory, solving the infeasibilities (elastic programming), etc.

The most efficient and robust nonlinear optimization algorithms use combinations of the active-set methods (local models based on sequential linear or quadratic programming), or interior point methods with globalization strategies (augmented Lagrangian, penalty and merit function, or filter methods), and refining mechanisms (line-search or trust-region methods), all of them in a frame where advanced computational linear algebra techniques are used. The most advanced nonlinear optimization algorithms implementing the above concepts are illustrated in the chapters of this book.

1.3 Structure of the Book

The book is structured in 21 chapters. Chapter 2 presents certain aspects of nonlinear optimization modeling using algebraic oriented languages. In Chapter 3 the main points of the GAMS technology are detailed. Chapter 4 introduces a number of 18 continuous nonlinear optimization applications which are used to see the performances of the optimization algorithms described in this book. These applications are taken from Andrei (2013b). Chapter 5 is dedicated to the mathematical theory of nonlinear optimization. Here the core is the KKT optimality conditions which characterize optimal points and provide the basis for most algorithms. The theory, algorithms, and some applications of simple bound optimization problems are discussed in Chapter 6. The applications in this chapter are taken from MINPACK-2 test problem collection (Averick, Carter, Moré, & Xue, 1992). The remaining chapters describe the most important continuous nonlinear optimization algorithms imbedded in GAMS: MINOS, KNITRO, CONOPT, SNOPT, and IPOPT, which implement the combination of methods presented above. To see the performances of these algorithms and to make comparisons, some other optimization algorithms *not integrated in GAMS* are described. They are based on concepts or methods like penalty barrier (SPENBAR), a sequential quadratic programming method using only the equality-constrained sub-problem (DONLP), a sequential quadratic programming algorithm with successive error restoration (NLPQLP), and filter methods with sequential linear programming (filterSD) and filter with sequential quadratic programming (filterSQP). *The motivation behind this approach is to illustrate the importance of designing the optimization algorithms by combining different concepts and strategies, as well as by using advanced computational linear algebra techniques which improve the numerical performances of the algorithms imbedded in the GAMS technology.* In the last chapter, a thorough numerical analysis and comparisons of the algorithms working under the GAMS technology are presented.

Notes and References

Continuous nonlinear optimization with engineering applications is practically present in any area of activity and is essential in running engineering complex structures and complex economy aggregates. Optimization is often called *mathematical programming*, a somewhat confusing term introduced by Tjalling Koopmans (1910–1985) and consolidated inter alia by Leonid Kantorovich (1912–1986), George Dantzig (1914–2005), and by some others (see Lenstra, Rinnooy Kan, & Schrijver, 1991; Grötschel, 2012). By optimization we understand the design and analysis of algorithms which fulfill the following requirements: *efficiency* (they should not require excessive computing time and storage), *robustness* (they should solve a wide variety of optimization problems from different starting points), and *accuracy* (they should be able to generate a solution with a specified precision without being sensitive to errors in data or to arithmetic rounding errors). The mathematical theory of optimization is found in many books: Luenberger (1973, 1984), Bazaraa, Sherali, and Shetty (1993), Bertsekas (1999), Boyd and Vandenberghe (2006), Nocedal and Wright (2006), Sun and Yuan (2006), Bartholomew-Biggs (2008), Luenberger and Ye (2008), and Andrei (2015), just to cite some of them. Now, the professional optimization algorithms work under the assistance of some systems of modeling and optimization like GAMS, AMPL, TOMLAB, etc. In this book the GAMS language and technology for modeling and solving continuous nonlinear optimization applications are chosen.

Chapter 2
Mathematical Modeling Using Algebraic Oriented Languages for Nonlinear Optimization

In the last two decades, significant research activities have taken place in the area of local and global optimization, including many theoretical, computational, and software contributions. The access to this advanced optimization software needs more and more sophisticated modeling tools. Algebraic oriented optimization modeling languages represent an important class of tools that facilitate the communication of optimization models to decision-making systems based on the optimization paradigm. In a wider context, the algebraic oriented modeling tools evolve toward fully integrated modeling and optimization management systems with access to databases, spreadsheets, and graphical user interfaces.

As already known, at the fundamentals of every mathematical model lie the conservation laws which are active in the domain to which the mathematical model belongs. Therefore, knowing these conservation laws is essential, and their usage leads to models satisfying the *adequacy to real principle*. However, besides the conservation laws, additional empirical knowledge, principles, different rules, or previous experience must be considered in order to develop a mathematical model of a given reality.

In this chapter, certain aspects concerning the mathematical modeling process in the context of mathematical modeling technologies based on the algebraic oriented languages are discussed.

2.1 Linguistic Models Versus Mathematical Models

In a common way, everyone is familiar with different representations of the surrounding world, the so-called mental models, used in each moment of their existence. The decisions made in different situations are not based on the real world but on our mental images of the real world, on our mental images of the relations among the components of the real world. The mental models represent our

N. Andrei, *Continuous Nonlinear Optimization for Engineering Applications in GAMS Technology*, Springer Optimization and Its Applications 121,
DOI 10.1007/978-3-319-58356-3_2

understanding of the part of the creation that we want to know. Having in view that the support of our thinking is the word, it follows that mental models are actually linguistic models. Therefore, in one way or another, the human beings set forth their understanding of the real world as a linguistic description expressed as a corpus of assertions (theorems) of the general form: *if . . . then . . . .*

The mental models have some advantages which recommend them to be used in our efforts to understand the world. A mental model is *flexible* in the sense that it can take into consideration a domain of information sensibly larger than the numerical one. Besides, a linguistic model may be quickly *adapted* to some new situations and can be *modified* as soon as some new information is available. The mental models are *filters* through which we can explain our experiences, evaluate, and select different actions. In a way, the greatest philosophical systems, political and economical doctrines, theories of physics, and literature itself are linguistic models.

However, the mental models have some disadvantages as well. They are not so easily understood by the others. Their interpretation is *dependent* on their creator. Besides, the hypothesis used to generate them is difficult to examine and even to accept. The ambiguities and contradictions contained in these types of models can remain undetected, unexplained, and therefore unsolved. It is quite natural to have difficulties in understanding the linguistic models suggested by the others. Surprisingly, we are not so good at developing and understanding our own mental models or at using them during the process of decision-making. Psychologists have shown that we are able to consider only a very small number of factors in decision-making. In other words, the mental models we use in decision-making are *extremely simple*. These are often imperfect because we frequently persist in error when deducing the consequences from the suppositions on which they are based. These models often express what we would like to happen and not what actually happens in reality.

But the greatest *imperfection* of the mental models lies in the fact that these intellectual developments do not satisfy the criteria of *completeness*, *minimality*, and *non-contradiction*. It is very likely that some very important assertions which change their significance may be omitted in a mental (linguistic) model. Clearly, some new assertions may be introduced, which are in contradiction with those we have previously considered in our reasoning. At the same time, in using the linguistic models, we are often faced with the very difficult problem of *word rectification*. This problem dramatically limits the usage of linguistic models with different groups of people. Finally, we must emphasize that during the linguistic model analysis and solving process, we are very likely to be confronted with the *circularity danger*. The problem of the circularity of formal systems was solved by Gödel (1931), who showed that expressing knowledge into a formal system and into logical formal systems, like those of Russel or of Zermelo-Fraenkel-Newmann, is an illusion. There are relatively simple assertions (theorems) which are impossible to be solved (decidable) in this kind of formal systems.

A mathematical model is a representation in mathematical symbols of the relations between the variables and the parameters belonging to the part of the creation we are interested in. The relations describing the mathematical model often include variables and their derivatives, thus expressing the *local character* of the model and that of *predictability* as well. The mathematical models have a number of advantages versus the linguistic models. The most important is that mathematical models do not have any imperfections which the linguistic models do. They are explicit in the sense that the hypothesis and the assumptions used in their development are public and subject to any criticism. Besides, the (logical) consequences after solving them are very well justified mathematically. Finally, the mathematical models are more comprehensive, being able to simultaneously consider an appreciable multitude of factors. But the most important characteristic of mathematical models is that they are written on the basis of the conservation laws. In this respect the Noether theorem shows that the conservation laws are direct consequences of different symmetries (Andrei, 2008b). For example, the conservation of energy is a consequence of temporal symmetry, while the conservation of momentum is a consequence of the symmetry of space. The entire physical world is depicted as being governed according to mathematical laws.

2.2 Mathematical Modeling and Computational Sciences

The theory and practice of mathematical modeling is based on computational sciences. Broadly, computational sciences consist in the usage of computer systems for analyzing and solving scientific problems. There is a sharp distinction between computer science and computational science. Computer science focuses on the building and the analysis of computers as well as on computability. On the other side, computational sciences emphasize the development of methods, techniques, and algorithms for solving mathematical models and centers on the convergence and complexity of algorithms.

The purpose of computational sciences is to understand the evolution of a real-world system by analyzing and solving the corresponding mathematical models associated to the considered system, using high-performance computers. By using mathematical symbols as well as mathematical accuracy and rigor, mathematical modeling has a central place in computational sciences. Mathematical modeling leads us to the maturation of the domain the mathematical model belongs to and also to the generation and the development of new mathematical methods and techniques able to solve new problems. The place of the mathematical model within the general scheme for modeling and solving a problem is shown in Figure 2.1.

Observe that a mathematical model is synthesized from the real world, thus arriving into an abstract world, where mathematical rigor is present through mathematical concepts. The model is solved using mathematical theories irrespective of the physical significance of the elements of the problem. The interpretation of the solution may lead us to some modifications of the model, which again determines our addressing to the real world.

Notice that in the abstract world of the modeler, we operate with concepts belonging to the domain the modeling process refers to. This is the question of

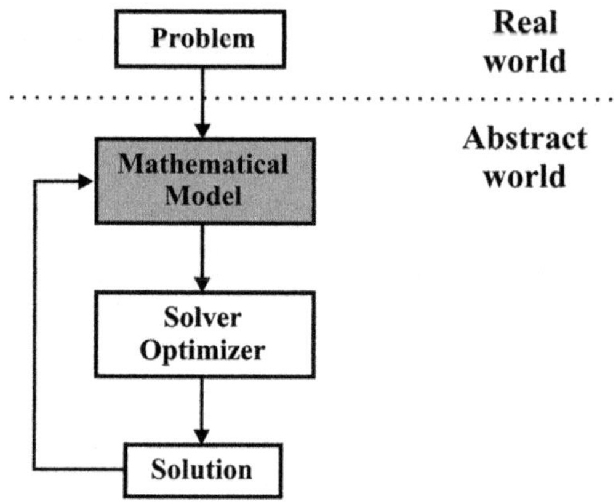

Figure 2.1 The process of modeling and solving a problem

the phenomenology and the conservation laws which characterize the domain of interest. At the same time, certain concepts from the theory of languages, of translators, and of compilers are used in order to develop informatics technologies able to elaborate, update, and maintain the mathematical models.

On the other hand, in the abstract world of algorithm developers, one operates with advanced mathematical concepts in order to generate optimization algorithms and studies of their convergence and complexity. Additionally, some advanced languages like Fortran or C++ are used here for implementing algorithms in computer programs.

It is quite clear that for the majority of problems, their mathematical models must be simplified. Therefore, we are faced with the following dilemma. *What is best: to consider an approximate (simplified) mathematical model of the problem and then try to get an exact solution as much as possible or to use a mathematical model as accurately as possible and then determine an approximate solution of it?* The practical experiences recommend that the second alternative gives better results.

2.3 Modeling Scheme

Formulating and developing a mathematical model of a given reality is one of the finest creative intellectual activities. A deep analysis of the modeling and solving process (see Figure 2.1) shows that mathematical modeling involves more effort and time for data analysis, verification, and documentation, as well as for updating and improving different variants of the model. The difficulties of the data management for mathematical modeling are coming from the fact that nowadays we do not have a clear methodology for mathematical modeling expressed as an algorithm or as a group of algorithms to be used in a general context. However, understanding the linguistic representation of the process that we need to represent through a mathematical model is crucial.

The development of a mathematical model is a complex process consisting in a mapping between the objects (together with the relations among them) from the real-world and the symbolic and mathematical objects. This process involves both a very good knowledge of the reality we want to represent in mathematical terms and a multitude of methods and techniques for mathematical modeling. The correctness of a model is usually established after its solving. As a result of intensive computational experiments, it is often necessary to reformulate the model by introducing some new algebraic or differential equations (which were initially ignored) and also the corresponding data. The mathematical modeling process is closely linked to the solving process of the model, supporting each other. These two processes interact in order to build up a mathematical object placed in the *perspective of the infinite similarity with reality*.

Let us now try to detail the modeling process. At the very beginning, for the process we want to represent into mathematical terms we have a *linguistic description*, actually a text expressed in the natural language which describes the process. It is worth saying that this linguistic description is a *scientific text*, not a literary one. Obviously, this description includes references to the important conservation laws which are active in the process domain under consideration. As a result of an abstracting creation activity from this linguistic description and using the conservation laws, we get an *algebraic-differential representation* of the process, a *mathematical model*. This activity mainly consists in identifying the variables and parameters of the process, together with the algebraic and differential relations among these entities by using principles, empirical knowledge, different rules, and some other additional knowledge of the process. Once having this algebraic-differential representation of the process, the next step is to develop an *internal representation* of it, a representation directly admitted by any professional solver, in our case an optimizer. This is done through the *external representation* of the model using the so-called algebraic oriented mathematical modeling languages. Therefore, the mathematical model has three forms of representation: an algebraic-differential one using mathematical symbols, an external one using an algebraic oriented mathematical modeling language (like GAMS, AMPL, ALLO, etc.), and an internal representation which is mainly an informatics description including different files and databases directly admitted by any solver (optimizer). Figure 2.2 presents the modeling scheme for nonlinear optimization based on algebraic oriented languages.

Suppose we have written the optimization mathematical model, i.e., we have specified the objective function, the constraints, and the simple bounds on variables. The problem now is how to transmit this model to the solver (optimizer) in order to get a solution. Since we consider mathematical models with thousands of variables and constraints, this problem appears to be quite challenging. An elegant and very efficient solution is to use algebraic oriented languages for achieving an external description of the model which can automatically be translated into the internal description of it by means of a translator associated to the language used in this

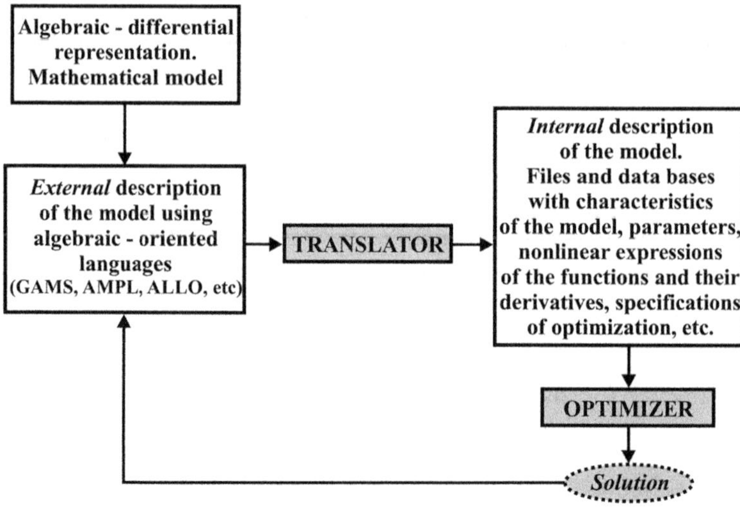

Figure 2.2 Modeling scheme of nonlinear optimization based on algebraic oriented languages

respect. For example, in case of linear programming problems, the internal representation of the model is given by the MPS form. Therefore, the translator associated to an algebraic oriented language (like GAMS, AMPL, ALLO, etc.) generates the MPS form of the model from its external description. In case of nonlinear optimization problems, the mechanism is more complex but mainly the same.

2.4 Algebraic Oriented Modeling Languages

Algebraic modeling languages represent a problem in a purely declarative way. Most of them include some computational facilities to manipulate the data as well as certain control structures. Many optimization models can be well represented declaratively in a very concise way, which also favors the insight of the model. Besides, a mathematical model is stated in a declarative way using mathematical equalities and inequalities. This gives a clear documentation of the model which can be parsed and interpreted by a compiler.

The algebraic languages are more declarative rather than imperative. A declarative language emphasizes *what is being computed* more than *how it is being computed*. Furthermore, the reasons for which we use a declarative presentation of models are *conciseness*, *insight*, and *documentation*.

It is important to emphasize that one of their strengths is the complete separation of the problem formulation from finding a solution which is supposed to be determined by an external program called *solver* or *optimizer*. This characteristic allows the modeler not only to separate the two main tasks of model formulation and model solving but also to easily switch among several solvers. This is very important indeed because it is known that a model instant can be solved using one method, while another instance is efficiently solvable by using another method. Another important advantage of algebraic oriented modeling languages is that they clearly separate between the model structure which only contains parameters, but not data, and the model instance in which the parameters are replaced by a specific set of data.

Therefore, the main features of the algebraic oriented modeling languages are:

- Purely declarative representation of the model
- Clear separation between formulation of the model and its solving
- Clear separation between the model structure and the model data

An algebraic oriented modeling language has a declarative and an algorithmic knowledge which are clearly separated. Either of them may be empty, meaning that the problem is represented as a purely declarative or as a purely algorithmic form. The declarative part consists of the basic building blocks of declarative knowledge including variables, parameters, constraints, sets, etc. On the other hand, the algorithmic part consists of the control structures describing (explicitly or implicitly) the computation of solving a problem.

The most important algebraic oriented languages used for solving nonlinear optimization problems are presented in Table 2.1.

What is characteristic of algebraic oriented languages is that they permit to the modeler to express the model in a way based on indices that belong to certain abstract entities, which are sets of evolution of indices, parameters, variables, and constraints. Furthermore, there is the possibility of grouping the entities in sets with similar characteristics to which we can make reference through indices as elements of certain sets. The grouping of entities (such as variables or constraints) allows their representation in a compact manner, very similar to the algebraic one. For instance, the mathematical formulation $\sum_{i \in I} x_i$ can be translated into the expression **`SUM(I,x(I))`** in the GAMS modeling language. Therefore, as can be seen, this leads to a problem formulation that is very close to the formulation using algebraic notations. The task of the translator associated to the modeling language is to expand this compact representation into one accessible to a solver, the so-called the *problem instantiation*, the one ready to be considered by an appropriate optimization solver. This expanding operation is realized by replicating every entity over the different elements of the set. This is referred to as a *set-indexing* ability of the algebraic oriented modeling languages. The modeler can define generic expressions that are indexed over several sets. Set indexing in such cases involves compound sets.

As already mentioned, the algebraic oriented modeling languages use essentially declarative statements as opposed to programming languages (Fortran, C++, etc.) in which the use of procedural statements dominates over the use of declarative ones. For instance, the few procedural statements used in algebraic oriented languages are **`read/write data`** and **`solve`** commands. In order to deal with more complicated nonlinear optimization models, these languages include **`if-then-else,`**

Table 2.1 Algebraic oriented modeling languages (partial list)

GAMS	*General Algebraic Modeling System* (www.gams.com) Development Research Center. The World Bank, 1818 H. Street, Washington D.C., USA. Brooke, A., Kendrick, D., Meeraus, A., Raman, R., Rosenthal, R.E., *GAMS A user's guide*. GAMS Development Corporation, December 1998
AMPL	*A Mathematical Programming Language* (www.ampl.com) Department of Industrial Engineering and Management Sciences. Northwestern University, Evanstone, Illinois 60,201, USA. Fourer, R., Gay, M., Kernighan, B.W., *AMPL: A modeling language for mathematical programming*. Second edition. Duxbury Press/ Brooks/Cole Publishing Company, 2002
ALLO	*A Language for Linear Optimization* (http://camo.ici.ro/projects/allo/allo.htm) Research Institute for Informatics – Bucharest. 8–10 Bdl. Mareşal Alexandru Averescu, sector 1, 011455 Bucharest – Romania. Andrei, N., *The ALLO language for linear programming*. ICI Technical Report No.1/2004, Bucharest. Andrei, N., *Criticism of the linear programming algorithms reasoning*. Academy Publishing House, Bucharest 2011. (see Chapter 17, Annex A6)
LPL	*A Structured Language for Modeling Linear Programs* (http://diuflx71.unifr.ch/lpl/mainmodel.html) Institut of Informatics, University of Fribourg, Regina Mundi, rue de Faucigny 2, CH-1700 Fribourg, Switzerland. Hürlimann, T., *The LPL modeling language: Highlights*. University of Fribourg, Switzerland, December 8, 2011
AIMMS	*The Modeling System* (www.aimms.com) Paragon Decision Technology B.V. P.O. Boxes 3277, 2001 DG Haarlem, The Netherlands Bisschop, J., Roelofs, M., *AIMMS – The user's guide*. Paragon Decision Technology, 1999
MPL	*Modeling System* (www.maximalsoftware.com) Maximal Software, Inc., 2111 Wilson Blvd. Suite 700, Arlington, VA, USA. Kristjansson, B., *MPL User manual*. Maximal Software Inc., Iceland, 1993
LINDO	*Powerful Library of Optimization Solvers and Mathematical Programming Tools* (http://www.lindo.com) Lindo Systems Inc., 1415 North Dayton Street, Chicago, Il, 60,642, USA. Schrage, L., *Optimization Modeling with LINDO. 5th edition*. Duxbury Press, 1997
MOSEL	*An Extensible Environment for Modeling and Programming Solutions* (www.fico.com/en/Company) FICO® Xpress Optimization Suite. 901 Marquette Avenue, Suite 3200, Minneapolis, MN 55402, USA. (formerly DASH) Colombani, Y., Heipcke, S., *Mosel: An Overview*, Xpress Team, FICO, Leamington Spa CV32 5YN, UK. (http://www.fico.com/xpress)
TOMLAB	*TOMLAB Optimization Environment* (tomopt.com/tomlab) 113 Cherry St Ste. 95,594. Seattle, WA 98104–2205, USA

`for, while` commands. Such commands enable the modeler to write certain solution algorithms directly in the modeling language (Bisschop & Meeraus, 1982; Fourer, 1983; Hürlimann, 1999; Kallrath & Wilson, 1997; Conejo, Castillo, Minguez, & Garcia-Bertrand, 2006; Castillo, Conejo, Pedregal, García, & Alguacil, 2001). The next chapter is dedicated to illustrate the main aspects and facilities of the GAMS algebraic oriented language.

Notes and References

This chapter on modeling for nonlinear optimization using algebraic oriented languages for continuous nonlinear optimization is based on developments presented in Andrei (2012, 2013b).

Chapter 3
Introduction to GAMS Technology

This chapter deals with an introduction to the General Algebraic Modeling System – GAMS. It is a high-level algebraic modeling technology elaborated by the GAMS Development Corporation.[1] It permits developing and solving the linear or nonlinear optimization models. In GAMS the optimization models can be introduced in a way similar to the one which presents them in research papers or books. GAMS has the ability to solve large-scale linear programming problems and integer linear programming problems, as well as to find local or global optima of nonlinear and mixed-integer programs of very different complexity.

The system is especially useful for handling and solving real, large, and complex optimization problems. It is available for use on personal computers, workstations, mainframes, and supercomputers.

GAMS is documented in a number of publications including (Andrei, 2011b, 2013b; Brooke, Kendrick, Meeraus, Raman, & Rosenthal, 1998; Jensen, 2006; Kalvelagen, 2001, 2002, 2003; Markusen, 2005; McCarl, 1998; McCarl, Meeraus, Van der Eijk, Bussieck, Dirkse, Steacy, & Nelissen, 2011; McKinney & Savitsky, 2003; Robichaud, 2010; Rosenthal, 2011[2]).

3.1 Basics of Modeling

As we know, optimization consists in the representation and the solving of optimization models. All the optimization models can be represented in a number of ways using algebraic and/or differential notations. They can be solved using a number of

[1] GAMS Software Corp. 1217 Potomac Street, NW, Washington, DC 20007, USA. GAMS Software GmbH, Eupener Strasse 135–137, 50,933 Cologne, Germany

[2] *GAMS – A User's Guide*. Tutorial by Richard E. Rosenthal, July 2011. GAMS Development Corporation, Washington, DC, USA

N. Andrei, *Continuous Nonlinear Optimization for Engineering Applications in GAMS Technology*, Springer Optimization and Its Applications 121,
DOI 10.1007/978-3-319-58356-3_3

methods which have different corresponding algorithms. Any optimization model has three main components: objective function, variables, and constraints. Conceptually, an optimization model consists of two main parts: its algebraic structure and the associated data instance. For example, the formal linear programming model has the following structure:

$$\begin{aligned} &\underset{x \in \mathbb{R}^n}{\text{minimize}} \ c^{\mathrm{T}} x \\ &\text{subject to:} \\ &Ax = b, \end{aligned} \tag{3.1}$$

and the associated data are the matrix $A \in \mathbb{R}^{m \times n}$ and the vectors b and c of corresponding dimensions. A nonlinear programming model is more complicated and has the following structure:

$$\begin{aligned} &\underset{x \in \mathbb{R}^n}{\text{minimize}} \ f(x) \\ &\text{subject to:} \\ &g(x) = 0, \\ &h(x) \geq 0, \\ &l \leq x \leq u, \end{aligned} \tag{3.2}$$

where $f: \mathbb{R}^n \to \mathbb{R}$, $g: \mathbb{R}^n \to \mathbb{R}^m$, and $h: \mathbb{R}^n \to \mathbb{R}^p$ are functions of different forms and l and u are simple bounds on variables. The associated data of this problem consists of the parameters entering into the algebraic expression of the functions of the problem and the bounds l and u. Of course, the optimization models can be of a more general form including mixed-integer linear programming, mixed-integer nonlinear programming, stochastic programming, mixed complementarity problems, problems with equilibrium constraints, generalized equilibrium, multiobjective programming, etc.

GAMS is a system for formulating and solving these types of general optimization models. It has a language which allows a high-level algebraic representation of mathematical optimization models, a compiler which is responsible for user interaction by compiling and executing user commands given in a GAMS source file, and a set of solvers to solve them. A text source file can implement a simple optimization model, or it can represent a large-scale model consisting of several interrelated sub-models. The solvers are stand-alone programs that implement different efficient optimization algorithms for different GAMS model types.

The GAMS technology can be called from the command line or through the integrated development environment (IDE) which is a Windows-based environment that facilitates both the modeling by interacting editors with the GAMS and solving the generated optimization problem.

Table 3.1 The basic components of a GAMS model

Inputs	Outputs
Sets	Echo print
Data: parameters, table, scalars	Reference maps
Variables	Equation listing
Bounds, initial values	Variables listing
Equations (equalities and inequalities)	Status report
Model and solve statements	Results
DISPLAY/PUT statement	

3.2 Structure of a GAMS Model

The GAMS technology uses general-purpose language containing elements of both ordinary programming languages data declaration or control structure and high-level modeling structures such as sets, variables, equations, etc. The basic components are listed in Table 3.1.

Some general characteristics of the GAMS language are as follows:

- A GAMS model is a collection of statements in the GAMS language. The only restriction on the ordering of the statements of a model in GAMS is that an entity of the model cannot be referenced before it is declared to exist.
- Every statement is terminated with a semicolon. The GAMS compiler does not distinguish between upper- and lowercase letters.
- Typographically, GAMS statements may be laid out in any style. Multiple lines per statement, multiple statements per line, and embedded blank lines are allowed.
- There are at least two possibilities to insert documentation within the GAMS language. Firstly, any line starting with an asterisk in column 1 is considered as a comments line by the GAMS compiler. Secondly, documentary texts can be inserted using "`$ontext`" and "`$offtext`" instructions.
- Generally, creation of the input components involves two steps: a declaration and an assignment or definition. Declaration means declaring the existence of something and giving it a name. Assignment or definition means giving a specific value or form.

3.3 Sets

In any complex optimization model, we use data, variables, and constraints which are indexed. The sets form the basis for indexing. For example, the following statements are valid in GAMS:

```
Set t time periods /1990*2001/;
Set m machines /mach1*mach5/;
```

These lines declare the two sets: `t` and `m`. The set `t` contains the elements from 1990 through 2001. Similarly, the `m` set contains the elements from mach1 through mach5. The asterisk indicates filling out the intervening elements. Observe that set elements are stored as character strings, so the elements of `t` are not numbers.

The ALIAS statement is a convenient way used to give another name to a previously declared set. In the following example, `Alias (t,tp);` the name `tp` is like `t`.

Most GAMS modeling elements (data, variables, etc.) can be indexed up to 10 indices. GAMS has several capabilities for generating dynamic sets, which acquire their members through the execution of set-theoretic and logical operations.

Operations can be performed on sets using the symbols + (set union), – (difference of sets), * (set intersection), and `not` (set complement operation).

3.4 Data

Numerical data in GAMS are introduced as *scalars*, *parameters*, and *tables*. Scalar is a single real number (no indices are allowed). Parameter is an indexed data collection of numbers up to ten indices. Table is a syntactically convenient way to declare and to initialize a parameter.

The following example shows the declaration and immediate initialization of a scalar: `scalar tax_rate /0.50/;`. Initialization can occur in a later assignment statement, as shown in the following example:

```
scalar pi;
pi = 2*arctan(inf);
```

Parameters are data sets indexed by one or more indices. Indices are sets which are previously declared. In the following example, two parameters are declared: one for the initial position and the other one for the initial velocity of a vehicle:

```
Parameters c_0(c) initial position /x 0, y 1000/
           v_0(c) initial velocity /x 13.23, y -1.288/;
```

Here `c` has been previously declared as a set.

A table declaration is a way to declare a multidimensional parameter. For example, to declare the three-dimensional matrix `a`, to be inverted, the following statements are introduced in a source GAMS text:

```
Set i /i1*i3/;
Alias(i,j);
Table a(i,j) matrix to be inverted
    i1 i2 i3
i1  1  5  4
i2  2  3  3
i3  1  1  7;
```

Observe that the values of `a` are given in the above statement under the appropriate heading. If in the table there are blank entries, they are interpreted as zeroes.

3.5 Variables

To declare the variables of an optimization model, GAMS uses the variable declarations. Each variable is given a name, a domain if appropriate and (optionally) a text. The variables can be continuous or discrete or some mixture of the two. Table 3.2 shows different declaration possibilities of variables.

The default bounds can be reset through the `.LO` and `.UP` attributes. Variables can have up to ten indices.

It is worth saying that terms like "variable" or "parameter" mean slightly different things in different contexts. In optimization a "variable" represents an unknown, a decision to be considered. When GAMS solves an optimization model, the solver finds appropriate values for the variables that satisfy the constraints of the model. In economic literature and practice, such variables are sometimes called *endogenous variables*, in contrast to *exogenous variables*, or *parameters*. The parameters (exogenous variables) may be assigned to some particular known values, and they are fixed for the duration of the solving process of the model.

In nonlinear optimization, it is very important for the analyst to help the solver by specifying as narrow a range as possible between lower and upper bound. It is also very helpful to specify an initial point from which the solver starts searching for the optimum. The lower and upper bounds of a variable are set automatically according to the variable's type (*free*, *positive*, *binary*, or *integer*). However, these bounds can be overwritten by the GAMS user using: `.LO` for lower bound, `.UP` for upper bound, `.FX` for fixing a variable to a given value, `.L` for initialization of a variable, and `.M` for marginal or dual value. The following sequence of statements illustrates these concepts:

```
positive variables y(i,j);
y.LO(i,j) = 5; y.UP(i,j) = 100;
y.FX(i,j) = 18;
```

Table 3.2 The different kinds of variables and their declarations

Variable declarations and types		Default bounds	
Variable	Continuous	−INF	+INF
Positive variable	Continuous	0	+INF
Binary variable	Discrete	0	1
Integer variable	Discrete	0	100
Semicont variable	Either 0 or in [LO;UP]	0	INF
Semiint variable	Either 0 or in {LO,LO + 1,...UP}	0	100

Observe that fixing a variable does not remove it from the model. Assigning a value to the `.FX` attribute is equivalent to setting the variable's lower and upper bounds to the same value. Variables also have a scaling attribute, `.SCALE`; see the User's Guide for details.

3.6 Equations

Equations in GAMS mean both equalities and inequalities. The power of GAMS is illustrated in the creation of the equality and inequality constraints that comprise the model under construction. This is because whenever a group of constraints (equalities or inequalities) has the same algebraic structure, all the members of the group are created simultaneously, not individually. The objective function and the constraints declarations are introduced by the keyword `EQUATIONS`. Like variables, GAMS requires that each constraint and the objective function have a name which will be active in the optimization model. Concerning the objective function, it is worth saying that in GAMS the model $\max c^{\mathrm{T}}x$ is rewritten as $\max Z, Z = c^{\mathrm{T}}x$, where Z is a variable unrestricted in sign. This variable can be named as you want, in this case Z. Observe that for the objective function, it is necessary to introduce a new unrestricted variable.

The declaration of equations is the same as for other GAMS entities. First comes the name of the equation being defined, followed by the domain, the domain restriction condition (optional), the symbol "`..`", the left-hand-side expression, the relational operator: `=L=`, `=E=` or `=G=`, and the right-hand-side expression. A less-or-equal inequality constraint is indicated by `=L=`. A greater-or-equal constraint is specified by `=G=`. An equality constraint and the objective function are indicated by `=E=`.

Constraints and the objective function definitions require the use of expressions and assignments. GAMS provides a very rich set of *operators* and built-in *functions* for forming expressions. The list of operators in GAMS grouped on their priority is presented in Table 3.3.

The current GAMS guide contains the most up-to-date list of intrinsic and extrinsic functions. A partial list of functions in GAMS is shown in Table 3.4.

Expressions are built up from numerical values, that is, *floating point constants*, *scalars*, and *parameter* and *table* elements. Numerical values may be joined using operators.

The `ord` function takes as an argument the name of a set and returns the ordinate value of the index relative to the set. The `card` function takes a set name and returns the cardinality of the set, i.e., the number of elements in the set.

The functions `SUM`, `PROD`, `SMAX`, and `SMIN` can be used in GAMS for defining simple or complex expressions. The format of these functions is that of an operator with two arguments. For example, the summation notation is as follows: `sum (index of summation, summand)`. A comma separates the two arguments, and if the first argument requires a comma, then it should be placed in parentheses.

Table 3.3 Operators in GAMS expressions, grouped by priority

Operator	Description
`$`	Conditional operator
`**`	Exponentiation, a^b only for $a > 0$
`*, /`	Multiplication, division
`+, −`	Addition, subtraction
`lt, <`	Less than
`gt, >`	Greater than
`eq, =`	Equals
`le, <=`	Less than or equal to
`ge, >=`	Greater than or equal to
`ne, <>`	Not equal to
`not`	Logical Not
`and`	Logical And
`or`	Logical Or
`xor`	Logical Exclusive Or

Table 3.4 Functions in GAMS

Function	Description
`ord`	Ordinate value of index
`card`	Cardinality of set
`sum`	Summation over set
`prod`	Product over set
`smin`	Minimum over set
`smax`	Maximum over set
`exp(x)`	Exponential, e^x
`log(x)`	Natural log (for x > 0)
`log10(x)`	Base-10 log (for x > 0)
`normal(x,y)`	Normal distribution; mean x, std.dev. y
`uniform(x,y)`	Uniform distribution in [x,y[
`abs(x)`	Absolute value
`ceil(x)`	Smallest integer greater than or equal to x
`floor(x)`	Largest integer less than or equal to x
`max(x,y,...)`	Maximum of arguments
`min(x,y,...)`	Minimum of arguments
`mod(x,y)`	Remainder (modulo)
`power(x,y)`	Power, y must be an integer
`round(x)`	Rounding to nearest integer
`sign(x)`	−1, 0 or 1 depending on the sign of x
`sqr(x)`	Square of x
`sqrt(x)`	Square root of x
`cos(x)`	Cos, x in radians
`sin(x)`	Sin, x in radians
`arctan(x)`	Arcus tangent, result in radians

The second argument can be any mathematical expression including another summation. As a simple example, $\sum_j x_{ij}$ is expressed as `sum(j, x(i,j));`. A more complex summation operation like $\sum_i \sum_j c_{ij} x_{ij}$ can be represented as `sum((i,j), c(i,j)*x(i,j));`. This expression can also be written as a nested summation as follows: `sum(i, sum(j, c(i,j)*x(i,j)));`. Products are defined in GAMS using exactly the same format as summation but replacing `SUM` by `PROD`. For example, `prod(j, x(i,j));` is $\prod_j x_{ij}$. The functions `SMAX` and `SMIN` represent the element-wise maximum and element-wise minimum over the second argument values, respectively.

The index set over which the operation is applied can be qualified using the `$`-operator. At any time, GAMS performs an indexed operation; a `$`-condition can be used to specify a subset of the indices over which the operation should be performed. The dollar operator can be introduced in a number of contexts as it is illustrated below:

1. *Conditionally execute an assignment.* The condition `x$(y gt 0) = 20;` says that $x = 20$ if the scalar y is greater than zero.
2. *Conditionally add a term in sum or other set operation.* The condition `z = sum (i$(y(i) gt 0),x(i));` says include the term for set element i only if $y(i) > 0$.
3. *Conditionally define an equation.* The condition `Eq1 $(rr gt 0) .. sum(i, a(i)*x(i))=e=1;` says that if rr > 0, then equation $\sum_i a_i x_i = 1$ is included into the model.
4. *Conditionally include a term into an equation.* The statement `Eqc .. xvar + yvar $(aa gt 0) =e= 1;` says that in equation Eqc, yvar is added to xvar only if aa > 0.

3.7 Model Declarations

The word "model" has a precise meaning in GAMS. Model declarations serve to collect the constraints and the variables that are part of the model and to name the model. The format of the declaration is the keyword `model` followed by the name of the model, followed by a list of equation names (without indices) enclosed in slashes. If all the constraints defined in the source file up to this point are part of the model, then one can simply write: `model name-model /all/;`.

Models have „attributes" which can be used to communicate information to and from a solver. Some of these attributes are set by the user and correspond to setting the values using an `OPTION` statement. Table 3.5 shows the most important options.

Other attributes are set as a result of executing a `SOLVE` statement. They can be used to test the result of solving a model and hence decide on further actions to take (see the User's Guide of GAMS).

As already seen, the variable suffix `.FX` will fix a variable, i.e., its lower and upper bounds will be set to the same value. However, the variable is still present

Table 3.5 The most important options in GAMS

Solver control	
reslim	Maximum number of CPU seconds to solve. Default 1000
iterlim	Maximum number of iterations to solve. Default 1000
Listing and printing control	
limrow = n	List n equations for each equation block (default 3)
limcol = n	List n variables for each variable block (default 3)
sysout = ON/OFF	Include solver output files into listing
solprint = ON/OFF	List the solution after each solve statement
decimals = d	Print numerical data with d decimals
Solver selection	
Model type = solver	Specifies the solver to be used for the model type, e.g., LP = BDMLP. Table 3.6 contains the list of model types

The arguments n and d indicate nonnegative integers

into the model even though it has only a single feasible value. The model attribute holdfixed when set to 1, as in the below sequence of declarations,

```
MODEL circle /all/;
Circle.holdfixed = 1;
```

will cause the values of all fixed variables to be substituted for the value throughout in the model. This can greatly reduce the complexity of the model, for instance, a nonlinear model can be converted to a linear one.

3.8 The SOLVE Statement and Model Types

Once the model has been declared and the equations assigned, we are ready to call the solver to get the solution. This is done with the SOLVE statement which has the following general form:

```
SOLVE model_name MINIMIZING obj_var USING model_type;
```

where model_name is the model to be solved, obj_var is the variable whose value should be minimized (we can also ask for MAXIMIZING the value), and model_type indicates the type of the model to be solved. Table 3.6 presents the GAMS model types. GAMS will select a default solver that is capable to solve the model, or a desired solver can be specified.

For example, OPTION LP = BDMLP; will impose GAMS to use the program BDMLP for solving linear programming models.

Table 3.6 GAMS model types

Model types		Kinds of variables and constraints
LP	Linear program	Linear
MIP	Mixed integer	Linear, discrete
RMIP	Relaxed MIP	As MIP, solved as an LP
NLP	Nonlinear program	Linear, nonlinear
DNLP	Discontinuous NLP	Linear, nonlinear, non-diff. Constr.
MINLP	Mixed-integer NLP	Linear, discrete, nonlinear
RMINLP	Relaxed MINLP	As MINLP, solved as a NLP
MCP	Mixed complementarity problem	Complementarity constraints
CNS	Constrained nonlinear system	LP or NLP without objective

GAMS incorporates all major commercial and academic state-of-the-art solution technologies for a broad range of problem types as presented in Table 3.6. The most important model types and the associated solvers are as follows:

- *LP* – In this case the model contains only linear constraints (the function objective is also linear), and the variables are continuous. The specialized linear programming solvers in GAMS are BDMLP, CPLEX, SCIP, XA, and XPRESS.
- *MIP* – In this case the model contains linear constraints, including the objective function, but the variables are discrete (integer or binary). These kinds of models are very time-consuming, especially when there are a large number of variables and constraints. A good practice is to introduce upper bounds on variables (the implicit upper bounds is 100). Also, the default iteration and resource limits of 1000 iterations and 1000 CPU seconds, respectively, may be changed. For mixed-integer linear programming, the specialized solvers are CPLEX, OSL, SCIP, XA, and XPRESS.
- *RMIP* – For solving a MIP optimization model while ignoring the integrality constraints, we use RMIP. For these models GAMS can use BDMLP solver.
- *NLP* – In this case the model contains nonlinear constraints and continuous variables. These models are easy to solve when the constraint set is convex and the objective function is convex (for a minimization problem) or concave (for a maximization problem). GAMS always gives locally optimal solutions. For nonlinear programming, GAMS may use: CONOPT, KNOPT, MINOS, SNOPT, MOSEK, PATHNLP, and LGO.
- *MINLP* – The model may contain nonlinear expressions and discrete variables. For mixed-integer nonlinear programming in GAMS, the following solvers can be used: AlphaECP, DICOPT, LINDOGlobal, SBB, BARON, and OQNLP.
- *DNLP* – The model may contain nonlinear constraints that are not differentiable. The following solvers can be used: CONOPT, MINOS, SNOPT, BARON, LGO, OQNLP, and MOSEK.

For stochastic programming in GAMS, the following solvers are active: DECIS and OSLSE. The mixed complementarity problems can be solved with MILES, PATH, and NLPEC. For global optimization, the following solvers can be used:

BARON, COIN-OR, GLOMIQO, LINDOGLOBAL, LGO, MSNLP, and OQNLP. Details on GAMS solvers are found on the web page of GAMS.

To control the execution of statements depending on a condition, GAMS has the following control statements: `IF`, `WHILE`, `FOR`, and `LOOP`.

The `IF` statement has the following forms:

```
IF(condition, executable-statements);
IF(condition, executable-statements.
  ELSE executable-statements);
IF(condition, executable-statements.
  ELSEIF condition, executable-statements ....
  ELSE executable-statements);
```

where `condition` is a conditional expression and `executable-statements` is a semicolon-separated list of executable statements. The `IF...ELSEIF...` variant allows any number of `ELSEIF` parts, while the `ELSE` part is optional. The body of the `IF` statement contains `SOLVE` statements. However, we cannot make declarations or define equations inside `IF` statement.

The iterative control structures allow for repeated execution of groups of statements until some condition is satisfied (`WHILE`) or for the under control of either a scalar parameter (`FOR`) or a set index (`LOOP`). The syntax of these iterative control structures is as follows:

```
WHILE(condition, executable-statements);
FOR(param=val1 TO/DOWNTO val2 BY val3,
   executable-statements);
LOOP(set-index, executable-statements);
```

where `condition` is a conditional expression and `executable-statements` is a semicolon-separated list of executable statements.

Inside the `WHILE`, `FOR`, and `LOOP` statements, we cannot make declarations or define equations. On the other hand, one can use these iterative control structures to control `SOLVE` statement.

The following example illustrates the above iterative control structures (`WHILE`, `FOR`, and `LOOP`) for solving the nonlinear program “design” from 20 random start points and reports all the solutions.

```
model design/all/;
scalar t /1/;
while((t le 20),
     x.l(j) = uniform(x.lo(j), x.up(j));
     solve design using nlp minimizing cost;
     dispay x.l, cost.l;
     t=t + 1;
);
```

Table 3.7 Possible values of modelstat

Modelstat	Description
1	Optimal solution
2	Locally optimal solution
3	Unbounded solution
4	Infeasible solution
5	Locally infeasible solution
6	Incomplete solution, infeasible
7	Incomplete solution, feasible
8	Integer solution
9	Incomplete solution, noninteger
10	Integer infeasible solution

```
model design/all/;
scalar t;
for(t = 1 to 20,
      x.l(j) = uniform(x.lo(j), x.up(j));
      solve design using nlp minimizing cost;
      dispay x.l, cost.l;
);
model design/all/;
set t /1*20/;
loop(t,
      x.l(j) = uniform(x.lo(j), x.up(j));
      solve design using nlp minimizing cost;
      dispay x.l, cost.l;
);
```

In order to use the above iterative control structures, it is convenient to be able to check the status of the last solving of the model. The model status is available as quantity “`model_name.modelstat.`” Table 3.7 presents the possible values of modelstat.

3.9 DISPLAY and PUT Statements

The easiest way to output data and results of optimization is the `DISPLAY` statement. Data (scalars and parameters) are specified without indices. To display variables and parameters, it is necessary to specify which attributes we wish to display. For example, `X.L` is the level values, `X.M` the marginal values, and `X.LO` and `X.UP` the bounds.

To control the output format, the `OPTION` statement may be used. For example, `OPTION DECIMALS` = 8 causes data to be displayed with 8 decimals. On the other hand, `OPTION X:8` does the same for data related to the symbol `X`.

Multidimensional data are displayed in a tabular format. The option `Matrix:d:r:c` causes GAMS to display using `d` decimals, listing the last `c` indices across the top, the next `r` indices in the left-most column, and any remaining indices in separate tables.

The purpose of `PUT` writing facility is to output individual items under a format control onto different files (other than the listing file). The basic structure of the `PUT` writing facility in its simplest form is:

```
file f_name(s);
put f_name;
put item(s);
```

where `f_name` represents the name used inside the GAMS model to refer to an external file. Items are any type of output, such as explanatory text, labels, parameters, and variable or equation values. In the above structure, the first line defines one or more files in which we intend to write. The second line assigns one of these defined files as the current file. The third line represents the actual writing of output items to the current file.

3.10 GAMS Output

The output from GAMS contains plenty of information for checking and understanding the model we wish to solve and for visualizing its solution. The main outputs from GAMS are echo print, error messages, reference maps, equation and column listings, model statistics, status reports, and solution reports. Details on all these outputs can be found in GAMS User's Guide (Rosenthal 2011).

The first section of output from GAMS running is an echo, i.e., a copy of the input file. In this copy GAMS puts line numbers on the left-hand side of the echo, for future reference.

In case the GAMS compiler encounters an error in the input file, it inserts a coded error message inside the echo print on the line immediately following the error. These messages always start with **** and contain a "$" directly below the point at which the compiler found the error. The $ is followed by a numerical code which is explained after the echo print.

The next section of output, which is the last if errors have been detected, is a pair of reference maps that contain summaries and analyses of the input file for purposes of debugging and documentation. The first reference map is a cross-reference map which is an alphabetical cross-referenced list of all the entities of the model. The second part of the reference map is a list of model entities grouped by their type and listed with their associated documentary text.

The equation and column listings are a product of the `SOLVE` command. The equation listing shows the specific instance of the model that is created when the current values of the sets and parameters are plugged into the general algebraic form of the model. A maximum of three specific equations for each generic equation are output by default. To change the default prior to the `SOLVE` command, the input statement `option limrow = r;` must be inserted, where `r` is the desired number. The column listing shows the coefficients of three specific variables for each generic variable. To change the default number of specific column printouts per generic variable, the above command can be extended as `option limrow = r, limcol = c;`, where `c` is the desired number of columns. In particular, these input statements are important when the user has trouble in identifying what is wrong with a model. The corresponding outputs will have terms fully collected and coefficients evaluated so that the user can immediately see if there were errors into the model.

Before invoking the solver, GAMS produces the model statistics section in which a group of statistics about the model's size as well as some other statistics describing its degree of nonlinearity are output. A typical example is as follows:

```
MODEL STATISTICS
BLOCKS OF EQUATIONS    3   SINGLE EQUATIONS  210
BLOCKS OF VARIABLES    3   SINGLE VARIABLES   41
NON ZERO ELEMENTS    839
```

The `BLOCK` counts refer to the number of generic equations and variables. The `SINGLE` counts refer to individual rows and columns in the specific model instance when being generated. For nonlinear models, some other statistics are issued to describe the degree of nonlinearity of the model.

After the model has been solved, GAMS issues a brief solve summary. The most important entries are `SOLVER STATUS` and `MODEL STATUS`. The status reports are preceded by the string `****`. The desired solver status is: `1 NORMAL COMPLETION`, but there are some other possibilities as illustrated in Table 3.8.

The model status is described in Table 3.7 where the usual linear programming termination states (`1 OPTIMAL`, `3 UNBOUNDED`, `4 INFEASIBLE`) and the most important `2 LOCALLY OPTIMAL` for nonlinear models are issued.

Table 3.8 Possible values of solve status

Solve status	Description
1	Normal completion
2	Iteration interrupt
3	Resource interrupt
4	Terminated by solver
5	Evaluation error limit
6	Capability problems
7	Licensing problems
8	User interrupt

The last section of output is the solution reports. The results of the optimization are presented in a standard mathematical programming output format in which the rows and columns are grouped and labeled according to names that are appropriate for the specific model just solved. In this format, there is a line of printout for each row and column giving: the lower limit, the level, the upper limit, and the marginal value. The single dots in the output represent zeroes. The entry EPS means "very small" but nonzero.

At the end of the solvers' solution report, there is a very important *report summary* giving the total number of non-optimal, infeasible, and unbounded rows and columns.

3.11 Basic Solver Usage

This short presentation illustrates several of the capabilities of GAMS for building and solving practical large-scale optimization models. GAMS has an enormous number of features and options which allow it to support the most real sophisticated mathematical programming applications in virtually all domains of activity. An advanced presentation of GAMS is given in (Rosenthal, 2011).

The solvers active in GAMS allow the user to set additional, solver-specific options. In most cases, it is not necessary to use any solver-specific options, the use of the generic GAMS options is sufficient. Since the solvers interpret the GAMS options in a consistent way, a GAMS option setting applies to all solvers, not just to a specific one. In *GAMS – The Solvers Manuals* (GAMS – The Solvers Manuals, 2005), plenty of details on the GAMS options and the solver option file are presented.

In GAMS, options exist in two forms: global and model specific. The option statement sets a global GAMS option like `option iterlim = 100;`, while the model suffix sets a GAMS option for an individual model: `mymodel.iterlim = 20;`. The main GAMS options for controlling solvers are given in Table 3.9.

To specify solver-specific options, it is necessary to use a solver option file. Two things are required for doing this: an option file having a proper name must be created, and the solver must be instructed to read and use this option file. To tell a solver to use an option file, the `optfile` model suffix can be set to a positive value.

```
model mymodel/all/;
mymodel.optfile = 1;
solve mymodel using nlp minimizing fobj;
```

The option file takes its name from the solver being used: *solvername.xxx*, where "*solvername*" is the name of the solver that is specified and the suffix "*xxx*" depends on the value to which the model suffix `optfile` has been set. If its value is 1, the suffix is *opt*. For example, the option file for KNITRO is called *knitro.opt*; for

Table 3.9 The main GAMS options

Option	Description
iterlim	Sets a limit on the simplex iterations performed by the solver
reslim	Set the time limit in seconds
optfile	If nonzero, the solver should read an option file
bratio	GAMS uses the bratio value to determine if an advanced basis exists
domlim	Sets the domain violation limit. Domain errors are evaluation errors in the nonlinear functions (e.g., $\sqrt{x}$ for $x < 0$)
sysout	If option sysout = on GAMS will echo all the solver messages to the GAMS listing file
workfactor	Specifies a factor to be applied to the solver-computed memory estimate
workspace	Specifies the amount (in MB) of memory the solver should allocate

CONOPT it is *conopt.opt*; etc. If one does not set the `.optfile` suffix to a nonzero value, no option file will be used even if there is one.

3.12 Running a Job

GAMS is a two-pass program. Firstly, the user must create a file with the extension *gms* which contains GAMS instructions. When this file is complete, it is submitted to GAMS. In turn, GAMS executes those instructions causing calculations to be done and solvers to use and generate a solution file of the execution results. There are two alternatives for submitting the job: the command line approach and the IDE approach.

The basic procedure for running command line GAMS is to create a file with a text editor and, when done, run it with DOS or UNIX or other operating system command line instructions like `GAMS myfile.gms`, where myfile.gms is the file to be run. This command line allows for a number of arguments, as it is explained in the GAMS Command Line Parameters chapter.

The integrated development environment IDE is designed to provide a Windows graphical interface to allow for editing, developing, debugging, and running the GAMS models in one single program. When the IDE is run, there is again the creation of a file with extension `.lst`, i.e., if the input file of GAMS is called "myfile.gms," then the output will be "myfile.lst."

3.13 Program Development

The GAMS programs are developed in stages. A typical style is to put the sets first, followed by tables and data manipulations, then equations, solving commands, and finally the assignments used to generate reports. As each piece of the model is built, it should be run and checked for errors by inserting diagnostic display and abort statements. By restarting, it is possible to work only on the piece of program under active development, thus minimizing computer costs and efforts in each of the developmental runs. This approach is useful when entering the statements needed for reporting. The solution phase is much more expensive than the reporting, but the report is likely to involve many details of content and layout that have to be tried several times before they are satisfactory. The optimization model can be solved and the results saved in work files. One can then restart from these work files when developing the report.

In many cases where solves are known to be difficult, it may be too risky to ask GAMS to process a job containing many solve statements. The risk is that if one solve does not proceed to normal completion, then the following one will be started from a bad initial point and much time and effort will be wasted. An alternative to this scenario is to request one solve at a time and save the work file. Then the output is carefully inspected before proceeding. If everything is normal, the job is restarted with the next solve request. If not, the previous solve can be repeated, probably with a different initial point, or continued if the cause of the trouble was an iteration limit or limits amount of time used by the solver, for example. This approach is common when doing repeated solves of a model that successively represent several consecutive time periods (see the application *Optimal temperature field in a rectangular area – DIFF* in Chapter 4). It uses work files in a sequential rather than a tree-structure way.

The majority of modeling examples involves a "base case," and the point of the study is to see how the system changes when the circumstances change, either naturally or by design. This is often done by making many different changes to the base case and separately considering the effects; it is called *what if* analysis. The base case can be saved using work files, and after that as many different scenarios can be run separately by restarting as necessary. Each scenario may involve only making a change in data or in bounds, solving the changed model, and reporting the solution. Some details on this multiple scenarios are given in (Brooke, Kendrick, Meeraus, & Raman, 2005).

Notes and References

Details on the GAMS technology can be found in *GAMS – A User's Guide* (2005) and *GAMS – The Solver Manuals* (2005). This chapter is based on the papers by Brooke, Kendrick, Meeraus, Raman, and Rosenthal (1998), McKinney and Savitsky (2003), Brooke, Kendrick, Meeraus, and Raman (2005), and Andrei (2013b).

Chapter 4
Applications of Continuous Nonlinear Optimization

A number of 18 real continuous nonlinear optimization applications are presented in this chapter. These are used for *numerical experiments and comparisons among the algorithms described in this book*. For each application, the mathematical model, its GAMS representation, and the solution are given. GAMS is a standard technology for modeling and solving large-scale linear and nonlinear optimization applications. It is characterized by a very powerful language and a large number of different advanced optimization algorithms which are imbedded in this technology. The syntax of the GAMS language is not too complicated and practically all types of difficult nonlinear optimization applications can be represented and solved. Additionally, the nonlinear optimal control application, by discretization, can also be represented and solved by GAMS. Therefore, in this chapter, we include both continuous nonlinear optimization applications and some optimal control problems from different areas of activity. Some applications are from mechanical, electrical, and chemical engineering, heat transfer and fluid dynamics, and economic development. Others are from optimal control. *The purpose is to present these applications in algebraic form and to see how all these can be represented and solved in GAMS*. The solutions of these applications are determined by the optimization algorithms imbedded in the GAMS technology (MINOS, CONOPT, KNITRO, SNOPT, IPOPT) and also by some other packages described in this book (SPENBAR, DONLP, NLPQLP, filterSD) which are not imbedded in GAMS.

There is a large diversity of continuous nonlinear optimization applications explained and solved in this chapter. Practically, nonlinear optimization applications are found in every domain of activity, but we refer here only to mechanical, electrical, and chemical engineering, to heat transfer and fluid dynamics, or to economic development. The mathematical models of these applications have different structures and complexities and are taken from: Bartholomew-Biggs (1976), Dembo (1976), Rijckaert (1973), Hock and Schittkowski (1981), Schittkowski (1987), Floudas and Pardalos (1990), McKinney and Savitsky (2003), Dolan, Moré and Munson (2004), Andrei (2003, 2011b, 2013a, 2013b, 2015), etc.

N. Andrei, *Continuous Nonlinear Optimization for Engineering Applications in GAMS Technology*, Springer Optimization and Its Applications 121,
DOI 10.1007/978-3-319-58356-3_4

4.1 Chemical Equilibrium (ELCH)

This application is described in Hock and Schittkowski (1981, pp.123), Andrei (1999, pp.803), and Andrei (2015, pp.974):

$$\min \sum_{j=1}^{10} x_j \left(c_j + \ln \frac{x_j}{x_1 + \cdots + x_{10}} \right)$$

subject to: (4.1)

$$x_1 + 2x_2 + 2x_3 + x_6 + x_{10} - 2 = 0,$$
$$x_4 + 2x_5 + x_6 + x_7 - 1 = 0,$$
$$x_3 + x_7 + x_8 + 2x_9 + x_{10} - 1 = 0.$$

The GAMS expression of this application is presented in Figure 4.1.

```
$Ontext
Application ELCH (Chemical equilibrium)
References:
[1] Hock, W., Schittkowski,K., Test examples for nonlinear
programming codes. Lecture Notes in Economics and Mathematical
Systems, Springer Verlag, Berlin, 1981. (Problem 112, page 121)
[2] Andrei, N., Critica Ratiunii Algoritmilor de Optimizare cu
Restrictii. Editura Academiei Romane - Bucuresti, 2015.
ISBN: 978-973-27-2527-6. pp. 972-974.
[3] Andrei, N., Modele, Probleme de Test si Aplicatii de
programare Matematica. Editura Tehnica, Bucuresti, 2003.
(Aplicatia A15, pagina 362)
$Offtext

scalars
c1 /-6.089/,         c2 /-17.164/,
c3 /-34.054/,        c4 /-5.914/,
c5 /-24.721/,        c6 /-14.986/,
c7 /-24.100/,        c8 /-10.708/,
c9 /-26.662/,        c10/-22.179/;

Variables x1,x2,x3,x4,x5,x6,x7,x8,x9,x10, obj;
Equations e1,e2,e3,e;

e1.. x1+2*x2+2*x3+x6+x10 - 2 =e= 0;
e2.. x4+2*x5+x6+x7 - 1 =e= 0;
e3.. x3+x7+x8+2*x9+x10 - 1 =e= 0;
e..  obj =e=  x1*(c1+log(x1/(x1+x2+x3+x4+x5+x6+x7+x8+x9+x10)))+
              x2*(c2+log(x2/(x1+x2+x3+x4+x5+x6+x7+x8+x9+x10)))+
              x3*(c3+log(x3/(x1+x2+x3+x4+x5+x6+x7+x8+x9+x10)))+
              x4*(c4+log(x4/(x1+x2+x3+x4+x5+x6+x7+x8+x9+x10)))+
```

Figure 4.1 GAMS expression of application ELCH

```
                x5*(c5+log(x5/(x1+x2+x3+x4+x5+x6+x7+x8+x9+x10)))+
                x6*(c6+log(x6/(x1+x2+x3+x4+x5+x6+x7+x8+x9+x10)))+
                x7*(c7+log(x7/(x1+x2+x3+x4+x5+x6+x7+x8+x9+x10)))+
                x8*(c8+log(x8/(x1+x2+x3+x4+x5+x6+x7+x8+x9+x10)))+
                x9*(c9+log(x9/(x1+x2+x3+x4+x5+x6+x7+x8+x9+x10)))+

x10*(c10+log(x10/(x1+x2+x3+x4+x5+x6+x7+x8+x9+x10)));

x1.lo = 1.e-6;  x2.lo = 1.e-6;
x3.lo = 1.e-6;  x4.lo = 1.e-6;
x5.lo = 1.e-6;  x6.lo = 1.e-6;
x7.lo = 1.e-6;  x8.lo = 1.e-6;
x9.lo = 1.e-6;  x10.lo = 1.e-6;

*Initial point
x1.l =  0.1;  x2.l = 0.1;
x3.l =  0.1;  x4.l = 0.1;
x5.l =  0.1;  x6.l = 0.1;
x7.l =  0.1;  x8.l = 0.1;
x9.l =  0.1;  x10.l = 0.1;

Model elch /all/;

OPTION NLP=KNITRO;
OPTION SYSOUT=ON;

Solve elch using nlp minimizing obj;

file out /elch.dat/
put out;
put x1.l:15:7, x2.l:15:7, x3.l:15:7, x4.l:15:7, x5.l:15:7 /;
put x6.l:15:7, x7.l:15:7, x8.l:15:7, x9.l:15:7, x10.l:15:7 /;
put obj.l:15:7/;
* End elch
```

Figure 4.1 (continued)

The solution of the application is as follows:

Nr.	C_j	Initial point	Lower bound	Solution
1	–6.089	0.1	1.e–6	0.0406685
2	–17.164	0.1	1.e–6	0.1477301
3	–34.054	0.1	1.e–6	0.7831533
4	–5.914	0.1	1.e–6	0.001414229
5	–24.721	0.1	1.e–6	0.4852466
6	–14.986	0.1	1.e–6	0.0006931799
7	–24.100	0.1	1.e–6	0.02739941
8	–10.708	0.1	1.e–6	0.01794741
9	–26.662	0.1	1.e–6	0.03731418
10	–22.179	0.1	1.e–6	0.09687152

$$f(x_0) = -20.96029 \qquad f(x^*) = -47.761090859$$

4.2 Optimization of an Alkilation Process (ALKI)

This application is taken from Hock and Schittkowski (1981, pp.123), Andrei (1999, pp.803), and Andrei (2015, pp.974):

$$\min(5.04x_1 + 0.035x_2 + 10x_3 + 3.36x_5 - 0.063x_4x_7)$$

subject to: (4.2)

$$\begin{aligned}
&35.82 - 0.222x_{10} - bx_9 \geq 0,\\
&-133 + 3x_7 - ax_{10} \geq 0,\\
&-35.82 + 0.222x_{10} + bx_9 + (1/b - b)x_9 \geq 0,\\
&133 - 3x_7 + ax_{10} + (1/a - a)x_{10} \geq 0,\\
&1.12x_1 + 0.13167x_1x_8 - 0.00667x_1x_8^2 - ax_4 \geq 0,\\
&57.425 + 1.098x_8 - 0.038x_8^2 + 0.325x_6 - ax_7 \geq 0,\\
&-1.12x_1 - 0.13167x_1x_8 + 0.00667x_1x_8^2 + ax_4 + (1/a - a)x_4 \geq 0,\\
&-57.425 - 1.098x_8 + 0.038x_8^2 - 0.325x_6 + ax_7 + (1/a - a)x_7 \geq 0,\\
&1.22x_4 - x_1 - x_5 = 0,\\
&98000x_3/(x_4x_9 + 1000x_3) - x_6 = 0,\\
&(x_2 + x_5)/x_1 - x_8 = 0,
\end{aligned}$$

where $a = 0.99$ and $b = 0.9$.

The GAMS expression of this application is presented in Figure 4.2.

```
$Ontext
Application ALKI (Alkilation process)
References:
[1] Hock, W., Schittkowski, K., Test examples for nonlinear
programming codes. Lecture Notes in Economics and Mathematical
Systems, Springer Verlag, Berlin, 1981. (Problem 114, page 123)
[2] Andrei, N., Critica Ratiunii Algoritmilor de Optimizare cu
Restrictii. Editura Academiei Romane - Bucuresti, 2015.
ISBN: 978-973-27-2527-6. pp. 974-976.
[3] Andrei, N., Modele, Probleme de Test si Aplicatii de
programare Matematica. Editura Tehnica, Bucuresti, 2003.
(Aplicatia A19, pagina 369)
$Offtext

scalars a /0.99/, b /0.9/;
Variables x1,x2,x3,x4,x5,x6,x7,x8,x9,x10, obj;
Equations e1,e2,e3,e4,e5,e6,e7,e8,e9,e10,e11, e;
e1.. 35.82-0.222*x10-b*x9 =g= 0;
e2.. -133+3*x7-a*x10 =g= 0;

e3.. -35.82+0.222*x10+b*x9+x9*(1/b-b) =g= 0;
e4.. 133-3*x7+a*x10+x10*(1/a-a) =g= 0;

e5.. 1.12*x1+0.13167*x1*x8-0.00667*x1*x8*x8-a*x4 =g= 0;
```

Figure 4.2 GAMS expression of application ALKI

```
e6.. 57.425+1.098*x8-0.038*x8*x8+0.325*x6-a*x7 =g= 0;

e7.. -1.12*x1-0.13167*x1*x8+0.00667*x1*x8*x8+a*x4+x4*(1/a-a)
     =g= 0;
e8.. -57.425-1.098*x8+0.038*x8*x8-0.325*x6+a*x7+x7*(1/a-a)
     =g= 0;

e9.. 1.22*x4-x1-x5 =e= 0;
e10.. 98000*x3/(x4*x9+1000*x3)-x6 =e= 0;
e11.. (x2+x5)/x1-x8 =e= 0;

e..  obj =e= 5.04*x1+0.035*x2+10*x3+3.36*x5-0.063*x4*x7;

* Bounds on variables
x1.lo = 0.00001;  x2.lo = 0.00001;
x3.lo = 0.00001;  x4.lo = 0.00001;
x5.lo = 0.00001;  x6.lo = 85;
x7.lo = 90;       x8.lo = 3;
x9.lo = 1.2;      x10.lo = 145;

x1.up = 2000;     x2.up = 16000;
x3.up = 120;      x4.up = 5000;
x5.up = 2000;     x6.up = 93;
x7.up = 95;       x8.up = 12;
x9.up = 4;        x10.up = 162;

*Initial point
x1.l =  1745;      x2.l = 12000;
x3.l =  110;       x4.l = 3048;
x5.l =  1974;      x6.l = 89.2;
x7.l =  92.8;      x8.l = 8;
x9.l =  3.6;       x10.l = 145;

Model alki /all/;
alki.optfile=1;
option nlp=knitro;
option sysout=on;

Solve alki using nlp minimizing obj;

file out /alki.dat/
put out;
put x1.l:15:7, x2.l:15:7, x3.l:15:7, x4.l:15:7, x5.l:15:7 /;
put x6.l:15:7, x7.l:15:7, x8.l:15:7, x9.l:15:7, x10.l:15:7 /;
put obj.l:15:7/;
* End alki
```

Figure 4.2 (continued)

The solution of the application is as follows:

Nr.	Initial point	Lower bound	Solution	Upper bound
1	1745	0.00001	1698.094	2000
2	12,000	0.00001	15818.51	16,000
3	110	0.00001	54.10228	120
4	3048	0.00001	3031.226	5000
5	1974	0.00001	2000	2000
6	89.2	85	90.11548	93
7	92.8	90	95	95
8	8	3	10.49324	12
9	3.6	1.2	1.561636	4
10	145	145	153.5354	162

$$f(x_0) = -872.3874 \qquad f(x^*) = -1768.80696$$

4.3 Optimal Design of a Reactor as a Geometric Programming Problem (PREC)

This application is presented in Dembo (1976), Rijckaert (1973), Andrei (1999, pp.801), and Hock and Schittkowski (1981; pp.113):

$$\min\left(0.4x_1^{0.67}x_7^{-0.67} + 0.4x_2^{0.67}x_8^{-0.67} - x_1 - x_2 + 10\right)$$

subject to: (4.3)

$$\begin{aligned}
&1 - 0.0588x_5x_7 - 0.1x_1 \geq 0,\\
&1 - 0.0588x_6x_8 - 0.1x_1 - 0.1x_2 \geq 0,\\
&1 - 4x_3x_5^{-1} - 2x_3^{-0.71}x_5^{-1} - 0.0588x_3^{-1.3}x_7 \geq 0,\\
&1 - 4x_4x_6^{-1} - 2x_4^{-0.71}x_6^{-1} - 0.0588x_4^{-1.3}x_8 \geq 0,\\
&0.4x_1^{0.67}x_7^{-0.67} + 0.4x_2^{0.67}x_8^{-0.67} - x_1 - x_2 + 10 \geq 1,\\
&0.4x_1^{0.67}x_7^{-0.67} + 0.4x_2^{0.67}x_8^{-0.67} - x_1 - x_2 + 10 \leq 4.2.
\end{aligned}$$

The GAMS expression of this application is presented in Figure 4.3.

```
$Ontext
Application PREC (A variant of optimal reactor design problem)
References:
[1] Dembo, R.S., A set of geometric programming test problems
and their solutions. Mathematical Programming, vol. 10, 1976,
pp. 192-213.
[2] Floudas, C.A., Pardalos, P.M., et al. Handbook of Test
Problems in Local and Global Optimization. Kluwer Academic
Publishers, Dordrecht, 1999. Section 7.2.4, (Test problem 4,
pp. 91-92)
[3] Andrei, N., Nonlinear Optimization Applications using the
GAMS Technology. Spriger Science + Business Media New York,
2013. (Application 7.1, Chapter 7, pp. 161-162)
$Offtext

Variables x1,x2,x3,x4,x5,x6,x7,x8, obj;

Equations e1,e2,e3,e4, eobj;

* Constraints:

e1.. 0.0588*x5*x7 + 0.1*x1 =l= 1;

e2.. 0.0588*x6*x8 + 0.1*x1 + 0.1*x2 =l= 1;

e3.. 4*x3/x5 + 2*(x3**(-0.71))/x5 + 0.0588*x7*(x3**(-1.3))
     =l= 1;

e4.. 4*x4/x6 + 2*(x4**(-0.71))/x6 + 0.0588*x8*(x4**(-1.3))
     =l= 1;
* Objective function to be minimized:
eobj.. obj =e= 0.4*(x1**(0.67))*(x7**(-0.67)) +
               0.4*(x2**(0.67))*(x8**(-0.67)) + 10 - x1 - x2;

* Bounds on variables:
x1.lo = 0.1;   x1.up = 10;
x2.lo = 0.1;   x2.up = 10;
x3.lo = 0.1;   x3.up = 10;
x4.lo = 0.1;   x4.up = 10;
x5.lo = 0.1;   x5.up = 10;
x6.lo = 0.1;   x6.up = 10;
x7.lo = 0.1;   x7.up = 10;
x8.lo = 0.1;   x8.up = 10;
Model reactor /all/;
reactor.optfile=1;
option nlp=knitro;
option sysout=on;
Solve reactor minimizing obj using nlp;
file out /prec.dat/
put out;
put x1.l:15:7, x2.l:15:7, x3.l:15:7, x4.l:15:7 /;
put x5.l:15:7, x6.l:15:7, x7.l:15:7, x8.l:15:7 /;
put obj.l:15:7/;
* End PREC
```

Figure 4.3 GAMS expression of application PREC

The solution of the application is as follows:

Nr.	Initial point	Lower bound	Solution	Upper bound
1	6	0.1	6.465115	10
2	3	0.1	2.232708	10
3	0.4	0.1	0.6673975	10
4	0.2	0.1	0.5957564	10
5	6	0.1	5.932676	10
6	6	0.1	5.527235	10
7	1	0.1	1.013322	10
8	0.5	0.1	0.4006682	10

$$f(x_0) = 3.657366 \qquad f(x^*) = 3.951163508$$

4.4 Cost Minimization of a Transformer Design (TRAFO)

The objective function of this application represents the worth of the transformer, including the operating cost. The constraints refer to the rating of the transformer and to the allowable transmission loss. The variables x_1, x_2, x_3, and x_4 are physical dimensions of winding and core. The variables x_5 and x_6 are the magnetic flux density and the current density, respectively (Price, 1983; Ballard Jelinek & Schinzinger, 1974). The mathematical model is described in Hock and Schittkowski (1981) and Andrei (2003, Application A1, pp. 344):

$$\begin{aligned}\min\ & 0.0204x_1x_4(x_1+x_2+x_3)+0.0187x_2x_3(x_1+1.57x_2+x_4)+\\ & 0.0607x_1x_4x_5^2(x_1+x_2+x_3)+0.0437x_2x_3x_6^2(x_1+1.57x_2+x_4)\end{aligned}$$

subject to: (4.4)

$$\begin{aligned}& 0.001x_1x_2x_3x_4x_5x_6-2.07\geq 0,\\ & 1-0.00062x_1x_4x_5^2(x_1+x_2+x_3)\\ & -0.00058x_2x_3x_6^2(x_1+1.57x_2+x_4)\geq 0,\end{aligned}$$

where the variables are bounded as $x_i \geq 0$, $i = 1, \ldots, 6$.

The GAMS expression of this model is presented in Figure 4.4.

The solution of this application is:

$$\begin{aligned}& x_1^* = 5.3326663, \quad x_2^* = 4.6567441, \quad x_3^* = 10.4329919,\\ & x_4^* = 12.0823063, \quad x_5^* = 0.7526074, \quad x_6^* = 0.8786509.\end{aligned}$$

```
$ONTEXT
Transformer design
References:
[1] Andrei, N., Nonlinear Optimization Applications using the
GAMS Technology, Springer Optimization and Its Applications 81,
Springer Science+Business Media, New York, 2013. (Application
6.1, pp.137-139)
[2] Hock, W., Schittkowski, K., Test Examples for Nonlinear
Programming Codes. Springer Verlag, New York, 1981. (Problem
93. pp. 108)
$OFFTEXT
VARIABLES
         x1, x2, x3, x4, x5, x6,
         objval    objective function variable;
FREE VARIABLES    objval;
EQUATIONS
         f     Objective function
         g1
         g2;
f  .. objval
  =e=0.0204*x1*x4*(x1+x2+x3)+0.0187*x2*x3*(x1+1.57*x2+x4)+
                   0.0607*x1*x4*POWER(x5,2)*(x1+x2+x3)+
                   0.0437*x2*x3*POWER(x6,2)*(x1+1.57*x2+x4);
g1  .. 0.001*x1*x2*x3*x4*x5*x6 =g= 2.07;
g2  .. -0.00062*x1*x4*POWER(x5,2)*(x1+x2+x3)-
        0.00058*x2*x3*POWER(x6,2)*(x1+1.57*x2+x4) =g= -1;
* Bounds on variables:
x1.LO = 0;
x2.LO = 0;
x3.LO = 0;
x4.LO = 0;
x5.LO = 0;
x6.LO = 0;
* Initial point:
x1.L=5.4;
x2.L=4.4;
x3.L=11.02;
x4.L=11.82;
x5.L=0.702;
x6.L=0.852;
MODEL trafo /ALL/;
$onecho >bench.opt
  solvers conopt knitro snopt
$offecho
trafo.optfile=1;
option nlp=bench;
SOLVE trafo USING NLP MINIMIZING objval;
file out /traf.dat/
put out
put x1.l:15:7,  x2.l:15:7,  x3.l:15:7,  x4.l:15:7, x5.l:15:7,
x6.l:15:7  /;
put objval.l:15:7 /;
* End trafo
```

Figure 4.4 GAMS expression of application TRAFO

4.5 Optimization of a Multi-spindle Automatic Lathe (LATHE)

The optimization of a multi-spindle automatic lathe is to minimize a nonlinear objective subject to 15 nonlinear constraints with 10 variables. It is taken from (Schittkowski, 1987, pp.195) and has the following mathematical expression:

$$\min\left(-20000\frac{0.15x_1+14x_2-0.06}{0.002+x_1+60x_2}\right)$$

subject to: (4.5)

$$\begin{aligned}
&x_1-\frac{0.75}{x_3x_4}\geq 0,\\
&x_1-\frac{x_9}{x_4x_5}\geq 0,\\
&x_1-\frac{x_{10}}{x_4x_6}-\frac{10}{x_4}\geq 0,\\
&x_1-\frac{0.19}{x_4x_7}-\frac{10}{x_4}\geq 0,\\
&x_1-\frac{0.125}{x_4x_8}\geq 0,\\
&10000x_2-0.00131x_9x_5^{0.666}x_4^{1.5}\geq 0,\\
&10000x_2-0.001038x_{10}x_6^{1.6}x_4^{3}\geq 0,\\
&10000x_2-0.000223x_7^{0.666}x_4^{1.5}\geq 0,\\
&10000x_2-0.000076x_8^{3.55}x_4^{5.66}\geq 0,\\
&10000x_2-0.000698x_3^{1.2}x_4^{2}\geq 0,\\
&10000x_2-0.00005x_3^{1.6}x_4^{3}\geq 0,\\
&10000x_2-0.00000654x_3^{2.42}x_4^{4.17}\geq 0,\\
&10000x_2-0.000257x_3^{0.666}x_4^{1.5}\geq 0,\\
&30-2.003x_4x_5-1.885x_4x_6-0.184x_4x_8-2x_4x_3^{0.803}\geq 0,\\
&x_9+x_{10}-0.255=0,
\end{aligned}$$

where the variables are bounded as:

$$0\leq x_1\leq 10,\quad 0\leq x_2\leq 0.1,\quad 0.5e-4\leq x_3\leq 0.0081,\quad 10\leq x_4\leq 1000,$$
$$0.5e-4\leq x_5\leq 0.0017,\quad 0.5e-4\leq x_6\leq 0.0013,\quad 0.5e-4\leq x_7\leq 0.0027,$$
$$0.5e-4\leq x_8\leq 0.002,\quad 0.5e-4\leq x_9\leq 1,\quad 0.5e-4\leq x_{10}\leq 1.$$

The GAMS expression of this application is presented in Figure 4.5.

The solution of the problem is:

$$x_1^*=0.1472722,\quad x_2^*=0.1,\quad x_3^*=0.0081,\quad x_4^*=628.7173075,\quad x_5^*=0.0017,$$
$$x_6^*=0.0011816,\quad x_7^*=0.0027,\quad x_8^*=0.00135,\quad x_9^*=0.1574074,$$
$$x_{10}^*=0.0975926.$$

```
$Ontext
Optimization of a Multi-spindle Automatic Lathe.
References:
[1] Andrei, N., Nonlinear Optimization Applications using the
GAMS Technology, Springer Optimization and Its Applications 81,
Springer Science+Business Media, New York, 2013. (Application
5.13, pp.90-94)
[2] Schittkowski, K., More test examples for nonlinear
programming codes. Lecture Notes in Economics and Mathematical
Systems, Springer-Verlag, Berlin, 1987. (Problem 376, pp. 195)
$Offtext

Variables x1, x2, x3, x4, x5, x6, x7, x8, x9, x10, obj;
Equations e1, e2, e3, e4, e5, e6, e7, e8, e9, e10, e11, e12,
          e13,e14,e15, eobj;

* Objective function to be minimized:
eobj.. obj =e= -20000*(0.15*x1+14*x2-0.06)/(0.002+x1+60*x2);

* Constraints:
e1.. x1 - 0.75/(x3*x4) =g= 0;
e2.. x1 - x9/(x4*x5) =g= 0;
e3.. x1 - x10/(x4*x6) - 10/x4 =g= 0;
e4.. x1 - 0.19/(x4*x7) - 10/x4 =g= 0;
e5.. x1 - 0.125/(x4*x8) =g= 0;
e6.. 10000*x2 - 0.00131*x9*(x5**0.666)*(x4**1.5) =g= 0;
e7.. 10000*x2 - 0.001038*x10*(x6**1.6)*(x4**3) =g= 0;
e8.. 10000*x2 - 0.000223*(x7**0.666)*(x4**1.5) =g= 0;
e9.. 10000*x2 - 0.000076*(x8**3.55)*(x4**5.66) =g= 0;
e10.. 10000*x2 - 0.000698*(x3**1.2)*(x4**2) =g= 0;
e11.. 10000*x2 - 0.00005*(x3**1.6)*(x4**3) =g= 0;
e12.. 10000*x2 - 0.00000654*(x3**2.42)*(x4**4.17) =g= 0;
e13.. 10000*x2 - 0.000257*(x3**0.666)*(x4**1.5) =g= 0;
e14.. 30 - 2.003*x4*x5 - 1.885*x4*x6 - 0.184*x4*x8 -
          2*x4*(x3**0.803) =g= 0;
e15.. x9 + x10 - 0.255 =e= 0;

* Bounds on variables;
x1.lo  = 0;          x1.up = 10;
x2.lo  = 0;          x2.up = 0.1;
x3.lo  = 0.5e-4;     x3.up = 0.0081;
x4.lo  = 10;         x4.up = 1000;
x5.lo  = 0.5e-4;     x5.up = 0.0017;
x6.lo  = 0.5e-4;     x6.up = 0.0013;
x7.lo  = 0.5e-4;     x7.up = 0.0027;
x8.lo  = 0.5e-4;     x8.up = 0.002;
x9.lo  = 0.5e-4;     x9.up = 1;
x10.lo = 0.5e-4;     x10.up = 1;
```

Figure 4.5 GAMS expression of application LATHE

```
Model lathe /all/;
$onecho >bench.opt
  solvers conopt knitro minos snopt
$offecho
lathe.optfile=1;
option nlp=bench
Solve lathe minimizing obj using nlp;
file out /lathel.dat/
put out;
put x1.l:15:7,  x2.l:15:7,  x3.l:15:7,  x4.l:15:7 /;
put x5.l:15:7,  x6.l:15:7,  x7.l:15:7,  x8.l:15:7 /;
put x9.l:15:7,  x10.l:15:7 /;
put /"Objective ="  obj.l:15:7 /;
* End lathe
```

Figure 4.5 (continued)

4.6 Static Power Scheduling (PPSE)

This application involves two electrical generators connected into a net with three nodes. Variables x_1 and x_2 are the real power outputs from the generators, x_3 and x_4 represent the reactive power outputs, and x_5, x_6, and x_7 are the voltage magnitudes at the nodes of the electrical network. Finally, x_8 and x_9 are the voltage phase angles at two of the nodes. The constraints of the model, other than the simple bounds on variables, are the real and reactive power balance equations, the constraint stating that the power flowing into a node must balance the power flowing out. The mathematical model, described in Bartholomew-Biggs (1976), Hock and Schittkowski (1981, pp. 116), Andrei (2003, pp. 347), and Andrei (2013b, pp. 144), is as follows:

$$\min\left(3000x_1 + 1000x_1^3 + 2000x_2 + 666.667x_2^3\right)$$

subject to: (4.6)

$$\begin{aligned}
&0.4 - x_1 + 2Cx_5^2 + x_5x_6(D\sin(-x_8) - C\cos(-x_8)) + \\
&\qquad x_5x_7(D\sin(-x_9) - C\cos(-x_9)) = 0, \\
&0.4 - x_2 + 2Cx_6^2 + x_5x_6(D\sin(x_8) - C\cos(x_8)) + \\
&\qquad x_6x_7(D\sin(x_8 - x_9) - C\cos(x_8 - x_9)) = 0, \\
&0.8 + 2Cx_7^2 + x_5x_7(D\sin(x_9) - C\cos(x_9)) + \\
&\qquad x_6x_7(D\sin(x_9 - x_8) - C\cos(x_9 - x_8)) = 0, \\
&0.2 - x_3 + 2Dx_5^2 + x_5x_6(C\sin(-x_8) + D\cos(-x_8)) - \\
&\qquad x_5x_7(C\sin(-x_9) + D\cos(-x_9)) = 0, \\
&0.2 - x_4 + 2Dx_6^2 - x_5x_6(C\sin(x_8) + D\cos(x_8)) - \\
&\qquad x_6x_7(C\sin(x_8 - x_9) + D\cos(x_8 - x_9)) = 0, \\
&-0.337 + 2Dx_7^2 - x_5x_7(C\sin(x_9) + D\cos(x_9)) - \\
&\qquad x_6x_7(C\sin(x_9 - x_8) + D\cos(x_9 - x_8)) = 0,
\end{aligned}$$

where $C = \sin(0.25)48.4/50.176$ and $D = \cos(0.25)48.4/50.176$. The simple bounds on variables are $x_i \geq 0$, $i = 1, 2$ and $0.90909 \leq x_i \leq 1, 0909$, for $i = 5, 6, 7$.

```
$Ontext
Static power scheduling.
References:
[1] Andrei, N., Critica Ratiunii Algoritmilor de Optimizare cu
Restrictii. Editura Academiei Romane - Bucuresti, 2015.
ISBN: 978-973-27-2527-6. pp. 978-981.
[2] Andrei, N., Nonlinear Optimization Applications using the
GAMS Technology, Springer Optimization and Its Applications 81,
Springer Science+Business Media, New York, 2013. (Application
6.4, pp.144-147)
[3] Bartholomew-Biggs, M.C., A numerical comparison between two
approaches to nonlinear programming problems. Technical Report
No.77, 1976. Hatfield: Numerical Optimization Centre.
[4] Hock, W., Schittkowski, K., Test examples for nonlinear
programming codes. Lecture Notes in Economics and Mathematical
Systems. Vol. 187, Springer-Verlag, Berlin, 1981. (Problem 107,
pp. 116)
$Offtext
Parameters C, D;

C = sin(0.25)*48.4/50.176;
D = cos(0.25)*48.4/50.176;

* Variables x1 and x2 are the real power outputs from two
* generators, x3 and x4 are the reactive power outputs; x5,
* x6 and x7 are the voltage magnitudes at three nodes of an
* electrical network; and x8 and x9 are the voltage phase
* angles at two of these nodes.

Variables x1, x2, x3, x4, x5, x6, x7, x8, x9, obj;

Equations e1, e2, e3, e4, e5, e6, eobj;

* Objective function:

eobj.. obj =e= 3000*x1 + 1000*x1**3 + 2000*x2 + 666.667*x2**3;

* Constraints:
* The constraints are the real and the reactive power balance
* equations at each node of the electrical network.

e1.. 0.4 - x1 + 2*C*x5**2 + x5*x6*(D*sin(-x8)-C*cos(-x8)) +
                x5*x7*(D*sin(-x9)-C*cos(-x9)) =e= 0;

e2.. 0.4 - x2 + 2*C*x6**2 + x5*x6*(D*sin(x8)-C*cos(x8)) +
                x6*x7*(D*sin(x8-x9)-C*cos(x8-x9)) =e= 0;

e3.. 0.8 + 2*C*x7**2 + x5*x7*(D*sin(x9)-C*cos(x9)) +
                x6*x7*(D*sin(x9-x8)-C*cos(x9-x8)) =e= 0;

e4.. 0.2 - x3 + 2*D*x5**2 - x5*x6*(C*sin(-x8)+D*cos(-x8)) -
                x5*x7*(C*sin(-x9)+D*cos(-x9)) =e= 0;

e5.. 0.2 - x4 + 2*D*x6**2 - x5*x6*(C*sin(x8)+D*cos(x8)) -
                x6*x7*(C*sin(x8-x9)+D*cos(x8-x9)) =e= 0;

e6.. -0.337 + 2*D*x7**2 - x5*x7*(C*sin(x9)+D*cos(x9)) -
                x6*x7*(C*sin(x9-x8)+D*cos(x9-x8)) =e= 0;

* Bounds on variables:
x1.lo = 0;
x2.lo = 0;
x5.lo = 0.90909;  x5.up = 1.0909;
x6.lo = 0.90909;  x6.up = 1.0909;
x7.lo = 0.90909;  x7.up = 1.0909;
```

Figure 4.6 GAMS expression of application PPSE

```
Model static /all/;
$onecho >bench.opt
  solvers conopt knitro minos snopt
$offecho
static.optfile=1;
option nlp=bench;

Solve static minimizing obj using nlp;

file out /static1.dat/
put out;
put x1.1:15:7,  x2.1:15:7,  x3.1:15:7,  x4.1:15:7, x5.1:15:7,
    x6.1:15:7,  x7.1:15:7,  x8.1:15:7,  x9.1:15:7  /;
put obj.1:15:7 /;
* End static
```

Figure 4.6 (continued)

The GAMS model of this application is given in Figure 4.6.
The solution is:

$$x_1^* = 0.6670128, \quad x_2^* = 1.0223847, \quad x_3^* = 0.2282871, \quad x_4^* = 0.1848218,$$
$$x_5^* = 1.0909, \quad x_6^* = 1.0909, \quad x_7^* = 1.069036, \quad x_8^* = 0.1066106,$$
$$x_9^* = -0.3387876.$$

4.7 Optimization of a Separation Process in a Membrane with three Stages (MSP3)

This application is taken from Dembo (1976), Hock and Schittkowski (1981, pp. 124), and Andrei (1999, pp.802):

$$\min(x_{11} + x_{12} + x_{13})$$

subject to: (4.7)

$$x_3 - x_2 \geq 0,$$
$$x_2 - x_1 \geq 0,$$
$$1 - 0.002x_7 + 0.002x_8 \geq 0,$$
$$x_{13} - 1.262626x_{10} + 1.231059x_3x_{10} \geq 0,$$
$$x_5 - 0.03475x_2 - 0.975x_2x_5 + 0.00975x_2^2 \geq 0,$$
$$x_6 - 0.03475x_3 - 0.975x_3x_6 + 0.00975x_3^2 \geq 0,$$

$$x_5x_7 - x_1x_8 - x_4x_7 + x_4x_8 \geq 0,$$

$$1 - 0.002(x_2x_9 + x_5x_8 - x_1x_8 - x_6x_9) - x_5 - x_6 \geq 0,$$

$$x_2x_9 - x_3x_{10} - x_6x_9 - 500x_2 + 500x_6 + x_2x_{10} \geq 0,$$

$$x_2 - 0.9 - 0.002(x_2x_{10} - x_3x_{10}) \geq 0,$$

$$x_4 - 0.03475x_1 - 0.975x_1x_4 + 0.00975x_1^2 \geq 0,$$

$$x_{11} - 1.262626x_8 + 1.231059x_1x_8 \geq 0,$$

$$x_{12} - 1.262626x_9 + 1.231059x_2x_9 \geq 0,$$

$$x_{11} + x_{12} + x_{13} \geq 50,$$

$$x_{11} + x_{12} + x_{13} \leq 250.$$

The solution of this application is as follows:

Nr.	Initial point	Lower bound	Solution	Upper bound
1	0.5	0.1	0.8037730	1
2	0.8	0.1	0.8999858	1
3	0.9	0.1	0.9709695	1
4	0.099	0.0001	0.09999994	0.1
5	0.899	0.1	0.1908132	0.9
6	0.5	0.1	0.4605417	0.9
7	489	0.1	574.0773	1000
8	80	0.1	74.0776	1000
9	650	500	500.0162	1000
10	450	0.1	0.1	500
11	149.9	1	20.23311	150
12	149.9	0.0001	77.34769	150
13	149.9	0.0001	0.006730541	150

$$f(x_0) = 449.700 \qquad f(x^*) = 97.587532422$$

Figure 4.7 presents the GAMS representation of this application.

```
$Ontext
Application MSP3 (3-stage membrane separation)
References:
[1] Hock, W., Schittkowski,K., Test examples for nonlinear
programming codes. Lecture Notes in Economics and Mathematical
Systems, Springer Verlag, Berlin, 1981. (Problem 116, page
124).
[2] Andrei, N., Critica Ratiunii Algoritmilor de Optimizare cu
Restrictii. Editura Academiei Romane - Bucuresti, 2015.
ISBN: 978-973-27-2527-6. pp. 981-983.
[3] Andrei, N., Modele, Probleme de Test si Aplicatii de
programare Matematica. Editura Tehnica, Bucuresti, 2003.
(Aplicatia A17, pagina 365-366).
$Offtext
Variables x1,x2,x3,x4,x5,x6,x7,x8,x9,x10,x11,x12,x13, obj;
Equations e1,e2,e3,e4,e5,e6,e7,e8,e9,e10,e11,e12,e13,e14,e15,
          e;
e1.. x3-x2 =g= 0;
e2.. x2-x1 =g= 0;
e3.. 1-0.002*x7+0.002*x8 =g=0;
e4.. x13-1.262626*x10+1.231059*x3*x10 =g= 0;
e5.. x5-0.03475*x2-0.975*x2*x5+0.00975*x2*x2 =g= 0;
e6.. x6-0.03475*x3-0.975*x3*x6+0.00975*x3*x3 =g= 0;
e7.. x5*x7-x1*x8-x4*x7+x4*x8 =g= 0;
e8.. 1-0.002*(x2*x9+x5*x8-x1*x8-x6*x9)-x5-x6 =g= 0;
e9.. x2*x9-x3*x10-x6*x9-500*x2+500*x6+x2*x10 =g= 0;
e10.. x2-0.9-0.002*(x2*x10-x3*x10) =g= 0;
e11.. x4-0.03475*x1-0.975*x1*x4+0.00975*x1*x1 =g= 0;
e12.. x11-1.262626*x8+1.231059*x1*x8 =g= 0;
e13.. x12-1.262626*x9+1.231059*x2*x9 =g= 0;
e14.. x11+x12+x13 =g= 50;
e15.. x11+x12+x13 =l= 250;
e..  obj =e= x11+x12+x13;
* Bounds on variavles
x1.lo = 0.1;  x2.lo = 0.1;
x3.lo = 0.1;  x4.lo = 0.0001;
x5.lo = 0.1;  x6.lo = 0.1;
x7.lo = 0.1;  x8.lo = 0.1;
x9.lo = 500;  x10.lo = 0.1;
x11.lo = 1;   x12.lo = 0.0001;
x13.lo =0.0001;
x1.up = 1;      x2.up = 1;
x3.up = 1;      x4.up = 0.1;
x5.up = 0.9;    x6.up = 0.9;
x7.up = 1000;   x8.up = 1000;
x9.up = 1000;   x10.up = 500;
x11.up = 150;   x12.up = 150;
x13.up = 150;
*Initial point
x1.l =  0.5;      x2.l = 0.8;
x3.l =  0.9;      x4.l = 0.099;
x5.l =  0.899;    x6.l = 0.5;
x7.l =  489;      x8.l = 80;
x9.l =  650;      x10.l = 450;
x11.l =  149.9;   x12.l = 149.9;
x13.l =  149.9;
```

Figure 4.7 GAMS expression of application MSP3

```
Model msp3 /all/;

option nlp=knitro;
option sysout=on;

Solve msp3 using nlp minimizing obj;

file out /msp3.dat/
put out;
put x1.l:15:7, x2.l:15:7, x3.l:15:7, x4.l:15:7, x5.l:15:7 /;
put x6.l:15:7, x7.l:15:7, x8.l:15:7, x9.l:15:7, x10.l:15:7 /;
put x11.l:15:7, x12.l:15:7, x13.l:15:7 /;
put obj.l:15:7/;
* End MSP3
```

Figure 4.7 (continued)

4.8 Optimization of a Separation Process in a Membrane with five Stages (MSP5)

This application is taken from Dembo (1976) and Andrei (2003, pp.367):

$$\min\{1.262626(x_{12}+x_{13}+x_{14}+x_{15}+x_{16}) - 1.231060(x_1x_{12}+x_2x_{13}+x_3x_{14}+x_4x_{15}+x_5x_{16})\}$$

subject to: (4.8)

$$x_6 - 0.03475x_1 - 0.975x_1x_6 + 0.00975x_1^2 \ge 0,$$

$$x_7 - 0.03475x_2 - 0.975x_2x_7 + 0.00975x_2^2 \ge 0,$$

$$x_8 - 0.03475x_3 - 0.975x_3x_8 + 0.00975x_3^2 \ge 0,$$

$$x_9 - 0.03475x_4 - 0.975x_4x_9 + 0.00975x_4^2 \ge 0,$$

$$x_{10} - 0.03475x_5 - 0.975x_5x_{10} + 0.00975x_5^2 \ge 0,$$

$$x_7x_{11} - x_6x_{11} - x_1x_{12} + x_6x_{12} \ge 0,$$

$$x_8 - x_7 - 0.002(x_7x_{12} + x_2x_{13} - x_8x_{13} - x_1x_{12}) \ge 0,$$

$$1 - x_8 - x_9 - 0.002(x_8x_{13} + x_3x_{14} - x_2x_{13} - x_9x_{14}) \ge 0,$$

$$x_3x_{14} - x_9x_{14} - x_4x_{15} - 500x_{10} + 500x_9 + x_8x_{15} \ge 0,$$

$$x_4x_{15} - x_5x_{16} - x_{10}x_{15} - 500x_4 + 500x_{10} + x_4x_{16} \ge 0,$$

$$x_4 - 0.002x_4x_{16} + 0.002x_5x_{16} - 0.9 \ge 0,$$

$$1 - 0.002x_{11} + 0.002x_{12} \ge 0,$$

$$x_{11} - x_{12} \geq 0,$$
$$x_5 - x_4 \geq 0,$$
$$x_4 - x_3 \geq 0,$$
$$x_3 - x_2 \geq 0,$$
$$x_2 - x_1 \geq 0,$$
$$x_{10} - x_9 \geq 0,$$
$$x_9 - x_8 \geq 0,$$
$$1.262626(x_{12} + x_{13} + x_{14} + x_{15} + x_{16})$$
$$-1.231060(x_1x_{12} + x_2x_{13} + x_3x_{14} + x_4x_{15} + x_5x_{16}) - 50 \geq 0,$$
$$-1.262626(x_{12} + x_{13} + x_{14} + x_{15} + x_{16})$$
$$+1.231060(x_1x_{12} + x_2x_{13} + x_3x_{14} + x_4x_{15} + x_5x_{16}) + 250 \geq 0,$$

Solution of the application is as follows:

Nr.	Initial point	Lower bound	Solution	Upper bound
1	0.8	0.1	0.8037651	0.9
2	0.83	0.1	0.8161088	0.9
3	0.85	0.1	0.9	0.9
4	0.87	0.1	0.9	0.9
5	0.91	0.9	0.9	1
6	0.09	0.0001	0.0999996	0.1
7	0.12	0.1	0.1070319	0.9
8	0.19	0.1	0.1908370	0.9
9	0.25	0.1	0.1908370	0.9
10	0.29	0.1	0.1908370	0.9
11	512	1	505.0219	1000
12	13.1	0.000001	5.046301	500
13	71.8	1	72.63798	500
14	640	500	500.0	1000
15	650	500	500.0	1000
16	5.7	0.000001	0.1042061e–5	500

$$f(x_0) = 284.6696 \qquad f(x^*) = 174.787136606$$

The GAMS representation of this application is given in Figure 4.8.

```
$Ontext
Application MSP5 (5-stage membrane separation)
References:
[1] Dembo, R.S., A set of geometric programming test problems
and their solutions. Mathematical Programming, vol. 10, 1976,
pp.192-213.
[2] Andrei, N., Critica Ratiunii Algoritmilor de Optimizare cu
Restrictii. Editura Academiei Romane - Bucuresti, 2015.
ISBN: 978-973-27-2527-6. pp. 983-987.
$Offtext

Variables x1,x2,x3,x4,x5,x6,x7,x8,x9,x10,x11,x12,
          x13,x14,x15,x16, obj;
Equations e1,e2,e3,e4,e5,e6,e7,e8,e9,e10,e11,e12,
          e13,e14,e15,e16,e17,e18,e19,e20,e21, e;

e1.. x6- 0.03475*x1- 0.975*x1*x6+0.00975*x1*x1 =g= 0;
e2.. x7- 0.03475*x2- 0.975*x2*x7+0.00975*x2*x2 =g= 0;
e3.. x8- 0.03475*x3- 0.975*x3*x8+0.00975*x3*x3 =g=0;
e4.. x9- 0.03475*x4- 0.975*x4*x9+0.00975*x4*x4 =g= 0;
e5.. x10-0.03475*x5-0.975*x5*x10+0.00975*x5*x5 =g= 0;

e6.. x7*x11-x6*x11-x1*x12+x6*x12 =g= 0;

e7.. x8-x7-0.002*(x7*x12+x2*x13-x8*x13-x1*x12) =g= 0;

e8.. 1-x8-0.002*(x8*x13+x3*x14-x2*x13-x9*x14)-x9 =g= 0;

e9..  x3*x14-x9*x14-x4*x15-500*x10+500*x9+x8*x15 =g= 0;

e10.. x4*x15-x5*x16-x10*x15-500*x4+x4*x16+500*x10  =g= 0;

e11.. x4-0.002*(x4*x16-x5*x16)-0.9 =g= 0;

e12.. 1-0.002*x11+0.002*x12 =g= 0;

e13.. x11-x12 =g= 0;
e14.. x5-x4   =g= 0;
e15.. x4-x3   =g= 0;
e16.. x3-x2   =g= 0;
e17.. x2-x1   =g= 0;
e18.. x10-x9  =g= 0;
e19.. x9-x8   =g= 0;

e20.. 1.262626*(x12+x13+x14+x15+x16)-
      1.231060*(x1*x12+x2*x13+x3*x14+x4*x15+x5*x16) =g= 50;
e21.. 1.262626*(x12+x13+x14+x15+x16)-
      1.231060*(x1*x12+x2*x13+x3*x14+x4*x15+x5*x16) =l= 250;

e..  obj =e= 1.262626*(x12+x13+x14+x15+x16)-
             1.231060*(x1*x12+x2*x13+x3*x14+x4*x15+x5*x16);

* Bounds on variables
x1.lo = 0.1;     x2.lo = 0.1;
x3.lo = 0.1;     x4.lo = 0.1;
x5.lo = 0.9;     x6.lo = 0.0001;
x7.lo = 0.1;     x8.lo = 0.1;
x9.lo = 0.1;     x10.lo = 0.1;
x11.lo = 1;      x12.lo = 0.000001;
x13.lo = 1;      x14.lo = 500;
x15.lo = 500;    x16.lo = 0.000001;
```

Figure 4.8 GAMS expression of application MSP5

```
x1.up = 0.9;      x2.up = 0.9;
x3.up = 0.9;      x4.up = 0.9;
x5.up = 1;        x6.up = 0.1;
x7.up = 0.9;      x8.up = 0.9;
x9.up = 0.9;      x10.up = 0.9;
x11.up = 1000;    x12.up = 500;
x13.up = 500;     x14.up = 1000;
x15.up = 1000;    x16.up = 500;

*Initial point
x1.l =  0.8;       x2.l = 0.83;
x3.l =  0.85;      x4.l = 0.87;
x5.l =  0.91;      x6.l = 0.09;
x7.l =  0.12;      x8.l = 0.19;
x9.l =  0.25;      x10.l = 0.29;
x11.l = 512;       x12.l = 13.1;
x13.l = 71.8;      x14.l = 640;
x15.l = 650;       x16.l = 5.7;

Model msp5 /all/;

option nlp=knitro;
option sysout=on;

Solve msp5 using nlp minimizing obj;

file out /msp5.dat/
put out;

put x1.l:15:7,  x2.l:15:7,  x3.l:15:7,  x4.l:15:7,  x5.l:15:7
/;
put x6.l:15:7,  x7.l:15:7,  x8.l:15:7,  x9.l:15:7,  x10.l:15:7
/;
put x11.l:15:7, x12.l:15:7, x13.l:15:7, x14.l:15:7, x15.l:15:7
/;
put x16.l:15:7 /;
put obj.l:15:7 /;
* End MSP5
```

Figure 4.8 (continued)

4.9 Blending/Pooling with Five Feeds and Two Products (POOL) (Andrei, 1999, pp. 808; Andrei, 2003, pp. 382)

$$\min\big(1.1x_1 + 1.1x_2 + 1.1x_3 + 0.878x_4 + 0.878x_5 + 0.878x_6 + 0.878x_7 + 1.6x_8 + 1.6x_9 + 0.5x_{10} + 0.5x_{11} + 0.98x_{12} + 0.998x_{13}\big)$$

subject to: (4.9)

$$90x_1 + 89x_2 + 91x_3 - (x_1 + x_2 + x_3)x_{14} = 0,$$

$$87x_4 + 90.5x_5 + 89.5x_6 + 94x_7 - (x_4 + x_5 + x_6 + x_7)x_{15} = 0,$$

$$89x_8 + 89.1x_9 - (x_8 + x_9)x_{16} = 0,$$

$$89.2x_{10} + 89.3x_{11} + x_{14}x_{29} + x_{15}x_{30} - (x_{10} + x_{11} + x_{29} + x_{30})x_{17} = 0,$$

$$89.4x_{12} + 89.5x_{13} + x_{15}x_{31} + x_{16}x_{32} - (x_{12} + x_{13} + x_{31} + x_{32})x_{18} = 0,$$

$$86x_1 + 85.5x_2 + 86.5x_3 - (x_1 + x_2 + x_3)x_{19} = 0,$$

$$83.2x_4 + 86.9x_5 + 85x_6 + 89.8x_7 - (x_4 + x_5 + x_6 + x_7)x_{20} = 0,$$

$$85.1x_8 + 85.3x_9 - (x_8 + x_9)x_{21} = 0,$$

$$85.4x_{10} + 85.2x_{11} + x_{19}x_{29} + x_{20}x_{30} - (x_{10} + x_{11} + x_{29} + x_{30})x_{22} = 0,$$

$$85.5x_{12} + 85.6x_{13} + x_{20}x_{31} + x_{21}x_{32} - (x_{12} + x_{13} + x_{31} + x_{32})x_{23} = 0,$$

$$0.78x_1 + 0.8x_2 + 0.81x_3 - (x_1 + x_2 + x_3)x_{24} = 0,$$

$$0.77x_4 + 0.775x_5 + 0.78x_6 + 0.82x_7 - (x_4 + x_5 + x_6 + x_7)x_{25} = 0,$$

$$0.785x_8 + 0.79x_9 - (x_8 + x_9)x_{26} = 0,$$

$$0.787x_{10} + 0.776x_{11} + x_{24}x_{29} + x_{25}x_{30} - (x_{10} + x_{11} + x_{29} + x_{30})x_{27} = 0,$$

$$0.783x_{12} + 0.779x_{13} + x_{25}x_{31} + x_{26}x_{32} - (x_{12} + x_{13} + x_{31} + x_{32})x_{28} = 0,$$

$$x_1 + x_2 + x_3 - x_{29} = 0,$$

$$x_4 + x_5 + x_6 + x_7 - x_{30} - x_{31} = 0,$$

$$x_8 + x_9 - x_{32} = 0,$$

$$x_{10} + x_{11} + x_{29} + x_{30} - x_{33} = 0,$$

$$x_{12} + x_{13} + x_{31} + x_{32} - x_{34} = 0.$$

Solution of this application is as follows:

Nr.	Initial point	Lower bound	Solution	Upper bound
1	182	0	131.9361	11,000
2	140	0	132.5016	11,000
3	70	0	135.5624	11,000
4	1070	0	264.3126	11,000
5	47	0	273.9741	11,000
6	11	0	270.9078	11,000
7	12	0	290.8055	11,000
8	700	0	398.3295	11,000
9	60	0	401.6705	11,000
10	190	0	100.0567	11,000
11	0.1	0	99.94331	11,000
12	0.1	0	0	11,000
13	0.1	0	0	11,000
14	88	0	90.00765	11,000
15	88	0	90.33802	11,000
16	88	0	89.05021	11,000
17	88	0	89.98826	11,000
18	88	0	89.65119	11,000
19	88	0	86.00383	11,000
20	88	0	86.30969	11,000
21	88	0	85.20042	11,000
22	88	0	85.98542	11,000
23	88	0	85.71808	11,000
24	0.77	0	0.7967923	11,000
25	0.77	0	0.7869266	11,000
26	0.77	0	0.7875104	11,000
27	0.77	0	0.7897882	11,000
28	0.77	0	0.7872380	11,000
29	401	400	400	11,000
30	402	400	400	11,000
31	702	700	700	11,000
32	810	800	800	11,000
33	1001	1000	1000	11,000
34	1510	1500	1500	11,000

$$f(x_0) = 2743.368 \qquad f(x^*) = 2785.80$$

Figure 4.9 presents the GAMS expression of this application.

```
$Ontext
Application POOL (Blending/Pooling with five feeds and two
products)
References:
[1] Andrei, N., Critica Ratiunii Algoritmilor de Optimizare cu
Restrictii. Editura Academiei Romane - Bucuresti, 2015.
ISBN: 978-973-27-2527-6. pp. 987-991.
$Offtext
Variables x1,x2,x3,x4,x5,x6,x7,x8,x9,x10,x11,x12,
          x13,x14,x15,x16,x17,x18,x19,x20,x21,x22,
          x23,x24,x25,x26,x27,x28,x29,x30,x31,x32,
          x33,x34, obj;
Equations e1,e2,e3,e4,e5,e6,e7,e8,e9,e10,e11,e12,
          e13,e14,e15,e16,e17,e18,e19,e20, e;

e1.. 90*x1+89*x2+91*x3-(x1+x2+x3)*x14 =e= 0;
e2.. 87*x4+90.5*x5+89.5*x6+94*x7-(x4+x5+x6+x7)*x15 =e= 0;
e3.. 89*x8+89.1*x9-(x8+x9)*x16 =e=0;
e4.. 89.2*x10+89.3*x11+x29*x14+x30*x15-(x10+x11+x29+x30)*x17
     =e= 0;
e5.. 89.4*x12+89.5*x13+x31*x15+x32*x16-(x12+x13+x31+x32)*x18
     =e= 0;
e6.. 86*x1+85.5*x2+86.5*x3-(x1+x2+x3)*x19 =e= 0;
e7.. 83.2*x4+86.9*x5+85*x6+89.8*x7-(x4+x5+x6+x7)*x20 =e= 0;
e8.. 85.1*x8+85.3*x9-(x8+x9)*x21 =e= 0;
e9.. 85.4*x10+85.2*x11+x29*x19+x30*x20-(x10+x11+x29+x30)*x22
     =e= 0;

e10.. 85.5*x12+85.6*x13+x31*x20+x32*x21-(x12+x13+x31+x32)*x23
      =e= 0;
e11.. 0.78*x1+0.8*x2+0.81*x3-(x1+x2+x3)*x24 =e= 0;
e12.. 0.77*x4+0.775*x5+0.78*x6+0.82*x7-(x4+x5+x6+x7)*x25 =e= 0;
e13.. 0.785*x8+0.79*x9-(x8+x9)*x26 =e= 0;
e14.. 0.787*x10+0.776*x11+x29*x24+x30*x25-(x10+x11+x29+x30)*x27
      =e= 0;

e15.. 0.783*x12+0.779*x13+x31*x25+x32*x26-(x12+x13+x31+x32)*x28
      =e= 0;
e16.. x1+x2+x3-x29 =e= 0;
e17.. x4+x5+x6+x7-x30-x31 =e= 0;
e18.. x8+x9-x32 =e= 0;
e19.. x10+x11+x29+x30-x33 =e= 0;
e20.. x12+x13+x31+x32-x34 =e= 0;

e..  obj =e= 1.1*x1+1.1*x2+1.1*x3 +
             0.878*x4+0.878*x5+0.878*x6+0.878*x7 +
             1.6*x8+1.6*x9+0.5*x10+0.5*x11+0.98*x12+0.98*x13;

* Bounds on variables
x1.lo = 0;        x2.lo = 0;
x3.lo = 0;        x4.lo = 0;
x5.lo = 0;        x6.lo = 0;
x7.lo = 0;        x8.lo = 0;
x9.lo = 0;        x10.lo = 0;
x11.lo = 0;       x12.lo = 0;
x13.lo = 0;       x14.lo = 0;
x15.lo = 0;       x16.lo = 0;
x17.lo = 0;       x18.lo = 0;
x19.lo = 0;       x20.lo = 0;
```

Figure 4.9 GAMS expression of application POOL

```
x21.lo = 0;        x22.lo = 0;
x23.lo = 0;        x24.lo = 0;
x25.lo = 0;        x26.lo = 0;
x27.lo = 0;        x28.lo = 0;
x29.lo = 40;       x30.lo = 400;
x31.lo = 700;      x32.lo = 800;
x33.lo = 1000;     x34.lo = 1500;
x1.up = 11000;     x2.up = 11000;
x3.up = 11000;     x4.up = 11000;
x5.up = 11000;     x6.up = 11000;
x7.up = 11000;     x8.up = 11000;
x9.up = 11000;     x10.up = 11000;
x11.up = 11000;    x12.up = 11000;
x13.up = 11000;    x14.up = 11000;
x15.up = 11000;    x16.up = 11000;
x17.up = 11000;    x18.up = 11000;
x19.up = 11000;    x20.up = 11000;
x21.up = 11000;    x22.up = 11000;
x23.up = 11000;    x24.up = 11000;
x25.up = 11000;    x26.up = 11000;
x27.up = 11000;    x28.up = 11000;
x29.up = 11000;    x30.up = 11000;
x31.up = 11000;    x32.up = 11000;
x33.up = 11000;    x34.up = 11000;

*Initial point
x1.l =  182;      x2.l = 140;
x3.l =  70;       x4.l = 1070;
x5.l =  47;       x6.l = 11;
x7.l =  12;       x8.l = 700;
x9.l =  60;       x10.l = 190;
x11.l = 0.1;      x12.l = 0.1;
x13.l = 0.1;      x14.l = 88;
x15.l = 88;       x16.l = 88;
x17.l = 88;       x18.l = 88;
x19.l = 88;       x20.l = 88;
x21.l = 88;       x22.l = 88;
x23.l = 88;       x24.l = 0.77;
x25.l = 0.77;     x26.l = 0.77;
x27.l = 0.77;     x28.l = 0.77;
x29.l = 401;      x30.l = 402;
x31.l = 702;      x32.l = 810;
x33.l = 1001;     x34.l = 1510;

Model pool /all/;
option nlp=knitro;
option sysout=on;

Solve pool using nlp minimizing obj;
file out /pool.dat/
put out;
put x1.l:15:7,   x2.l:15:7,   x3.l:15:7,   x4.l:15:7,   x5.l:15:7
/;
put x6.l:15:7,   x7.l:15:7,   x8.l:15:7,   x9.l:15:7,   x10.l:15:7
/;
put x11.l:15:7, x12.l:15:7, x13.l:15:7, x14.l:15:7, x15.l:15:7
put x16.l:15:7  x17.l:15:7, x18.l:15:7, x19.l:15:7, x20.l:15:7
/;
put x21.l:15:7  x22.l:15:7, x23.l:15:7, x24.l:15:7, x25.l:15:7
/;
```

Figure 4.9 (continued)

```
put x26.l:15:7  x27.l:15:7, x28.l:15:7, x29.l:15:7, x30.l:15:7
/;
put x31.l:15:7  x32.l:15:7, x33.l:15:7, x34.l:15:7 /;
put obj.l:15:7 /;
* End POOL
```

Figure 4.9 (continued)

4.10 Distribution of Electrons on a Sphere (DES)

The problem is described in Dolan, Moré, and Munson (2004, pp. 5) (see also Andrei (2001, pp. 292) and Andrei (2003, pp. 348)). Given n_p electrons, find the equilibrium state distribution (of minimal Coulomb potential) of the electrons positioned on a conducting sphere. The problem, known as the Thomson problem (raised in 1904), consists in finding the lowest energy configuration of n_p point charges on a conducting sphere. This is an important problem in physics and chemistry, which determines a minimal structure with respect to atomic positions.

If (x_i, y_i, z_i) are the positions of the n_p points (electrons), then the potential energy is.

$$\sum_{i=1}^{n_p-1} \sum_{j=i+1}^{n_p} \left(\left(x_i - x_j\right)^2 + \left(y_i - y_j\right)^2 + \left(z_i - z_j\right)^2 \right)^{-\frac{1}{2}}, \tag{4.10a}$$

which must be minimized subject to the n_p constraints:

$$x_i^2 + y_i^2 + z_i^2 = 1, \quad i = 1, \ldots, n_p. \tag{4.10b}$$

The problem has a multitude of local minima at which the objective value is relatively close to the objective value at the global minimum. Also, the number of local minima grows exponentially with n_p. Therefore, determining the global minimum is a difficult task, and solvers are usually expected to find only a local minimum.

The GAMS representation of this application is presented in Figure 4.10.

For $n_p = 50$ the value of the potential energy is $1,055.1823147$.

```
$ontext
Electron
Given n electrons, find the equilibrium state distribution (of
minimal Coulomb potential) of the electrons positioned on a
conducting sphere.
[1] Andrei, N., Critica Ratiunii Algoritmilor de Optimizare cu
Restrictii. Editura Academiei Romane - Bucuresti, 2015.
ISBN: 978-973-27-2527-6. pp. 1001-1003.
$offtext
$if     set n  $set np %n%
$if not set np $set np 50
Set i        electrons /i1 * i%np%/
    ut(i,i) upper triangular part;

Alias (i,j);
    ut(i,j)$(ord(j) > ord(i)) = yes;

Variables   x(i) x-coordinate of the electron
            y(i) y-coordinate of the electron
            z(i) z-coordinate of the electron
            potential Coulomb potential;
Equations   obj     objective
            ball(i) points on unit ball;
obj.. potential =e=
  sum{ut(i,j), 1.0/sqrt(sqr(x[i]-x[j]) + sqr(y[i]-y[j]) +
                         sqr(z[i]-z[j]))};

ball(i).. sqr(x(i)) + sqr(y(i)) + sqr(z(i)) =e= 1;

* Set the starting point to a quasi-uniform distribution
* of electrons on a unit sphere
scalar pi;
pi = 2*arctan(inf);

parameter theta(i), phi(i);
theta(i) = 2*pi*uniform(0,1);
phi(i)   = pi*uniform(0,1);

x.l(i) = cos(theta(i))*sin(phi(i));
y.l(i) = sin(theta(i))*sin(phi(i));
z.l(i) =               cos(phi(i));

model electron /all/;

electron.iterlim = 500000;
option reslim=5000;

$onecho >bench.opt
solvers conopt knitro minos
$offecho
electron.optfile=1;
electron.workfactor=20;
option nlp=bench;
solve electron using nlp minimizing potential;
* End electron
```

Figure 4.10 GAMS expression of application DES

4.11 Hanging Chain (HANG)

Find the chain of uniform density of length L suspended between two points with minimal potential energy. This is a classical problem, known as the dog problem (Cesari, 1983). It was suggested by Mittelmann and it is described in (Dolan, Moré, & Munson, 2004, pp. 9). The problem is to determine a function $x(t)$, the shape of the chain that minimizes the potential energy:

$$\int_0^1 x(t)\left(1+x'(t)^2\right)^{1/2} dt,$$

subject to the constraints on the length of the chain:

$$\int_0^1 \left(1+x'(t)^2\right)^{1/2} dt = L,$$

as well as the end conditions $x(0)=a$ and $x(1)=b$, where a and b are given constants.

Another formulation of this problem is possible by introducing a control variable $u(t)=x_1'(t)$ and the potential energy function:

$$x_2(t)=\int_0^t x_1(s)\left(1+u(s)^2\right)^{1/2} ds.$$

This formulation leads to minimizing the total potential energy $x_2(1)$ subject to the differential equations:

$$x_1'(t)=u,$$
$$x_2'(t)=x_1(t)\left(1+u(t)^2\right)^{1/2},$$
$$x_3'(t)=\left(1+u(t)^2\right)^{1/2}.$$

By introducing n discretization points by means of a uniform time step of length $h=1/(n+1)$, then the discrete variant of the problem is

$$\min h\sum_{i=1}^{n+1}\left(\frac{x_i+x_{i-1}}{2}\right)\sqrt{1+\left(\frac{x_i-x_{i-1}}{h}\right)^2} \tag{4.11a}$$

subject to

$$\sum_{i=1}^{n+1}\sqrt{1+\left(\frac{x_i-x_{i-1}}{h}\right)^2}=\frac{L}{h}, \tag{4.11b}$$

where $x_0=a$ and $x_{n+1}=b$ (Bondarenko, Bortz, & Moré, 1999).

The GAMS representation of this application uses the variant of integration of the above system of differential equations with $a=1$, $b=3$, and $L=4$. Figure 4.11 gives the GAMS expression of this problem.

For $n_h=400$ the form of the chain is shown in Figure 4.12.

```
$ontext
Find the chain (of uniform density) of length L suspended
between two points with minimal potential energy.
[1] Andrei, N., Critica Ratiunii Algoritmilor de Optimizare cu
Restrictii. Editura Academiei Romane - Bucuresti, 2015.
ISBN: 978-973-27-2527-6. pp. 1019-1022.
$offtext

$if     set n  $set nh %n%
$if not set nh $set nh 1000

set nh /i0 * i%nh%/;
alias(nh,i);

scalars L  length of the suspended chain        / 4 /
        a  height of the chain at t=0 (left)    / 1 /
        b  height of the chain at t=1 (left)    / 3 /
        tf ODEs defined in [0 tf]               / 1 /
        h  uniform interval length
        n  number of subintervals
        tmin;

if (b>a, tmin = 0.25 else tmin = 0.75);
n = card(nh) - 1;
h = tf/n;

*Variables:
Variables
  x(i)   height of the chain
  u(i)   derivative of x
  energy potential energy ;

x.fx('i0')     = a;
x.fx('i%nh%') = b;

x.l(i) = 4*abs(b-a)*((ord(i)-1)/n)*(0.5*((ord(i)-1)/n)
                     - tmin) + a;
u.l(i) = 4*abs(b-a)*(((ord(i)-1)/n) - tmin);

* Equations:
equations obj, x_eqn(i), length_eqn ;

obj.. energy =e=
       0.5*h*sum(nh(i+1), x(i)*sqrt(1+sqr(u(i))) +
                          x(i+1)*sqrt(1+sqr(u(i+1))));

x_eqn(i+1).. x(i+1) =e= x(i) + 0.5*h*(u(i)+u(i+1));

length_eqn.. 0.5*h*sum(nh(i+1), sqrt(1+sqr(u(i))) +
                                sqrt(1+sqr(u(i+1)))) =e= L;
model chain /all/;

chain.optfile=1;
chain.workspace=120;
option nlp=conopt

solve chain using nlp minimizing energy;

*file res /chain.dat/;
*put res
*loop(i, put x.l(i):10:5, put/)
* End hang
```

Figure 4.11 GAMS expression of application HANG

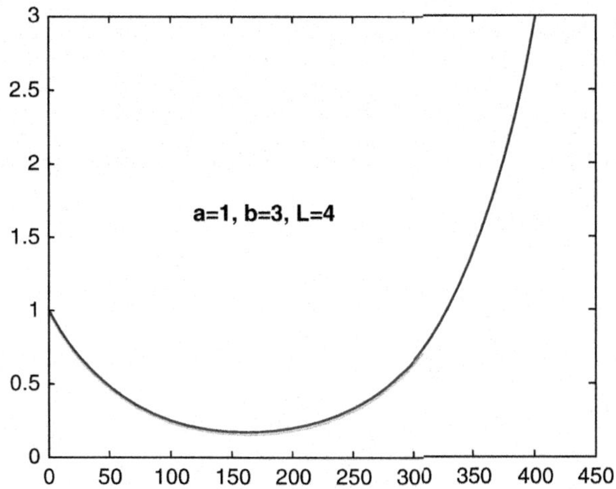

Figure 4.12 Hanging chain of minimal potential energy of length $L=4$

4.12 Determine the Optimal Mixing Policy of Two Catalysts Along the Length of a Tubular Plug Flow Reactor Involving Several Reactions (CAT)

This application is described in von Stryk (1999) and Dolan, Moré, and Munson (2004, pp. 33) and has the following form:

$$x_1'(t) = u(t)(10x_2(t) - x_1(t)), \tag{4.12a}$$

$$x_2'(t) = u(t)(x_1(t) - 10x_2(t)) - (1 - u(t))x_2(t). \tag{4.12b}$$

The initial conditions are $x_1(0)=1$ and $x_2(t)=0$. The control variable $u(t)$ represents the mixing ratio of the catalysts and satisfies the bounds:

$$0 \leq u(t) \leq 1. \tag{4.13}$$

The problem is to minimize:

$$-1 + x_1(t_f) + x_2(t_f), \quad t_f = 1. \tag{4.14}$$

A uniform partition of the interval [0, 1] with n_h subintervals is considered and the equations discretized in a natural way. As initial point, we take $u=0$, $x_1=1$, and $x_2=0$ evaluated in the grid points. The GAMS expression of this application is given in Figure 4.13.

Figure 4.14 presents the time evolution of the control variable $u(t)$ given by CONOPT and MINOS, respectively. We see the bang-bang character of this evolution, which is imposed by the presence of bounds on the control variable.

Figure 4.15 presents the evolution of variables $x_1(t)$ and $x_2(t)$ given by CONOPT.

```
$ontext
Determine the optimal policy of two catalysts along the length
of a tubular plug flow reactor involving several reactions.
References:
[1] Andrei, N., Nonlinear Optimization Applications using the
GAMS Technology, Springer Optimization and Its Applications 81,
Springer Science+Business Media, New York, 2013. (Application
7.11, pp.201-205)
$offtext

$if     set n  $set nh %n%
$if not set nh $set nh 800

Set nh number of subintervals / 0*%nh% /;
Alias (nh,i);

Scalar tf Final time                  /1/
       x1_0 Initial condition for x1  /1/
       x2_0 Initial condition for x2  /0/
       alpha smoothing parameter      /0/
       h;
       h = tf/%nh%;

Variable u(nh), x1(nh), x2(nh), obj;

Positive variable u; u.up(nh) = 1;

Equations defobj objective function
          ode1(nh)
          ode2(nh);
defobj.. obj =e= -1 + x1['%nh%'] + x2['%nh%'] +
                  alpha*h*sum{nh(i+1), sqr(u[i+1] - u[i])};
ode1(nh(i+1)).. x1[i+1] =e= x1[i] +
                  (h/2)*(u[i]*(10*x2[i]-x1[i]) +
                        u[i+1]*(10*x2[i+1]-x1[i+1]));

ode2(nh(i+1)).. x2[i+1] =e= x2[i] +
                  (h/2)*(u[i]*(x1[i]-10*x2[i]) -
                        (1-u[i])*x2[i] +
                         u[i+1]*(x1[i+1]-10*x2[i+1]) -
                        (1-u[i+1])*x2[i+1]);
* Initial point
x1.l[nh] = 1;

x1.fx['0'] = x1_0;
x2.fx['0'] = x2_0;

model catmix /all/;

$onecho >bench.opt
  solvers conopt knitro minos
$offecho
catmix.optfile=1;
option nlp=bench;
solve catmix minimizing obj using nlp;
*file res1 /catmix.dat/;
*put res1
*loop(nh, put x1.l(nh):10:5, put/)
* End of model catmix
```

Figure 4.13 GAMS expression of application CAT

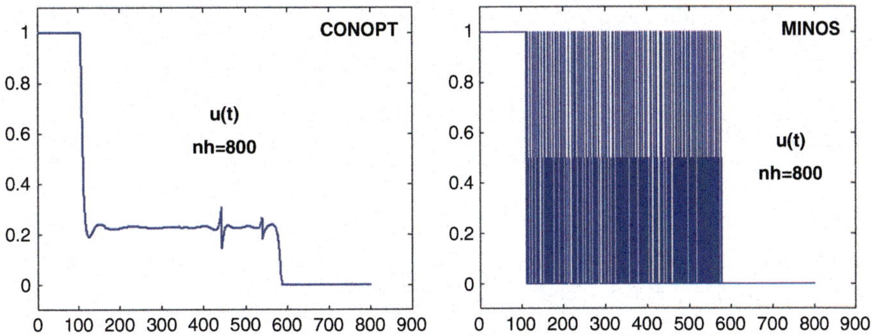

Figure 4.14 Evolution of $u(t)$ given by CONOPT and MINOS

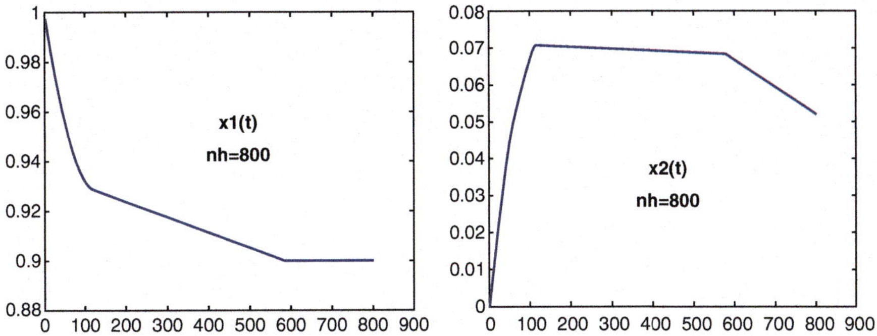

Figure 4.15 Evolution of $x_1(t)$ and $x_2(t)$ given by CONOPT

4.13 Optimal Control of a Continuous Stirred-Tank Chemical Reactor (CSTC)

Let us consider a continuous stirred-tank chemical reactor where the flow of a coolant through a coil inserted in the reactor is to control the first-order, irreversible exothermic reaction taking place in the reactor (Lapidus & Luus, 1967). The state variables are $x_1(t)$ and $x_2(t)$, which represent the deviation from the steady-state temperature and the deviation from the steady-state concentration. $u(t)$ is the normalized control variable representing the effect of the coolant flow on the chemical reaction. The state equations are:

$$x_1'(t) = -2(x_1(t) + 0.25) + (x_2(t) + 0.5)\exp\left(\frac{25x_1(t)}{x_2(t) + 2}\right) - (x_1(t) + 0.25)u(t), \tag{4.15a}$$

$$x_2'(t) = (0.5 - x_2(t)) - (x_2(t) + 0.5)\exp\left(\frac{25x_1(t)}{x_2(t) + 2}\right). \tag{4.15b}$$

The initial conditions are $x_1(0) = 0.05$ and $x_2(0) = 0$. The performance measure to be minimized is

$$J = \int_0^{0.78} \left(x_1^2(t) + x_2^2(t) + Ru^2(t)\right)\mathrm{d}t, \tag{4.16}$$

showing that the desired objective is to maintain the temperature and the concentration close to their steady-state values, without expending large amounts of control effort. Here R is a weighting factor that we arbitrarily shall select as 0.2.

The above optimal control problem can be represented as a nonlinear programming problem, thus avoiding the backward in time solution of the Riccati equation. To solve the problem, we approximate the state differential equations by differences, while the integral term in the performance measure by summation. The GAMS expression of this application is as in Figure 4.16.

Figures 4.17 and 4.18 show the time evolution of the state variables and of the control for a discretization of the time interval $[0, t_f]$ in 100 subintervals.

If for $t \in [0, 0.78]$ the control variable is bounded as $0.2 \le u(t) \le 1.2$, then the time evolution of the variables and of the control is as in Figures 4.19 and 4.20, respectively.

```
$Ontext
Optimal control of a Continuous Stirred-Chemical Reactor.
Linear Quadratic tracking problem solved as a General Nonlinear
Programming Problem instead of the usual Backward Riccati
Matrix Recursion.
References:
[1] Lapidus, L., Luus, R., The control of nonlinear systems:
Part II: convergence by combined first and second variations.
A.I.Ch. E. Journal, 1967, pp.108-113.
[2] Kirk, D.E., Optimal Control Theory. An introduction.
Prentice Hall Inc., Englewood Cliffs, New Jersey, 1970.
(Example 6.2-2, pp. 338-342)
[3] Andrei, N., Nonlinear Optimization Applications using the
GAMS Technology, Springer Optimization and Its Applications 81,
Springer Science+Business Media, New York, 2013. (Application
7.12, pp.202-209)
$Offtext

$if     set n   $set nh %n%
$if not set nh $set nh 100

set nh Number of subintervals / 0*%nh% /;
alias (nh,k);

Scalar    tf final time /0.78/
          rk penalty control / 0.20 /
          x1_0 initial value for x1 / 0.05/
          x2_0 initial value for x2 / 0.00 /
          h;
          h=tf/%nh%;
```

Figure 4.16 GAMS expression of application CSTC

```
Variables x1(nh), x2(nh) state variables
          u(nh) control variable
          obj criterion ;

Equations criterion   criterion definition
          state1(nh)  state equation 1
          state2(nh)  state equation 2;

criterion..
      obj =e= .5*sum(nh(k), (2*x1(k)*x1(k)+2*x2(k)*x2(k) +
               u(k)*rk*u(k)));

state1(nh(k+1))..
x1[k+1] =e=
      x1(k)+(h/2)*(-2*(x1(k)+0.25)+(x2(k)+0.5)*exp(25*x1(k)/
           (x1(k)+2))-(x1(k)+0.25)*u(k)
      -2*(x1(k+1)+0.25)+(x2(k+1)+0.5)*exp(25*x1(k+1)/
           (x1(k+1)+2))-(x1(k+1)+0.25)*u(k+1)) ;

state2(nh(k+1))..
x2[k+1] =e=
      x2(k)+(h/2)*(0.5-x2(k)- (x2(k)+0.5)*exp(25*x1(k)/
           (x1(k)+2))+
      0.5-x2(k+1)-(x2(k+1)+0.5)*exp(25*x1(k+1)/
           (x1(k+1)+2)));

*Initial point
x1.l[nh]=0;
x2.l[nh]=0;
u.l[nh]=0;

x1.fx ['0'] = x1_0;
x2.fx ['0'] = x2_0;

Model control2 /all/;
$onecho >bench.opt
  solvers conopt knitro minos snopt
$offecho

control2.optfile=1;
option nlp=bench

Solve control2 minimizing obj using nlp;

file stat1 /state1.dat/;
file stat2 /state2.dat/;
file cont  /control.dat/;
put stat1;
loop(nh, put x1.l(nh):10:5,',', put/)
put stat2;
loop(nh, put x2.l(nh):10:5,',', put/)
put cont;
loop(nh, put u.l(nh):10:5,',', put/)
* End of model cstc
```

Figure 4.16 (continued)

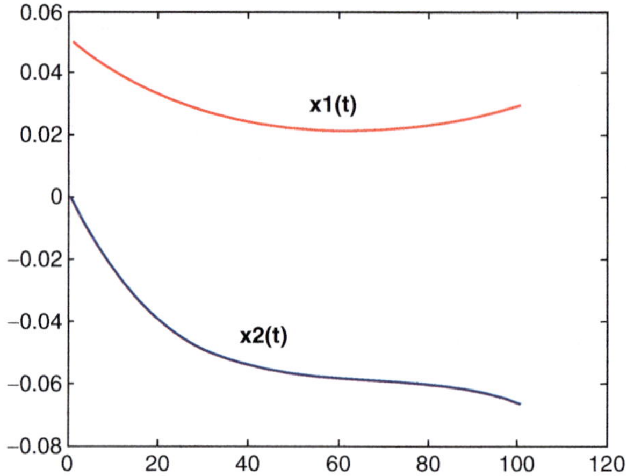

Figure 4.17 Evolution of $x_1(t)$ and $x_2(t)$

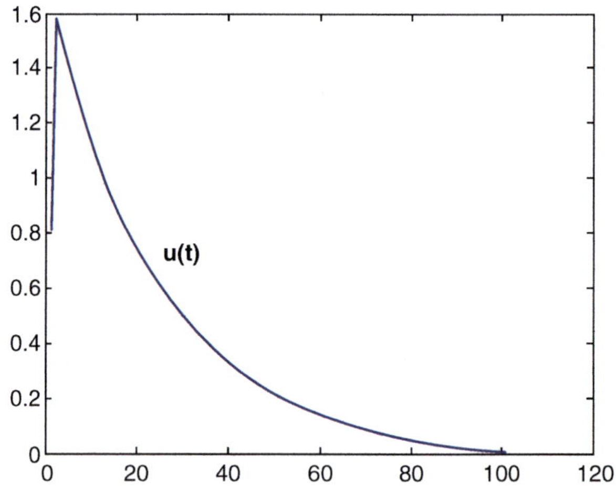

Figure 4.18 Evolution of $u(t)$

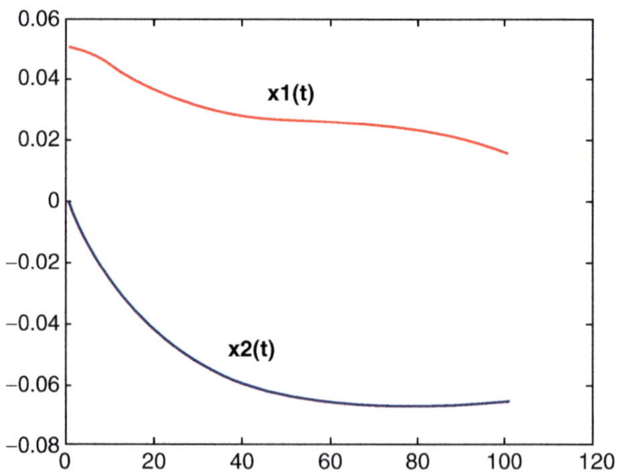

Figure 4.19 Evolution of $x_1(t)$ and $x_2(t)$

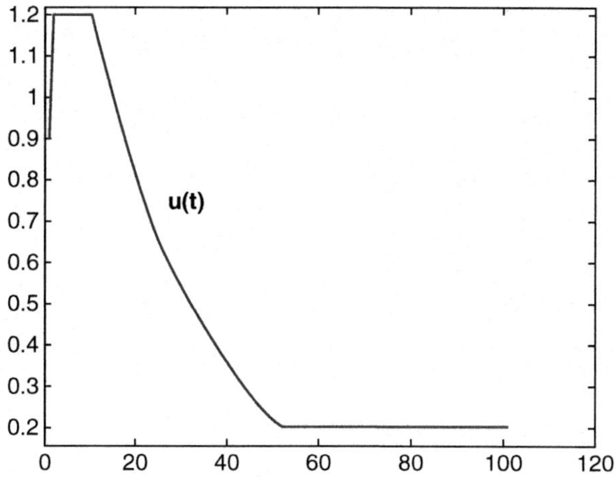

Figure 4.20 Evolution of $u(t)$

4.14 Optimal Temperature Field in a Rectangular Area (DIFF)

This application considers the problem of determining the stationary temperature field in a rectangular area with heterogeneous thermal conductivity and with some source points of heat inside the area as well as fixed temperatures on the boundaries of the solution domain (McKinney & Savitsky, 2003). Therefore, consider an area with the thermal conductivity $V[Wm^{-1}K^{-1}]$ heterogeneously and symmetrically distributed in the area of interest, i.e., there is a symmetric zone at the center of the solution domain, where V becomes larger than in the rest of the area. Then the heat transportation equation is

$$\frac{\partial(c\rho T)}{\partial t} = \nabla.(V\nabla(c\rho T)) + I, \tag{4.17}$$

where c is the specific heat capacity of substance (cal/kg degree), ρ is the density of substance (kg/m^3), T is temperature of substance (degree), V is the thermal conductivity (m^2/sec), I is the point source of heat (cal/m^{3}sec), and t is time in seconds. On the boundary of the considered area, the temperature is fixed as.

$$T = T_0(x,y)^0 \text{ for } (x,y) \text{ on the boundary.}$$

In steady-state case, no characteristics of the problem change in time. Therefore, the transportation equation becomes:

$$\nabla.(V\nabla(c\rho T)) + I = 0. \tag{4.18}$$

Now, if c and ρ do not vary spatially, the above equation can be written as:

$$\nabla.(V\nabla T) + \frac{I}{c\rho} = 0. \tag{4.19}$$

To solve this equation, the differential operators must be approximated by algebraic analogs, and a conservative scheme of solving the resulting system of algebraic equations must be used. For this, the rectangular area is subdivided into some number of small rectangles with sides parallel to the x and the y axes. Thus, a number of intersection points known as nodes are obtained. Using this system of nodes, an algebraic analogs for the differential operators $\nabla.()$ and $\nabla()$ can be constructed.

If a and B are differential functions, then:

$$\nabla.(a\nabla B) = a\nabla.(\nabla B) + \nabla a.\nabla B.$$

Using this formula, we get the following equation for the heat transportation:

$$V\nabla.(\nabla T) + \nabla T.\nabla V + \frac{I}{c\rho} = 0. \tag{4.20}$$

Using the finite difference in two dimensions, this equation can be discretized as.

$$V_{i,j}\frac{\left(T_{i+1,j} - 2T_{i,j} + T_{i-1,j}\right)}{(\Delta x)^2} + V_{i,j}\frac{\left(T_{i,j+1} - 2T_{i,j} + T_{i,j-1}\right)}{(\Delta y)^2}$$

$$+\frac{\left(V_{i+1,j} - V_{i-1,j}\right)}{2(\Delta x)}\frac{\left(T_{i+1,j} - T_{i-1,j}\right)}{2(\Delta x)} + \frac{\left(V_{i,j+1} - V_{i,j-1}\right)}{2(\Delta y)}\frac{\left(T_{i,j+1} - T_{i,j-1}\right)}{2(\Delta y)} + \frac{I}{c\rho} = 0,$$

for $i = 1, \ldots, n_x$ and $j = 1, \ldots, n_y$ where n_x and n_y are the numbers of the discretization points along the sides of the area, respectively.

Consider the case in which, on the boundary of the area, the temperature is zero, excepting the cells (I1,J10), (I1,J11), (I10,J1), and (I11,J1), where the temperature is about 100 °C. If $n_x = 20$ and $n_y = 20$, then the GAMS representation of the stationary temperature field in the considered area is as in Figure 4.21 (Andrei, 2004).

Figure 4.22 presents the solution to the heat transportation problem with two fixed boundary conditions.

Now, let us consider the situation in which an additional source of heat is placed in the center of cell (I5,J5) of value 10000(cal/m^{3}sec). The distribution of temperature and the constant level curves are as in Figure 4.23.

Observe that the symmetry of the problem and the boundary conditions determine the symmetry of the solution. To get the solution of the problem in this case, observe that the instruction `SUPPLY('I5','J5'):=10000;` is introduced in the GAMS representation of the problem.

```
$ontext
Stationary temperature field in a rectangular area.
Determination of the stationary temperature field in a
rectangular area with heterogeneous thermal conductivity and
some source points of heat inside the area as well as fixed
temperature on the boundaries of the solution domain.
References:
[1] McKinney, D.C., Savitsky, A.G., Basic optimization models
for water and energy management. June 1999 (revision 6,
February 2003).
[2] Andrei, N., Nonlinear Optimization Applications using the
GAMS Technology, Springer Optimization and Its Applications 81,
Springer Science+Business Media, New York, 2013. (Application
8.1, pp.223-229)

$offtext

   set x /i1*i20/
   set y /j1*j20/;

   set inside(x,y);
       inside(x,y)                    = yes;
       inside(x,y)$(ord(x)=1)         = no;
       inside(x,y)$(ord(x)=card(x))   = no;
       inside(x,y)$(ord(y)=1)         = no;
       inside(x,y)$(ord(y)=card(y))   = no;

* Temperature supply determination
   parameter supply(x,y);
             supply(x,y)          := 0;
*             supply('i17','j17')  := 15000;
*             supply('i5','j5')    := 10000;

   scalar dx /0.1/;
   scalar dy /0.1/;
   scalar c /1./;
   scalar rho /1./;

   TABLE v(x,y)
     j1  j2  j3  j4  j5  j6  j7  j8  j9 j10 j11 j12 j13 j14 j15
j16 j17 j18 j19 j20
i1  0.5 0.5 0.5 0.5 0.5 0.5 0.5 0.5 0.5 0.5 0.5 0.5 0.5 0.5 0.5
0.5 0.5 0.5 0.5 0.5
i2  0.5 0.5 0.5 0.5 0.5 0.5 0.5 0.5 0.5 0.5 0.5 0.5 0.5 0.5 0.5
0.5 0.5 0.5 0.5 0.5
i3  0.5 0.5 0.5 0.5 0.5 0.5 0.5 0.5 0.5 0.5 0.5 0.5 0.5 0.5 0.5
0.5 0.5 0.5 0.5 0.5
i4  0.5 0.5 0.5 0.5 0.5 0.5 0.5 0.5 0.5 0.5 0.5 0.5 0.5 0.5 0.5
0.5 0.5 0.5 0.5 0.5
i5  0.5 0.5 0.5 0.5 0.5 0.5 0.5 0.5 0.5 0.5 1.0 1.0 0.5 0.5 0.5
0.5 0.5 0.5 0.5 0.5
i6  0.5 0.5 0.5 0.5 0.5 0.5 0.5 0.5 0.5 0.5 1.0 1.0 0.5 0.5 0.5
0.5 0.5 0.5 0.5 0.5
i7  0.5 0.5 0.5 0.5 0.5 0.5 0.5 0.5 0.5 0.5 1.0 1.0 0.5 0.5 0.5
0.5 0.5 0.5 0.5 0.5
```

Figure 4.21 GAMS expression of application DIFF

```
i8  0.5 0.5 0.5 0.5 0.5 0.5 0.5 0.5 0.5 0.5 1.0 1.0 0.5 0.5 0.5
0.5 0.5 0.5 0.5 0.5
i9  0.5 0.5 0.5 0.5 0.5 0.5 0.5 0.5 0.5 0.5 1.0 1.0 1.0 1.0 0.5
0.5 0.5 0.5 0.5 0.5
i10 0.5 0.5 0.5 0.5 1.0 1.0 1.0 1.0 1.0 1.0 1.0 1.0 1.0 1.0 0.5
0.5 0.5 0.5 0.5 0.5
i11 0.5 0.5 0.5 0.5 1.0 1.0 1.0 1.0 1.0 1.0 1.0 1.0 1.0 1.0 0.5
0.5 0.5 0.5 0.5 0.5
i12 0.5 0.5 0.5 0.5 0.5 0.5 0.5 0.5 0.5 1.0 1.0 1.0 1.0 1.0 1.0
0.5 0.5 0.5 0.5 0.5
i13 0.5 0.5 0.5 0.5 0.5 0.5 0.5 0.5 0.5 1.0 1.0 1.0 1.0 1.0 1.0
1.0 0.5 0.5 0.5 0.5
i14 0.5 0.5 0.5 0.5 0.5 0.5 0.5 0.5 0.5 0.5 0.5 0.5 1.0 1.0 1.0
1.0 1.0 0.5 0.5 0.5
i15 0.5 0.5 0.5 0.5 0.5 0.5 0.5 0.5 0.5 0.5 0.5 0.5 0.5 1.0 1.0
1.0 1.0 1.0 0.5 0.5
i16 0.5 0.5 0.5 0.5 0.5 0.5 0.5 0.5 0.5 0.5 0.5 0.5 0.5 0.5 1.0
1.0 1.0 1.0 0.5 0.5
i17 0.5 0.5 0.5 0.5 0.5 0.5 0.5 0.5 0.5 0.5 0.5 0.5 0.5 0.5 0.5
1.0 1.0 1.0 0.5 0.5
i18 0.5 0.5 0.5 0.5 0.5 0.5 0.5 0.5 0.5 0.5 0.5 0.5 0.5 0.5 0.5
0.5 0.5 0.5 0.5 0.5
i19 0.5 0.5 0.5 0.5 0.5 0.5 0.5 0.5 0.5 0.5 0.5 0.5 0.5 0.5 0.5
0.5 0.5 0.5 0.5 0.5
i20 0.5 0.5 0.5 0.5 0.5 0.5 0.5 0.5 0.5 0.5 0.5 0.5 0.5 0.5 0.5
0.5 0.5 0.5 0.5 0.5

  Variables
        obj
        t(x,y);
* Variable bounds
   t.lo(x,y)  = 0;
   t.up(x,y)  =  200;

* Initial values
   t.l(x,y) = 0;

* Boundary conditions
   t.fx(x,y)$(ord(x)=1)         = 0.0;
   t.fx(x,y)$(ord(x)=card(x))  = 0.0;
   t.fx(x,y)$(ord(y)=1)         = 0.0;
   t.fx(x,y)$(ord(y)=card(y))  = 0.0;

   t.fx('i1','j10') = 100;
   t.fx('i1','j11') = 100;
   t.fx('i10','j1') = 100;
   t.fx('i11','j1') = 100;

* Equations of the model
   equation
        temp(x,y)
        ben;

 temp(x,y)$(inside(x,y)).. supply(x,y)/c/rho +
      v(x,y)*(t(x-1,y) - 2*t(x,y) + t(x+1,y))/dx/dx +
      v(x,y)*(t(x,y-1) - 2*t(x,y) + t(x,y+1))/dy/dy +
```

Figure 4.21 (continued)

```
        (v(x,y+1) - v(x,y-1))*(t(x,y+1)-t(x,y-1))/dy/dy/2.0 +
        (v(x+1,y) - v(x-1,y))*(t(x+1,y)-t(x-1,y))/dx/dx/2.0
                    =e= 0;
*     ben.. obj =e= sum(x,t(x,'j2'));
      ben.. obj =e= 0;

   model Diffusion1 /all/;

$onecho >bench.opt
  solvers conopt knitro minos snopt
$offecho
Diffusion1.optfile=1;
option nlp=bench;

   solve Diffusion1 using nlp minimizing obj;

* Put the solution
file res1 / Diffusion1.txt/
file res2 / Diffusion1.dat/
put res1;
loop(x, put x.tl:6; loop(y, put t.l(x,y):6:2; ); put /;) put /;
          put " inside 20 " /;
loop(x, put x.tl:6; loop(y, put inside(x,y):6; )put /;);
          put "inside 20 " /;
loop(x, put x.tl:6; loop(y, put supply(x,y):6; ); put /;);
put res2;
loop(x, loop(y, put t.l(x,y):4:0; ); put /;); put /;
* End Diffusion1
```

Figure 4.21 (continued)

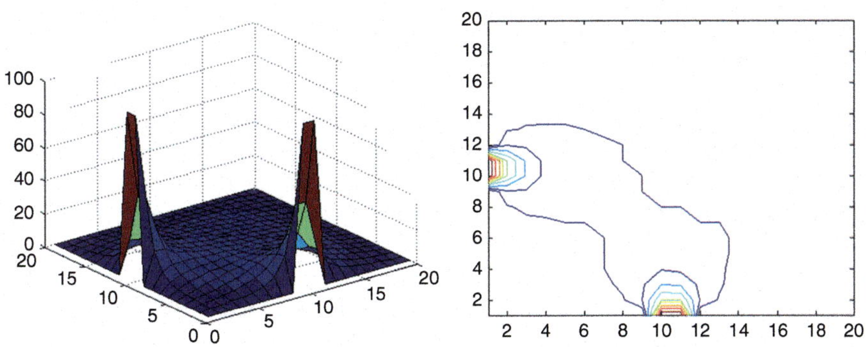

Figure 4.22 Solution to the heat transportation problem with two fixed boundary conditions

In the following, let us consider another situation in which, in the center of the cell (I17,J17) an additional source of value 15,000 is introduced. Figure 4.24 shows the distribution of temperature in the rectangular area.

In the following, we shall consider a *time-dependent temperature field in a rectangular area with heterogeneous thermal conductivity and a point source of*

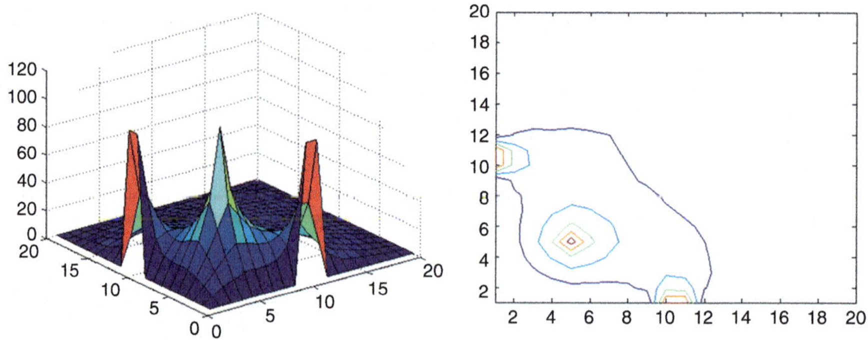

Figure 4.23 Solution to the heat transportation problem with two fixed boundary conditions and one heat source on cell (I5,J5)

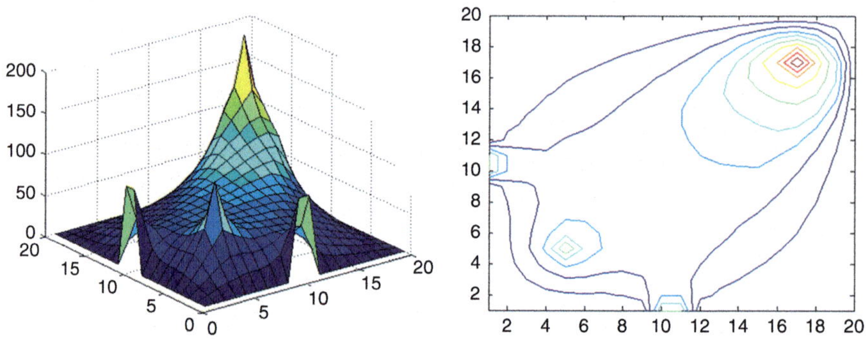

Figure 4.24 Solution to the heat transportation problem with two fixed boundary conditions and two heat sources

heat and heat flows through the border of the solution domain. Let a point source of heat begin to heat the area in point (I10, J18) of value 10,000 (cal/m^{3}sec). The heat from this source is distributed over the solution domain according to the heat transportation equation (4.19). The solution of the time-dependent temperature field consists in calculating the changes in the temperature field inside the solution domain at one time step on the basis of the previous temperature field. After the current period temperature field has been computed from the previous time period, it is saved in order to be used in the next time step.

In GAMS, this method of solving the time-dependent temperature field is implemented by using the LOOP operator as in Figure 4.25. Inside the body of the LOOP operator, the SOLVE command is repeatedly executed until the maximum number of cycles has been completed.

Observe that each pass through the loop executes the GAMS solver which determines the temperature field for the current moment *t*. This is determined from the values of the temperature field corresponding to the previous time step *Past_T(x,y)*.

```
$ontext
Time dependent temperature field in a rectangular area.
Determination of the time dependent temperature field in a
rectangular area with heterogeneous thermal conductivity and a
source points of heat inside the area as well as heat flows
temperature through the borders of the solution domain.
References:
[1] McKinney, D.C., Savitsky, A.G., Basic optimization models
for water and energy management. June 1999 (revision 6,
February 2003).
[2] Andrei, N., Nonlinear Optimization Applications using the
GAMS Technology, Springer Optimization and Its Applications 81,
Springer Science+Business Media, New York, 2013. (Application
8.1, pp.229-233)

$offtext

   set time   /t1*t25/;
   set x      /i1*i20/;
   set y      /j1*j20/;
   set inside(x,y);
       inside(x,y)                   = yes;
       inside(x,y)$(ord(x)=1)        = no;
       inside(x,y)$(ord(x)=card(x))  = no;
       inside(x,y)$(ord(y)=1)        = no;
       inside(x,y)$(ord(y)=card(y))  = no;

* Parameters determination
   scalar dx step for space in x direction      /0.1/;
   scalar dy step for space in y direction      /0.1/;
   scalar dt step for time                      /0.1/;
   scalar c                                     /1./;
   scalar rho                                   /1./;
   scalar heat accumulation in one time step    /0.0/;

* Temperature supply determination
   parameter supply(x,y);
             supply(x,y)            := 0;
             supply('i10','j18')    := 100/dx/dy;
   parameter past_T(x,y);
             past_T(x,y)            :=0;

   TABLE v(x,y)
     j1  j2  j3  j4  j5  j6  j7  j8  j9 j10 j11 j12 j13 j14 j15
j16 j17 j18 j19 j20
i1   0.5 0.5 0.5 0.5 0.5 0.5 0.5 0.5 0.5 0.5 0.5 0.5 0.5 0.5 0.5
0.5 0.5 0.5 0.5 0.5
i2   0.5 0.5 0.5 0.5 0.5 0.5 0.5 0.5 0.5 0.5 0.5 0.5 0.5 0.5 0.5
0.5 0.5 0.5 0.5 0.5
i3   0.5 0.5 0.5 0.5 0.5 0.5 0.5 0.5 0.5 0.5 0.5 0.5 0.5 0.5 0.5
```

Figure 4.25 GAMS expression of the time-dependent temperature field DIFF2

```
0.5 0.5 0.5 0.5 0.5
i4  0.5 0.5 0.5 0.5 0.5 0.5 0.5 0.5 0.5 0.5 0.5 0.5 0.5 0.5 0.5
0.5 0.5 0.5 0.5 0.5
i5  0.5 0.5 0.5 0.5 0.5 0.5 0.5 0.5 0.5 0.5 1.0 1.0 0.5 0.5 0.5
0.5 0.5 0.5 0.5 0.5
i6  0.5 0.5 0.5 0.5 0.5 0.5 0.5 0.5 0.5 0.5 1.0 1.0 0.5 0.5 0.5
0.5 0.5 0.5 0.5 0.5
i7  0.5 0.5 0.5 0.5 0.5 0.5 0.5 0.5 0.5 0.5 1.0 1.0 0.5 0.5 0.5
0.5 0.5 0.5 0.5 0.5
i8  0.5 0.5 0.5 0.5 0.5 0.5 0.5 0.5 0.5 0.5 1.0 1.0 0.5 0.5 0.5
0.5 0.5 0.5 0.5 0.5
i9  0.5 0.5 0.5 0.5 0.5 0.5 0.5 0.5 0.5 0.5 1.0 1.0 1.0 1.0 0.5
0.5 0.5 0.5 0.5 0.5
i10 0.5 0.5 0.5 0.5 1.0 1.0 1.0 1.0 1.0 1.0 1.0 1.0 1.0 1.0 0.5
0.5 0.5 0.5 0.5 0.5
i11 0.5 0.5 0.5 0.5 1.0 1.0 1.0 1.0 1.0 1.0 1.0 1.0 1.0 1.0 0.5
0.5 0.5 0.5 0.5 0.5
i12 0.5 0.5 0.5 0.5 0.5 0.5 0.5 0.5 0.5 1.0 1.0 1.0 1.0 1.0 1.0
0.5 0.5 0.5 0.5 0.5
i13 0.5 0.5 0.5 0.5 0.5 0.5 0.5 0.5 0.5 1.0 1.0 1.0 1.0 1.0 1.0
1.0 0.5 0.5 0.5 0.5
i14 0.5 0.5 0.5 0.5 0.5 0.5 0.5 0.5 0.5 0.5 0.5 0.5 1.0 1.0 1.0
1.0 1.0 0.5 0.5 0.5
i15 0.5 0.5 0.5 0.5 0.5 0.5 0.5 0.5 0.5 0.5 0.5 0.5 0.5 1.0 1.0
1.0 1.0 1.0 0.5 0.5
i16 0.5 0.5 0.5 0.5 0.5 0.5 0.5 0.5 0.5 0.5 0.5 0.5 0.5 0.5 1.0
1.0 1.0 1.0 0.5 0.5
i17 0.5 0.5 0.5 0.5 0.5 0.5 0.5 0.5 0.5 0.5 0.5 0.5 0.5 0.5 0.5
1.0 1.0 1.0 0.5 0.5
i18 0.5 0.5 0.5 0.5 0.5 0.5 0.5 0.5 0.5 0.5 0.5 0.5 0.5 0.5 0.5
0.5 0.5 0.5 0.5 0.5
i19 0.5 0.5 0.5 0.5 0.5 0.5 0.5 0.5 0.5 0.5 0.5 0.5 0.5 0.5 0.5
0.5 0.5 0.5 0.5 0.5
i20 0.5 0.5 0.5 0.5 0.5 0.5 0.5 0.5 0.5 0.5 0.5 0.5 0.5 0.5 0.5
0.5 0.5 0.5 0.5 0.5

* Model description
   variables
     obj     objective variable
     t(x,y)  field of temperature
     Q       temperature on boundaries ;

* Variable bounds
     t.lo(x,y)   =   0.0;
     t.up(x,y)   =   200.0;

* Initial values
     t.l(x,y) = 0.0;
     Q.l      = 0.0;

* Equations of the model
   equation
     temp(x,y) main equation of heat transport
     f1(x,y)   boundary computation dt:dn
     f2(x,y)   boundary computation dt:dn
     f3(x,y)   boundary computation dt:dn
```

Figure 4.25 (continued)

```
     f4(x,y)    boundary computation dt:dn
     fp1(x,y)   boundary computation t
     fp2(x,y)   boundary computation t
     fp3(x,y)   boundary computation t
     fp4(x,y)   boundary computation t
     eobj       objective equation ;

   temp(x,y)$(inside(x,y)).. t(x,y) - past_t(x,y) =e=
     dt*(supply(x,y)/c/rho +
       ( (v(x+1,y) + v(x,y))*(t(x+1,y) - t(x,y))-
         (v(x,y) + v(x-1,y))*(t(x,y) - t(x-1,y))) /dx/dx/2.0 +
         ( (v(x,y+1) + v(x,y))*(t(x,y+1) - t(x,y)) -
         (v(x,y-1) = v(x,y))*(t(x,y) - t(x,y-1))) /dy/dy/2.0);

   f1(x,y)$((ord(x) = 1))..          t(x+1,y)-t(x,y) =g= 0;
   f2(x,y)$((ord(x) = card(x)))..   t(x,y)-t(x-1,y) =l= 0;
   f3(x,y)$((ord(y) = 1))..          t(x,y+1)-t(x,y) =g= 0;
   f4(x,y)$((ord(y) = card(y)))..   t(x,y)-t(x,y-1) =l= 0;

   fp1(x,y)$((ord(x) = 1))..          t(x,y) =e= Q;
   fp2(x,y)$((ord(x) = card(x)))..   t(x,y) =e= Q;
   fp3(x,y)$((ord(y) = 1))..          t(x,y) =e= Q;
   fp4(x,y)$((ord(y) = card(y)))..   t(x,y) =e= Q;

   eobj.. obj =e= Q;

model Diffusion2 /all/;

*$onecho >bench.opt
*  solvers conopt knitro minos snopt
*$offecho
Diffusion2.optfile=1;
*option nlp=bench;

* Output file
*  file res1 / Diffusion2.txt/
  file res2 / Diffusion2.dat/
*  put res1;
  put res2;

* Cycle of computation
loop (time,
    solve Diffusion2 using nlp minimizing obj;
    put " Time interval = ";
    put time.tl:20; put /;
*   loop(x, put x.tl:6;
*       loop(y, put t.l(x,y):6:2; ); put /;) put /;

    loop(x,loop(y, put t.l(x,y):8:2;); put /;) ; put /;

    heat = heat +sum( (x,y), t.l(x,y)-past_t(x,y))*dx*dy;
    past_t(x,y) = t.l(x,y);
);
* End Diffusion2
```

Figure 4.25 (continued)

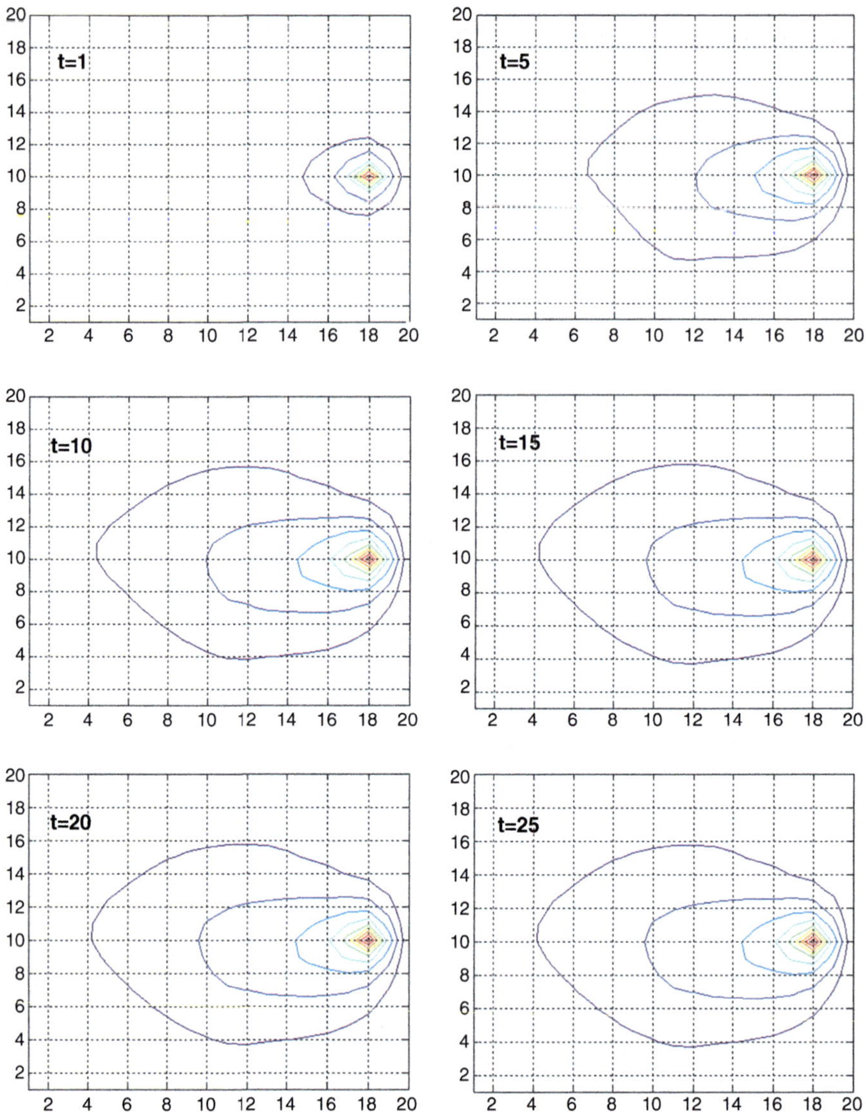

Figure 4.26 Solution of the transient heat transportation problem at six time periods

Figure 4.26 presents the solution of the transient heat transportation problem at a number of some periods: *t1*, *t5*, *t10*, *t15*, *t20*, and *t25*. A stationary solution to the transient equation is found by using a large number of iterations under stationary boundary conditions. After a number of iterations, the influence of the initial conditions becomes very small and $\partial T/\partial t \rightarrow 0$. In Figure 4.26, observe that the field temperature converges to a stationary solution after about 15–20 iterations.

4.15 Stationary Flow of an Incompressible Fluid in a Rectangular Area (FLOW/FLOWO)

The equations that describe the flow of a fluid are the conservation of linear momentum and the conservation of mass (McKinney & Savitsky, 2003). The conservation of linear momentum of a viscous, incompressible fluid in horizontal plane area in x direction is:

$$\frac{\partial V_x}{\partial t} + V_x \frac{\partial V_x}{\partial x} + V_y \frac{\partial V_y}{\partial y} = -\frac{1}{\rho} \frac{\partial P}{\partial x} + \mu \nabla^2 V_x, \tag{4.21}$$

and in y direction:

$$\frac{\partial V_y}{\partial t} + V_y \frac{\partial V_y}{\partial x} + V_y \frac{\partial V_y}{\partial y} = -\frac{1}{\rho} \frac{\partial P}{\partial y} + \mu \nabla^2 V_y. \tag{4.22}$$

The equation of conservation of mass (continuity equation) for the fluid is

$$\frac{\partial V_x}{\partial x} + \frac{\partial V_y}{\partial y} = 0, \tag{4.23}$$

where V_x is x component of velocity [m/sec], V_y is y component of velocity [m/sec], P is pressure [pa], μ is the kinematic viscosity [m^2/sec], and ρ is the density of the fluid [kg/m^3].

Observe that these equations are valid in the case when the viscosity is constant and the fluid weight acts perpendicularly to the $x - y$ plane. Therefore, the gravitational force is not present in these equations.

With these preparatives, let us consider a stationary flow of an incompressible fluid in a rectangular area with given inflow of fluid on the borders of the solution domain (McKinney & Savitsky, 2003). In this application, the zones of inflow and outflow of water are on two opposite sides. On the zone of inflow of water, the boundary conditions are $V_x = 0.5$ [m/sec] and $V_y = 0$ [m/sec]. On the zone of outflow of water, the boundary conditions are $\frac{\partial V_x}{\partial x} = 0$ and $V_y = 0$. On the other sides of the rectangle, the so-called non-slip conditions are taken: $V_x = 0$ [m/sec] and $V_y = 0$ [m/sec].

To compute the flow of water, a finite difference analog of the differential equations will be considered. The solution domain is divided into small rectangles by means of a grid of parallel lines and to each intersection point the indices (i, j) are associated. At each point, the fluid velocity and the pressure are computed. The fluid velocity is not computed in points, but on the lines connecting the points. On lines parallel to the x axis, the V_x is computed. On the other hand, on lines parallel to the y axis, we compute V_y. Therefore, the continuity equation has the following finite difference analog:

$$\frac{V_{x,i,j} - V_{x,i-1,j}}{\Delta x} + \frac{V_{y,i,j} - V_{y,i,j-1}}{\Delta y} = 0. \tag{4.24}$$

The pressure is computed at the node points (i, j) of each line, where the pressure gradient is computed from the velocity:

$$\left.\frac{\partial P}{\partial x}\right|_{i,j} = \frac{P_{i+1,j} - P_{i,j}}{\Delta x}, \tag{4.25}$$

$$\left.\frac{\partial P}{\partial y}\right|_{i,j} = \frac{P_{i,j+1} - P_{i,j}}{\Delta y}. \tag{4.26}$$

The diffusive terms from (4.21) and (4.22) are computed as:

$$\mu \nabla^2 V_x \approx \mu \left(\frac{V_{x,i+1,j} - 2V_{x,i,j} + V_{x,i-1,j}}{(\Delta x)^2} + \frac{V_{x,i,j+1} - 2V_{x,i,j} + V_{x,i,j-1}}{(\Delta y)^2} \right), \tag{4.27}$$

$$\mu \nabla^2 V_y \approx \mu \left(\frac{V_{y,i+1,j} - 2V_{y,i,j} + V_{y,i-1,j}}{(\Delta x)^2} + \frac{V_{y,i,j+1} - 2V_{y,i,j} + V_{y,i,j-1}}{(\Delta y)^2} \right). \tag{4.28}$$

Since there are errors introduced by the algebraic approximations, it is not always possible to approximate differentials by finite differences. Analyzing the system of equations (4.21) and (4.22), we see that these equations contain velocity components connected by the pressure gradient ∇P. If P is not connected with boundary conditions, then all the approximation errors will leave the solution domain through the borders. In equation (4.23), the velocity vector is in strong contact with the boundary conditions. Therefore, in the calculations of the continuity equation we expect big errors to appear at each point. To skip over this difficulty, the continuity equation is written in a slightly different (relaxed) form as

$$\frac{\partial V_x}{\partial x} + \frac{\partial V_y}{\partial y} = \delta, \tag{4.29}$$

where δ is the error in the approximation of differentials, which must be minimized. The final system of equations which have to be solved has the following form:

$$-\frac{1}{\rho}\frac{\partial P}{\partial x} + \mu \nabla^2 V_x = 0, \tag{4.30}$$

$$-\frac{1}{\rho}\frac{\partial P}{\partial y} + \mu \nabla^2 V_y = 0, \tag{4.31}$$

$$\frac{\partial V_x}{\partial x} + \frac{\partial V_y}{\partial y} = \delta. \tag{4.32}$$

The finite difference analog of (4.30), (4.31), and (4.32) are:

$$\frac{1}{\rho}\frac{P_{i+1,j}-P_{i,j}}{\Delta x}=\mu\left(\frac{V_{x,i+1,j}-2V_{x,i,j}+V_{x,i-1,j}}{(\Delta x)^2}+\frac{V_{x,i,j+1}-2V_{x,i,j}+V_{x,i,j-1}}{(\Delta y)^2}\right),$$

$$\frac{1}{\rho}\frac{P_{i,j+1}-P_{i,j}}{\Delta y}=\mu\left(\frac{V_{y,i+1,j}-2V_{y,i,j}+V_{y,i-1,j}}{(\Delta x)^2}+\frac{V_{y,i,j+1}-2V_{y,i,j}+V_{y,i,j-1}}{(\Delta y)^2}\right),$$

$$\frac{V_{x,i,j}-V_{x,i-1,j}}{\Delta x}+\frac{V_{y,i,j}-V_{y,i,j-1}}{\Delta y}=\delta_{i,j}.$$

In the following, consider a discretization of the rectangular area in which the x axis is discretized in 15 intervals and the y axis in 20, where $\Delta x = \Delta y = 1$, (Figure 4.27) (McKinney & Savitsky, 2003).

Let the fluid density be equal to 1000 [Kg/m^3]. The kinematic viscosity of the fluid is 0.005 [m^2/sec]. The flow of the fluid is parallel to the x axis. Now, consider that in the interval [u5, u15] on the y axis the following boundary conditions are active: $V_x = 0.5$ [m/sec] and $V_y = 0$ [m/sec]. On the opposite side, the boundary conditions are $\frac{\partial V_x}{\partial x} = 0$ and $V_y = 0$. On the rest of the intervals of the discretization, the boundary conditions are $V_x = 0$ [m/sec] and $V_y = 0$ [m/sec].

With these details, the GAMS representation of this application is presented in Figure 4.28.

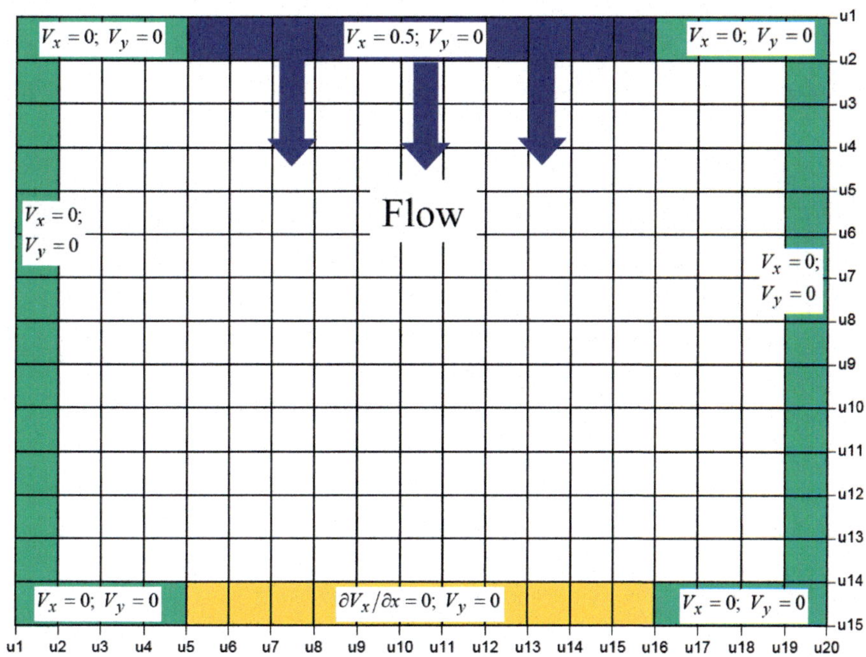

Figure 4.27 Flow domain and its discretization

```
$ontext
Stationary flow of an incompressible fluid in a rectangular
area.
References:
[1] McKinney, D.C., Savitsky, A.G., Basic optimization models
for water and energy management. June 1999 (revision 6,
February 2003).
http://www.ce.utexas.edu/prof/mckynney/ce385d/papers/GAMS-
Tutorial.pdf (pp. 130-136)
[2] Andrei, N., Nonlinear Optimization Applications using the
GAMS Technology, Springer Optimization and Its Applications 81,
Springer Science+Business Media, New York, 2013. (Application
8.2, pp.229-245)

$offtext

set X /u1*u15/;
set Y /u1*u20/;
set Yout(Y) /u5*u16/;

* Determination of zone for water movement equation
Set Inside(X,Y);
    Inside(X,Y)                      = yes;
    Inside(X,Y)$(ord(X)=1)           = no;
    Inside(X,Y)$(ord(X)=card(X))     = no;
    Inside(X,Y)$(ord(Y)=1)           = no;
    Inside(X,Y)$(ord(Y)=card(Y))     = no;

* Scalars and Parameters
scalar dx  step space in x direction /1/;
scalar dy  step space in y direction /1/;
scalar r   density of the fluid      /1000/;
parameter m(X,Y) kinematic viscosity ;
          m(X,Y) := 0.5;

* Variables definition
Variables
 obj        objective variable
 D(X,Y)     error
 P(X,Y)     pressure
 Vx(X,Y)    x-direction velocity
 Vy(X,Y)    y-direction velocity
 Vdx(Y)     delta Vx at the outlet;

* Variable bounds and initialization
  Vx.up(X,Y)     =  1.5;
  Vx.lo(X,Y)     = -1.5;
  Vx.l(X,Y)      =  0.5;
```

Figure 4.28 GAMS expression of application FLOW

```
  Vy.up(X,Y)      =  1.0;
  Vy.lo(X,Y)      = -1.0;
  Vy.l(X,Y)       =  0.0;
  D.lo(X,Y)       = 0.0;
  D.up(X,Y)       = 0.0000005;
  Vy.l(X,Y)       = 0.0;
  Vy.l(X,Y)$(inside(X,Y)) = 0.0000001;
  P.up(X,Y)       =  2000;
  P.lo(X,Y)       = -2000;
  P.l(X,Y)        =  0.000001;

*Boundary conditions
  Vx.lo('u1',Y)    =0.5;
  Vx.fx('u15',Y)   =0.5;
  Vx.fx(X,'u1')    = 0;
  Vx.fx(X,'u20')   = 0;
  Vx.fx('u1','u2') = 0;
  Vx.fx('u1','u3') = 0;
  Vx.fx('u1','u4') = 0;
  Vx.fx('u1','u17') = 0;
  Vx.fx('u1','u18') = 0;
  Vx.fx('u1','u19') = 0;
  Vx.fx('u15','u2')=0;
  Vx.fx('u15','u3')=0;
  Vx.fx('u15','u4')=0;
  Vx.fx('u15','u17')=0;
  Vx.fx('u15','u18')=0;
  Vx.fx('u15','u19')=0;
  Vy.fx('u1',Y) = 0;
  Vy.fx(X,'u1') = 0;
  Vy.fx(X,'u20') =0;
  Vy.fx('u15',Y)=0;

Equations
     For_Vx(X,Y)
     For_Vy(X,Y)
     Div_Vxy(X,Y)
     Vx_Vx(Y)
     eobj  objective function ;

For_Vx(X,Y)$(Inside(X,Y))..
   (P(X+1,Y)-P(X,Y))/(r*dx) =e=
    m(x,Y)*((Vx(X+1,Y)-2*Vx(X,Y)+Vx(x-1,Y))/(dx*dy)
   +        (Vx(X,Y+1)-2*Vx(X,Y)+Vx(X,Y-1))/(dx*dy));
For_Vy(X,Y)$(Inside(X,Y))..
   (P(X,Y+1)-P(X,Y))/(r*dy) =e=
    m(X,Y)*((Vy(X+1,Y)-2*Vy(X,Y)+Vy(X-1,Y))/(dx*dy)
   +        (Vy(X,Y+1)-2*Vy(X,Y)+Vy(X,Y-1))/(dy*dx));
Div_Vxy(X,Y)$((ord(X) > 1) $ (ord(Y) > 1))..
    (Vx(X-1,Y)-Vx(X,Y))/dx + (Vy(X,Y-1)-Vy(X,Y))/dy =e=
        D(X,Y);
Vx_Vx(Y).. Vdx(Y) =e= Vx('u15',Y)-Vx('u14',Y);

eobj.. obj =e= SUM(Y$Yout(Y), Vdx(Y)*Vdx(Y))
              +SUM((X,Y),(D(X,Y)*D(X,Y)));
```

Figure 4.28 (continued)

```
Model flow /all/;
$onecho >bench.opt
  solvers conopt knitro minos snopt
$offecho
flow.optfile=1;
option nlp=bench;
Solve flow using nlp minimizing obj;
* Put the solution
  file res1 /pressure.dat/
  put res1;
  loop(X, loop(Y, put P.l(x,y):9:1; ); put /;); put /;
  file res2 / vx.dat/
  put res2;
  loop(X, loop(Y, put Vx.l(x,y):9:4; ); put /;); put /;
  file res3 / vy.dat/
  put res3;
  loop(X, loop(Y, put Vy.l(x,y):5:1; ); put /;); put /;
  file res4 /err.dat/
  put res4
  loop(X, loop(Y, put D.l(x,y):5:1; ); put /;); put /;
* End flow
```

Figure 4.28 (continued)

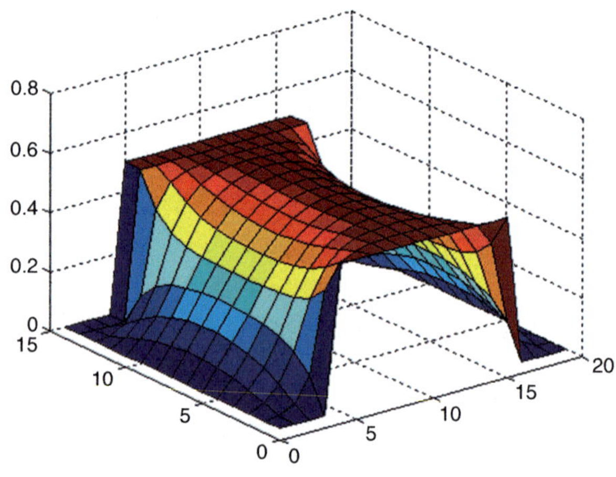

Figure 4.29 Velocity in x direction

The error in the approximation through finite differences is found in the file err.dat. If this error is visualized, we see that it is zero. Besides, notice that the optimal value of the objective function is also zero. Figure 4.29 presents the evolution of the velocity in direction x.

Figure 4.30 shows three cross sections of velocity V_x in x direction.

*In the following, we shall compute the stationary flow of water in a rectangular area in the presence of an obstacle (*Figure 4.31*).* As above, let us consider a rectangular area. On the top edge of the area, an inflow of water with velocity of 0.5 m/sec occurs.

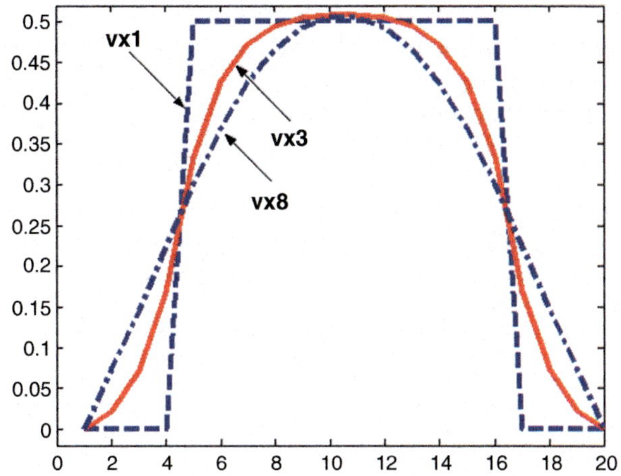

Figure 4.30 Three cross sections of velocity in *x* direction

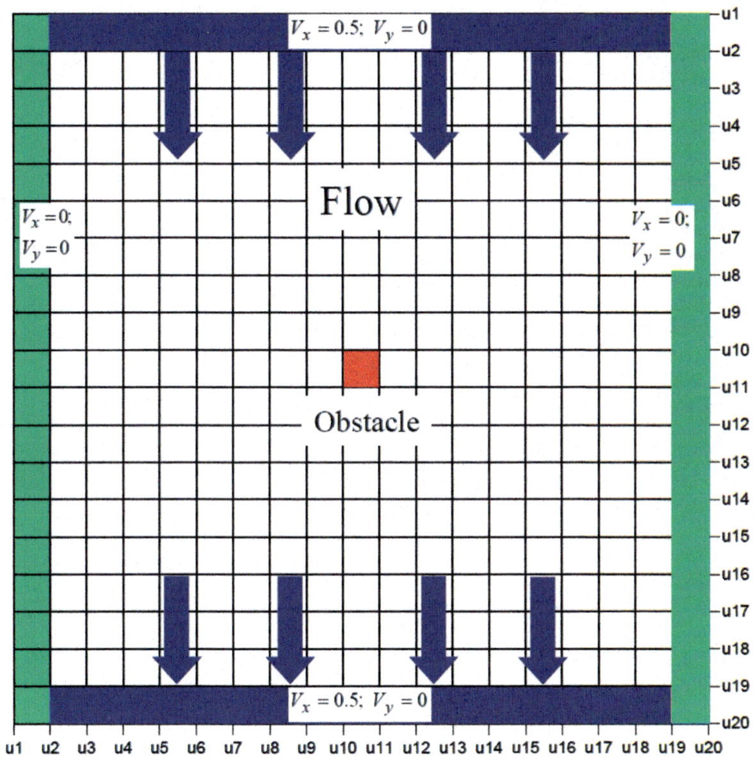

Figure 4.31 Flow domain with an obstacle

The same flow leaves the area through the bottom edge of the domain. The boundary conditions are the same as in the previous application, but in the middle of the domain, there is a symmetrically located obstacle around which the fluid is compelled to flow as represented in Figure 4.31 (McKinney & Savitsky, 2003).

The equations of flow in this case are the same as before. However, the additional instructions with the definition of the obstacle in the middle of the solution domain must be introduced. The GAMS representation of the model in this case is represented in Figure 4.32.

```
$ontext
Stationary flow of an incompressible fluid in a rectangular
area in the presence of an obstacle.
References:
[1] McKinney, D.C., Savitsky, A.G., Basic optimization models
for water and energy management. June 1999 (revision 6,
February 2003). (pp. 136-142)
[2] Andrei, N., Nonlinear Optimization Applications using the
GAMS Technology, Springer Optimization and Its Applications 81,
Springer Science+Business Media, New York, 2013. (Application
8.2, pp.229-245)

$offtext

set X /u1*u20/;
set Y /u1*u20/;

* Determination of zone for water movement equation

Set vyside(X,Y);
Set vxside(X,Y);
    vxside(X,Y)                     = yes;
    vxside(X,Y)$(ord(X)=1)          = no;
    vxside(X,Y)$(ord(X)=card(X))    = no;
    vxside(X,Y)$(ord(Y)=1)          = no;
    vxside(X,Y)$(ord(Y)=card(Y))    = no;
    vyside(X,Y) = vxside(X,Y);
    vxside('u10','u10')                = no;
    vxside('u10','u11')                = no;

* Scalars and Parameters
scalar dx  step space in x direction   /1/;
scalar dy  step space in y direction   /1/;
scalar r   density of the fluid        /1000/;

parameter m(X,Y) kinematic viscosity ;
          m(X,Y) := 0.5;

* Variables definition
variables
 obj         objective variable
 D(X,Y)      error
 P(X,Y)      pressure
 Vx(X,Y)     x-direction velocity
```

Figure 4.32 GAMS expression of flow with an obstacle FLOWOBS

```
 Vy(X,Y)     y-direction velocity  ;

* Variable bounds and initialization
  Vx.l(X,Y)      =  0.0;
  Vy.l(X,Y)      =  0.0;
  Vy.l(x,y)$(vxside(x,y)) = 0.0;

  D.lo(X,Y)      = 0.0;
  D.up(X,Y)      = 7.50;

  P.up(X,Y)      =  20000;
  P.lo(X,Y)      = -20000;
  P.l(X,Y)       =  0.000001;

*Boundary conditions
  Vx.fx('u1',Y)    = 0.5;
  Vx.fx('u20',Y)   = 0.5;
  Vx.fx(X,'u1')    = 0;
  Vx.fx(X,'u20')   = 0;
  Vy.fx('u1',Y)    = 0;
  Vy.fx('u20',Y)   = 0;
  Vy.fx(X,'u1')    = 0;
  Vy.fx(X,'u20')   = 0;

* Obstacle description
  vx.fx('u10','u10') = 0;
  vx.fx('u10','u11') = 0;

Equations
     For_Vx(X,Y)
     For_Vy(X,Y)
     Div_Vxy(X,Y)
     eobj  objective function ;

For_Vx(X,Y)$(vxside(X,Y))..
   (P(X+1,Y)-P(X,Y))/(r*dx) =e=
     m(x,Y)*((Vx(X+1,Y)-2*Vx(X,Y)+Vx(x-1,Y))/(dx*dy)
    +         (Vx(X,Y+1)-2*Vx(X,Y)+Vx(X,Y-1))/(dx*dy));

For_Vy(X,Y)$(vxside(X,Y))..
   (P(X,Y+1)-P(X,Y))/(r*dy) =e=
     m(X,Y)*((Vy(X+1,Y)-2*Vy(X,Y)+Vy(X-1,Y))/(dx*dy)
    +         (Vy(X,Y+1)-2*Vy(X,Y)+Vy(X,Y-1))/(dy*dx));

*--
*For_Vx(X,Y)$(Vxside(X,Y))..
*    Vx(X,Y)*(Vx(X+1,Y)-Vx(X-1,Y))/(2*dx) +
*    0.25*(Vy(X+1,Y-1)+Vy(X+1,Y)+Vy(X,Y-1)+Vy(X,Y)) *
*          (Vx(X,Y+1)-Vx(X,Y-1))/(2*dy) +
*          (P(X+1,Y)-P(X,Y))/(r*dx)
*    =e=
*    m(X,Y)*((Vx(X+1,Y)-2*Vx(X,Y)+Vx(X-1,Y))/(dx*dx) +
*            (Vx(X,Y+1)-2*Vx(X,Y)+Vx(X,Y-1))/(dy*dy));
*
*For_Vy(X,Y)$(Vyside(X,Y))..
*    0.25*(Vx(X-1,Y+1)+Vx(X-1,Y)+Vx(X,Y+1)+Vx(X,Y)) *
```

Figure 4.32 (continued)

```
*            (Vy(X+1,Y)-Vy(X-1,Y))/(2*dy) +
*           Vy(X,Y)*(Vy(X,Y+1)-Vy(X,Y-1))/(2*dy) +
*            (P(X,Y+1)-P(X,Y))/(r*dy)
*      =e=
*      m(X,Y)*((Vy(X+1,Y)-2*Vy(X,Y)+Vy(X-1,Y))/(dx*dx) +
*              (Vy(X,Y+1)-2*Vy(X,Y)+Vy(X,Y-1))/(dy*dy));
*--

Div_Vxy(X,Y)$((ord(X) > 1) $ (ord(Y) > 1))..
    (Vx(X,Y)-Vx(X-1,Y))/dx + (Vy(X,Y)-Vy(X,Y-1))/dy =e=
        D(X,Y);

eobj.. obj =e= SUM((X,Y),(D(X,Y)*D(X,Y)));

Model flowobs /all/;

*$onecho >bench.opt
*  solvers conopt knitro minos snopt
*$offecho

flowobs.optfile=1;
*option nlp=bench;

Solve flowobs using nlp minimizing obj;

* Put the solution
  file res1 /pressure.dat/
  put res1;
  loop(X, loop(Y, put P.l(x,y):10:4; ); put /;); put /;

  file res2 / vx.dat/
  put res2;
  loop(X, loop(Y, put Vx.l(x,y):9:5; ); put /;); put /;

  file res3 / vy.dat/
  put res3;
  loop(X, loop(Y, put Vy.l(x,y):5:1; ); put /;); put /;

  file res4 /err.dat/
  put res4
  loop(X, loop(Y, put D.l(x,y):5:1; ); put /;); put /;
* End flowobs
```

Figure 4.32 (continued)

Figure 4.33 presents the evolution of the velocity V_x in direction x.

Figure 4.34 shows three cross sections of velocity V_x in x direction.

Observe the asymmetry of the approximations. This is because the velocities are not determined at the cell centers, but on the cell faces.

If the inertial terms of the equations are considered, then these terms can be computed using the "marker and cell" method as is described in Peyret and Taylor (1985):

Figure 4.33 Velocity in x direction

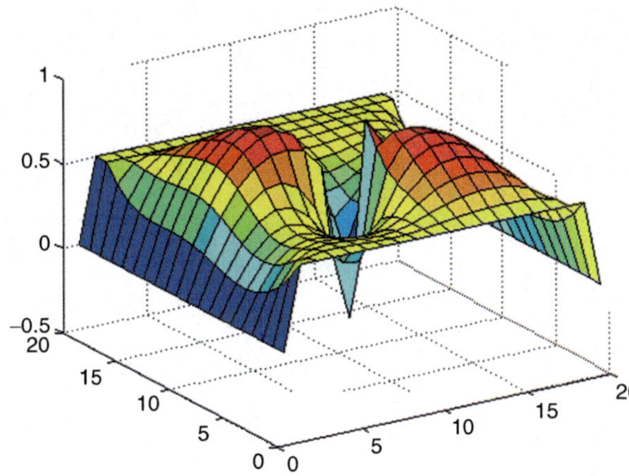

Figure 4.34 Three cross sections of velocity in x direction

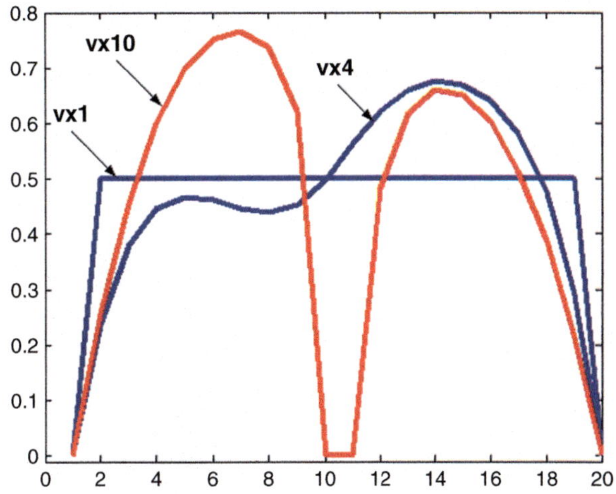

$$V_x\frac{\partial V_x}{\partial x}+V_y\frac{\partial V_x}{\partial y}\cong V_{x,i,j}\frac{V_{x,i+1,j}-V_{x,i,j}}{2\Delta x}+\frac{V_{y,i+1,j}+V_{y,i,j}+V_{y,i,j-1}+V_{y,i+1,j-1}}{4}\frac{V_{x,i,j+1}-V_{x,i,j-1}}{2\Delta y},$$

$$V_x\frac{\partial V_y}{\partial x}+V_y\frac{\partial V_y}{\partial y}\cong V_{y,i,j}\frac{V_{y,i,j+1}-V_{y,i,j-1}}{2\Delta y}+\frac{V_{x,i,j}+V_{x,i,j+1}+V_{x,i-1,j+1}+V_{x,i-1,j}}{4}\frac{V_{y,i+1,j}-V_{y,i-1,j}}{2\Delta x}.$$

In Figure 4.32, these terms can be included in the model by removing the asterisks from the GAMS code and by commenting the old equations corresponding

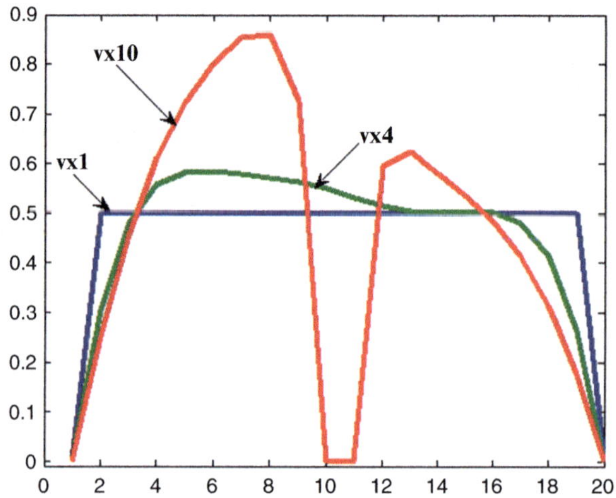

Figure 4.35 Three cross sections of velocity in x direction with inertial effects

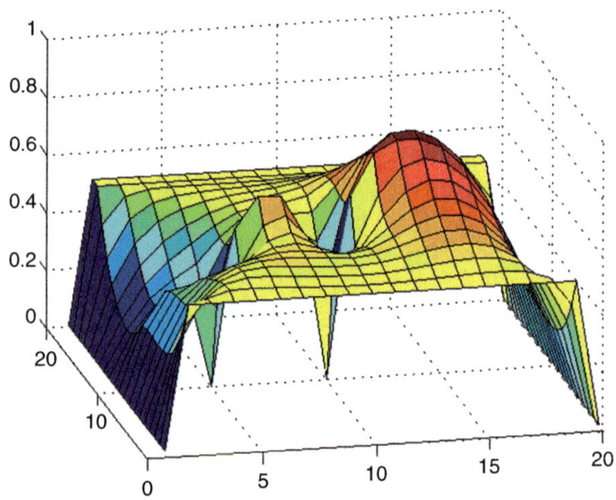

Figure 4.36 Velocity in x direction for two obstacles

to the stationary flow of fluid without an obstacle. Figure 4.35 presents three cross sections of velocity in x direction with inertial terms.

It is worth seeing the solution of the flow in a rectangular area with *two* obstacles (u10,u5) and (u10,u10). Figure 4.36 shows the velocity in x direction.

Figure 4.37 shows three cross sections of velocity V_x in x direction.

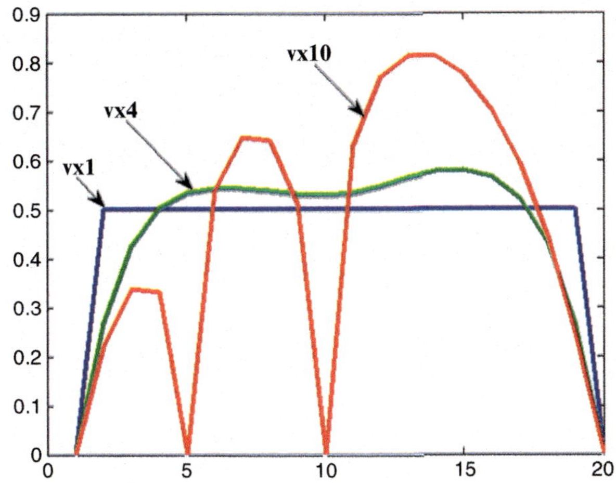

Figure 4.37 Three cross sections of velocity in x direction

4.16 Fed-Batch Fermenter for Penicillin Production (PENICI)

This problem deals with the dynamic optimization of a fed-batch fermenter for production of penicillin through anaerobic glucose fermentation. The dynamic optimization of this process with fixed final time was studied by Banga, Alonso, and Singh (1997), Cuthrell and Biegler (1989), Luus (1993), and Larrosa (2008). The problem consists in maximizing the total amount of penicillin produced using the feed rate of substrate as the control variable. The mathematical model of this application is

$$\max\left(y_2\left(t_f\right)y_4\left(t_f\right)\right)$$

subject to (4.33)

$$\frac{dy_1}{dt} = h_1 y_1 - \frac{u y_1}{500 y_4},$$

$$\frac{dy_2}{dt} = h_2 y_1 - 0.01 y_2 - \frac{u y_2}{500 y_4},$$

$$\frac{dy_3}{dt} = -\frac{h_1 y_1}{0.47} - \frac{h_2 y_1}{1.2} - \frac{0.029 y_1 y_3}{0.0001 + y_3} + \frac{u}{y_4}\left(1 - \frac{y_3}{500}\right),$$

$$\frac{dy_4}{dt} = \frac{u}{500},$$

where:

$$h_1 = \frac{0.11 y_3}{0.006 y_1 + y_3},$$

$$h_2 = \frac{0.0055 y_3}{0.0001 + y_3(1 + 10 y_3)}.$$

In this model, y_1 represents the concentration of biomass, y_2 is the penicillin concentration, y_3 is the concentration of substrate, and y_4 is the fermenter volume (in L). The initial conditions are $y_1(0)=1.5$, $y_2(0)=0.1$, $y_3(0)=0.1$, and $y_4(0)=7$. The final product is destined to human consumption. Therefore, the concentrations of the present species are subject to the following path constraints:

$$0 \le y_1 \le 40, \quad 0 \le y_3 \le 25, \quad 0 \le y_4 \le 10.$$

The control variable is bounded as $0 \le u \le 50$, and the total process time is fixed as $t_f = 100$ h.

For solving this application as well as some other optimal control examples, let us discuss some aspects of the numerical solutions of the ordinary differential equations:

$$y' = f(t, y), \;\; t_0 \le t \le t_f, \;\; y(t_0) = y_0. \tag{4.34}$$

One of the simplest methods for solving (4.34) is the *Euler's method*. For this, the interval $[t_0, t_f]$ is divided by the *mesh-points* $t_k = t_0 + kh$, $k=0, \ldots, N$, where $h=(t_f - t_0)/N$ is the *step size* and N is a positive integer. Now, consider that for each k, we seek a numerical approximation y_k to $y(t_k)$, the value of the analytical solution at the mesh point t_k. The Euler's method proceeds by integrating the differential equation (4.34) between two consecutive points t_k and t_{k+1}:

$$y(t_{k+1}) = y(t_k) + \int_{t_k}^{t_{k+1}} f(t, y(t))\mathrm{d}t, \quad k = 0, \ldots, N-1, \tag{4.35}$$

and then by applying the numerical integration rule

$$\int_{t_k}^{t_{k+1}} g(t)\mathrm{d}t \approx hg(t_k), \tag{4.36}$$

called the *rectangle rule*, with $g(t) = f(t, y(t))$, to get:

$$y(t_{k+1}) \approx y(t_k) + hf(t_k, y(t_k)), \;\; k = 0, \ldots, N-1, \;\; y(t_0) = y_0.$$

This simple derivation motivates the definition of *Euler's method* as:

$$y_{k+1} = y_k + hf(t_k, y_k), \quad k = 0, \ldots, N-1. \tag{4.37}$$

This idea can be very easily generalized by replacing the rectangle rule in the derivation of the Euler method with one-parameter family of integration rule of the form:

$$\int_{t_k}^{t_{k+1}} g(t)\mathrm{d}t \approx h[(1-\theta)g(t_k) + \theta g(t_{k+1})],$$

where $\theta \in [0, 1]$ is a parameter. Now, applying this in (4.35) with $g(t) = f(t, y(t))$ we get:

$$y(t_{k+1}) \approx y(t_k) + h[(1-\theta)f(t_k, y(t_k)) + \theta f(t_{k+1}, y(t_{k+1}))],$$
$$k = 0, \ldots, N-1, \ \ y(t_0) = y_0.$$

Therefore, this motivates the introduction of the following one-parameter family of methods: given that y_0 is specified by (4.34), define:

$$y_{k+1} = y_k + h\left[(1-\theta)f(t_k, y_k) + \theta f\left(t_{k+1}, y_{k+1}\right)\right],$$
$$k = 0, \ldots, N-1. \tag{4.38}$$

known as the $\theta-$*method*. Observe that for $\theta = 0$ we recover the Euler's method. Now, for $\theta = 1$ and y_0 given by (4.34), we get:

$$y_{k+1} = y_k + hf\left(t_{k+1}, y_{k+1}\right), \quad k = 0, \ldots, N-1. \tag{4.39}$$

known as the *implicit* or *backward Euler method*, since unlike the Euler's method defined by (4.37), (4.39) requires the solution of an implicit equation in order to determine y_{k+1}, given y_k. In this context, (4.37) is called the *explicit* or *forward Euler method*. For $\theta = 1/2$ in (4.38), we get another interesting computational scheme: given that y_0 is specified by (4.34), y_{k+1} are computed as:

$$y_{k+1} = y_k + \frac{1}{2}h\left[f(t_k, y_k) + f\left(t_{k+1}, y_{k+1}\right)\right], \ k = 0, \ldots, N-1, \tag{4.40}$$

known as the *trapezoidal rule method*. Like the backward Euler method, the trapezoidal rule is implicit: in order to compute y_{k+1} a nonlinear system of algebraic equations must be solved.

Each of these methods is *consistent* with the ordinary differential equation (4.34). That is, if we plug solutions to the exact equation into the numerical method, we get a *small local error*. For example, the forward Euler has consistency of order 1, and the trapezoidal rule has second-order consistency.

While the trapezoidal rule method leads to more accurate approximations, the forward Euler method is less convenient from the computational viewpoint since it requires the solution of implicit equations at each mesh point x_{k+1} to get y_{k+1}. To skip over this difficulty, an attractive computational scheme lies in using the forward Euler method to compute an initial crude approximation for $y(t_{k+1})$ and then use this value within the trapezoidal rule to get a more accurate approximation of $y(t_{k+1})$. The resulting numerical method is.

$$y_{k+1} = y_k + \frac{1}{2}h[f(t_k, y_k) + f(t_{k+1}, y_k + hf(t_k, y_k))], \ k = 0, \ldots, N-1, \tag{4.41}$$

known as *improved Euler method*. This is an explicit one-step computational scheme.

Now, using the trapezoidal rule (4.40), the GAMS representation of this application is given in Figure 4.38.

```
$Ontext
Optimal control of fed-batch fermented for Penicillin
Production. The dynamic optimization of a fed-batch fermented
for the production of penicillin through anaerobic glucose
fermentation.
References:
[1] Larrosa, J.A.E., New heuristics for global optimization of
complex bioprocesses. Ph.D. Thesis, Universidade de Vigo,
Departamento de Enxeñería Química, Vigo, 2008.
[2] Andrei, N., Nonlinear Optimization Applications using the
GAMS Technology, Springer Optimization and Its Applications 81,
Springer Science+Business Media, New York, 2013. (Application
12.2, pp.288-299)
The trapezoid rule method for solving ordinary differential
equations.
$Offtext

$if     set n  $set nh %n%
$if not set nh $set nh 100

set nh Number of subintervals / 0*%nh% /;
alias (nh,k);

Scalar    tf   final time /100/
          y1_0 initial value for y1 / 1.5/
          y2_0 initial value for y2 / 0.1 /
          y3_0 initial value for y3 / 0.1 /
          y4_0 initial value for y4 / 7 /
          a    a parameter in objective function /0.0/
          h ;
          h=tf/%nh%;

* Variables definition.
Variables y1(nh)  biomass
          y2(nh)  penicillin
          y3(nh)  substrate concentration
          y4(nh)  fermenter volume
          u(nh)   control variable
          h1(nh)
          h2(nh)
          obj     criterion ;

* Equations definition.
Equations eobj         criterion definition
          state1(nh)   state equation 1
          state2(nh)   state equation 2
          state3(nh)   state equation 3
          state4(nh)   state equation 4
          eh1(nh)
          eh2(nh) ;
eobj..
 obj =e= y2['%nh%']*y4['%nh%'] +
        h*a*sum(nh(k+1), ((u(k+1)-u(k))*(u(k+1)-u(k))) );

 eh1(nh(k)).. h1(k) =e=
              0.11*y3(k)/(0.006*y1(k)+y3(k));
 eh2(nh(k)).. h2(k) =e=
              0.0055*y3(k)/(0.0001+y3(k)*(1+10*y3(k)));
```

Figure 4.38 GAMS expression of application PENICI

```
* The trapezoidal rule method
state1(nh(k+1))..
y1[k+1] =e= y1(k)+
    (h/2)*( h1(k)*y1(k)       - u(k)*y1(k)/(500*y4(k)) +
            h1(k+1)*y1(k+1)   - u(k+1)*y1(k+1)/(500*y4(k+1)) ) ;

state2(nh(k+1))..
y2[k+1] =e= y2(k)+
    (h/2)*( h2(k)*y1(k) - 0.01*y2(k) -  u(k)*y2(k)/(500*y4(k)) +
            h2(k+1)*y1(k+1) -0.01*y2(k+1) -
            u(k+1)*y2(k+1)/(500*y4(k+1)) );

state3(nh(k+1))..
y3[k+1] =e= y3(k)+
    (h/2)*( -h1(k)*y1(k)/0.47 - h2(k)*y1(k)/1.2 -
             0.029*y1(k)*y3(k)/(0.0001+y3(k)) +
             (u(k)/y4(k))*(1-y3(k)/500) -
             h1(k+1)*y1(k+1)/0.47 - h2(k+1)*y1(k+1)/1.2 -
             0.029*y1(k+1)*y3(k+1)/(0.0001+y3(k+1)) +
             (u(k+1)/y4(k+1))*(1-y3(k+1)/500) );

state4(nh(k+1))..
y4[k+1] =e= y4(k) + (h/2)*(u(k)/500 + u(k+1)/500);

*Initial point.
y1.l[nh]=1.5;
y2.l[nh]=0.1;
y3.l[nh]=0.1;
y4.l[nh]=7;
u.l[nh] =10;

y1.fx ['0'] = y1_0;
y2.fx ['0'] = y2_0;
y3.fx ['0'] = y3_0;
y4.fx ['0'] = y4_0;

*Bounds.
y1.lo(nh) = 0;    y1.up(nh) = 40;
y3.lo(nh) = 0;    y3.up(nh) = 25;
y4.lo(nh) = 0;    y4.up(nh) = 10;
u.lo(nh)  = 0;    u.up(nh)  = 50;

Model penicillin /all/;

$onecho >bench.opt
  solvers conopt knitro minos snopt
$offecho
penicillin.optfile=1;
option nlp=bench
Solve penicillin maximizing obj using nlp;
file stat1 /pen1.dat/;
file stat2 /pen2.dat/;
file stat3 /pen3.dat/;
file stat4 /pen4.dat/;
file cont  /pen.dat/;
put stat1;
```

Figure 4.38 (continued)

```
loop(nh, put y1.l(nh):10:5,',', put/)
put stat2;
loop(nh, put y2.l(nh):10:5,',', put/)
put stat3;
loop(nh, put y3.l(nh):10:5,',', put/)
put stat4;
loop(nh, put y4.l(nh):10:5,',', put/)
put cont;
loop(nh, put u.l(nh):10:5,',', put/)
* End of penicillin
```

Figure 4.38 (continued)

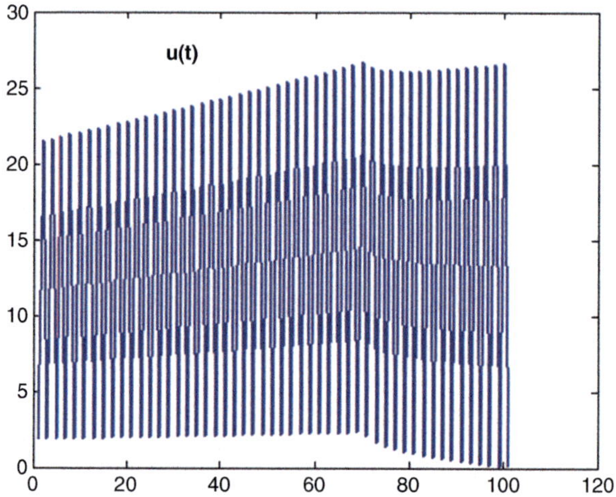

Figure 4.39 Evolution of the control $u(t)$

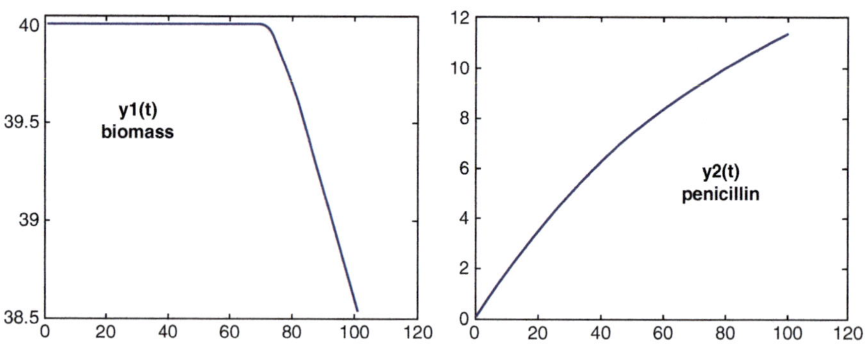

Figure 4.40 State variables $y_1(t)$ and $y_2(t)$

Figure 4.39 presents the evolution of the control $u(t)$. Observe the bang-bang character of this evolution.

Figure 4.40 presents the evolution of the state variables $y_1(t)$ (biomass) and $y_2(t)$ (penicillin).

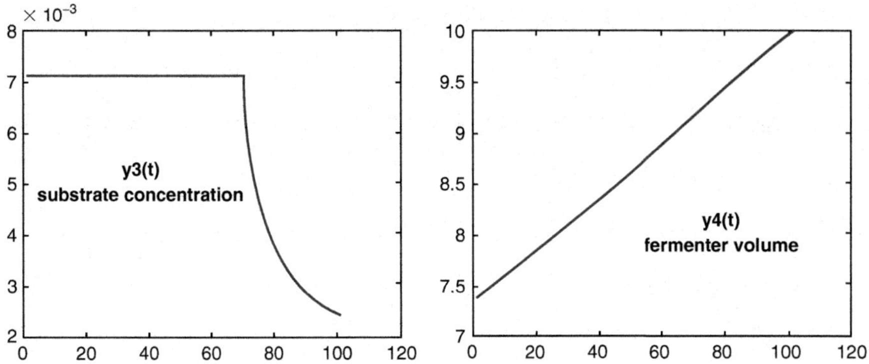

Figure 4.41 State variables $y_3(t)$ and $y_4(t)$

Figure 4.41 presents the time evolution of the state variables $y_3(t)$ (substrate concentration) and $y_4(t)$ (volume).

It is interesting to see the solution of this application when the backward Euler method is used in its GAMS representation. Figure 4.42 presents the time evolution of the variables associated to this application.

Observe the differences between these two methods for solving the ordinary differential equations used in the GAMS technology. The trapezoidal rule method is more accurate, better spotting the characteristics of the optimal control $u(t)$ as well as of the state variables $y_i(t), i = 1, \ldots, 4$.

4.17 A Standard Linear Lumped Parameter System (CONT)

For a linear dynamic system, it is desired to find the optimal control which will drive the state vector from its initial state to the origin in minimum time. The mathematical model of this application is as follows:

$$\min \sum_{i=1}^{4} x_i^2(t)$$

subject to: (4.42)

$$\frac{dx_1}{dt} = -0.5x_1 + 5x_2, \qquad \frac{dx_2}{dt} = -5x_1 - 0.5x_2 + u,$$
$$\frac{dx_3}{dt} = -0.6x_3 + 10x_4, \qquad \frac{dx_4}{dt} = -10x_3 - 0.6x_4 + u,$$
$$-1 \le u(t) \le 1,$$

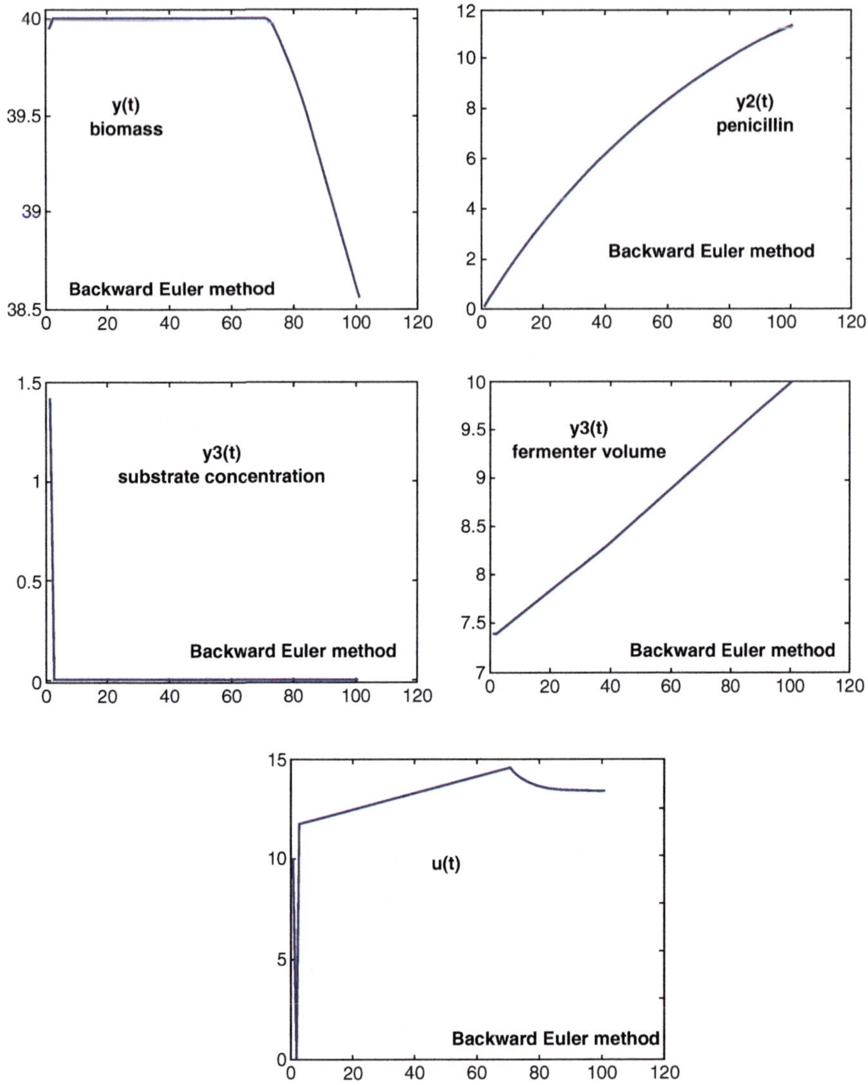

Figure 4.42 Time evolution of variables (backward Euler method)

with initial conditions: $x(t_0) = [10 \quad 10 \quad 10 \quad 10]$ and the final time $t_f = 4.2$ (Nishida, Liu, Lapidus, & Hiratsuka, 1976; Irizarry, 2005).

The GAMS representation of this application is presented in Figure 4.43.

Figures 4.44 and 4.45 present the time evolution of the state variables and the control of this application, respectively.

```
$Ontext
A standard linear lumped parameter system (CONT).
For a linear dynamic system it is desired to find the optimal
control which will drive the state vector from its initial
state to the origin in minimum time.

References:
[1] Nishida, N., Liu, Y.A., Lapidus, L., Niratsuka, S., An
effective computational algorithm for suboptimal and/or bang-
bang control.
AIChE Journal, vol.22, No.3, May 1976, pp.505-513.
[2] Andrei, N., Nonlinear Optimization Applications using the
GAMS Technology, Springer Optimization and Its Applications 81,
Springer Science+Business Media, New York, 2013. (Application
12.8, pp.316-322)
$Offtext

$if     set n  $set nh %n%
$if not set nh $set nh 500

set nh Number of subintervals / 0*%nh% /;
alias (nh,k);

Scalar    tf   final time             / 4.2  /
          x1_0 initial value for x1 / 10.0 /
          x2_0 initial value for x2 / 10.0 /
          x3_0 initial value for x3 / 10.0 /
          x4_0 initial value for x4 / 10.0 /
          h ;
          h=tf/%nh%;

Variables x1(nh)
          x2(nh)
          x3(nh)
          x4(nh)
          u(nh)    control variable
          obj      criterion ;
Equations eobj         criterion definition
          state1(nh)  state equation 1
          state2(nh)  state equation 2
          state3(nh)  state equation 3
          state4(nh)  state equation 4;

eobj..
 obj =e= x1['%nh%']*x1['%nh%'] + x2['%nh%']*x2['%nh%'] +
        x3['%nh%']*x3['%nh%'] + x4['%nh%']*x4['%nh%'];

state1(nh(k+1))..
x1[k+1] =e= x1(k) +
   (h/2)*( -0.5*x1(k) + 5*x2(k) - 0.5*x1(k+1) + 5*x2(k+1) ) ;

state2(nh(k+1))..
x2[k+1] =e= x2(k)+
   (h/2)*(  -5*x1(k)   -   0.5*x2(k)   +   u(k)   -   5*x1(k+1)   -
0.5*x2(k+1) + u(k+1) );

state3(nh(k+1))..
x3[k+1] =e= x3(k)+
   (h/2)*( -0.6*x3(k) + 10*x4(k) - 0.6*x3(k+1) + 10*x4(k+1) );
```

Figure 4.43 GAMS expression of application CONT

```
state4(nh(k+1))..
x4[k+1] =e= x4(k)+
   (h/2)*(  -10*x3(k)  -  0.6*x4(k)  +  u(k)  -   10*x3(k+1)  -
0.6*x4(k+1) + u(k+1) );

*Initial point
x1.l[nh]=10.0;
x2.l[nh]=10.0;
x3.l[nh]=10.0;
x4.l(nh)=10.0;
u.l(nh) =0.0;

x1.fx ['0'] = x1_0;
x2.fx ['0'] = x2_0;
x3.fx ['0'] = x3_0;
x4.fx ['0'] = x4_0;

*Bounds
u.lo(nh)  = -1.0;    u.up(nh)  = 1.0;

Model control4 /all/;

option reslim=60000;

*$onecho >bench.opt
* solvers knitro
*$offecho
control4.optfile=1;
*option nlp=bench

control4.iterlim=80000;
Solve control4 minimizing obj using nlp;

file stat1 /x1.dat/;
file stat2 /x2.dat/;
file stat3 /x3.dat/;
file stat4 /x4.dat/;
file cont  /u.dat/;

put stat1;
loop(nh, put x1.l(nh):10:5,',', put/)
put stat2;
loop(nh, put x2.l(nh):10:5,',', put/)
put stat3;
loop(nh, put x3.l(nh):10:5,',', put/)
put stat4;
loop(nh, put x4.l(nh):10:5,',', put/)
put cont;
loop(nh, put u.l(nh):10:5,',', put/)
* End of control4
```

Figure 4.43 (continued)

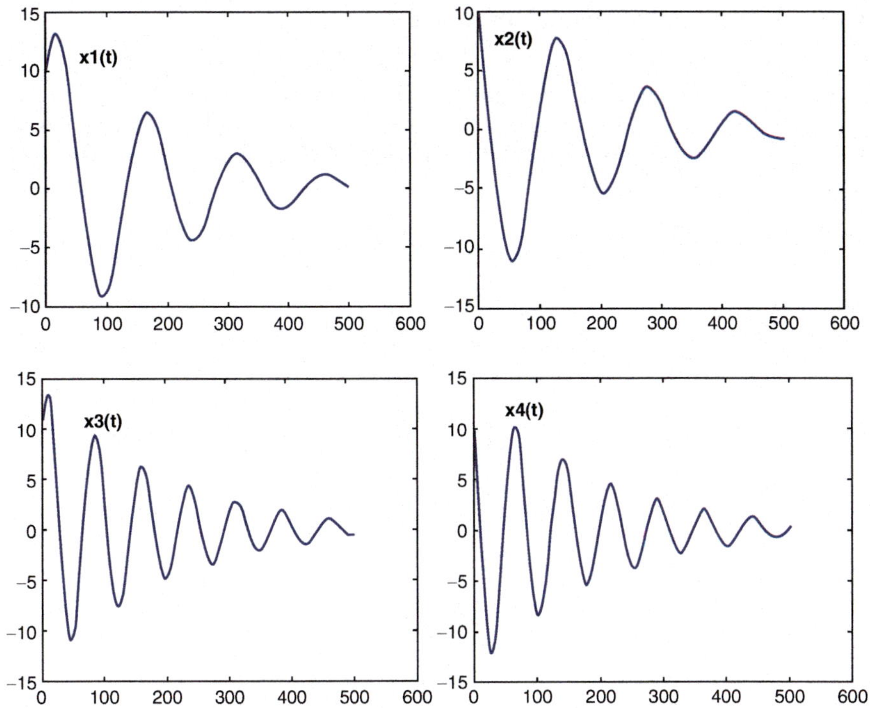

Figure 4.44 Time evolution of the state variables

Figure 4.45 Time evolution of control $u(t)$

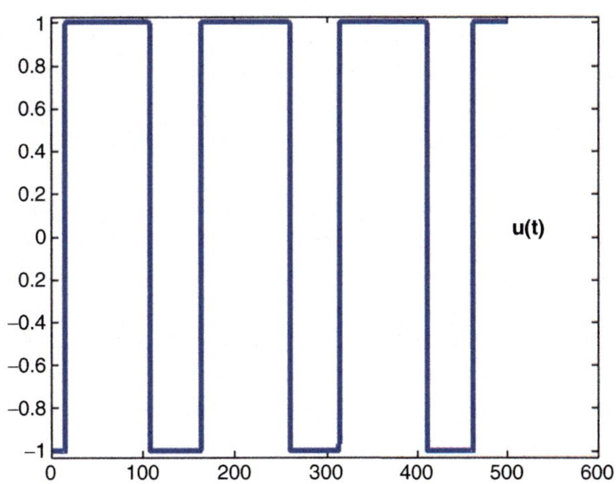

4.18 Van der Pol Oscillator (POL)

One of the classical equations of nonlinear dynamics was formulated by van der Pol (1927). It was a model for an electrical circuit with a triode. A brief description of this circuit is given in Strogatz (1994). The van de Pol oscillator has the following mathematical expression:

$$\max_{u(t)} y_3(t_f)$$

subject to (4.43)

$$\begin{aligned}
\frac{dy_1}{dt} &= \left(1 - y_2^2\right) - y_2 + u, \\
\frac{dy_2}{dt} &= y_1, \\
\frac{dy_3}{dt} &= y_1^2 + y_2^2 + u^2, \\
&-0.3 \le u(t) \le 1.0,
\end{aligned} \tag{4.43}$$

with the initial conditions $y(0) = [0, 1, 0]$ and the final time $t_f = 5$.

The GAMS representation of this application is given in Figure 4.46.

Figures 4.47, 4.48, and 4.49 show the evolution of the state variables and of the control for this application.

Some comments are as follows. We presented a limited number of 18 continuous nonlinear optimization applications. More optimization applications are found in (Andrei, 2003, 2011b, 2013b). Observe that these are continuous optimization problems. Continuous nonlinear optimization problems are easier to solve because the smoothness of the functions makes it possible to use objective and constraint information in a particular point x to deduce information about the function's behavior at all points close to x. This is done by the Taylor series representation, which is a major tool in the theory of nonlinear optimization.

At the same time, observe that the applications described in this chapter classify in two classes. The first one is the *static nonlinear optimization* and the second one is *dynamic nonlinear optimization*, often called *optimal control*. The strategy for solving optimal control problems is to "kill" the dynamics, i.e., to transform the dynamic nonlinear optimization problem into a static one. In this way, we get a large-scale optimization problem for which the Jacobian of the constraints has a block diagonal or a block triangular structure. Some optimizers (e.g., CONOPT) can take into consideration this structure to solve the problem in an efficient and robust manner.

The nonlinear optimization applications included in this chapter were selected to illustrate the capabilities of GAMS as a language for nonlinear optimization modeling and also to see the performances of the optimizers imbedded in GAMS

```
$Ontext
Van der Pol oscillator.
[1] Andrei, N., Nonlinear Optimization Applications using the
GAMS Technology, Springer Optimization and Its Applications 81,
Springer Science+Business Media, New York, 2013. (Application
12.5, pp.305-311)
$Offtext

$if     set n  $set nh %n%
$if not set nh $set nh 1000

set nh Number of subintervals / 0*%nh% /;
alias (nh,k);

Scalar    tf   final time            / 5 /
          y1_0 initial value for y1 / 0.0 /
          y2_0 initial value for y2 / 1.0 /
          y3_0 initial value for y3 / 0.0 /
          h ;
          h=tf/%nh%;
Variables y1(nh)
          y2(nh)
          y3(nh)
          u(nh)   control variable
          obj     criterion ;
Equations eobj          criterion definition
          state1(nh)  state equation 1
          state2(nh)  state equation 2
          state3(nh)  state equation 3  ;
eobj..
 obj =e= y3['%nh%'] ;

state1(nh(k+1))..
y1[k+1] =e= y1(k)+

   (h/2)*( (1-y2(k)*y2(k)) - y2(k) + u(k) +
           (1-y2(k+1)*y2(k+1)) - y2(k+1) + u(k+1)) ;

state2(nh(k+1))..
y2[k+1] =e= y2(k)+ (h/2)*(y1(k) + y1(k+1));

state3(nh(k+1))..
y3[k+1] =e= y3(k)+
   (h/2)*(y1(k)*y1(k) + y2(k)*y2(k) + u(k)*u(k) +
          y1(k+1)*y1(k+1) + y2(k+1)*y2(k+1) + u(k+1)*u(k+1));

*Initial point
y1.l[nh]=0.0;
y2.l[nh]=1.0;
y3.l[nh]=0.0;
u.l[nh] =0.0;
y1.fx ['0'] = y1_0;
y2.fx ['0'] = y2_0;
y3.fx ['0'] = y3_0;

*Bounds
u.lo(nh)  =-0.3;    u.up(nh)  = 1;

Model pol /all/;
*$onecho >bench.opt
*  solvers conopt knitro minos snopt
```

Figure 4.46 GAMS expression of application POL

```
*$offecho
pol.optfile=1;
*option nlp=bench
Solve pol maximizing obj using nlp;
file stat1 /pol1.dat/;
file stat2 /pol2.dat/;
file stat3 /pol3.dat/;
file cont  /pol.dat/;
put stat1;
loop(nh, put y1.l(nh):10:5,',', put/)
put stat2;
loop(nh, put y2.l(nh):10:5,',', put/)
put stat3;
loop(nh, put y3.l(nh):10:5,',', put/)
put cont;
loop(nh, put u.l(nh):10:5,',', put/)
* End pol
```

Figure 4.46 (continued)

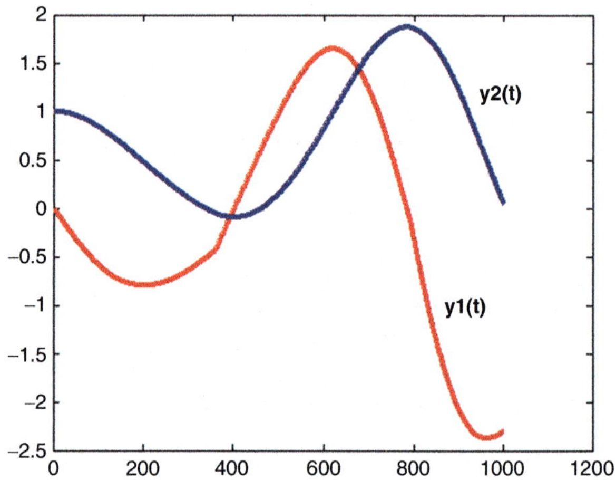

Figure 4.47 Evolution of $y_1(t)$ and $y_2(t)$

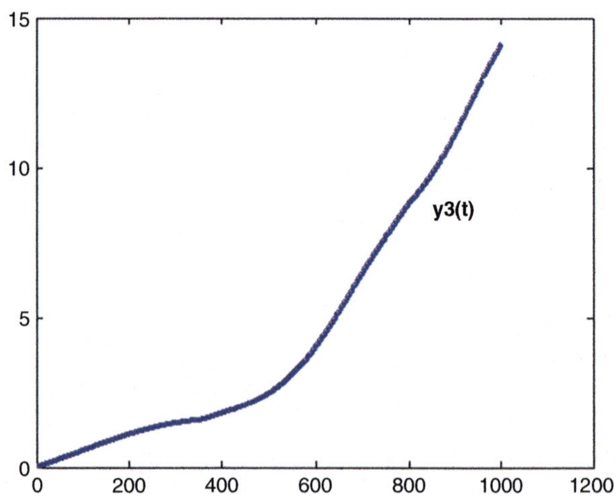

Figure 4.48 Evolution of $y_3(t)$

Figure 4.49 Control $u(t)$

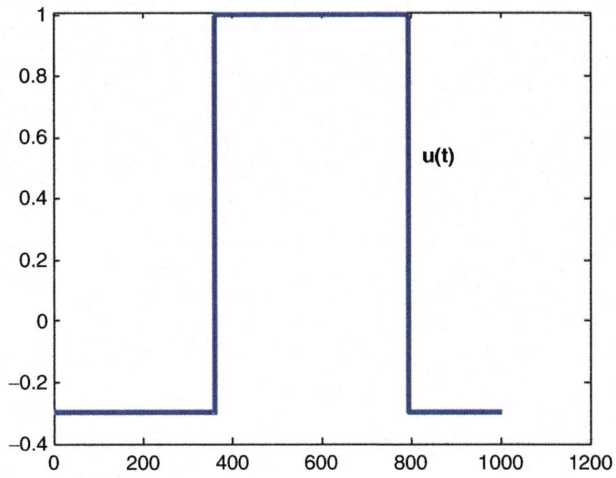

and discussed in this book (CONOPT, IPOPT, KNITRO, MINOS, and SNOPT). Observe the application *Optimal temperature field in a rectangular area* (*DIFF*), where the LOOP statement (see Section 3.8) is used to compute the time-dependent temperature field in a rectangular area with heterogeneous thermal conductivity and a point source of heat and heat flows through the border of the solution domain. Another interesting application is *Stationary flow of an incompressible fluid in a rectangular area* (*FLOW*/*FLOWO*), where different integration methods (without or with inertial terms) are presented. Remarkable is also the application *Fed-batch fermenter for penicillin production* (*PENICI*), where some aspects of the numerical solution of ordinary differential equations and their implementation in GAMS are discussed.

Notes and References

The description of all these 18 nonlinear optimization applications in the GAMS technology is taken from Andrei (2013b, 2015).

Chapter 5
Optimality Conditions for Continuous Nonlinear Optimization

The optimization problems considered in this book involve minimization or maximization of a function of several real variables subject to one or more constraints. The constraints may be nonnegativity of variables, simple bounds on variables, and equalities or inequalities as functions of these variables. These problems are known as continuous nonlinear constrained optimization or nonlinear programming.

The purpose of this chapter is to introduce the main concepts and the fundamental results in nonlinear optimization, known as optimality conditions. Plenty of very good books dedicated to these problems are known in literature: Luenberger (1973), Gill, Murray, and Wright (1981), Peressini, Sullivan, and Uhl (1988), Bazaraa, Sherali, and Shetty (1993), Bertsekas (1999), Boyd and Vandenberghe (2006), Nocedal and Wright (2006), Sun and Yuan (2006), Chachuat (2007), Andrei (2009, 2015), etc.

The general continuous nonlinear optimization problem is expressed as

$$\begin{aligned} &\min f(x) \\ &\text{subject to:} \\ &c_i(x) \leq 0, \quad i = 1, \ldots, m, \\ &h_i(x) = 0, \quad i = 1, \ldots, p, \end{aligned} \tag{5.1}$$

where $x \in \mathbb{R}^n, f: \mathbb{R}^n \to \mathbb{R}$, $c_i: \mathbb{R}^n \to \mathbb{R}$, $i = 1, \ldots, m$, and $h_i: \mathbb{R}^n \to \mathbb{R}$, $i = 1, \ldots, p$, are continuously differentiable functions. Usually, the function f is called the *objective function*. Each of the constraints $c_i(x) \leq 0$, $i = 1, \ldots, m$, is called an *inequality constraint*, and each $h_i(x) = 0$, $i = 1, \ldots, p$, is called an *equality constraint*. Often, (5.1) is called a *nonlinear program*. A vector x satisfying all the equality and inequality constraints is called a *feasible solution* (*point*) to the problem (5.1). Define

N. Andrei, *Continuous Nonlinear Optimization for Engineering Applications in GAMS Technology*, Springer Optimization and Its Applications 121,
DOI 10.1007/978-3-319-58356-3_5

$$X = \{x : c_i(x) \leq 0, i = 1, \ldots, m, \quad h_i(x) = 0, i = 1, \ldots, p\}$$

as the *feasible region* (or *feasible domain*).

In this chapter we are interested to specify what is meant by optimality for the general nonlinear optimization problem and to give conditions under which a solution for the problem (5.1) exists. Both necessary and sufficient conditions for optimality are presented, starting with unconstrained problems and continuing with problems with inequality constraints, equality constraints, and finally for general nonlinear optimization problems with equality and inequality constraints. The key to understand the nonlinear optimization is the Karush-Kuhn-Tucker (KKT) optimality conditions. This is a major result which identifies an algebraic system of equalities and inequalities, which corresponds to the solution to any nonlinear optimization problem. This system can often be used to develop algorithms for computing a solution for the problem or to get some additional information about the sensitivity of the minimum value of the problem subject to changes in the constraints. In general, many optimization algorithms can be interpreted as methods for numerically solving the KKT nonlinear system of equations.

In the mathematical optimization, the KKT conditions are first-order necessary conditions for a solution in nonlinear optimization to be optimal, provided that some regularity conditions are satisfied. For problems with inequality constraints, the KKT approach generalizes the method of Lagrange multipliers, which allows only equality constraints. For the development of the KKT optimality conditions, three possible approaches can be used. One is based on the *separation and support theorems* from the convex set theory. Another one uses the *penalty functions* and the third one comes from the *theory of Lagrange multipliers*. Each of these approaches has its own virtues and provides its own insights on the KKT Theorem. In this text we consider the optimality conditions for the continuous nonlinear optimization (the mathematical programming), using the formalism of Lagrange.

5.1 General Concepts in Nonlinear Optimization

In the following, we shall present some definitions and results used in the context of nonlinear programming. At the same time, we shall define a particular class of nonlinear programs, the convex programming. In this section, $X \subset \mathbb{R}$ denotes a nonempty set of real numbers.

Definition 5.1 *(Upper bound, lower bound). A real number α is called an upper bound for X if $x \leq \alpha$ for all $x \in X$. The set X is said to be bounded above if it has an upper bound. Similarly, a real number α is called a lower bound for X if $x \geq \alpha$ for all $x \in X$. The set X is said to be bounded below if it has a lower bound.*

Definition 5.2 *(Least upper bound, Greatest lower bound). A real number α is called the least upper bound (or supremum, or* sup*) of X, if (i) α is an upper bound* *for X; and (ii) there does not exist an upper bound for X that is strictly smaller than α. The supremum, if it exists, is unique and is denoted by* sup X. *A real number α is called the greatest lower bound (or infimum, or* inf*) of X, if (i) α is a lower bound for X; and (ii) there does not exist a lower bound for X that is strictly greater than α. The infimum, if it exists, is unique and is denoted by* inf X.

It is worth saying that for sups and infs, the following equivalent definition is useful.

Definition 5.3 *(Supremum, infimum). The supremum of X, provided it exists, is the least upper bound for X,* i.e., *a real number α satisfying: (i)* $z \le \alpha$ *for any* $z \in X$*; (ii) for any* $\overline{\alpha} < \alpha$*, there exists* $z \in X$ *such that* $z > \overline{\alpha}$*. Similarly, the infimum of X, provided it exists, is the greatest lower bound for X,* i.e., *a real number α satisfying: (i)* $z \ge \alpha$ *for any* $z \in X$*; (ii) for any* $\overline{\alpha} > \alpha$*, there exists* $z \in X$ *such that* $z < \overline{\alpha}$*.*

Definition 5.4 *(Maximum, minimum). The maximum of a set X is its largest element if such an element exists. The minimum of a set X is its smallest element if such an element exists.*

The key differences between the *supremum* and the *maximum* concepts are as follows. If a set has a maximum, then the maximum is also a supremum for this set, but the converse is not true. A finite set has always a maximum which is also its supremum, but an infinite set does not need to have a maximum. The supremum of a set X does not need to be an element of the set X itself, but the maximum of X must always be an element of X.

Concerning the existence of infima and suprema in $\mathbb{R}$, fundamental is the *axiom of completeness. If a nonempty subset of real numbers has an upper bound, then it has a least upper bound. If a nonempty set of real numbers has a lower bound, it has a greatest lower bound.* In other words, the completeness axiom guarantees that, for any nonempty set of real numbers that is bounded above, a supremum exists (in contrast to the maximum, which may or may not exist).

Let us consider the minimization problem:

$$\min\{f(x) : x \in X\}, \tag{5.2}$$

where $X \subset \mathbb{R}^n$ represents the *feasible set*. Any point $x \in X$ is a *feasible point* or an *admissible point*. Any point $x \in \mathbb{R}^n \backslash X$ is called to be *infeasible*.

Definition 5.5 *(Global minimum, strict global minimum). A point* $x^* \in X$ *is said to be a global minimum of f on X if* $f(x) \ge f(x^*)$ *for any* $x \in X$*. A point* $x^* \in X$ *is said to be a strict global minimum of f on X if* $f(x) > f(x^*)$ *for any* $x \in X$ *with* $x \ne x^*$*.*

Definition 5.6 *(Global maximum, strict global maximum). A point* $x^* \in X$ *is said to be a global maximum of f on X if* $f(x) \le f(x^*)$ *for any* $x \in X$*. It is a strict global maximum of f on X if* $f(x) < f(x^*)$ *for any* $x \in X$ *with* $x \ne x^*$*.*

The point x^* is called an *optimal solution* of the optimization problem. The real number $f(x^*)$ is known as the *optimal value* of the objective function subject to the constraints $x \in X$.

Observe the distinction between the minimum/maximum and the infimum/supremum. The value $\min\{f(x): x \in X\}$ must be attained at one or more points $x \in X$. On the other hand, the value $\inf\{f(x): x \in X\}$ does not necessarily have to be attained at any points $x \in X$. However, if a minimum (maximum) exists, then its optimal value equals the infimum (supremum).

If a minimum exists, it is not necessarily unique. That is, there may be a finite number, or even an infinite number of feasible points x^* that satisfy the inequality $f(x) \geq f(x^*)$ for any $x \in X$. The notation

$$\operatorname{argmin}\{f(x) : x \in X\} \triangleq \{x \in X : f(x) = \inf\{f(x) : x \in X\}\}$$

is reserved for the set of the minima of function f on X, that is, a set in $\mathbb{R}^n$.

Definition 5.7 *(Local minimum, strict local minimum). A point $x^* \in X$ is said to be a local minimum of f on X if there exists $\varepsilon > 0$ such that $f(x) \geq f(x^*)$ for any $x \in B(x^*, \varepsilon) \cap X$, where $B(x^*, \varepsilon)$ is the open ball centered at x^* of radius ε. Similarly, a point $x^* \in X$ is said to be a strict local minimum of f on X if there exists $\varepsilon > 0$ such that $f(x) > f(x^*)$ for any $x \in B(x^*, \varepsilon) \backslash \{x^*\} \cap X$.*

Definition 5.8 *(Local maximum, strict local maximum). A point $x^* \in X$ is said to be a local maximum of f on X if there exists $\varepsilon > 0$ such that $f(x) \leq f(x^*)$ for any $x \in B(x^*, \varepsilon) \cap X$. Similarly, it is said to be a strict local maximum of f on X if there exists $\varepsilon > 0$ such that $f(x) < f(x^*)$ for any $x \in B(x^*, \varepsilon) \backslash \{x^*\} \cap X$.*

A fundamental problem in optimizing a function on a given set is whether a minimum or a maximum point exists in the given set. This result is known as the theorem of Weierstrass. It shows that if X is nonempty, closed, and bounded, and f is continuous on X, then a minimum of f on X exists.

Theorem 5.1 *(Weierstrass). Let X be a nonempty and compact set. Assume that $f: X \to \mathbb{R}$ is continuous on X. Then, the problem $\min\{f(x): x \in X\}$ attains its minimum.*

Proof. If f is continuous on X and X is both closed and bounded, it follows that f is bounded below on X. Now, since X is nonempty, from the axiom of completeness, there exists a greatest lower bound $\alpha = \inf\{f(x): x \in X\}$. Let $0 < \varepsilon < 1$ and consider the set $X_k = \{x \in X: \alpha \leq f(x) \leq \alpha + \varepsilon^k\}$, $k = 1, 2, \ldots$ By the definition of the infimum, for each k it follows that $X_k \neq \varnothing$. Therefore, a sequence of points $\{x_k\} \subset X$ can be constructed by selecting a point x_k for each $k = 1, 2, \ldots$ Since X is bounded, there exists a convergent subsequence $\{x_k\}_K \subset X$ indexed by the set $K \subset \mathbb{N}$ with x^* as its limit. Since X is closed it follows that $x^* \in X$. By continuity of f on X, since $\alpha \leq f(x_k) \leq \alpha + \varepsilon^k$, we have $\alpha = \lim_{k \to \infty, k \in K} f(x_k) = f(x^*)$. Therefore, there exists a solution $x^* \in X$ so that $f(x^*) = \alpha = \inf\{f(x): x \in X\}$, i.e., x^* is a minimizing solution. ♦

All the hypotheses of this theorem are important. The feasible set must be *nonempty*; otherwise there are no feasible points at which the minimum is attained. The feasible set must be *closed*, i.e., it must contain its boundary points. The objective function must be *continuous* on the feasible set; otherwise the limit at a point may not exist or it may be different from the value of the function at that point. Finally, the feasible set must be *bounded*; otherwise even continuous functions can be unbounded on the feasible set.

Definition 5.9 *(Convex program). Let C be a convex set in* $\mathbb{R}^n$ *and let* $f: C \to \mathbb{R}^n$ *be a convex function on C. Then,* $\min\{f(x): x \in C\}$ *is called a convex optimization problem or a convex program.*

The fundamental result in convex programming is the following theorem.

Theorem 5.2 *Let* x^* *be a local minimum of a convex program. Then,* x^* *is also a global minimum.*

Proof. If x^* is a local minimum, then there exists $\varepsilon > 0$ such that $f(x) \geq f(x^*)$ for any $x \in B(x^*, \varepsilon)$. Now, suppose that x^* is not a global minimum. Then there exists $y \in C$ such that $f(y) < f(x^*)$. Let $\lambda \in (0, 1)$ be chosen such that the point $z = \lambda y + (1 - \lambda)x^* \in B(x^*, \varepsilon)$. By convexity of C, $z \in C$. Therefore,

$$f(z) \leq \lambda f(y) + (1 - \lambda)f(x^*) < \lambda f(x^*) + (1 - \lambda)f(x^*) = f(x^*),$$

which is a contradiction, since x^* is a local minimum. ♦

5.2 Optimality Conditions for Unconstrained Optimization

Let us consider the problem of minimizing a function $f(x)$ without constraints on the variables $x \in \mathbb{R}^n$:

$$\min\{f(x) : x \in \mathbb{R}^n\}.$$

For a given point $x \in \mathbb{R}^n$, the optimality conditions determine whether or not a point is a local or a global minimum of f. To formulate the optimality conditions, it is necessary to introduce some concepts which characterize an improving direction along which the values of the function f decrease.

Definition 5.10 *(Descent direction). Suppose that* $f: \mathbb{R}^n \to \mathbb{R}$ *is continuous at* x^**. A vector* $d \in \mathbb{R}^n$ *is a descent direction for f at* x^* *if there exists* $\delta > 0$ *such that* $f(x^* + \lambda d) < f(x^*)$ *for any* $\lambda \in (0, \delta)$*. The cone of descent directions at* x^**, denoted by* $C_{dd}(x^*)$*, is given by:*

$$C_{dd}(x^*) = \{d: \text{there exists } \delta > 0 \text{ such that } f(x^* + \lambda d) < f(x^*), \text{ for any } \lambda \in (0, \delta)\}.$$

Assume that f is a differentiable function. To get an algebraic characterization for a descent direction for f at x^*, let us define the set:

$$C_0(x^*) = \left\{d: \nabla f(x^*)^T d < 0\right\}.$$

The following result shows that every $d \in C_0(x^*)$ is a descent direction at x^*.

Proposition 5.1 *(Algebraic characterization of a descent direction). Suppose that $f: \mathbb{R}^n \to \mathbb{R}$ is differentiable at x^*. If there exists a vector d such that $\nabla f(x^*)^{\mathrm{T}} d < 0$, then d is a descent direction for f at x^**, i.e., $C_0(x^*) \subseteq C_{dd}(x^*)$.

Proof. Since *f* is differentiable at x^*, it follows that

$$f(x^* + \lambda d) = f(x^*) + \lambda \nabla f(x^*)^{\mathrm{T}} d + \lambda \|d\| o(\lambda d),$$

where $\lim_{\lambda \to 0} o(\lambda d) = 0$. Therefore:

$$\frac{f(x^* + \lambda d) - f(x^*)}{\lambda} = \nabla f(x^*)^{\mathrm{T}} d + \|d\| o(\lambda d).$$

Since $\nabla f(x^*)^{\mathrm{T}} d < 0$ and $\lim_{\lambda \to 0} o(\lambda d) = 0$, it follows that there exists a $\delta > 0$ such that $\nabla f(x^*)^{\mathrm{T}} d + \|d\| o(\lambda d) < 0$ for all $\lambda \in (0, \delta)$. ◆

Theorem 5.3 *(First-order necessary conditions for a local minimum). Suppose that $f: \mathbb{R}^n \to \mathbb{R}$ is differentiable at x^*. If x^* is a local minimum, then $\nabla f(x^*) = 0$.*

Proof. Suppose that $\nabla f(x^*) \neq 0$. If we consider $d = -\nabla f(x^*)$, then $\nabla f(x^*)^{\mathrm{T}} d = -\|\nabla f(x^*)\|^2 < 0$. By Proposition 5.1 there exists a $\delta > 0$ such that for any $\lambda \in (0, \delta)$, $f(x^* + \lambda d) < f(x^*)$. But this is in contradiction with the assumption that x^* is a local minimum for *f*. ◆

Observe that the above necessary condition represents a system of *n* algebraic nonlinear equations. All the points x^* which solve the system $\nabla f(x) = 0$ are called *stationary points*. Clearly, the stationary points need not all be local minima. They could very well be local maxima or even saddle points. In order to characterize a local minimum, we need more restrictive necessary conditions involving the Hessian matrix of the function *f*.

Theorem 5.4 *(Second-order necessary conditions for a local minimum). Suppose that $f: \mathbb{R}^n \to \mathbb{R}$ is twice differentiable at point x^*. If x^* is a local minimum, then $\nabla f(x^*) = 0$ and $\nabla^2 f(x^*)$ is positive semidefinite.*

Proof. Consider an arbitrary direction *d*. Then, using the differentiability of *f* at x^* we get

$$f(x^* + \lambda d) = f(x^*) + \lambda \nabla f(x^*)^{\mathrm{T}} d + \frac{1}{2} \lambda^2 d^{\mathrm{T}} \nabla^2 f(x^*) d + \lambda^2 \|d\|^2 o(\lambda d),$$

where $\lim_{\lambda \to 0} o(\lambda d) = 0$. Since x^* is a local minimum, $\nabla f(x^*) = 0$. Therefore:

$$\frac{f(x^* + \lambda d) - f(x^*)}{\lambda^2} = \frac{1}{2} d^{\mathrm{T}} \nabla^2 f(x^*) d + \|d\|^2 o(\lambda d).$$

Since x^* is a local minimum, for λ sufficiently small, $f(x^* + \lambda d) \geq f(x^*)$. For $\lambda \to 0$ it follows from the above equality that $d^{\mathrm{T}} \nabla^2 f(x^*) d \geq 0$. Since *d* is an arbitrary direction, it follows that $\nabla^2 f(x^*)$ is positive semidefinite. ◆

In the above theorems, we have presented the *necessary* conditions for a point x^* to be a local minimum, i.e., these conditions *must be satisfied* at every local minimum solution. However, a point satisfying these necessary conditions need not be a local minimum. In the following theorems, the *sufficient* conditions for a global minimum are given, provided that the objective function is convex on $\mathbb{R}^n$ (see Appendix A).

Theorem 5.5 *(First-order sufficient conditions for a strict local minimum). Suppose that $f: \mathbb{R}^n \to \mathbb{R}$ is differentiable at x^* and convex on $\mathbb{R}^n$. If $\nabla f(x^*) = 0$, then x^* is a global minimum of f on $\mathbb{R}^n$.*

Proof. Since f is convex on $\mathbb{R}^n$ and differentiable at x^* then from the property of convex functions given by Proposition A4.5, it follows that for any $x \in \mathbb{R}^n$ $f(x) \geq f(x^*) + \nabla f(x^*)^T(x - x^*)$. But x^* is a stationary point, i.e., $f(x) \geq f(x^*)$ for any $x \in \mathbb{R}^n$. ♦

The following theorem gives the second-order sufficient conditions characterizing a local minimum point for those functions which are strictly convex in a neighborhood of the minimum point.

Theorem 5.6 (*Second-order sufficient conditions for a strict local minimum*). *Suppose that $f: \mathbb{R}^n \to \mathbb{R}$ is twice differentiable at point x^*. If $\nabla f(x^*) = 0$ and $\nabla^2 f(x^*)$ is positive definite, then x^* is a local minimum of f.*

Proof. Since f is twice differentiable, for any $d \in \mathbb{R}^n$, we can write

$$f(x^* + d) = f(x^*) + \nabla f(x^*)^T d + \frac{1}{2} d^T \nabla^2 f(x^*) d + \|d\|^2 o(d),$$

where $\lim_{d \to 0} o(d) = 0$. Let λ be the smallest eigenvalue of $\nabla^2 f(x^*)$. Since $\nabla^2 f(x^*)$ is positive definite, it follows that $\lambda > 0$ and $d^T \nabla^2 f(x^*) d \geq \lambda \|d\|^2$. Therefore, since $\nabla f(x^*) = 0$, we can write:

$$f(x^* + d) - f(x^*) \geq \left[\frac{\lambda}{2} + o(d)\right] \|d\|^2.$$

Since $\lim_{d \to 0} o(d) = 0$, then there exists a $\eta > 0$ such that $|o(d)| < \lambda/4$ for any $d \in B(0, \eta)$. Hence

$$f(x^* + d) - f(x^*) \geq \frac{\lambda}{4} \|d\|^2 > 0$$

for any $d \in B(0, \eta) \backslash \{0\}$, i.e., x^* is a strict local minimum of function f. ♦

If we assume f to be twice continuously differentiable, we observe that, since $\nabla^2 f(x^*)$ is positive definite, $\nabla^2 f(x^*)$ is positive definite in a small neighborhood of x^* and so f is strictly convex in a small neighborhood of x^*. Therefore, x^* is a strict local minimum, that is, it is the unique global minimum over a small neighborhood of x^*.

5.3 Optimality Conditions for Problems with Inequality Constraints

In the following, we shall discuss the nonlinear optimization problems with inequality constraints:

$$\min f(x)$$

subject to:
$$x \in X, \tag{5.3}$$

where X is a general set. Further on, we will be more specific and will define the problem as to minimize $f(x)$ subject to $c(x) \leq 0$, where $c(x)$ is the vector of constraint functions.

Definition 5.11 *(Feasible direction). Let X be a nonempty set in $\mathbb{R}^n$. A nonzero vector $d \in \mathbb{R}^n$ is a feasible direction at $x^* \in cl(X)$ if there exists a $\delta > 0$ such that $x^* + \eta d \in X$ for any $\eta \in (0, \delta)$. Moreover, the cone of feasible directions at x^*, denoted by $C_{fd}(x^*)$, is given by*

$$C_{fd}(x^*) \triangleq \{d \neq 0, \text{ there is } \delta > 0 \text{ such that } x^* + \eta d \in X, \text{ for any } \eta \in (0, \delta)\}.$$

Clearly, a small movement from x^* along the direction $d \in C_{fd}(x^*)$ leads to feasible points. On the other hand, a similar movement along a direction $d \in C_0(x^*)$ (see Definition 5.10) leads to solutions which improve the value of the objective function. The following theorem, which gives a geometrical interpretation of the local minima, shows that a necessary condition for local optimality is that *every improving direction is not a feasible direction.*

Theorem 5.7 *(Geometric necessary condition for a local minimum). Let X be a nonempty set in $\mathbb{R}^n$ and let $f: \mathbb{R}^n \to \mathbb{R}$ be a differentiable function. Suppose that x^* is a local minimum of the problem* (5.3). *Then, $C_0(x^*) \cap C_{fd}(x^*) = \varnothing$.*

Proof. Suppose that there exists a nonzero vector $d \in C_0(x^*) \cap C_{fd}(x^*)$. By Proposition 5.1 of the algebraic characterization of a descent direction, there exists $\delta_1 > 0$ such that $f(x^* + \eta d) < f(x^*)$ for any $\eta \in (0, \delta_1)$. On the other hand, by Definition 5.11 of feasible direction, there exists $\delta_2 > 0$ such that $x^* + \eta d \in X$ for any $\eta \in (0, \delta_2)$. Therefore, there exists $x \in B(x^*, \eta) \cap X$ such that $f(x^* + \eta d) < f(x^*)$, for every $\eta \in (0, \min\{\delta_1, \delta_2\})$, which contradicts the assumption that x^* is a local minimum of f on X (see Definition 5.7). ◆

So far, we have obtained a geometric characterization of the optimality condition for the problem (5.3) given by Theorem 5.7 where $C_{fd}(x^*)$ is the cone of feasible directions. To get a practical optimality condition, implementable in computer programs, we need to convert this geometric condition into an algebraic one. For this, we introduce the concept of *active constraints* at x^* and define a cone

$C_{ac}(x^*) \subseteq C_{fd}(x^*)$ in terms of the gradients of these active constraints. Now, we specify the feasible set X as

$$X \triangleq \{x : c_i(x) \leq 0, \ \ i = 1, \ldots, m\}, \tag{5.4}$$

where $c_i : \mathbb{R}^n \to \mathbb{R}$, $i = 1, \ldots, m$, are continuous functions. Define the vector $c(x) = [c_1(x), \ldots, c_m(x)]$.

Definition 5.12 *(Active constraint, active set). Let $c_i : \mathbb{R}^n \to \mathbb{R}$, $i = 1, \ldots, m$, which define the feasible set $X = \{x : c_i(x) \leq 0, i = 1, \ldots, m\}$, and consider $x^* \in X$ a feasible point. For each $i = 1, \ldots, m$, the constraint c_i is said to be active or binding at x^* if $c_i(x^*) = 0$. It is said to be inactive at x^* if $c_i(x^*) < 0$. The set*

$$A(x^*) \triangleq \{i : c_i(x^*) = 0\},$$

denotes the set of active constraints at x^*.

The following proposition gives an algebraic characterization of a feasible direction showing the relation between a cone $C_{ac}(x^*)$ expressed in terms of the gradients of the active constraints and the cone of the feasible directions.

Proposition 5.2 *(Algebraic characterization of a feasible direction). Let $c_i : \mathbb{R}^n \to \mathbb{R}$, $i = 1, \ldots, m$, be differentiable functions and consider the feasible set $X = \{x : c_i(x) \leq 0, i = 1, \ldots, m\}$. For any feasible point $x^* \in X$, we have:*

$$C_{ac}(x^*) \triangleq \left\{d : \nabla c_i(x^*)^{\mathrm{T}} d < 0, \ \ i \in A(x^*)\right\} \subseteq C_{fd}(x^*).$$

Proof. Suppose that $C_{ac}(x^*)$ is a nonempty set. Let $d \in C_{ac}(x^*)$. Observe that $\nabla c_i(x^*)^{\mathrm{T}} d < 0$ for each $i \in A(x^*)$. Therefore, by Proposition 5.1, the algebraic characterization of a descent direction, it follows that d is a descent direction for c_i at x^*, i.e., there exists $\delta_2 > 0$ such that $c_i(x^* + \eta d) < c_i(x^*) = 0$ for any $\eta \in (0, \delta_2)$ and for any $i \in A(x^*)$. On the other hand, since c_i is differentiable at x^*, it follows that it is continuous at x^*. Therefore, since $c_i(x^*) < 0$ and c_i is continuous at x^* for each $i \notin A(x^*)$, there exists $\delta_1 > 0$ such that $c_i(x^* + \eta d) < 0$ for any $\eta \in (0, \delta_1)$ and for any $i \notin A(x^*)$. Besides, for all $\eta \in (0, \min\{\delta_1, \delta_2\})$, the points $x^* + \eta d \in X$. Therefore, by Definition 5.11 of feasible direction, $d \in C_{fd}(x^*)$. ◆

Remark 5.1 From the Theorem 5.7 we know that $C_0(x^*) \cap C_{fd}(x^*) = \varnothing$. But, from Proposition 5.2 we have that $C_{ac}(x^*) \subseteq C_{fd}(x^*)$. Therefore $C_0(x^*) \cap C_{ac}(x^*) = \varnothing$, for any local optimal solution x^*. ◆

The above geometric characterization of the local optimal solution (see Theorem 5.7) holds either at the interior points $\operatorname{int} X \triangleq \{x \in \mathbb{R}^n : c_i(x) < 0, i = 1, \ldots, m\}$, or at the boundary points. For the interior points, any direction is feasible, and the necessary condition $C_0(x^*) \cap C_{ac}(x^*) = \varnothing$ reduces to the very well-known condition $\nabla f(x^*) = 0$, which is identical to the necessary optimal condition for the unconstrained optimization (see Theorem 5.3).

It is important to notice that the condition $C_0(x^*) \cap C_{ac}(x^*) = \varnothing$ can be satisfied by nonoptimal points, i.e., this condition is *necessary but not sufficient* for a point x^* to be a local minimum of the function f on X. For example, at any point x^* for which $\nabla c_i(x^*) = 0$, for an arbitrary index $i \in A(x^*)$ the condition $C_0(x^*) \cap C_{ac}(x^*) = \varnothing$ is trivially satisfied.

In the following, in order to get an algebraic necessary optimality condition to be used in numerical computation, we want to transform the geometric necessary optimality condition $C_0(x^*) \cap C_{ac}(x^*) = \varnothing$ to a statement in terms of the gradient of the objective function and the gradient of the constraints. Thus, the first-order optimality conditions, known as the *Karush-Kuhn-Tucker (KKT) necessary conditions*, are obtained. In order to formulate the KKT conditions, we need to introduce the concept of *regular point* and of *KKT point.*

Definition 5.13 *(Regular point – inequality constraints). Let $c_i : \mathbb{R}^n \to \mathbb{R}$, $i = 1, \dots, m$, be differentiable functions, and consider the feasible set $X = \{x \in \mathbb{R}^n : c_i(x) \le 0, i = 1, \dots, m\}$. A point $x^* \in X$ is a regular point if the gradient vectors $\nabla c_i(x^*)$, $i \in A(x^*)$, are linear independent,* i.e.:

$$\mathrm{rank}[\nabla c_i(x^*), i \in A(x^*)] = \mathrm{card}(A(x^*)).$$

Definition 5.14 *(KKT point). Let $f : \mathbb{R}^n \to \mathbb{R}$ and $c_i : \mathbb{R}^n \to \mathbb{R}$, $i = 1, \dots, m$, be differentiable functions. Consider the problem* $\min\{f(x) : c(x) \le 0\}$. *If a point $(x^*, \mu^*) \in \mathbb{R}^n \times \mathbb{R}^m$ satisfies the algebraic conditions:*

$$\nabla f(x^*) + (\mu^*)^T \nabla c(x^*) = 0, \tag{5.5}$$

$$\mu^* \ge 0, \tag{5.6}$$

$$c(x^*) \le 0, \tag{5.7}$$

$$(\mu^*)^T c(x^*) = 0. \tag{5.8}$$

then (x^*, μ^*) is called a KKT point.

In the Definition 5.14, the scalars μ_i, $i = 1, \dots, m$, are called the *Lagrange multipliers*. The first condition (5.5) is known as the *primal feasibility* condition. The conditions (5.6) and (5.7) are known as *dual feasibility* conditions. The last condition (5.8), expressed as $\mu_i^* c_i(x^*) = 0$, $i = 1, \dots, m$, are the *complementarity slackness* (or *transversality*) conditions.

We are now in the position to present the KKT necessary conditions for the optimality of the nonlinear optimization problem with inequality constraints. For this, a very useful result is given by the Theorem of Gordan (see Theorem A4.2). This is extensively used in the derivation of the optimality conditions of linear and nonlinear programming problems.

Theorem 5.8 *(KKT necessary conditions). Let $f : \mathbb{R}^n \to \mathbb{R}$ and $c_i : \mathbb{R}^n \to \mathbb{R}$, $i = 1, \dots, m$, be differentiable functions. Consider the problem* $\min\{f(x) : c(x) \le 0\}$. *If x^* is a local minimum and a regular point of the constraints, then there exists a unique vector μ^* such that (x^*, μ^*) is a KKT point.*

Proof. We know that x^* is an optimal solution for the problem $\min\{f(x) : c(x) \leq 0\}$. Therefore, using the Remark 5.1, no direction $d \in \mathbb{R}^n$ exists such that $\nabla f(x^*)^T d < 0$ and $\nabla c_i(x^*)^T d < 0$, for any $i \in A(x^*)$ are simultaneously satisfied. Now, let $A \in \mathbb{R}^{(\text{card}(A(x^*))+1)\times n}$ be the matrix whose rows are $\nabla f(x^*)^T$ and $\nabla c_i(x^*)^T$, $i \in A(x^*)$. By Gordan's theorem (see Theorem A4.2), there exists a nonzero vector $p = [u_0, u_1, \ldots, u_{\text{card}(A(x^*))}] \geq 0$ in $\mathbb{R}^{\text{card}(A(x^*))+1}$ such that $A^T p = 0$. Therefore,

$$u_0 \nabla f(x^*) + \sum_{i \in A(x^*)} u_i \nabla c_i(x^*) = 0,$$

where $u_0 \geq 0$ and $u_i \geq 0$ for $i \in A(x^*)$ and $[u_0, u_1, \ldots, u_{\text{card}(A(x^*))}]$ is not the vector zero. Considering $u_i = 0$ for all $i \notin A(x^*)$, the following conditions are obtained:

$$\begin{aligned}
&u_0 \nabla f(x^*) + u^T \nabla c(x^*) = 0, \\
&u^T c(x^*) = 0, \\
&u_0 \geq 0, \quad u \geq 0, \\
&(u_0, u) \neq (0, 0),
\end{aligned}$$

where u is the vector with components u_i for $i = 1, \ldots, m$, some of them being $u_0, u_1, \ldots, u_{\text{card}(A(x^*))}$ and the others being zero. Observe that $u_0 \neq 0$, because otherwise the assumption that the gradient of the active constraints is linear independent at x^* is not satisfied. Now, considering the vector μ^* as the vector u whose components are divided by u_0, we get that (x^*, μ^*) is a KKT point. ♦

The above theorem shows the importance of the active constraints. A major difficulty in applying this result is that we do not know in advance which constraints are active and which are inactive at the solution of the problem. In other words, we do not know the active set. The majority of algorithms for solving this optimization problem with inequalities face this difficulty of identifying the active set. Of course, the idea of investigating all the possible active sets of a problem in order to get the points satisfying the KKT conditions is usually impractical.

Remark 5.2 *(Constrained qualification).* Observe that *not every local minimum is a KKT point.* For a local minimum x^* to be a KKT point, an additional condition must be introduced on the behavior of the constraints. Such a condition is known as the *constraint qualification.* Observe that the first-order Taylor series expansion of the functions defining the problem about x is used to form an approximate problem in which both the objective and the constraints are linear. If, near the current point x the linearization is fundamentally different from the feasible set, then the linear approximation of the problem does not yield useful information about the original problem. Therefore, certain assumptions about the nature of the constraints c_i that are active at x must be introduced to ensure that near x, the linearized approximation is similar to the feasible set. Given the point x and the active set $A(x)$, the *linear independence constraint qualification* (LICQ) holds if the *set of active constraint gradients* $\{\nabla c_i(x), i \in A(x)\}$ *is linear independent.* In Theorem 5.8 such a constraint qualification is that x^* is a regular point, which is also known as the *linear independence constraint qualification*. The Lagrange multipliers are guaranteed to be unique in Theorem 5.8 if LICQ holds. In general, if LICQ holds, none of the active constraint gradients can be zero. Another (weaker) constraint qualification is the *Mangasarian-Fromovitz constraint qualification* (MFCQ). It requires that there exists (at least) one direction $d \in C_{ac}(x^*)$, i.e., such that $\nabla c_i(x^*)^T d < 0$, for each $i \in A(x^*)$. The MFCQ is weaker than LICQ, i.e., the Lagrange multipliers are guaranteed to be unique if LICQ holds, while this uniqueness property may be lost under MFCQ. Finally, observe that the constraint qualifications are *sufficient* conditions for the linear approximation of the problem to be adequate; they are not necessary conditions. ♦

In the following theorem we present a sufficient condition which guarantees that any KKT point of an inequality constrained nonlinear optimization problem is a global minimum of the problem. Of course, this result is obtained under the convexity hypothesis.

Theorem 5.9 *(KKT sufficient conditions). Let $f : \mathbb{R}^n \to \mathbb{R}$ and $c_i : \mathbb{R}^n \to \mathbb{R}$, $i = 1, \ldots, m$, be convex and differentiable functions. Consider the problem $\min\{f(x) : c(x) \le 0\}$. If (x^*, μ^*) is a KKT point, then x^* is a global minimum of the problem.*

Proof. Let us define the function $L(x) \triangleq f(x) + \sum_{i=1}^{m} \mu_i^* c_i(x)$. Since f and c_i, $i = 1, \ldots, m$, are convex functions and $\mu_i^* \ge 0$, $i = 1, \ldots, m$, it follows that L is also convex. Now, the dual feasibility conditions determine that $\nabla L(x^*) = 0$. Therefore, by Theorem 5.5, x^* is a global minimum for L on $\mathbb{R}^n$, i.e., $L(x) \ge L(x^*)$ for any $x \in \mathbb{R}^n$. Therefore, for any x such that $c_i(x) \le c_i(x^*) = 0$, $i \in A(x^*)$, it follows that:

$$f(x) - f(x^*) \ge - \sum_{i \in A(x^*)} \mu_i^* [c_i(x) - c_i(x^*)] \ge 0.$$

On the other hand, the set $\{x \in \mathbb{R}^n : c_i(x) \le 0, \ i \in A(x^*)\}$ contains the feasible set $\{x \in \mathbb{R}^n : c_i(x) \le 0, i = 1, \ldots, m\}$. Therefore, x^* is a global minimum for the problem with inequality constraints. ♦

5.4 Optimality Conditions for Problems with Equality Constraints

In this section, the nonlinear optimization problem with equality constraints is considered:

$$\begin{aligned} &\min f(x) \\ \text{subject to:} \quad & \\ &h_i(x) = 0, \quad i = 1, \ldots, p, \end{aligned} \tag{5.9}$$

where $f : \mathbb{R}^n \to \mathbb{R}$ and $h_i : \mathbb{R}^n \to \mathbb{R}$, $i = 1, \ldots, p$, are continuously differentiable functions. The functions $h_i(x) = 0$, $i = 1, \ldots, p$, of the above problem define the vector $h(x) = [h_1(x), \ldots h_p(x)]$. If x^* satisfies the constraints from (5.9), i.e., $h_i(x^*) = 0$, $i = 1, \ldots, p$, it is said to be *feasible*. Otherwise, it is called *infeasible*.

The optimality of x^* can be seen as a balance between the function minimization and the constraint satisfaction. A move away from x^* cannot be made without either violating a constraint or increasing the value of the objective function. Formally, this can be stated as the following proposition.

Proposition 5.3 *(Balance between objective function and constraints). If x^* is a solution of* (5.9) *and $x^* + \delta x$ is a nearby point, then:*

1. *If $f(x^* + \delta x) < f(x^*)$ then $h_i(x^* + \delta x) \neq 0$ for some i.*
2. *If $h_1(x^* + \delta x) = \ldots = h_p(x^* + \delta x) = 0$ then $f(x^* + \delta x) \geq f(x^*)$.* ◆

In order to establish the optimality conditions for the nonlinear optimization problems with equality constraints, we need to introduce some relevant concepts. An equality constraint $h(x) = 0$ defines in $\mathbb{R}^n$ a set which can be viewed as a *hypersurface*. When there are p equality constraints $h_i(x) = 0$, $i = 1, \ldots, p$, then their intersection defines a (possible empty) set:

$$X \triangleq \{ x \in \mathbb{R}^n : h_i(x) = 0, i = 1, \ldots, p \}.$$

If the functions defining the equality constraints are differentiable, then the set X is said to be a *differentiable manifold* or a *smooth manifold*.

Now, in any point on a differentiable manifold, the *tangent set* can be defined as follows. A *curve* η on a manifold X is a continuous application $\eta : I \subset \mathbb{R} \to X$, i.e., a family of points $\eta(t) \in X$ continuously parameterized by t in the interval $I \subset \mathbb{R}$. Clearly, a curve passes through the point x^* if $x^* = \eta(t^*)$ for some $t^* \in I$. The *derivative* of a curve at t^*, if it exists, is defined in a classical manner as:

$$\dot{\eta}(t^*) \triangleq \lim_{\xi \to 0} \frac{\eta(t^* + \xi) - \eta(t^*)}{\xi}.$$

A curve is *differentiable* or *smooth* if a derivative exists for each $t \in I$.

Definition 5.15 *(Tangent set). Let X be a differentiable manifold in $\mathbb{R}^n$ and a point $x^* \in X$. Consider the collection of all the continuously differentiable curves on X passing through x^*. Then, the collection of all the vectors tangent to these curves at x^* is the tangent set to X at x^*, denoted by $T_X(x^*)$.*

Definition 5.16 *(Regular point – equality constraints). Let $h_i : \mathbb{R}^n \to \mathbb{R}$, $i = 1, \ldots, p$, be differentiable functions on $\mathbb{R}^n$, and consider the set $X \triangleq \{x \in \mathbb{R}^n : h_i(x) = 0, i = 1, \ldots, p\}$. A point $x^* \in X$ is a regular point if the gradient vectors $\nabla h_i(x^*)$, $i = 1, \ldots, p$, are linearly independent,* i.e.:

$$\operatorname{rank}\left[\nabla h_1(x^*), \ldots, \nabla h_p(x^*)\right] = p. \tag{5.10}$$

If the constraints are regular in the sense of the above definition, then X is a subspace of dimension $n-p$. In this case $T_X(x^*)$ is a subspace of dimension $n-p$, called *tangent space*. At regular points the tangent space can be characterized in terms of the gradients of the constraints (Luenberger, 1973).

Proposition 5.4 *(Algebraic characterization of a tangent space). Let $h_i : \mathbb{R}^n \to \mathbb{R}$, $i=1, \ldots, p$, be differentiable functions on $\mathbb{R}^n$, and consider the set $X \triangleq \{x \in \mathbb{R}^n : h_i(x)=0, i=1, \ldots, p\}$. At a regular point $x^* \in X$, the tangent space is such that:*

$$T_X(x^*) = \left\{d : \nabla h(x^*)^{\mathrm{T}} d = 0\right\}. \tag{5.11}$$

Proof. Let $T_X(x^*)$ be the tangent space at x^* and $M(x^*) = \{d : \nabla h(x^*)^{\mathrm{T}} d = 0\}$. Consider any curve $\eta(t)$ passing through x^* at $t=t^*$, having derivative $\dot{\eta}(t^*)$ such that $\nabla h(x^*)^{\mathrm{T}} \dot{\eta}(t^*) \neq 0$. Since such a curve would not lie on X, it follows that $T_X(x^*) \subset M(x^*)$. Now to prove that $T_X(x^*) \supset M(x^*)$, we must show that if $d \in M(x^*)$, then there is a curve on X passing through x^* with derivative d. In order to construct such a curve, we consider the equations:

$$h\left(x^* + td + \nabla h(x^*)^{\mathrm{T}} u(t)\right) = 0,$$

where for fixed t the vector $u(t) \in \mathbb{R}^p$ is unknown. Observe that we have a nonlinear system of p equations with p unknowns, continuously parameterized by t. At $t=0$ there is a solution $u(0)=0$. The Jacobian matrix of the above system with respect to u at $t=0$ is the matrix $\nabla h(x^*) \nabla h(x^*)^{\mathrm{T}}$ which is nonsingular, since $\nabla h(x^*)$ is of full rank if x^* is a regular point. Thus, by the implicit function theorem (see Theorem A2.4), there is a continuous solution $u(t)$ for $-a \leq t \leq a$. The curve $\eta(t) = x^* + td + \nabla h(x^*)^{\mathrm{T}} u(t)$ by construction is a curve on X. By differentiating the above nonlinear system with respect to t at $t=0$, we get:

$$0 = \frac{\mathrm{d}}{\mathrm{d}t} h(\eta(t))\Big|_{t=0} = \nabla h(x^*)^{\mathrm{T}} d + \nabla h(x^*) \nabla h(x^*)^{\mathrm{T}} \dot{u}(0).$$

By definition of d we have $\nabla h(x^*)^{\mathrm{T}} d = 0$. Therefore, since $\nabla h(x^*) \nabla h(x^*)^{\mathrm{T}}$ is nonsingular, it follows that $\dot{u}(0) = 0$. Therefore, $\dot{\eta}(0) = d + \nabla h(x^*)^{\mathrm{T}} \dot{u}(0) = d$ and the constructed curve has derivative d at x^*. ♦

The Method of Lagrange Multipliers

Let us now present the optimality conditions for the nonlinear optimization problems with equality constraints by using the method of Lagrange multipliers. The idea is to restrict the search of a minimum of (5.9) to the manifold $X \triangleq \{x \in \mathbb{R}^n : h_i(x) = 0, i = 1, \ldots, p\}$. The following theorem gives the geometric necessary condition for a local minimum of a nonlinear optimization problem with equality constraints. It is shown that the tangent space $T_X(x^*)$ at a regular local minimum point x^* is *orthogonal* to the gradient of the objective function at x^*.

Theorem 5.10 *(Geometric necessary condition for a local minimum). Let f: $\mathbb{R}^n \to \mathbb{R}$ and $h_i : \mathbb{R}^n \to \mathbb{R}$, $i=1, \ldots, p$, be continuously differentiable functions. Suppose the x^* is a local minimum point of the problem* $\min\{f(x) : h(x)=0\}$. *Then, $\nabla f(x^*)$ is orthogonal to the tangent space $T_X(x^*)$,* i.e.*:*

$$C_0(x^*) \cap T_X(x^*) = \varnothing.$$

Proof. Assume that there exists a $d \in T_X(x^*)$ such that $\nabla f(x^*)^{\mathrm{T}} d \neq 0$. Let $\eta : I = [-a, a] \to X$, $a > 0$, be any smooth curve passing through x^*, with $\eta(0) = x^*$ and $\dot{\eta}(0) = d$. Also let φ be the function defined as $\varphi(t) \triangleq f(\eta(t))$ for any $t \in I$. Since x^* is a local minimum of f on $X \triangleq \{x \in \mathbb{R}^n : h(x) = 0\}$, by Definition 5.7, it follows that there exists $\delta > 0$ such that $\varphi(t) = f(\eta(t)) \geq f(x^*) = \varphi(0)$ for any $t \in B(0, \delta) \cap I$. Therefore, $t^* = 0$ is an unconstrained local minimum point for φ, and:

$$0 = \nabla\varphi(0) = \nabla f(x^*)^{\mathrm{T}} \dot{\eta}(0) = \nabla f(x^*)^{\mathrm{T}} d.$$

But, this is in contradiction with the assumption that $\nabla f(x^*)^{\mathrm{T}} d \neq 0$. ♦

The conclusion of this theorem is that if x^* is a regular point of the constraints $h(x) = 0$ and a local minimum point of f subject to these constraints, then all $d \in \mathbb{R}^n$ satisfying $\nabla h(x^*)^{\mathrm{T}} d = 0$ must also satisfy $\nabla f(x^*)^{\mathrm{T}} d = 0$.

The following theorem shows that this property that $\nabla f(x^*)$ is orthogonal to the tangent space implies that $\nabla f(x^*)$ is a linear combination of the gradients of $h_i(x^*)$, $i = 1, \ldots, p$, at x^*. This relation leads to the introduction of Lagrange multipliers and the Lagrange function.

Theorem 5.11 *(First-order necessary optimality conditions). Let f: $\mathbb{R}^n \to \mathbb{R}$ and $h_i : \mathbb{R}^n \to \mathbb{R}$, $i=1, \ldots, p$, be continuously differentiable functions. Consider the problem* $\min\{f(x) : h(x)=0\}$. *If x^* is a local minimum, and it is a regular point of the constraints, then there exists a unique vector $\lambda^* \in \mathbb{R}^p$ such that*

$$\nabla f(x^*) + \nabla h(x^*)^{\mathrm{T}} \lambda^* = 0. \tag{5.12}$$

Proof. Since x^* is a local minimum of f on $X = \{x \in \mathbb{R}^n : h(x) = 0\}$, by Theorem 5.10, it follows that $C_0(x^*) \cap T_X(x^*) = \varnothing$, i.e., the system.

$$\nabla f(x^*)^{\mathrm{T}} d < 0, \ \nabla h(x^*)^{\mathrm{T}} d = 0$$

is inconsistent. Now, consider the following two sets:

$$C_1 \triangleq \left\{(z_1, z_2) \in \mathbb{R}^{p+1} : z_1 = \nabla f(x^*)^{\mathrm{T}} d, \;\; z_2 = \nabla h(x^*)^{\mathrm{T}} d\right\},$$
$$C_2 \triangleq \left\{(z_1, z_2) \in \mathbb{R}^{p+1} : z_1 < 0, \;\; z_2 = 0\right\}.$$

Observe that C_1 and C_2 are convex sets and $C_1 \cap C_2 = \varnothing$. Therefore, by the separation of two convex sets, given by Proposition A4.2, there exists a nonzero vector $(\mu, \lambda) \in \mathbb{R}^{p+1}$ ($\mu \in \mathbb{R}$, $\lambda \in \mathbb{R}^p$) such that for any $d \in \mathbb{R}^n$ and for any $(z_1, z_2) \in C_2$:

$$\mu \nabla f(x^*)^{\mathrm{T}} d + \lambda^{\mathrm{T}} \left[\nabla h(x^*)^{\mathrm{T}} d\right] \geq \mu z_1 + \lambda^{\mathrm{T}} z_2.$$

Now, considering $z_2 = 0$ and having in view that z_1 can be made an arbitrary large negative number, it follows that $\mu \geq 0$. Additionally, considering $(z_1, z_2) = (0, 0)$, we must have $[\mu \nabla f(x^*) + \lambda^{\mathrm{T}} \nabla h(x^*)]^{\mathrm{T}} d \geq 0$, for any $d \in \mathbb{R}^n$. In particular, letting $d = -[\mu \nabla f(x^*) + \lambda^{\mathrm{T}} \nabla h(x^*)]$, it follows that $-\|\mu \nabla f(x^*) + \lambda^{\mathrm{T}} \nabla h(x^*)\|^2 \geq 0$, and thus:

$$\mu \nabla f(x^*) + \lambda^{\mathrm{T}} \nabla h(x^*) = 0, \;\; \text{with} \;\; (\mu, \lambda) \neq (0, 0).$$

Observe that $\mu > 0$, for otherwise the above relation would contradict the assumption that $\nabla h_i(x^*)$, $i = 1, \ldots, p$, are linear independent. The conclusion of the theorem follows letting $\lambda^* = \lambda/\mu$ and noting that the linear independence assumption implies the uniqueness of the λ^*. ◆

Remark 5.3 The first-order necessary optimality conditions given by the Theorem 5.11 together with the constraints of the problem (5.9):

$$\nabla f(x^*) + \nabla h(x^*)^{\mathrm{T}} \lambda^* = 0, \tag{5.13a}$$
$$h(x^*) = 0, \tag{5.13b}$$

represent a total of $n+p$ nonlinear equations in the variables (x^*, λ^*). These conditions determine, at least locally, a unique solution (x^*, λ^*). However, as in the unconstrained case, a solution to the first-order necessary optimality conditions does not have to be a local minimum of the problem (5.9). ◆

Definition 5.17 *(Lagrange multipliers). The scalars* $\lambda_1^*, \ldots, \lambda_p^*$ *in* (5.12) *are called the Lagrange multipliers.*

Definition 5.18 *(Constraint normals). The vectors* $\nabla h_1(x), \ldots, \nabla h_p(x)$ *are called the constraint normals.*

The condition (5.12) shows that $\nabla f(x^*)$ is linearly dependent on the constraint normals. Therefore, a constrained minimum occurs when the gradients of the objective function and the constraints interact in such a way that any reduction in f can only be obtained by violating the constraints (see Proposition 5.3).

Definition 5.19 *(Lagrange function – Lagrangian). The function $L : \mathbb{R}^n \times \mathbb{R}^p \to \mathbb{R}$ associated to the nonlinear optimization problem* (5.9) *is defined as:*

$$L(x, \lambda) \triangleq f(x) + \lambda^{\mathrm{T}} h(x). \tag{5.14}$$

Remark 5.4 *(Regularity assumption).* For a local minimum to satisfy the above first-order necessary conditions (5.13) and, in particular, for unique Lagrange multipliers to exist, it is necessary that the equality constraints $h_i(x) = 0$, $i = 1, \ldots, p$, satisfy a regularity condition. As already seen, for a local minimum of an inequality constrained nonlinear optimization problem to be a KKT point, a constrained qualification is needed. For the equality constrained nonlinear optimization problems, the condition that *the minimum point is a regular point* corresponds *to linear independence constrained qualification.* ♦

If x^* is a regular local minimum of (5.9), then the first-order necessary optimality conditions (5.13) can be rewritten as:

$$\nabla_x L(x^*, \lambda^*) = 0, \tag{5.15a}$$

$$\nabla_\lambda L(x^*, \lambda^*) = 0. \tag{5.15b}$$

Observe that the second condition (5.15b) is a restatement of the constraints. The solution of the optimization problem (5.9) corresponds to a saddle point of the Lagrangian.

Theorem 5.12 *(Second-order necessary optimality conditions). Let $f : \mathbb{R}^n \to \mathbb{R}$ and $h_i : \mathbb{R}^n \to \mathbb{R}$, $i = 1, \ldots, p$, be continuously differentiable functions. Consider the problem* $\min\{f(x) : h(x) = 0\}$. *If x^* is a local minimum and it is a regular point of the constraints, then there exists a unique vector $\lambda^* \in \mathbb{R}^p$ such that*

$$\nabla f(x^*) + \nabla h(x^*)^{\mathrm{T}} \lambda^* = 0, \tag{5.16}$$

and

$$d^{\mathrm{T}} \left(\nabla^2 f(x^*) + \nabla^2 h(x^*)^{\mathrm{T}} \lambda^* \right) d \geq 0 \tag{5.17}$$

for any $d \in \mathbb{R}^n$ such that $\nabla h(x^*)^{\mathrm{T}} d = 0$.

Proof. The first condition $\nabla f(x^*) + \nabla h(x^*)^{\mathrm{T}} \lambda^* = 0$ follows from the Theorem 5.11. Now we concentrate to the second condition. Let x^* be a regular point and consider d an arbitrary direction from $T_X(x^*)$, i.e., $\nabla h(x^*)^{\mathrm{T}} d = 0$. Let $\eta : I = [-a, a] \to X$, $a > 0$, be an arbitrary twice-differential curve passing through x^* with $\eta(0) = x^*$ and $\dot{\eta}(0) = d$. Consider φ a function defined as $\varphi(t) \triangleq f(\eta(t))$, for any $t \in I$. Since x^* is a local minimum of f on $X \triangleq \{x \in \mathbb{R}^n : h(x) = 0\}$, it follows that $t^* = 0$ is an unconstrained local minimum point for φ. Therefore, by Theorem 5.4, it follows that

$$\nabla^2\varphi(0) - \dot{\eta}(0)^{\mathrm{T}}\nabla^2 f(x^*)\dot{\eta}(0) + \nabla f(x^*)^{\mathrm{T}}\ddot{\eta}(0) \geq 0.$$

On the other hand, differentiating the relation $h(\eta(t))^{\mathrm{T}}\lambda = 0$ twice, we get:

$$\dot{\eta}(0)^{\mathrm{T}}\left(\nabla^2 h(x^*)^{\mathrm{T}}\lambda\right)\dot{\eta}(0) + \left(\nabla h(x^*)^{\mathrm{T}}\lambda\right)^{\mathrm{T}}\ddot{\eta}(0) = 0.$$

Now, adding the last two relations we obtain:

$$d^{\mathrm{T}}\left(\nabla^2 f(x^*) + \nabla^2 h(x^*)^{\mathrm{T}}\lambda^*\right)d \geq 0,$$

which must hold for every d such that $\nabla h(x^*)^{\mathrm{T}}d = 0$. ♦

The above theorem says that if $T_X(x^*)$ is the tangent space to X at x^*, then the matrix $\nabla^2_{xx}L(x^*, \lambda^*) = \nabla^2 f(x^*) + \nabla^2 h(x^*)^{\mathrm{T}}\lambda^*$ is positive semidefinite on $T_X(x^*)$.

Remark 5.5 *(Feasible directions and second-order conditions).* An n-vector d is said to be a *feasible direction* at x^* if $\nabla h(x^*)d = 0$, where $\nabla h(x^*)$ is the Jacobian of the constraints at x^*. Let us assume that d is a feasible direction normalized so that $\|d\| = 1$. Considering the Taylor's expansion:

$$h(x^* + \varepsilon d) = h(x^*) + \varepsilon\nabla h(x^*)d + O\left(\|\varepsilon d\|^2\right),$$

then $h(x^* + \varepsilon d) = O(\varepsilon^2)$. Therefore, a move away from x^* along d keeps the constraints satisfied to first-order accuracy. In particular, if all the constraints in (5.9) are linear, then $x^* + \varepsilon d$ is a feasible point for all $\varepsilon > 0$. On the other hand, if any of the $h_i(x)$ in (5.9) are nonlinear, then d is a direction *tangential* to the constraints at x^*. It is easy to see that the condition (5.12) implies that, for any feasible direction d,

$$d^{\mathrm{T}}\nabla f(x^*) = 0.$$

To distinguish a *minimum* from a *maximum* or a *saddle point*, the second-order optimality condition must be used. These conditions can be stated as:

1. If the constraint functions h_i are all linear, the second-order condition that guarantees x^* is a minimum of problem (5.9) is

$$d^{\mathrm{T}}\nabla^2 f(x^*)d > 0$$

for any feasible direction d.

2. If the constraint functions h_i are nonlinear, the second-order condition that guarantees x^* is a minimum of problem (5.9) is

$$d^{\mathrm{T}}\nabla^2 L(x^*, \lambda^*)d > 0$$

for any feasible direction d. ♦

Remark 5.6 *(Eigenvalues in Tangent Space).* Geometrically, the restriction of the matrix $\nabla^2_{xx}L(x^*,\lambda^*)$ to $T_X(x^*)$ corresponds to the projection $P_{T_X(x^*)}\left[\nabla^2_{xx}L(x^*,\lambda^*)\right]$. A vector $y \in T_X(x^*)$ is an *eigenvector* of the projection $P_{T_X(x^*)}\left[\nabla^2_{xx}L(x^*,\lambda^*)\right]$ if there is a real number ν such that:

$$P_{T_X(x^*)}\left[\nabla^2_{xx}L(x^*,\lambda^*)\right]y = \nu y.$$

The real number ν is called the *eigenvalue* of $P_{T_X(x^*)}\left[\nabla^2_{xx}L(x^*,\lambda^*)\right]$. To obtain a matrix representation for $P_{T_X(x^*)}\left[\nabla^2_{xx}L(x^*,\lambda^*)\right]$, it is necessary to introduce a basis of the tangent subspace $T_X(x^*)$. It is best to introduce an orthonormal basis, say $E=[e_1, \ldots, e_{n-p}]$. Any vector $y \in T_X(x^*)$ can be written as $y=Ez$, where $z \in \mathbb{R}^{n-p}$. Now, $\nabla^2_{xx}L(x^*,\lambda^*)Ez$ represents the action of $\nabla^2_{xx}L(x^*,\lambda^*)$ on such a vector. To project this result back into $T_X(x^*)$ and to express the result in terms of the basis $E=[e_1, \ldots, e_{n-p}]$, it is necessary to multiply by E^{T}. Therefore, $E^{\mathrm{T}}\nabla^2_{xx}L(x^*,\lambda^*)Ez$ is the vector whose components give the representation in terms of the basis E. The $(n-p)\times(n-p)$ matrix $E^{\mathrm{T}}\nabla^2_{xx}L(x^*,\lambda^*)E$ is the matrix representation of $\nabla^2_{xx}L(x^*,\lambda^*)$ restricted to $T_X(x^*)$. *The eigenvalues of* $\nabla^2_{xx}L(x^*,\lambda^*)$ *restricted to* $T_X(x^*)$ *can be determined by computing the eigenvalues of* $E^{\mathrm{T}}\nabla^2_{xx}L(x^*,\lambda^*)E$. These eigenvalues are independent of the particular choice of the basis E. ♦

Recall that the conditions given in Theorems 5.11 and 5.12 are necessary conditions. These must hold at each local minimum point. However, a point satisfying these conditions may not be a local minimum. As in the unconstrained case, it is possible to derive second-order conditions for constrained optimization problems. The following theorem provides sufficient conditions for a stationary point of the Lagrange function to be a local minimum.

Theorem 5.13 *(Second-order sufficient conditions). Let* $f: \mathbb{R}^n \to \mathbb{R}$ *and* $h_i: \mathbb{R}^n \to \mathbb{R}$, $i=1, \ldots, p$, *be twice continuously differentiable functions. Consider the problem* $\min\{f(x): h(x)=0\}$. *If* x^* *and* λ^* *satisfy:*

$$\nabla_x L(x^*,\lambda^*) = 0, \tag{5.18a}$$

$$\nabla_\lambda L(x^*,\lambda^*) = 0 \tag{5.18b}$$

and

$$y^{\mathrm{T}}\nabla^2_{xx}L(x^*,\lambda^*)y > 0 \tag{5.19}$$

for any $y \neq 0$ *such that* $\nabla h(x^*)^{\mathrm{T}}y = 0$, *then* x^* *is a strict local minimum.*

Proof. Consider the augmented Lagrange function:

$$\bar{L}(x,\lambda) = f(x) + \lambda^{\mathrm{T}}h(x) + \frac{c}{2}\|h(x)\|^2,$$

where c is a scalar. Clearly,

$$\nabla_x \bar{L}(x,\lambda) = \nabla_x L(x,\bar{\lambda}),$$

$$\nabla_{xx}^2 \bar{L}(x,\lambda) = \nabla_{xx}^2 L(x,\bar{\lambda}) + c\nabla h(x)^{\mathrm{T}} \nabla h(x),$$

where $\bar{\lambda} = \lambda + ch(x)$. Since (x^*,λ^*) satisfy the sufficient conditions, by the Theorem A4.3, we obtain that $\nabla_x \bar{L}(x^*,\lambda^*) = 0$ and $\nabla_{xx}^2 \bar{L}(x^*,\lambda^*) > 0$, for sufficiently large c. $\bar{L}$ being positive definite at (x^*,λ^*), it follows that there exist $\rho > 0$ and $\delta > 0$ such that

$$\bar{L}(x,\lambda^*) \geq \bar{L}(x^*,\lambda^*) + \frac{\rho}{2}\|x - x^*\|^2$$

for $\|x - x^*\| < \delta$. Besides, since $\bar{L}(x,\lambda^*) = f(x)$ when $h(x) = 0$, we get

$$f(x) \geq f(x^*) + \frac{\rho}{2}\|x - x^*\|^2$$

if $h(x) = 0$, $\|x - x^*\| < \delta$, i.e., x^* is a strict local minimum. ♦

Sensitivity: Interpretation of the Lagrange Multipliers

The ith Lagrange multiplier can be viewed as the measuring of the sensitivity of the objective function with respect to the ith constraint, i.e., how much the optimal value of the objective function would change if that constraint was perturbed.

At the very beginning let us consider $p = 1$, i.e., the problem (5.9) has one constraint $h_1(x) = 0$. Now, suppose that x^* is a local solution of the problem:

$$\min\{f(x) : h_1(x) = 0\},$$

and consider the *perturbed* problem:

$$\min\{f(x) : h_1(x) = \delta\},$$

where δ is a known scalar. If the solution of the perturbed problem is $x^* + u$, then by using the Taylor's expansion, a first-order estimate of the optimum function value is $f(x^* + u) \approx f(x^*) + u^{\mathrm{T}}\nabla f(x^*)$. But, the optimality condition for the original problem given by (5.12) states that $\nabla f(x^*) = -\lambda_1^* \nabla h_1(x^*)$, where λ_1^* is the Lagrange multiplier. Hence:

$$f(x^* + u) \approx f(x^*) - \lambda_1^* u^{\mathrm{T}} \nabla h_1(x^*).$$

Since $x^* + u$ solves the perturbed problem, it follows that $h_1(x^* + u) = \delta$, hence $h_1(x^*) + u^{\mathrm{T}}\nabla h_1(x^*) \approx \delta$. But, $h_1(x^*) = 0$. Therefore, $u^{\mathrm{T}}\nabla h_1(x^*) \approx \delta$, that is:

$$f(x^* + u) - f(x^*) \approx -\delta\lambda_1^*.$$

In other words, *the Lagrange multiplier is an approximate measure of the change in the objective function that will occur if a unit amount is added to the right-hand side of the constraint.* In general, we have the following theorem.

Theorem 5.14 *(Interpretation of the Lagrange multipliers). Consider the family of problems* $\min\{f(x) : h(x) = w\}$, *where* $f : \mathbb{R}^n \to \mathbb{R}$ *and* $h : \mathbb{R}^n \to \mathbb{R}^p$ *are twice continuously differentiable. Suppose for* $w = 0$ *there is a local solution* x^* *that is a regular point and that, together with its associated Lagrange multiplier vector* λ, *satisfies the second-order sufficient conditions for a strict local minimum. Then, for every* $w \in \mathbb{R}^p$ *in a region containing* 0, *there is a* $x(w)$, *depending continuously on* w, *such that* $x(0) = x^*$ *and such that* $x(w)$ *is a local minimum of the problem. Furthermore:*

$$\nabla_w f(x(w))|_{w=0} = -\lambda.$$

Proof. Consider the system of equations:

$$\nabla f(x) + \nabla h(x)^{\mathrm{T}}\lambda = 0,$$
$$h(x) = w.$$

By hypothesis, when $w = 0$, there is a solution x^*, λ^* to this system. The Jacobian matrix of this system at this solution is:

$$\begin{bmatrix} L(x^*) & \nabla h(x^*)^{\mathrm{T}} \\ \nabla h(x^*) & 0 \end{bmatrix},$$

where $L(x^*) = \nabla^2 f(x^*) + \nabla^2 h(x^*)^{\mathrm{T}}\lambda^*$.

Since x^* is a regular point and $L(x^*)$ is positive definite on $\{y : \nabla h(x^*)^{\mathrm{T}} y = 0\}$, it follows that this matrix is nonsingular. Thus, by the implicit function theorem (see Theorem A2.4), there is a solution $x(w), \lambda(w)$ to the system which is twice continuously differentiable. Therefore:

$$\nabla_w f(x(w))|_{w=0} = \nabla f(x^*)^{\mathrm{T}} \nabla_w x(0),$$
$$\nabla_w h(x(w))|_{w=0} = \nabla h(x^*) \nabla_w x(0).$$

But, since $h(x^*) = w$ it follows that $\nabla h(x^*)\nabla_w x(0) = I$. On the other hand, from $\nabla f(x) + \nabla h(x)^{\mathrm{T}}\lambda = 0$, it follows that $\nabla_w f(x(w))|_{w=0} = -\lambda$. ♦

5.5 Optimality Conditions for General Problems

We now present a generalization of Theorems 5.8, 5.11, 5.12, and 5.13 to nonlinear optimization problems with equality and inequality constraints:

$$\min f(x)$$

subject to: (5.20)

$$c_i(x) \le 0, \quad i = 1, \dots, m,$$
$$h_j(x) = 0, \quad j = 1, \dots, p,$$

where $x \in \mathbb{R}^n$, $f: \mathbb{R}^n \to \mathbb{R}$, $c_i: \mathbb{R}^n \to \mathbb{R}$, $i = 1, \dots, m$, and $h_j: \mathbb{R}^n \to \mathbb{R}$, $j = 1, \dots, p$, are continuously differentiable functions. Define the vectors: $c(x) = [c_1(x), \dots, c_m(x)]$ and $h(x) = [h_1(x), \dots, h_p(x)]$.

Remark 5.7 *(Discarding the inactive constraints).* Let us consider the nonlinear optimization problem $\min\{f(x) : c_i(x) \le 0, \ i = 1, \dots, m\}$. Suppose that x^* is a local minimum point for this problem. Clearly, x^* is also a local minimum of the above problem, where the inactive constraints $c_i(x) \le 0$, $i \notin A(x^*)$ have been *discarded.* Therefore, *the inactive constraints at x^* can be ignored in the statement of the optimality conditions.* On the other hand, the active constraints can be treated as equality constraints at a local minimum point. Hence, x^* is also a local minimum point to the equality constrained problem:

$$\min\{f(x) : c_i(x) = 0, \ i \in A(x^*)\}$$

The difficulty is that we do not know the set of the active constraints at x^*.

From Theorem 5.11 it follows that if x^* is a regular point, there exists a unique Lagrange multiplier vector $\mu^* \in \mathbb{R}^m$ such that:

$$\nabla f(x^*) + \sum_{i \in A(x^*)} \mu_i^* \nabla c_i(x^*) = 0.$$

Now, assigning zero Lagrange multipliers to the inactive constraints, we get:

$$\nabla f(x^*) + \nabla c(x^*)^{\mathrm{T}} \mu^* = 0,$$
$$\mu_i = 0, \quad i \notin A(x^*).$$

Clearly, the last condition can be rewritten as $\mu_i^* c_i(x^*) = 0$, $i = 1, \dots, m$.

It remains to show that $\mu > 0$. For this, assume that $\mu_q < 0$ for some $q \in A(x^*)$. Now, let $\mathrm{A} \in \mathbb{R}^{(m+1) \times n}$ be the matrix whose rows are $\nabla f(x^*)$ and $\nabla c_i(x^*)$, $i = 1, \dots, m$. Since x^* is a regular point, it follows that the Lagrange multiplier vector μ^* is unique. Therefore, the condition $\mathrm{A}^{\mathrm{T}} y = 0$ can only be satisfied by $y^* = \gamma(1 \ \ \mu^*)^{\mathrm{T}}$ with $\gamma \in \mathbb{R}$. But, $\mu_q < 0$. Therefore, by Gordan's Theorem A4.2 there exists a direction $\bar{d} \in \mathbb{R}^n$ such that $\mathrm{A}\bar{d} < 0$. In other words, $\bar{d} \in C_0(x^*) \cap C_{ac}(x^*) \neq \varnothing$, which contradicts the hypothesis that x^* is a local minimum of the problem. All these results represent the KKT optimality conditions as stated by Theorem 5.8. Although this development is straightforward, it is somewhat limited by the regularity-type assumption at the optimal solution. ♦

Definition 5.20 *(Regular point – general case). Let $c_i : \mathbb{R}^n \to \mathbb{R}, i = 1, \ldots, m$, and $h_j : \mathbb{R}^n \to \mathbb{R}, j = 1, \ldots, p$, be continuously differentiable functions. Consider the set $X = \{x \in \mathbb{R}^n : c_i(x) \le 0, i = 1, \ldots, m,\ h_j(x) = 0, j = 1, \ldots, p\}$. A point $x^* \in X$ is a regular point of the constraints from* (5.20) *if the gradients $\nabla c_i(x^*)$, $i \in A(x^*)$ and $\nabla h_j(x^*), j = 1, \ldots, p$, are linearly independent.*

Definition 5.20 introduces the linear independence constraint qualification (LICQ) for general nonlinear optimization problems, i.e., the gradients of the *active* inequality constraints and the gradients of the equality constraints are all linearly independent at x^*. Another constraint qualification is the *linear constraint qualification* (LCQ), i.e., $c_i(x), i = 1, \ldots, m$, and $h_j : \mathbb{R}^n \to \mathbb{R}$, $j = 1, \ldots, p$, are affine functions. Another one is the *Slater condition* for a convex problem, i.e., there exists a point $\bar{x}$ such that $c_i(\bar{x}) < 0$, $i = 1, \ldots, m$, and $h(\bar{x}) = 0$.

We emphasize that the constraint qualification ensures that the linearized approximation to the feasible set X captures the essential shape of X in a neighborhood of x^*.

Theorem 5.15 *(First- and second-order necessary conditions). Let $f : \mathbb{R}^n \to \mathbb{R}$, $c_i : \mathbb{R}^n \to \mathbb{R}$, $i = 1, \ldots, m$, and $h_i : \mathbb{R}^n \to \mathbb{R}$, $i = 1, \ldots p$, be twice continuously differentiable functions. Consider the problem* $\min\{f(x) : c(x) \le 0,\ h(x) = 0\}$. *If x^* is a local minimum for this problem and it is a regular point of the constraints, then there exist unique vectors $\mu^* \in \mathbb{R}^m$ and $\lambda^* \in \mathbb{R}^p$ such that*

$$\nabla f(x^*) + \nabla c(x^*)^{\mathrm{T}} \mu^* + \nabla h(x^*)^{\mathrm{T}} \lambda^* = 0, \tag{5.21a}$$

$$\mu^* \ge 0, \tag{5.21b}$$

$$c(x^*) \le 0, \tag{5.21c}$$

$$h(x^*) = 0, \tag{5.21d}$$

$$(\mu^*)^{\mathrm{T}} c(x^*) = 0, \tag{5.21e}$$

and

$$y^{\mathrm{T}} \left(\nabla^2 f(x^*) + \nabla^2 c(x^*)^{\mathrm{T}} \mu^* + \nabla^2 h(x^*)^{\mathrm{T}} \lambda^* \right) y \ge 0, \tag{5.22}$$

for all $y \in \mathbb{R}^n$ such that $\nabla c_i(x^*)^{\mathrm{T}} y = 0$, $i \in A(x^*)$ and $\nabla h(x^*)^{\mathrm{T}} y = 0$.

Proof. Observe that since $\mu^* \ge 0$ and $c(x^*) \le 0$, (5.21e) is equivalent to the statement that a component of μ^* is nonzero only if the corresponding constraint is active. Since x^* is a minimum point over the constraint set, it is also a minimum over the subset of that set defined by setting the active constraints to zero. Therefore, for the resulting equality constrained problem defined in a neighborhood of x^*, there are Lagrange multipliers. Hence, (5.21a) holds with $\mu_i^* = 0$ if $c_i(x^*) \ne 0$.

It remains to show that $\mu^* \ge 0$. This is a little more elaborate. Suppose that for some $k \in A(x^*)$, $\mu_k^* < 0$. Let $\bar{X}$ and $\bar{T}$ be the surface and the tangent space, respectively, defined by all the other active constraints at x^*. By the regularity assumptions, there is a d such that $d \in \bar{T}$ and $\nabla c_k(x^*)^{\mathrm{T}} d < 0$. Let $\eta(t)$ be a curve on $\bar{X}$ passing through x^* at $t = 0$ with $\dot{\eta}(0) = d$. Then, for small $t \geq 0$, it follows that $\eta(t)$ is feasible and

$$\frac{\mathrm{d}f}{\mathrm{d}t}(\eta(t))\bigg|_{t=0} = \nabla f(x^*)^{\mathrm{T}} d < 0$$

by (5.21a), which contradicts the fact that x^* is a minimum point. ♦

The conditions (5.21) are known as the *Karush-Kuhn-Tucker conditions* or *KKT conditions*. The conditions (5.21e) written as $\mu_i^* c_i(x^*) = 0$, $i = 1, \ldots, m$, are the *complementary conditions*. They show that either constraint i is active or the corresponding Lagrange multiplier $\mu_i^* = 0$, or possibly both. For a given nonlinear optimization problem (5.20), and a solution point x^*, there may be many Lagrange multipliers (μ^*, λ^*) for which the conditions (5.21) and (5.22) are satisfied. However, when x^* is a regular point (the LICQ is satisfied), the optimal (μ^*, λ^*) is unique.

The KKT conditions motivate the following definition which *classifies* constraints according to whether or not their corresponding Lagrange multiplier is zero.

Definition 5.21 *(Strongly active (binding) – weakly active constraints). Let x^* be a local solution to the problem* (5.20) *and the Lagrange multipliers* (μ^*, λ^*) *which satisfy the KKT conditions* (5.21). *We say that an inequality constraint $c_i(x)$ is strongly active or binding if $i \in A(x^*)$ and the corresponding Lagrange multiplier $\mu_i^* > 0$. We say that $c_i(x)$ is weakly active if $i \in A(x^*)$ and the corresponding Lagrange multiplier $\mu_i^* = 0$.*

The nonnegativity condition (5.21b) on the Lagrange multiplier for the inequality constraints ensures that the function $f(x)$ will not be reduced by a move of any of the binding constraints at x^* to the interior of the feasible region.

A special case of complementarity is important because it introduces the concept of *degeneracy* in optimization.

Definition 5.22 *(Strict complementarity). Let x^* be a local solution to the problem* (5.20) *and the Lagrange multipliers* (μ^*, λ^*) *which satisfy the KKT conditions* (5.21). *We say that the strict complementarity holds if exactly one of μ_i^* and $c_i(x^*)$ is zero for each index $i = 1, \ldots, m$. In other words, we have $\mu_i^* > 0$ for each $i \in A(x^*)$.*

Usually, satisfaction of strict complementarity is beneficial for algorithms and makes it easier to determine the active set $A(x^*)$ so that convergence is more rapid.

Remark 5.8 *(Degeneracy).* A property that causes difficulties for some optimization algorithms is degeneracy. This concept refers to the following two situations:

- The gradients of the active constraints $\nabla c_i(x^*)$, $i \in A(x^*)$, are linearly dependent at the solution point x^*. Linear dependence of the gradients of the active constraints can cause difficulties during the computation of the step direction because certain matrices that must be factorized become rank deficient.

- Strict complementarity fails to hold, that is, there is some index $i \in A(x^*)$ such that all the Lagrange multipliers satisfying the KKT conditions (5.21) have $\mu_i^* = 0$. In the case when the problem contains weakly active constraints, it is difficult for an algorithm to determine whether these constraints are active at the solution. For some optimization algorithms (active-set algorithms and gradient projection algorithms), the presence of weakly active constraints can cause the algorithm to zigzag as the iterates move on and off the weakly constraints along the successive iterations. ♦

Theorem 5.16 *(Second-order sufficient conditions). Let* $f : \mathbb{R}^n \to \mathbb{R}$, $c_i : \mathbb{R}^n \to \mathbb{R}$, $i = 1, \ldots, m$, *and* $h_j : \mathbb{R}^n \to \mathbb{R}$, $j = 1, \ldots p$, *be twice continuously differentiable functions. Consider the problem* $\min\{f(x) : c(x) \le 0,\ \ h(x) = 0\}$. *If there exist* x^*, μ^* *and* λ^* *satisfying the KKT conditions* (5.21) *and*

$$y^{\mathrm{T}} \nabla_{xx}^2 L(x^*, \mu^*, \lambda^*) y > 0,$$

for all $y \neq 0$ *such that*

$$\nabla c_i(x^*)^{\mathrm{T}} y = 0, \quad i \in A(x^*) \ \ \text{with} \ \ \mu_i^* > 0, \tag{5.23a}$$

$$\nabla c_i(x^*)^{\mathrm{T}} y \le 0, \quad i \in A(x^*) \ \ \text{with} \ \ \mu_i^* = 0, \tag{5.23b}$$

$$\nabla h(x^*)^{\mathrm{T}} y = 0, \tag{5.23c}$$

where $L(x, \mu, \lambda) = f(x) + \mu^{\mathrm{T}} c(x) + \lambda^{\mathrm{T}} h(x)$, *then* x^* *is a strict local minimum of the problem.*

Proof. The theorem says that the Hessian of the Lagrangian is positive definite on the critical cone $C(x^*, \mu^*, \lambda^*)$ defined by (5.23) for x^*,μ^* and λ^* satisfying the KKT conditions (5.21).

Assume that x^* is not a strict local minimum, and let $\{y_k\}$ be a sequence of feasible points converging to x^* such that $f(y_k) \le f(x^*)$. Consider y_k of the form $y_k = x^* + \delta_k s_k$ with $\delta_k > 0$ and $\|s_k\| = 1$. Assume that $\delta_k \to 0$ and $s_k \to s^*$. Clearly, $\nabla f(x^*)^{\mathrm{T}} s^* \le 0$ and $\nabla h_j(x^*) s^* = 0$ for $j = 1, \ldots, p$.

On the other hand, for each active constraint c_i, we have $c_i(y_k) - c_i(x^*) \le 0$. Therefore, $\nabla c_i(x^*)^{\mathrm{T}} s^* \le 0$.

If $\nabla c_i(x^*)^{\mathrm{T}} s^* = 0$, for all $i \in \{l : c_l(x^*) = 0,\ \mu_l^* > 0\}$, then the proof is similar to that in Theorem 5.13. If $\nabla c_i(x^*)^{\mathrm{T}} s^* < 0$ for at least one $i \in \{l : c_l(x^*) = 0, \mu_l^* > 0\}$, then

$$0 \ge \nabla f(x^*)^{\mathrm{T}} s^* = -\lambda^{\mathrm{T}} \nabla h(x^*) s^* - \mu^{\mathrm{T}} \nabla c(x^*) s^* > 0,$$

which represents a contradiction. ♦

The KKT sufficient conditions for convex programming with inequality constraints given in Theorem 5.9 can immediately be generalized to nonlinear optimization problems with convex inequalities and affine equalities.

Theorem 5.17 *(KKT sufficient conditions for general problems). Let* $f: \mathbb{R}^n \to \mathbb{R}$ *and* $c_i: \mathbb{R}^n \to \mathbb{R}$, $i = 1, \dots, m$, *be convex and differentiable functions. Also, let* $h_i: \mathbb{R}^n \to \mathbb{R}$, $i = 1, \dots, p$, *be affine functions. Consider the problem* $\min f(x)$ *subject to* $x \in X \triangleq \{x \in \mathbb{R}^n : c(x) \leq 0,\ h(x) = 0\}$. *If* (x^*, μ^*, λ^*) *satisfies the KKT conditions* (5.21), *then* x^* *is a global minimum for* f *on* X. ♦

Sensitivity: Interpretation of the Lagrange Multipliers for General Problems
As we have already seen, Theorem 5.14 presents an interpretation of the Lagrange multipliers for nonlinear optimization problems with equality constraints. Each Lagrange multiplier tells us something about the *sensitivity* of the optimal objective function value $f(x^*)$ with respect to the corresponding constraint. Clearly, for an inactive constraint $i \notin A(x^*)$, the solution x^* and the function value $f(x^*)$ are independent of whether this constraint is present or not. If we slightly perturb c_i by a tiny amount, it will still be inactive and therefore x^* will still be a local solution of the optimization problem. Since $\mu_i^* = 0$ from (5.21e), the Lagrange multiplier shows that the constraint i has no importance in the system of the constraints. Otherwise, as in Theorem 5.14, the following theorem can be presented.

Theorem 5.18 *(Interpretation of the Lagrange multipliers for general problems). Consider the family of problems* $\min\{f(x) : c(x) \leq v,\ h(x) = w\}$, *where* $f: \mathbb{R}^n \to \mathbb{R}$, $c: \mathbb{R}^n \to \mathbb{R}^m$ *and* $h: \mathbb{R}^n \to \mathbb{R}^p$ *are twice continuously differentiable. Suppose for* $v = 0$, $w = 0$ *there is a local solution* x^* *that is a regular point and that, together with its associated Lagrange multiplier* $\mu^* \geq 0, \lambda^*$, *satisfies the second-order sufficient conditions for a strict local minimum. Then, for every* $(v, w) \in \mathbb{R}^{m+p}$, *in a region containing* $(0, 0) \in \mathbb{R}^{m+p}$, *there is a solution* $x(v, w)$ *continuously depending on* (v, w), *such that* $x(0, 0) = x^*$ *and such that* $x(v, w)$ *is a local minimum of the problem. Furthermore:*

$$\nabla_v f(x(v, w))|_{0,0} = -\mu^*.$$
$$\nabla_w f(x(v, w))|_{0,0} = -\lambda^*.$$
♦

Notes and References
Plenty of books and papers are dedicated to the theoretical developments of optimality conditions for continuous nonlinear optimization. Many details and properties of the theoretical aspects of the optimality conditions can be found in Bertsekas (1999), Boyd and Vandenberghe (2006), Nocedal and Wright (2006), Sun and Yuan (2006), Bartholomew-Biggs (2008), etc. The content of this chapter is taken from the books by Chachuat (2007), Bazaraa, Sherali, and Shetty (1993), Luenberger (1973, 1984), and Luenberger and Ye (2008). As for the optimality conditions for problems with inequality constraints, the material is inspired by Bazaraa, Sherali, and Shetty (1993). The derivation of the necessary and sufficient optimality conditions for problems with equality constraints follows the developments presented by Luenberger (1973). The sensitivity analysis and interpretation of the Lagrange multipliers for nonlinear optimization is taken from Luenberger

(1973). We have not treated here the duality and the saddle point optimality condition of the Lagrangian, but these can be found, for example, in Bazaraa, Sherali, and Shetty (1993) or Nocedal and Wright (2006).

The KKT conditions were originally named after Harold W. Kuhn (1925–2014), and Albert W. Tucker (1905–1995), who first published them in 1951 (Kuhn and Tucker, 1951). Later on, scholars discovered that the necessary conditions for this problem had been stated by William Karush (1917–1997) in his master's thesis in 1939 (Karush, 1939). Another approach of the optimality conditions for the nonlinear optimization problem was given in 1948 by Fritz John (1910–1994) (John, 1948). See also Cottle (2012).

Chapter 6
Simple Bound Constraints Optimization

The simple bound constraints optimization is a class of nonlinear optimization problems with a special structure, found in many real practical applications. The mathematical model of these problems is as follows:

$$\begin{aligned} &\min f(x) \\ \text{subject to:} \quad & \\ & x \in X = \{x \in \mathbb{R}^n, \;\; l_i \leq x_i \leq u_i, \;\; i = 1, \ldots, n\}. \end{aligned} \tag{6.1}$$

The function $f: \mathbb{R}^n \to \mathbb{R}$ is supposed to be at least twice continuously differentiable. The set X defined by the real numbers $-\infty < l_i \leq u_i < +\infty$, $i = 1, \ldots, n$, which represents the *bounds on the variables*, is the *feasibility domain* of the problem (6.1). Suppose that for any $i = 1, \ldots, n$, $l_i \leq u_i$, that is, the feasibility domain X is nonempty. Since X is a compact set, it follows that (6.1) has always a local optimal solution (see Theorem 5.1). The inequalities $-\infty < l_i \leq x_i \leq u_i < +\infty$, $i = 1, \ldots, n$, are called *simple bounds*. The ith simple bound constraint is *active* at $x \in X$ if $x_i = l_i$, or $x_i = u_i$. In the first case, if $x_i = l_i$, we say that the variable x_i is assigned to its *lower bound*. In the second one, if $x_i = u_i$, the variable x_i is assigned to its *upper bound*. The *set of active constraints* at the point $x \in X$ is denoted by$A(x)$. The *set of inactive constraints* at $x \in X$ is denoted by $I(x)$. If $l_i = u_i$, then the ith component of x is *fixed* to the value $l_i = u_i$.

This problem is often a sub-problem of the augmented Lagrangian or of the penalty computational schemes for solving general constrained optimization (Facchinei & Lucidi, 1992a, 1992b; Conn, Gould, & Toint, 1997; Nocedal & Wright, 2006; Sun & Yuan, 2006). Therefore, the development of numerical algorithms to efficiently solve (6.1), especially for large-scale problems, is important in both theory and practice.

The purpose of this chapter is to present this class of problems, the main aspects of the optimality conditions, and the main computational methods for solving these problems. For this, we follow the developments given by Kelley (1999), together

N. Andrei, *Continuous Nonlinear Optimization for Engineering Applications in GAMS Technology*, Springer Optimization and Its Applications 121,
DOI 10.1007/978-3-319-58356-3_6

with some details from Bertsekas (1976, 1982a, 1982b, 1999), Lin and Moré (1999), Birgin and Martínez (2001), Birgin, Martínez, and Raydan (2000, 2001), Byrd, Lu, and Nocedal, (1995), Byrd, Lu, Nocedal, and Zhu, (1995) and Hager and Zhang (2006a, 2006b). At the same time, we present some computational results, comparisons among the algorithms, and some nonlinear optimization applications with simple bounds. From the multitude of algorithms dedicated to this problem, we insist on the spectral projected gradient method, on the limited memory BFGS with simple bounds, and on the truncated Newton with simple bounds.

6.1 Necessary Conditions for Optimality

As we know, for a continuously differentiable function of one variable, the necessary conditions for unconstrained optimality at x^* are simply $f'(x^*)=0$, and if f is twice continuously differentiable, $f''(x^*)\geq 0$. For a simple bound constrained problem in which the variables restrict the domain of f to an interval $[a,b]$, the necessary condition must be changed in order to admit the possibility for the minimizer to be one of the endpoints of the interval $[a,b]$. If $x^*=a$ is a local minimizer, then $f(x)\geq f(a)$ for all $a\leq x$ sufficiently near a. Therefore, $f'(a)\geq 0$. Nothing can be said about f''. Similarly, if $x^*=b$ is a local minimizer, then $f'(b)\leq 0$. Hence, all three possibilities, $x^*=a$, $x^*=b$, and $a<x^*<b$, can be expressed by the following theorem:

Theorem 6.1 *Let f be a continuously differentiable function of one variable on the interval $[a,b]$. Let x^* be a local minimum of f on $[a,b]$. Then*

$$f'(x^*)(x-x^*)\geq 0 \textit{ for all } x\in[a,b] \tag{6.2}$$

and if f is twice continuously differentiable on $[a,b]$,

$$f''(x^*)(x^*-a)(b-x^*)\geq 0. \tag{6.3}$$

A point $x^*\in X$ is *stationary* for problem (6.1) if:

$$\nabla f(x^*)^{\mathrm{T}}(x-x^*)\geq 0 \text{ for all } x\in X. \tag{6.4}$$

As in the unconstrained case, *stationary points* are said to satisfy the *first-order necessary conditions.*

In order to present the second-order necessary conditions for problem (6.1), the *reduced Hessian* is introduced. Let f be twice differentiable at $x\in X$. The reduced Hessian $\nabla^2_R f(x)$ is the matrix:

$$\left(\nabla_R^2 f(x)\right)_{ij} = \begin{cases} \delta_{ij}, & i \in A(x),\ j \in A(x), \\ \left(\nabla^2 f(x)\right)_{ij}, & \text{otherwise}. \end{cases} \tag{6.5}$$

Theorem 6.2 *Let f be twice Lipschitz continuously differentiable and let* x^* *be the solution of the problem* (6.1). *Then the reduced Hessian* $\nabla_R^2 f(x^*)$ *is positive semidefinite.*

Proof. Assume that at point x there are t inactive indices and $n-t$ active indices. With this, the vector $x \in X$ can be partitioned by reordering the variables as $x=[z,y]$, where z corresponds to the inactive indices and y to the active ones. Then the map $\phi(z)=f(z,y^*)$ has an unconstrained local minimizer at $z^* \in \mathbb{R}^t$ and hence $\nabla^2\phi$ is positive semidefinite. But the reduced Hessian can be written as

$$\nabla_R^2 f(x^*) = \begin{bmatrix} \nabla^2\phi(x^*) & \\ & I \end{bmatrix}$$

if the variables are partitioned as above. ♦

Let P be the *projection* onto X, i.e., the map that takes x to the nearest point (in the l_2 norm) in X to x. Then:

$$P(x)_i = \begin{cases} l_i, & \text{if } (x)_i \le l_i, \\ (x)_i, & \text{if } l_i < (x)_i < u_i, \\ u_i, & \text{if } (x)_i \ge u_i. \end{cases} \tag{6.6}$$

The following theorem, proved in Kelley (1999), states the necessary condition for optimality.

Theorem 6.3 *Let f be continuously differentiable. A point* $x^* \in X$ *is stationary for problem* (6.1) *if and only if*

$$x^* = P(x^* - \alpha\nabla f(x^*)) \tag{6.7}$$

for all $\alpha \ge 0$. ♦

In a Lagrangian formalism, supposing that no l_i is $-\infty$ and no u_i is $+\infty$, $i=1,\dots,n$, then the KKT conditions for x^* to solve the problem (6.1) are

$$\begin{aligned} &\nabla f(x^*) - \lambda^* + \mu^* = 0, \\ &\lambda^* \ge 0, \quad (l-x^*)^T\lambda^* = 0, \\ &\mu^* \ge 0, \quad (x^*-u)^T\mu^* = 0, \\ &l \le x^* \le u, \end{aligned}$$

where $\lambda^*, \mu^* \in \mathbb{R}^n$ are the KKT multipliers. In this context, strict complementarity is said to hold at the KKT point (x^*,λ^*,μ^*) if $x_i^* = l_i$ implies $\lambda_i^* > 0$ and $x_i^* = u_i$

implies $\mu_i^* > 0$. Another equivalent way to present the KKT conditions is as follows:

$$\begin{aligned} & l \le x^* \le u, \\ & l_i < x_i^* < u_i \quad \Rightarrow \quad \nabla f(x^*)_i = 0, \\ & x_i^* = l_i \quad \Rightarrow \quad \nabla f(x^*)_i \ge 0, \\ & x_i^* = u_i \quad \Rightarrow \quad \nabla f(x^*)_i \le 0. \end{aligned}$$

If the projected gradient $\nabla_P f(x)$ is defined by

$$\nabla_P f(x^*)_i \triangleq \begin{cases} \min\{0, \nabla f(x^*)_i\}, & \text{if} \quad x_i^* = l_i, \\ \nabla f(x^*)_i, & \text{if} \quad l_i < x_i^* < u_i, \\ \max\{0, \nabla f(x^*)_i\}, & \text{if} \quad x_i^* = u_i, \end{cases}$$

then the first-order necessary conditions for optimality can be written as:

$$l \le x^* \le u, \quad \nabla_P f(x^*) = 0.$$

6.2 Sufficient Conditions for Optimality

The sufficient conditions are formulated by using the definition of the reduced Hessian. Observe that if x^* is stationary, $i \in I(x^*)$ and e_i is a unit vector in the ith coordinate direction, then $x^* \pm te_i \in X$ for all t sufficiently small. Since

$$\frac{df(x^* \pm te_i)}{dt} = \pm \nabla f(x^*)^T e_i \ge 0,$$

it follows that:

$$(\nabla f(x^*))_i = 0 \text{ for all } i \in I(x^*). \tag{6.8}$$

In order to formulate the sufficient conditions for optimality, the concept of nondegenerate stationary point is introduced. *A point $x^* \in X$ is a nondegenerate stationary point for problem* (6.1) *if x^* is a stationary point and:*

$$(\nabla f(x^*))_i \ne 0 \textit{ for all } i \in A(x^*). \tag{6.9}$$

If x^* is a solution of the problem (6.1), *then x^* is a nondegenerate local minimizer.*

Nondegeneracy is important in the formulation of sufficient conditions and in the design of termination criteria. Let S be an arbitrary set of indices. Then define:

$$P_S(x)_i = \begin{cases} x_i, & i \in S, \\ 0, & i \notin S. \end{cases} \tag{6.10}$$

The following theorem, proved in Kelley (1999), gives the sufficiency conditions associated to problem (6.1).

Theorem 6.4 *Let $x^* \in X$ be a nondegenerate stationary point for problem* (6.1). *Let f be twice continuously differentiable in a neighborhood of x^* and assume that the reduced Hessian at x^* is positive definite. Then x^* is a solution of problem* (6.1) *and hence a nondegenerate local minimizer.*

Proof. Let $x \in X$ and define $\varphi(t) = f(x^* + t(x - x^*))$. We prove that either (i) $\varphi'(0) > 0$, or (ii) $\varphi'(0) = 0$, $\varphi''(0) > 0$. Consider $e = x - x^*$. Observe that $\varphi'(0) = \nabla f(x^*)^T e = \nabla f(x^*)^T (P_A(e) + P_I(e))$. But from the stationarity, we have $\nabla f(x^*)^T P_I(e) = 0$. Now, if $P_A(e) \neq 0$, then from nondegeneracy it follows that $\nabla f(x^*)^T P_A(e) > 0$, i.e., (i) holds. On the other hand, if $P_A(e) = 0$, then

$$\varphi''(0) = (x - x^*)^T P_I \left(\nabla^2 f(x^*)\right) P_I (x - x^*) = (x - x^*)^T \nabla_R^2 f(x^*)(x - x^*) > 0,$$

which prove (ii). ♦

6.3 Methods for Solving Simple Bound Optimization Problems

One of the methods for solving simple bound optimization problems is the *active-set method.* Polyak (1969) proposed an extension of the conjugate gradient method to this class of problems with quadratic objective. In this method, the conjugate gradient method is used to explore a face of the feasible set and the negative gradient is used to leave a face. At every iteration, the algorithm of Polyak only added or dropped one constraint. Dembo and Tulowitzki (1983) proposed an algorithm which could add and drop many constraints at an iteration. Later, Yang and Tolle (1991) further developed Polyak's algorithm to obtain a finite termination even when the problem was degenerate at the local minimizer. Another variant of this algorithm, which includes a rigorous convergence analysis, was given by Wright (1990). Moré and Toraldo (1991) improved this algorithm by proposing a gradient projection method to identify a working face of the feasible set, followed by a conjugate gradient method to explore the face. For a simple bound optimization problem at point $x \in X$, the following two sets, $L(x) = \{i : x_i = l_i\}$ and $U(x) = \{i : x_i = u_i\}$, can be defined. Clearly, $A(x) = L(x) \cup U(x)$. If x^* is a local minimizer of f in X, then x^* is a local minimizer of $f(x)$ subject to $x_i = l_i$, $i \in L(x^*)$ and $x_i = u_i$, $i \in U(x^*)$. The active-set methods aim at predicting $L(x^*)$ and $U(x^*)$ by using in an iterative way the disjoint sets $L, U \subseteq \{1, \ldots, n\}$. The idea of the active-set method is as follows: given the sets L and U, the following problem:

$$\min f(x)$$

subject to

$$x_i = l_i, \ \ i \in L \text{ and } x_i = u_i, \ \ i \in U,$$

is approximately solved as an unconstrained optimization problem over the variables x_i, $i \notin A = L \cup U$. For the quadratic programming case, for which f is quadratic, the active-set methods were developed by Coleman and Hulbert (1989). The main difficulty with the active-set method is the selection of the sets L and U in an iterative process either by adding variables which violate one of their bounds or by removing those for which further progress is predicted.

An *active-set algorithm* was developed by Hager and Zhang (2006a) which consists of a nonmonotone gradient projection step, an unconstrained optimization step, and a set of rules for branching between steps. The implementation of this algorithm uses the cyclic Barzilai-Borwein algorithm for the gradient projection step and the conjugate gradient algorithm CG-DESCENT by Hager and Zhang (2005) for the unconstrained optimization step. An attractive feature of this algorithm is that the search directions are always sufficient descent directions and when the objective function is strongly convex quadratic, the convergence of the algorithm is achieved in a finite number of iterations, even when the strict complementarity slackness does not hold (see Definition 5.14) (Hager & Zhang, 2006a).

Other methods for solving these problems is the *gradient projected* (Rosen, 1960; Levitin & Polyak, 1966; Bertsekas, 1976). These methods are extensions of the steepest descent method to deal with convex constraints. The iteration is computed as $x_{k+1} = P_X(x_k - \alpha_k \nabla f(x_k))$, where $P_X(v)$ projects v onto X and α_k is the stepsize. In the case of simple bounds, $P_X(v)$ is trivial to compute. Bertsekas, (1976) showed that the method has the following important feature. For nondegenerate problems (i.e., for those problems for which removing one or more active constraints necessarily changes the solution), the optimal "face" of active constraints will be determined in a finite number of iterations. If the active faces visited by consecutive iterates of the gradient projected algorithm are identical, then the Newton-like method should be used to investigate this face. Therefore, the *projected Newton method* is generated (Bertsekas, 1982a). For the quadratic case, this idea was considered by Moré and Toraldo (1991). In this case, if the initial point is sufficiently near a nondegenerate local minimizer x^*, then the search direction is computed by using the reduced Hessian. The resulting projected Newton method will take full steps, i.e., $\alpha_k = 1$ along the iterations. On the other hand, if the initial point is far from x^* and the reduced Hessian is not symmetric and positive definite, then the line search may fail. Clearly, this possibility of indefiniteness is the main weakness in any line-search method which uses $\nabla^2 f$ when far from the minimizer. To accelerate the convergence of these methods, more research has been developed on Newton and trust-region methods. For nondegenerate bound optimization problems, the superlinear and quadratic convergence was established by Conn, Gould, and Toint (1988a, 1988b, 1991) and Facchinei, Júdice, and Soares (1998). For degenerate problems, these convergence results were established by Facchinei, Lucidi, and Palagi (2002), Friedlander, Martínez, and Santos (1994), Lescrenier (1991), and Lin and Moré (1999). Computing the Newton step can be expensive. Therefore, approximation techniques, such as sparse or incomplete Cholesky factorization, have been suggested in order to reduce the computational expense.

A *gradient projection method with limited memory BFGS* matrix to approximate the Hessian of the objective function was developed by Byrd, Lu, Nocedal, and Zhu (1995). In this algorithm, the gradient projection method is used to determine a set of active constraints at each iteration. This is a line-search algorithm (as opposed to trust region) that uses the limited memory BFGS matrices. Another algorithm that uses limited memory quasi-Newton methods to update the inactive variables and a projected gradient method to update the active variables was given by Ni and Yuan (1997).

A *trust-region Newton method* which uses a gradient projection to generate a Cauchy step, a preconditioned conjugate gradient method with an incomplete Cholesky factorization to generate a direction, and a projected search to compute the step is the TRON method (Lin & Moré, 1999). Using the projected searches allows TRON to examine the faces of the feasible domain by generating a small number of minor iterations.

The *affine-scaling interior point method* by Coleman and Li (1994, 1996, 1997) represents a different approach, which is related to the trust-region method. Some developments of this strategy are presented in Dennis, Heinkenschloss, and Vicente (1998), Heinkenschloss, Ulbrich, and Ulbrich (1999), and Ulbrich, Ulbrich, and Heinkenschloss (1999). These methods are characterized by a reformulation of the necessary optimality conditions obtained by multiplication with a scaling matrix. The resulting system is solved by using the Newton method. In another way of research, Zhang (2004) proposes an interior point approach for solving this linear system.

As we have already said, a projected Newton-like method for the solution of (6.1) is the *truncated Newton method* for large-scale box constrained optimization given by Facchinei, Lucidi, and Palagi (2002). This method is based on the works of Facchinei and Lucidi (1992a, 1992b): At each iteration k, the estimates $L(x_k)$ and $U(x_k)$ of the variables that supposedly will, respectively, be at their lower and upper bounds at the solution are defined. At the same time, the estimate $F(x_k)$ of the variables which we believe to be free is computed. This partition of variables suggests performing an unconstrained minimization in the space of free variables, a typical approach in the active-set methods. For a locally fast convergent method, an obvious choice for the unconstrained minimization in the subspace of free variables is the Newton method. In order to compensate the loss of curvature information that we have in the subspace of those variables that are active but with zero multiplier, Facchinei, Lucidi, and Palagi (2002) introduced a correction term in the right-hand term of the Newton system. Other approaches also based on the truncated Newton method are those given by Schlick and Fogelson (1992a, 1992b) and Nash (1984a, 1984b, 1985).

A *nonmonotone spectral projected gradient* (SPG) for solving simple bound optimization problems is the one given by Birgin, Martínez, and Raydan (2000, 2001). This algorithm combines the projected gradient method of Bertsekas (1976) with two new features in optimization. The first one is based on the nonmonotone line search developed by Grippo, Lampariello, and Lucidi (1986). The second one uses the spectral step length introduced by Barzilai and Borwein (1988) and further analyzed by Raydan (1993, 1997).

Another approach we mention here is the *filter-trust-region method* for simple bound constrained optimization given by Sainvitu and Toint (2006). The algorithm combines the filter-trust-region algorithm of Gould, Sainvitu, and Toint (2005) with a gradient projection method.

Finally, a *direct method* for solving (6.1) is the BOBYQA method by Powell (2009), where a quadratic approximation of the objective function is minimized at each iteration. Recent advances in simple bound constrained optimization have been presented by Hager and Zhang (2006b).

In the following, we shall present the spectral projected gradient method (SPG) by Birgin , Martínez, and Raydan (2000, 2001), the limited memory BFGS algorithm with gradient projection (L-BFGS-B) by Byrd, Lu, Nocedal, and Zhu (1995), and truncated Newton with simple bounds (TNBC) by Nash (1984a, 1984b, 1985) for solving simple bound constrained optimization problems.

6.4 Spectral Projected Gradient Method (SPG)

The gradient projected algorithm and its spectral variant are natural extensions of the steepest descent algorithm to simple bound constrained problems (Bertsekas, 1976). Let us consider a current iteration x. The new iteration is computed as

$$x_{+} = P(x - \alpha \nabla f(x)),$$

where α is the stepsize computed by the Armijo rule or by some other line-search procedures. The gradient projected algorithm determines a sequence of iterations which satisfies the simple bounds of the problem (6.1) and achieves a sufficient decrease of the function f. For $\alpha > 0$ define:

$$x(\alpha) = P(x - \alpha \nabla f(x)). \tag{6.11}$$

The sufficient decrease condition for line search in simple bound constrained problems is expressed as

$$f(x(\alpha)) - f(x) \leq \frac{-\rho}{\alpha} \|x - x(\alpha)\|^2, \tag{6.12}$$

where ρ is a positive parameter ($\rho = 10^{-4}$ (Dennis & Schnabel, 1983). Therefore, the general gradient projected algorithm can be presented as:

Algorithm 6.1 Gradient projected algorithm

1.	Select an initial point x_0, and a value for the parameter $\beta \in (0, 1)$. Set $k=1$
2.	Test a criterion for stopping the iterations
3.	Compute $f(x_k)$ and $\nabla f(x_k)$
4.	(Armijo rule) Find the least integer m such that (6.12) holds for $\alpha = \beta^m$
5.	Set $x_{k+1} = x(\alpha)$, $k = k+1$ and go to step 2 ♦

The spectral projected gradient method is a method of projected gradient which includes two ingredients. The first one is an extension of globalization techniques used in unconstrained optimization, based on the nonmonotone line search by Grippo, Lampariello, and Lucidi (1986) to the simple bound optimization problem. The second one consists in using the spectral stepsize introduced by Barzilai and Borwein (1988) and analyzed by Raydan (1993). Mainly, the scaled gradient projected algorithm is a version of the gradient projected algorithm by Bertsekas, which uses the Armijo rule along a nonlinear trajectory of projections.

Consider the problem (6.1) where the function f is defined and has continuous partial derivatives on an open set that contains X. Assume that f is bounded from below on X. Consider a point $x \in \mathbb{R}^n$. Then the *orthogonal projection* of x onto X is $P_X(x)$ with:

$$(P_X(x))_i = \max\{l_i, \min\{x_i, u_i\}\}, \quad i = 1, \ldots, n. \tag{6.13}$$

For any $x \in X$, the algorithm uses the *spectral projected gradient* defined as

$$g_t(x) = [P_X(x - t\nabla f(x)) - x], \tag{6.14}$$

where $t > 0$ is a spectral selection of the stepsize. Observe that zeroing the spectral projected gradient is equivalent to the optimality conditions of the first-order. Therefore, the norm of the spectral projected gradient is a criterion for stopping the iterations.

The algorithm starts with an initial point $x_0 \in \mathbb{R}^n$ and uses an integer $m \geq 1$ used in nonmonotone line search, a small parameter $\alpha_{\min} > 0$, a large parameter $\alpha_{\max} > \alpha_{\min}$, a small parameter $\varepsilon_1 > 0$, a sufficient decrease parameter $\gamma \in (0, 1)$, and safeguarding parameters $0 < \sigma_1 < \sigma_2 < 1$ used as protection in quadratic interpolation. Initially, $\alpha_0 \in [\alpha_{\min}, \alpha_{\max}]$ is arbitrary. The step length is computed by quadratic or by cubic interpolation.

Algorithm 6.2 Spectral projected gradient (SPG)

1.	*Initialization.* Consider some numerical values for the above described parameters, as well as the initial point $x_0 \in \mathbb{R}^n$. If $x_0 \notin X$, then set $x_0 = P_X(x_0)$. Set $k = 0$
2.	Compute $f(x_k)$ and $\nabla f(x_k)$
3.	Compute the *gradient projected* $P_X(x_k - \nabla f(x_k))$

(continued)

4.	Compute the *initial spectral step length* as: $$\alpha=\min\left\{\alpha_{\max},\max\left\{\alpha_{\min},\frac{1}{\lVert P_X(x_k-\nabla f(x_k))-x_k\rVert_\infty}\right\}\right\}.$$
5.	Test for stopping the iterations: If $$\lVert P_X(x_k-\nabla f(x_k))-x_k\rVert_\infty\le\varepsilon_1,$$ stop; otherwise continue with step 6
6.	Compute the *spectral direction of the projected gradient* as: $$d=P_X(x_k-\alpha\nabla f(x_k))-x_k$$
7.	*Initialization of the nonmonotone line search.* Set $\alpha=1$ and compute $x_{k+1}=x_k+\alpha d$. Compute $f(x_{k+1})$
8.	If $$f(x_{k+1})\le\max_{0\le j\le\min\{k,m\}}\{f(x_{k-j})\}+\gamma\alpha(d^T\nabla f(x_k)),$$ then go to step 11; otherwise continue with step 9 where the quadratic or cubic interpolation is implemented
9.	*Quadratic interpolation.* If $\alpha\le0.1$, then set $\alpha=\alpha/2$, otherwise compute: $$a_t=-\frac{\alpha^2\left(d^T\nabla f(x_k)\right)}{2\left(f(x_{k+1})-f(x_k)-\alpha\left(d^T\nabla f(x_k)\right)\right)}.$$ If $a_t<\sigma_1$ or $a_t>\sigma_2\alpha$, then set $a_t=a_t/2$. Set $\alpha=a_t$. Compute $x_{k+1}=x_k+\alpha d$ and go to step 8 *Cubic interpolation.* If $\alpha=1$, then compute $$\alpha_t=-\frac{\alpha^2\left(d^T\nabla f(x_k)\right)}{2\left(f(x_{k+1})-f(x_k)-\alpha\left(d^T\nabla f(x_k)\right)\right)}$$ and go to step 10. Otherwise, compute: $$p=f(x_{k+1})-f(x_k)-\alpha(d^T\nabla f(x_k)),\quad q=f_p-f(x_k)-\alpha_p(d^T\nabla f(x_k)),$$ $$a=\left(\frac{p}{\alpha^2}-\frac{q}{\alpha_p^2}\right)\frac{1}{\alpha-\alpha_p},\qquad b=\left(\frac{-p\alpha_p}{\alpha^2}+\frac{q\alpha}{\alpha_p^2}\right)\frac{1}{\alpha-\alpha_p},$$ $$r=b^2-3a(d^T\nabla f(x_k)).$$ If $a=0$, then set

(continued)

	$$\alpha_t = \frac{-\left(d^{\mathrm{T}} \nabla f(x_k)\right)}{2b},$$ otherwise $$\alpha_t = \frac{-b+\sqrt{r}}{3a}.$$ If $\alpha_t > \alpha/2$, then set $\alpha_t = \alpha/2$.
10.	Set $\alpha_p = \alpha$, $f_p = f(x_{k+1})$. If $\alpha_t \le \alpha/10$, then set $\alpha = \alpha/10$, otherwise $\alpha = \alpha_t$. Compute $x_{k+1} = x_k + \alpha d$, $f(x_{k+1})$ and continue with step 8
11.	Compute $s_k = x_{k+1} - x_k$, $y_k = \nabla f(x_{k+1}) - \nabla f(x_k)$ and determine the spectral step length. If $y_k^{\mathrm{T}} s_k \le 0$, then set $\alpha = \alpha_{\max}$, otherwise $$\alpha = \min\left\{\alpha_{\max}, \max\left\{\alpha_{\min}, \frac{s_k^{\mathrm{T}} s_k}{y_k^{\mathrm{T}} s_k}\right\}\right\},$$ $k = k+1$ and continue with step 5 ♦

The algorithm is based on the spectral projected gradient direction $P_X(x_k - \alpha_k \nabla f(x_k)) - x_k$, where α_k is the safeguarded inverse Rayleigh quotient $\left(s_k^{\mathrm{T}} s_k\right)/\left(y_k^{\mathrm{T}} s_k\right)$. Observe that $\left(s_k^{\mathrm{T}} s_k\right)/\left(y_k^{\mathrm{T}} s_k\right)$ is a Rayleigh quotient corresponding to the average Hessian matrix $\int_0^1 \nabla^2 f(x_k + t s_k) dt$.

The line search uses the protected quadratic interpolation or the cubic interpolation, implemented in steps 9 and 10, according to a value of a parameter which we do not explicit here. The protection of the quadratic interpolation is given by the parameters σ_1 and σ_2, which are initialized in step 1 of the algorithm. The quadratic interpolation acts when the minimum of the one-dimensional quadratic function $q(.)$ such that $q(0) = f(x_k)$, $q(\alpha) = f(x_k + \alpha d_k)$ and $\nabla q(0) = d_k^{\mathrm{T}} \nabla f(x_k)$, lies outside of the interval $[\sigma_1, \sigma_2 \alpha]$ and not when this minimum lies outside of the interval $[\sigma_1 \alpha, \sigma_2 \alpha]$, as usually implemented. This means that, when the interpolation tends to reject 90% (for $\sigma_1 = 0.1$) of the original search interval (let say [0,1]), then we consider that its prediction is not reliable and a more conservative bisection is preferred. On the other hand, the cubic interpolation considers an approximation of the minimizing function f by a cubic polynomial and uses four interpolation conditions based on the function values and its derivatives, each of them computed in two different points.

The convergence of this algorithm is proved by Birgin, Martínez, and Raydan (1999).

Theorem 6.5 *The algorithm SPG is well defined and any accumulation point of the sequence it generates is a constrained stationary point for the problem* (6.1). ♦

The proof is based on the results of Bertsekas (1999), which refer to the Armijo rule in the context of projections. Mainly, from the convexity of the domain X, it follows that for any $x \in X$ and $t \in (0, \alpha_{\max}]$ the following results are true:

(i) $\nabla f(x)^T g_t(x) \le -\frac{1}{t}\|g_t(x)\|_2^2 \le -\frac{1}{\alpha_{max}}\|g_t(x)\|_2^2$,
(ii) $g_t(\bar{x})$ is zero if and only if $\bar{x}$ is a stationary point.

These results are used in Theorem 6.5 to prove that any accumulation point $\bar{x}$ of the sequence $\{x_k\}$ generated by the algorithm SPG is a stationary point, i.e., for any $x \in X$, it follows that $\nabla f(\bar{x})^T(x-\bar{x}) \ge 0$.

Numerical Study (Quadratic Interpolation Versus Cubic Interpolation)
The SPG algorithm with both quadratic and cubic interpolation was implemented in double precision Fortran, compiled with f77 (default compiler settings) and run on a Workstation Intel Pentium 4 with 1.8 GHz. We selected a number of 80 large-scale simple bound optimization test functions in generalized or extended form presented in Andrei (2013a). For each test function, we considered 10 numerical experiments with the number of variables increasing as $n = 1000, 2000, \ldots, 10000$. Therefore, a set of 800 simple bound optimization problems is obtained. Each problem is solved with SPG in two variants: SPG with quadratic interpolation (SPGp) and SPG with cubic interpolation (SPGc). The parameters used in SPG were initialized as $m = 10$, $\alpha_{min} = 10^{-3}$, $\alpha_{max} = 10^3, \gamma = 10^{-4}$, $\sigma_1 = 0.1$, $\sigma_2 = 0.9$, $\varepsilon_1 = 0$. The algorithms we compare in these numerical experiments find local solutions. Therefore, the comparisons of algorithms are given in the following context. Let f_i^{SPGp} and f_i^{SPGc} be the optimal value found by SPGp and SPGc, for problem $i = 1, \ldots, 800$, respectively. We say that, in the particular problem *i*, the performance of SPGp with quadratic interpolation was better than the performance of SPGc with cubic interpolation if

$$\left| f_i^{SPGp} - f_i^{SPGc} \right| < 10^{-3} \tag{6.15}$$

and the number of iterations (*#iter*), or the number of function-gradient evaluations (*#fg*), or the CPU time of SPGp were less than the number of iterations, or the number of function-gradient evaluations, or the CPU time corresponding to SPGc, respectively.

Table 6.1 presents the number of problems out of 800 used in this numerical study for which SPGp versus SPGc achieved the minimum number of iterations (*#iter*), the minimum number of function and its gradient evaluations (*#fg*), and the minimum CPU computing time (*cpu*).

From Table 6.1, comparing SPGp versus SPGc, for example, subject to the number of iterations, we see that SPGp was better in 236 problems (i.e., it achieved

Table 6.1 Performance of the algorithms. SPG with quadratic interpolation (SPGp) versus SPG with cubic interpolation (SPGc)

	SPGp	SPGc	=
#iter	236	167	273
#fg	128	250	298
cpu	201	170	305

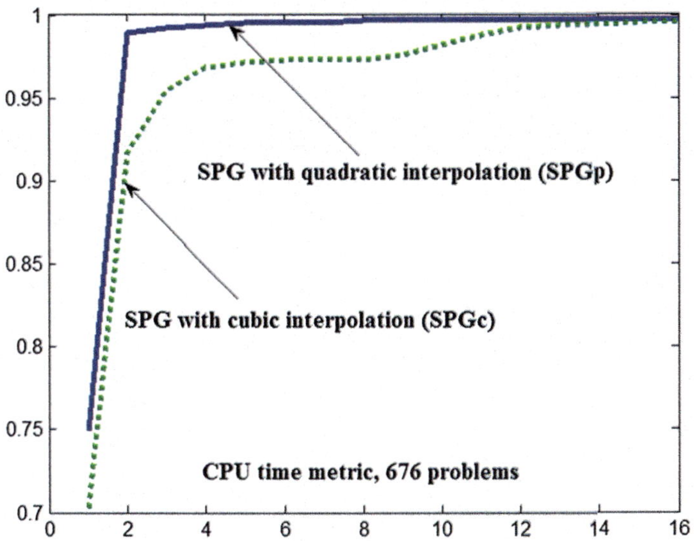

Figure 6.1 SPG: quadratic interpolation versus cubic interpolation

the minimum number of iterations in solving 236 problems). SPGc was better in 167 problems, and they achieved the same number of iterations in solving 273 problems. Therefore, subject to the number of iterations, we see that the SPGp is more efficient than SPGc.

Figure 6.1 shows the Dolan-Moré (2002) performance profiles of SPGp versus SPGc subject to the CPU time metric. In Figure 6.1, for every $\tau \geq 1$, the performance profile of these algorithms SPGp and SPGc gives the fraction of the test problems that each considered algorithmic variant has a performance within a factor of τ from the best. The left-hand side of the figure gives the percentage of the test problems for which an algorithm is the fastest; the right-hand side gives the percentage of the test problems that are successfully solved by these algorithms. Mainly, the left-hand side is a measure of the *efficiency* of an algorithm; the right-hand side is a measure of the *robustness* of an algorithm. Clearly, the top curve corresponds to the algorithm that solved the most problems in a time that was within a factor τ of the best time.

From Figure 6.1 we see that, subject to the CPU time metric, SPGp is more efficient than SPGc, but they have the same robustness for solving this set of simple bound constrained optimization problems. Both these algorithms find local optimal solutions. Out of 800 problems considered in this numerical study, only for 676 problems does the criterion (6.15) hold. The conclusion of this numerical study is that both these scaled gradient projected algorithms, with quadratic interpolation and with cubic one in line search, represent efficient and robust algorithms for solving a large variety of simple bound optimization problems.

6.5 L-BFGS with Simple Bounds (L-BFGS-B)

In this section we present a very efficient and reliable algorithm for solving the simple bound optimization problem (6.1) based on the limited memory BFGS update which approximates the Hessian of the function f. The algorithm is described by Byrd, Lu, Nocedal, and Zhu (1995) and is based on the developments given by Conn, Gould, and Toint (1988a) and Moré and Toraldo (1989). It uses the gradient projection method to determine a set of active constraints at each iteration. The main ingredient is the compact representation of the limited memory BFGS matrices described by Byrd, Nocedal, and Schnabel (1994).

Description of the Algorithm Suppose that at the current iterate x_k the following elements are known: the value of the objective function f_k, the gradient $g_k = \nabla f(x_k)$, and a positive definite limited memory approximation B_k of the Hessian $\nabla^2 f(x_k)$. Therefore, the following quadratic model of f can be formed:

$$m_k(x) = f(x_k) + g_k^T(x - x_k) + \frac{1}{2}(x - x_k)^T B_k(x - x_k). \tag{6.16}$$

The algorithm approximately minimizes $m_k(x)$ subject to the simple bounds $l_i \leq x_i \leq u_i,\ i = 1, \ldots, n$. This is done by using the gradient projection method to find a set of active constraints, followed by a minimization of $m_k(x)$ where the bounds are equality constraints.

For this, let us consider the piecewise linear path:

$$x(t) = P_X(x_k - tg_k, l, u)$$

obtained by projecting the steepest descent direction onto the feasible set X, where:

$$P_X(x, l, u)_i = \begin{cases} l_i, & x_i < l_i, \\ x_i, & l_i \leq x_i \leq u_i, \\ u_i, & x_i > u_i. \end{cases} \tag{6.17}$$

In the following, the generalized Cauchy point x^c is computed, which is defined as the first local minimizer of the univariate piecewise quadratic:

$$q_k(t) = m_k(x(t)).$$

All the variables whose value at point x^c is at the lower or at the upper bound and which form the active-set $A(x^c)$ are held fixed. With these, the following quadratic programming problem over the subspace of free variables:

$$\min\{m_k(x) : \ x_i = x_i^c, \ \forall i \in A(x^c)\}$$

subject to (6.18)

$$l_i \le x_i \le u_i, \ \forall i \notin A(x^c),$$

is considered. Firstly, an approximate solution of (6.18) is computed by ignoring the bounds on the free variables using x^c as the starting point. After an approximate solution $\bar{x}_{k+1}$ of the problem (6.18) has been obtained, the new iteration x_{k+1} is computed by line search along the direction $d_k = \bar{x}_{k+1} - x_k$ using the strong Wolfe line-search conditions:

$$f(x_{k+1}) \le f(x_k) + \rho\alpha_k g_k^T d_k, \tag{6.19}$$

$$\left|g_{k+1}^T d_k\right| \le \sigma\left|g_k^T d_k\right|, \tag{6.20}$$

where α_k is the step length and ρ and σ are positive parameters used in Wolfe line search (Wolfe, 1969, 1971). With all these elements, the gradient at x_{k+1} is evaluated, a new limited memory Hessian approximation B_{k+1} is computed, and the process is repeated.

Observe that the generalized Cauchy point x^c, which is a minimizer of $m_k(x)$ on the projected steepest descent direction, satisfies $m_k(x_k) > m_k(x^c)$ if the projected gradient is nonzero. Since $\bar{x}_{k+1}$ is on the path from x^c to the minimizer of (6.18), along which m_k decreases, it follows that the values of m_k at $\bar{x}_{k+1}$ are no larger than its value at x^c, i.e.:

$$f(x_k) = m_k(x_k) > m_k(x^c) \ge m_k(\bar{x}_{k+1}) = f(x_k) + g_k^T d_k + \frac{1}{2} d_k^T B_k d_k. \tag{6.21}$$

Therefore, if B_k is positive definite and d_k is not zero, the inequality (6.21) implies that $g_k^T d_k < 0$. In conclusion, since in our algorithm every Hessian approximation B_k is positive definite, it follows that the approximate solution $\bar{x}_{k+1}$ of the quadratic problem (6.18) defines a descent direction $d_k = \bar{x}_{k+1} - x_k$ for the objective function f.

Limited Memory BFGS Updates The limited memory BFGS matrices used in the algorithm are represented in the compact form Byrd, Nocedal, and Schnabel (1994). At every iteration x_k, the algorithm stores a small number, let us say m, of corrections pairs $\{s_i, y_i\}$, $i = k-1, \ldots, k-m$, where $s_k = x_{k+1} - x_k$ and $y_k = g_{k+1} - g_k$. These correction pairs contain information about the curvature of the function f, and in the frame of BFGS formula, they define the limited memory iteration matrix B_k. In the following, as it is described in Byrd, Nocedal, and Schnabel (1994), we shall represent these matrices without explicitly forming them.

Firstly, the following $n \times m$ correction matrices are formed:

$$Y_k = [y_{k-m}, \ldots, y_{k-1}], \quad S_k = [s_{k-m}, \ldots, s_{k-1}]. \tag{6.22}$$

Now, if θ is a positive scaling parameter and if the m correction pairs $\{s_i, y_i\}$, $i = k-1, \ldots, k-m$, satisfy the condition $y_i^T s_i > 0$, $i = k-1, \ldots, k-m$, then the matrix obtained by updating θI m-times using the BFGS updating formula and the pairs $\{s_i, y_i\}$, $i = k-1, \ldots, k-m$, can be written as

$$B_k = \theta I - W_k M_k W_k^{\mathrm{T}}, \tag{6.23}$$

where

$$W_k = [Y_k \ \ \theta S_k], \tag{6.24}$$

$$M_k = \begin{bmatrix} -D_k & L_k^{\mathrm{T}} \\ L_k & \theta S_k^{\mathrm{T}} S_k \end{bmatrix}^{-1}, \tag{6.25}$$

and where L_k and D_k are $m \times m$ matrices:

$$(L_k)_{ij} = \begin{cases} y_{k-m-1+j}^{\mathrm{T}} s_{k-m-1+i}, & i > j, \\ 0, & i \le j, \end{cases} \quad i,j = 1, \ldots, m, \tag{6.26}$$

$$D_k = \mathrm{diag}\left[y_{k-m}^{\mathrm{T}} s_{k-m}, \ldots, y_{k-1}^{\mathrm{T}} s_{k-1}\right]. \tag{6.27}$$

It is worth saying that since M_k is $2m \times 2m$ matrix and since m is chosen to be a small integer, the cost of computing the inverse in (6.25) is negligible. The compact representation (6.23) is very efficient in numerical computations involving the product of B_k times a vector, which often occurs in the algorithm.

In this context, as it is explained in Byrd, Nocedal, and Schnabel (1994), there is a similar representation of the inverse limited memory BFGS matrix H_k that approximates the inverse Hessian matrix:

$$H_k = \frac{1}{\theta} I + \bar{W}_k \bar{M}_k \bar{W}_k^{\mathrm{T}}, \tag{6.28}$$

where

$$\bar{W}_k = \begin{bmatrix} \frac{1}{\theta} Y_k & S_k \end{bmatrix}, \tag{6.29}$$

$$\bar{M}_k = \begin{bmatrix} 0 & -R_k^{-1} \\ -R_k^{-\mathrm{T}} & R_k^{-\mathrm{T}}\left(D_k + \frac{1}{\theta} Y_k^{\mathrm{T}} Y_k\right) R_k^{-1} \end{bmatrix}, \tag{6.30}$$

$$(R_k)_{ij} = \begin{cases} y_{k-m-1+j}^{\mathrm{T}} s_{k-m-1+i}, & i \le j, \\ 0, & i > j. \end{cases} \tag{6.31}$$

Since the bounds on variables possibly prevent the line search from satisfying the second Wolfe condition (6.20), there is no guarantee that the curvature condition $y_k^{\mathrm{T}} s_k > 0$ always holds. Therefore, in order to maintain the positive definiteness of the limited memory BFGS matrix, if the curvature condition $y_k^{\mathrm{T}} s_k > \varepsilon \|y_k\|^2$ is not satisfied for a small positive constant ε, then the correction pairs (s_k, y_k) are discarded.

The Generalized Cauchy Point A very important element in the frame of the L-BFGS-B algorithm is computing the generalized Cauchy point as the first local minimizer of the quadratic model along the piecewise linear path obtained by projecting points along the steepest descent direction, $x_k - tg_k$, onto the feasible domain. For this, define $x^0 = x_k$ and drop the index k, so that x, g, and B stand for x_k, g_k, and B_k, respectively. However, subscripts are used to denote the component of a vector; for example, g_i is the ith component of g. Superscripts will be used to represent iterates during the piecewise search for the Cauchy point.

To define the breakpoints in each coordinate direction, the following elements are computed:

$$t^i = \begin{cases} \left(x_i^0 - u_i\right)/g_i, & g_i < 0, \\ \left(x_i^0 - l_i\right)/g_i, & g_i > 0, \\ \infty, & g_i = 0, \end{cases} \quad i = 1, \ldots, n, \tag{6.32}$$

and sorted in increasing order to get the ordered set $\{t^j : t^j \le t^{j+1}, \ j = 1, \ldots, n\}$. In the following, a search along $P_X(x^0 - tg, l, u)$ is done, thus a piecewise linear path expressed as:

$$x_i(t) = \begin{cases} x_i^0 - tg_i, & t \le t^i, \\ x_i^0 - t^i g_i, & t > t^i, \end{cases} \quad i = 1, \ldots, n. \tag{6.33}$$

being obtained. Now, suppose that the interval $[t^{j-1}, t^j]$ is examined. Let us define the $(j-1)$-th breakpoint as $x^{j-1} = x(t^{j-1})$, such that on $[t^{j-1}, t^j]$ we have $x(t) = x^{j-1} + \Delta t d^{j-1}$, where $\Delta t = t - t^{j-1}$ and:

$$d_i^{j-1} = \begin{cases} -g_i, & t^{j-1} < t^i, \\ 0, & t^{j-1} \ge t^i, \end{cases} \quad i = 1, \ldots, n. \tag{6.34}$$

With this notation, on the line segment $[x(t^{j-1}), x(t^j)]$, the quadratic (6.16) can be written as

$$\begin{aligned} m(x) &= f + g^T\left(x - x^0\right) + \frac{1}{2}\left(x - x^0\right)^T B\left(x - x^0\right) \\ &= f + g^T\left(z^{j-1} + \Delta t d^{j-1}\right) + \frac{1}{2}\left(z^{j-1} + \Delta t d^{j-1}\right)^T B\left(z^{j-1} + \Delta t d^{j-1}\right), \end{aligned} \tag{6.35}$$

where

$$z^{j-1} = x^{j-1} - x^0. \tag{6.36}$$

Therefore, on the line segment $[x(t^{j-1}), x(t^j)]$, $m(x)$ can be written as a quadratic in Δt:

$$\begin{aligned}\widehat{m}(\Delta t) &= \left(f + g^{\mathrm{T}} z^{j-1} + \frac{1}{2}\left(z^{j-1}\right)^{\mathrm{T}} B z^{j-1}\right) + \left(g^{\mathrm{T}} d^{j-1} + \left(d^{j-1}\right)^{\mathrm{T}} B z^{j-1}\right)\Delta t \\ &\quad + \frac{1}{2}\left(\left(d^{j-1}\right)^{\mathrm{T}} B d^{j-1}\right)\Delta t^2 \\ &= f_{j-1} + f'_{j-1}\Delta t + \frac{1}{2} f''_{j-1}\Delta t^2,\end{aligned} \tag{6.37}$$

where

$$f_{j-1} = f + g^{\mathrm{T}} z^{j-1} + \frac{1}{2}\left(z^{j-1}\right)^{\mathrm{T}} B z^{j-1}, \tag{6.38}$$

$$f'_{j-1} = g^{\mathrm{T}} d^{j-1} + \left(d^{j-1}\right)^{\mathrm{T}} B z^{j-1}, \tag{6.39}$$

$$f''_{j-1} = \left(d^{j-1}\right)^{\mathrm{T}} B d^{j-1}. \tag{6.40}$$

From the equation $d\widehat{m}(\Delta t)/d\Delta t = 0$, we get $\Delta t^* = -f'_{j-1}/f''_{j-1}$. Since B is positive definite, it follows that this represents a minimzier, provided $t^{j-1} + \Delta t^*$ lies on $[t^{j-1}, t^j)$. Otherwise, the generalized Cauchy point lies at $x(t^{j-1})$ if $f'_{j-1} \geq 0$ and beyond or at $x(t^j)$ if $f'_{j-1} < 0$.

If, after exploring the interval $[t^{j-1}, t^j]$, the generalized Cauchy point has not been found, we set

$$x^j = x^{j-1} + \Delta t^{j-1} d^{j-1}, \quad \Delta t^{j-1} = t^j - t^{j-1}, \tag{6.41}$$

and update the directional derivatives f'_j and f''_j as the search moves to the next interval. As in Byrd, Lu, Nocedal, and Zhu (1995), let us assume that only one variable becomes active at t^j, and let us denote its index by b. Then, $t_b = t^j$ and we zero out the corresponding component of the search direction as

$$d^j = d^{j-1} + g_b e_b, \tag{6.42}$$

where e_b is the b-th unit vector. From (6.35) and (6.41) it follows that:

$$z^j = z^{j-1} + \Delta t^{j-1} d^{j-1}. \tag{6.43}$$

Therefore, using (6.39, 6.40, 6.42, and 6.43), we get

$$\begin{aligned}f'_j &= g^{\mathrm{T}} d^j + \left(d^j\right)^{\mathrm{T}} B z^j \\ &= g^{\mathrm{T}} d^{j-1} + g_b^2 + \left(d^{j-1}\right)^{\mathrm{T}} B z^{j-1} + \Delta t^{j-1}\left(d^{j-1}\right)^{\mathrm{T}} B d^{j-1} + g_b e_b^{\mathrm{T}} B z^j \\ &= f'_{j-1} + \Delta t^{j-1} f''_{j-1} + g_b^2 + g_b e_b^{\mathrm{T}} B z^j\end{aligned} \tag{6.44}$$

and

$$
\begin{aligned}
f_j'' &= \left(d^j\right)^{\mathrm{T}} B d^j \\
&= \left(d^{j-1}\right)^{\mathrm{T}} B d^{j-1} + 2 g_b e_b^{\mathrm{T}} B d^{j-1} + g_b^2 e_b^{\mathrm{T}} B e_b \\
&= f_{j-1}'' + 2 g_b e_b^{\mathrm{T}} B d^{j-1} + g_b^2 e_b^{\mathrm{T}} B e_b .
\end{aligned} \tag{6.45}
$$

As we can see, the only expensive computations in (6.44) and (6.45) are $e_b^{\mathrm{T}} B z^j$, $e_b^{\mathrm{T}} B d^{j-1}$, and $e_b^{\mathrm{T}} B e_b$, which can require $O(n)$ operations since B is a dense limited memory matrix. Therefore, it seems that the computation of the generalized Cauchy point could require $O(n^2)$ operations, since in the worst case n segments of the piecewise linear path can be examined. For large-scale problems, this cost would be prohibitive. However, using the limited memory BFGS formula (6.23) and (6.34), the updating formulae (6.44) and (6.45) become

$$
f_j' = f_{j-1}' + \Delta t^{j-1} f_{j-1}'' + g_b^2 + \theta g_b z_b^j - g_b w_b^{\mathrm{T}} M W^{\mathrm{T}} z^j, \tag{6.46}
$$

$$
f_j'' = f_{j-1}'' - \theta g_b^2 - 2 g_b w_b^{\mathrm{T}} M W^{\mathrm{T}} d^{j-1} - g_b^2 w_b^{\mathrm{T}} M w_b, \tag{6.47}
$$

where w_b^{T} stands for the b-th row of the matrix W. The only $O(n)$ operations remaining in (6.46) and (6.47) are $W^{\mathrm{T}} z^j$ and $W^{\mathrm{T}} d^{j-1}$. However, from (6.42) and (6.43), it follows that z^j and d^j are updated at every iteration by a simple computation. Therefore, if we store the two $2m$ vectors

$$
p^j \triangleq W^{\mathrm{T}} d^j = W^{\mathrm{T}} \left(d^{j-1} + g_b e_b\right) = p^{j-1} + g_b w_b, \tag{6.48}
$$

$$
c^j \triangleq W^{\mathrm{T}} z^j = W^{\mathrm{T}} \left(z^{j-1} + \Delta t^{j-1} d^{j-1}\right) = c^{j-1} + \Delta t^{j-1} p^{j-1}, \tag{6.49}
$$

then the updating of f_j' and f_j'' by means of the expressions

$$
f_j' = f_{j-1}' + \Delta t^{j-1} f_{j-1}'' + g_b^2 + \theta g_b z_b^j - g_b w_b^{\mathrm{T}} M c^j, \tag{6.50}
$$

$$
f_j'' = f_{j-1}'' - \theta g_b^2 - 2 g_b w_b^{\mathrm{T}} M p^{j-1} - g_b^2 w_b^{\mathrm{T}} M w_b, \tag{6.51}
$$

will require only $O(m^2)$ operations. If more than one variable becomes active at t^j, then the above updating process is repeated before examining the new interval $[t^j, t^{j+1}]$.

Observe that during the computation of the generalized Cauchy point the examination of the first segment of the projected steepest descent path requires $O(n)$ operations. All the subsequent segments require only $O(m^2)$ operations, where m is the number of corrections pairs stored in the limited memory matrix. Since m is usually small, say less than 10, the cost of examining all the segments after the first one is negligible. Besides, observe that it is not necessary to keep track of the $z^j \in \mathbb{R}^n$, since only the component z_b^j corresponding to the bound that has become active is needed to update f_j' and f_j''.

Algorithm 6.3 Computation of the generalized Cauchy point

1.	*Initialization*. Consider: x, l, u, g, and $B=\theta I-WMW^T$
2.	For $i=1,\ldots,n$, compute: $t_i=\begin{cases}(x_i-u_i)/g_i, & g_i<0,\\ (x_i-l_i)/g_i, & g_i>0,\\ \infty, & g_i=0,\end{cases}$ $\quad d_i=\begin{cases}-g_i, & t_i\neq 0,\\ 0, & t_i=0\end{cases}$
3.	Compute: $F=\{i:\ t_i>0\}$, $\quad p=W^Td$, $\quad c=0$, $f'=g^Td=-d^Td$, $\quad f''=\theta d^Td-d^TWMW^Td=-\theta f'-p^TMp$, $\Delta t_{\min}=-f'/f''$, $\quad t_{old}=0$, $\quad t=\min\{t_i,\ i\in F\}$, $b=i$ such that $t_i=t$ (remove b from F), $\Delta t=t$

4.	Examination of the subsequent segments:
	While $\Delta t_{\min}\geq\Delta t$ do
	$x_b^{cp}=\begin{cases}u_b, & d_b>0,\\ l_b, & d_b<0,\end{cases}$ $\quad z_b=x_b^{cp}-x_b$, $\quad c=c+\Delta tp$,
	$f'=f'+\Delta tf''+g_b^2+\theta g_bz_b-g_bw_b^TMc$,
	$f''=f''-\theta g_b^2-2g_bw_b^TMp-g_b^2w_b^TMw_b$,
	$p=p+g_bw_b$, $\quad d_b=0$, $\quad \Delta t_{\min}=-f'/f''$,
	$t_{old}=t$, $\quad t=\min\{t_i:\ i\in F\}$,
	$b=i$ such that $t_i=t$ (remove b din F),
	$\Delta t=t-t_{old}$.
	end while

5.	Compute: $\Delta t_{\min}=\max\{\Delta t_{\min},0\}$, $\quad t_{old}=t_{old}+\Delta t_{\min}$, $x_i^{cp}=x_i+t_{old}d_i$, for all i such that $t_i\geq t$, for all $i\in F$ with $t_i=t$, remove i from F, $c=c+\Delta t_{\min}p$ ♦

In the last step of Algorithm 6.3, the $2m$-vector c is updated so that in the end $c=W^T(x^c-x_k)$. This vector will be used to initialize the subspace minimization when the primal direct method or the conjugate gradient method are used.

Once the generalized Cauchy point x^c has been computed, an approximate minimum of the quadratic model m_k over the space of free variable is determined. In Byrd, Lu, Nocedal, and Zhu (1994, 1995) three approaches to minimize the model are considered: a *direct primal method based on the Sherman-Morrison-Woodbury formula*, a *primal iterative method using the conjugate gradient method*, and a *direct dual method using Lagrange multipliers*. In all these approaches, the minimization of m_k is done by ignoring the bounds. After that, an appropriate point truncates the move so as to satisfy the bound constraints. In the direct primal approach, $n-t$ variables are fixed at their bounds at the generalized Cauchy point x^c and the quadratic model (6.18) is solved over the subspace of the remaining t free variables, starting from x^c and imposing the free variables at the corresponding bounds from (6.18).

In the algorithm L-BFGS-B, the iterations are terminated when the projected gradient is small enough, i.e.:

$$\|P_X(x_k - g_k, l, u) - x_k\|_\infty < \varepsilon_g, \tag{6.52}$$

where $\varepsilon_g > 0$ is a parameter. Observe that $P_X(x_k - g_k, l, u) - x_k$ is the projected gradient. With these, a formal description of the L-BFGS-B algorithm can be presented as follows.

Algorithm 6.4 L-BFGS-B

1.	Consider an initial point x_0, as well as an integer m that determines the number of limited memory corrections stored (s_i, y_i), $i = 1, \ldots, m$. Select the values of the parameters $\varepsilon > 0$ and $\varepsilon_g > 0$
2.	If the convergence test (6.52) is satisfied, stop
3.	Using the Algorithm 6.3, compute the generalized Cauchy point
4.	Compute the search direction d_k by either the direct primal method, the conjugate gradient method, or the dual method
5.	Using the strong Wolfe line search (6.19) and (6.20), perform a line search along d_k subject to the bounds on the problem in order to compute a step length α_k
6.	Set $x_{k+1} = x_k + \alpha_k d_k$. Compute $f(x_{k+1})$ and $\nabla f(x_{k+1})$
7.	If y_k satisfies the curvature condition $y_k^T s_k > \varepsilon \|y_k\|^2$, then the correction pair (s_k, y_k) are added to the matrices S_k and Y_k. If more than m updates are stored, delete the oldest columns from S_k and Y_k and place the new ones instead
8.	Update $S_k^T S_k$, $Y_k^T Y_k$, L_k, R_k and set $\theta = y_k^T y_k / y_k^T s_k$
9.	Set $k = k + 1$ and continue with step 2 ♦

Some more details on the L-BFGS-B algorithm are given in Byrd, Lu, Nocedal, and Zhu (1994). In the following, we present a numerical study and comparisons versus the SPG algorithm.

Numerical Study (L-BFGS-B Versus SPG)

Consider the same set of 800 simple bound optimization problems used in the previous numerical experiment on the SPG algorithm with quadratic or cubic interpolation. L-BFGS-B implements two criteria of stopping the iterations:

$$\frac{f_k - f_{k+1}}{\max\{f_k, f_{k+1}, 1\}} \le \tau\varepsilon_m \text{ or } \|P_X(x_k - g_k, l, u) - x_k\|_\infty < \varepsilon_g,$$

where τ is a factor controlling the accuracy of the solution, ε_m is the precision of the machine, and ε_g is the tolerance on the projected gradient. If one of these two criteria is satisfied, then the iterations are stopped. The numerical values of the

Table 6.2 Performance of the algorithms. L-BFGS-B versus SPG with quadratic interpolation (SPGp)

	L-BFGS-B	SPGp	=
#iter	476	128	15
#fg	481	135	3
cpu	231	238	150

Table 6.3 Performances of the algorithms. L-BFGS-B versus SPG with cubic interpolation (SPGc)

	L-BFGS-B	SPGc	=
#iter	478	116	15
#fg	473	133	3
cpu	255	252	102

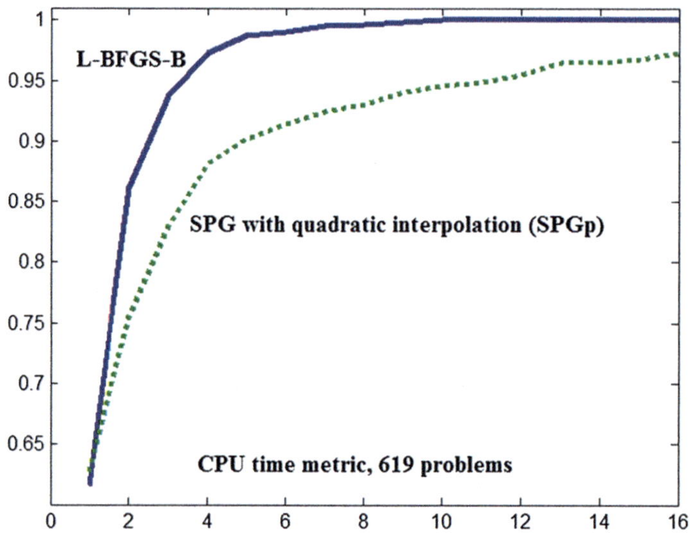

Figure 6.2 L-BFGS-B versus SPGp

parameter τ are as follows: $\tau = 1.d+12$ for small accuracy, $\tau = 1.d+7$ for medium accuracy, and $\tau = 1.d+1$ for high accuracy. In this numerical study, we have considered $\tau = 1.d+7$ and $\varepsilon_g = 1.d-5$. The number of correction pairs stored in computing the BFGS with limited memory is $m = 5$.

Table 6.2 presents the performances of L-BFGS-B versus SPG with quadratic interpolation (SPGp).

Table 6.3 presents the performances of L-BFGS-B versus SPG with cubic interpolation (SPGc).

Figures 6.2 and 6.3 show the performance profiles of L-BFGS-B ($m = 5$) versus SPG with quadratic interpolation (SPGp) and versus SPG with cubic interpolation (SPGc), respectively.

In the above Figures 6.2 and 6.3, we have the computational evidence that both algorithms are equally efficient, but subject to robustness, we see that L-BFGS-B is way more robust than both SPGp and SPGc. The algorithm SPG is more simple

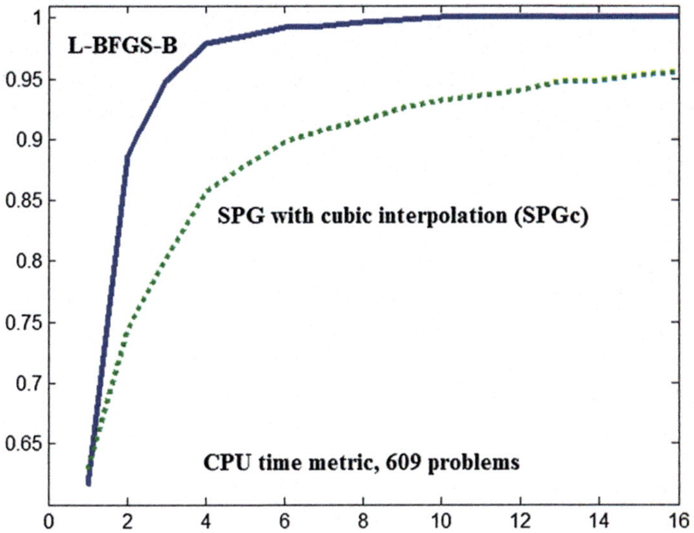

Figure 6.3 L-BFGS-B versus SPGc

than L-BFGS-B, at each iteration requiring a projection, the computation of the spectral gradient, and a nonmonotone linear search. On the other hand, L-BFGS-B is more complicated involving the computation of the generalized Cauchy point, the determination of the search direction, a linear search by strong Wolfe conditions (6.19) and (6.20), and the limited memory BFGS update.

6.6 Truncated Newton with Simple Bounds (TNBC)

This method, called TNBC, was elaborated by Nash (1984a, 1984b, 2000) and is especially developed to solve simple bound constrained optimization problems. The method is based on a line search with a classical active-set strategy for treating bounds. TNBC is a truncated Newton method (see Dembo and Steihaug (1983)) that uses the conjugate gradient projection method to obtain a new search direction. However, it computes only an approximation to the Newton direction because it is truncated before the solution to the subspace minimization problem is obtained. In other words, in the truncated Newton method, the current estimate of the solution is updated (i.e., a step is computed) by the approximate solving of the Newton system using an iterative algorithm. Therefore, the algorithm implements a double iterative method: an *outer iteration* for the nonlinear optimization problem and an *inner iteration* for the Newton system. The inner iteration is stopped or *truncated* before the solution to the Newton system is obtained. The motivation of this approach is that far away from the solution it is not necessary to solve the Newton system very accurately. Therefore, the Newton iterations can be truncated. The search direction

d_k is determined such that $\|\nabla^2 f(x_k)d_k + \nabla f(x_k)\| \leq \eta_k \|\nabla f(x_k)\|$, where $\{\eta_k\}$ is known as the "forcing sequence." The Hessian matrix with the second-order information is not given. Therefore, the Hessian vector product $B_k v$ for a given vector v required by the inner conjugate gradient algorithm is obtained by finite differencing as

$$\nabla^2 f(x_k)v \cong \frac{\nabla f(x_k + hv) \quad \nabla f(x_k)}{h},$$

where $h = \left(1 + \|x_k\|_2\right)\sqrt{\varepsilon}$ and ε is the relative machine precision. Each matrix vector product requires one gradient evaluation. The truncated Newton algorithm differs from the standard Newton approach mainly in its use of the parameter η_k which determines the accuracy with which the Newton system $\nabla^2 f(x_k)d_k = -\nabla f(x_k)$ is solved in order to obtain a search direction.

The conjugate gradient inner algorithm is preconditioned by a scaled two-step limited memory BFGS method with Powell's restarting strategy used to reset the preconditioner periodically (Nash, 1985).

The line search is based on cubic interpolation and is terminated when the strong Wolfe conditions are satisfied, as described in Gill and Murray (1979).

Consider the same set of 800 simple bound optimization problems used in the previous numerical experiments, where this time $n = 100, 200, \ldots, 1000$. Figures 6.4 and 6.5 present the performances of truncated Newton method in the implementation of Nash (1984b) versus L-BFGS-B (m=5) and SPGp and SPGc, respectively.

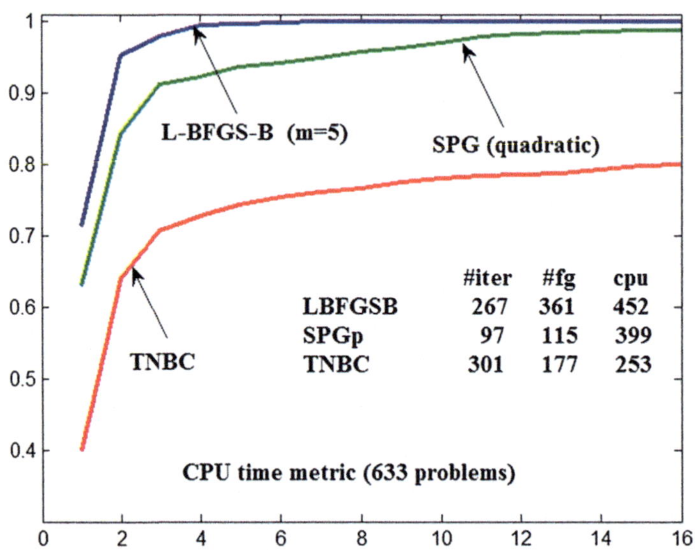

Figure 6.4 TNBC versus L-BFGS-B(m=5) and SPGp

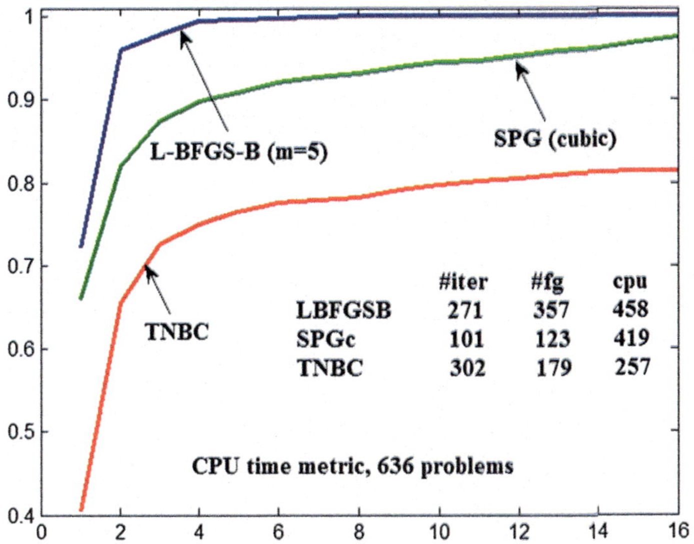

Figure 6.5 TNBC versus L-BFGS-B(m=5) and SPGc

6.7 Applications

This section is dedicated to the numerical performances of the above presented simple bounds optimization algorithms SPG and L-BFGS-B for solving five real industrial applications taken from the MINPACK-2 test problem collection (Averick, Carter, & Moré, 1991; Averick, Carter, Moré, & Xue, 1992). The mathematical models of these applications are expressed as unconstrained optimization problems. All these applications *without simple bounds* have been solved with the SCALCG algorithm, which implements a double BFGS preconditioned conjugate gradient (Andrei, 2007a, 2007b, 2007c, 2007d, 2007e, 2008a, 2010a) see also Andrei (2013c, 2014, 2016). In these numerical experiments for all the applications considered in this study, some simple bounds on the variables were introduced and solved with SPG and L-BFGS-B, respectively (Andrei, 2010b, 2010c).

Application A1 (Elastic-Plastic Torsion)
The description and the physical interpretation of the torsion problem are discussed in Glowinski (1984, pp.41–55). We follow the presentation of this problem from Averick, Carter, Moré, and Xue (1992). The elastic plastic torsion problem arises from the determination of the stress field on an infinitely long cylindrical bar. The infinite-dimensional version of this problem is of the following form:

$$\min\{q(v) \;:\; v \in K\},$$

where $q : K \to \mathbb{R}$ is a quadratic function:

$$q(v) = \frac{1}{2}\int_D \|\nabla v(x)\|^2 \mathrm{d}x - c\int_D v(x)\mathrm{d}x$$

for some constant c, and D is a bounded domain with smooth boundary. The convex set K is defined as

$$K = \left\{v \in H_0^1(D) : \ |v(x)| \le dist\left(x, \partial D\right), \ x \in D\right\},$$

where $dist(., \partial D)$ is the distance function to the boundary of D and $H_0^1(D)$ is the Hilbert space of all functions with compact support in D, such that v and $\|\nabla v\|^2$ belong to $L^2(D)$.

A finite element approximation to the torsion problem is obtained by triangulating D and then by replacing the minimization of q over $H_0^1(D)$ with the minimization of q over the set of piecewise linear functions that satisfy the constraints specified by K. The finite element approximation thus gives rise to a finite-dimensional minimization problem whose variables are the values of the piecewise linear function at the vertices of the triangulation.

In Averick, Carter, Moré, and Xue (1992), a finite element approximation to a minimization problem with a quadratic q of the general form:

$$q(v) = \frac{1}{2}\int_D w_q(x)\|\nabla v(x)\|^2 \mathrm{d}x - \int_D w_l(x)v(x)\mathrm{d}x \tag{6.53}$$

where $w_q : D \to \mathbb{R}$ and $w_l : D \to \mathbb{R}$ are functions defined on the rectangle D is described. In the torsion problem, $w_q = 1$ and $w_l = c$.

Let $D = (\xi_{1,l}, \xi_{1,u}) \times (\xi_{2,l}, \xi_{2,u})$ be a rectangle in $\mathbb{R}^2$. Vertices $z_{i,j} \in \mathbb{R}^2$ for a triangulation of D are obtained by choosing the grid spacing h_x and h_y and by defining the grid points:

$$z_{i,j} = (\xi_{1,l} + ih_x, \xi_{2,l} + jh_y), \quad 0 \le i \le n_x + 1, \quad 0 \le j \le n_y + 1,$$

such that $z_{n_x+1, n_y+1} = \left(\xi_{1,u}, \xi_{2,u}\right)$. The triangulation consists of the triangular elements T_L with vertices at $z_{i,j}$, $z_{i+1,j}$, and $z_{i,j+1}$ and the triangular elements T_U with vertices $z_{i,j}, z_{i-1,j}$ and $z_{i,j-1}$.

A finite element approximation to the torsion problem is obtained by minimizing q over the space of piecewise linear functions v with values $v_{i,j}$ at $z_{i,j}$. The approximation to the integral

$$\int_D w_q(x)\|\nabla v(x)\|^2 \mathrm{d}x$$

over the element T_L is the quadratic $q_{i,j}^L(v)$, where:

$$q_{i,j}^L(v) = \mu_{i,j}\left\{\left(\frac{v_{i+1,j} - v_{i,j}}{h_x}\right)^2 + \left(\frac{v_{i,j+1} - v_{i,j}}{h_y}\right)^2\right\},$$

$$\mu_{i,j} = \frac{h_x h_y}{6}\left\{w_q(z_{i,j}) + w_q(z_{i+1,j}) + w_q(z_{i,j+1})\right\}.$$

Similarly, the approximation over the element T_U is the quadratic $q_{i,j}^U(v)$, where:

$$q_{i,j}^U(v) = \lambda_{i,j}\left\{\left(\frac{v_{i-1,j} - v_{i,j}}{h_x}\right)^2 + \left(\frac{v_{i,j-1} - v_{i,j}}{h_y}\right)^2\right\},$$

$$\lambda_{i,j} = \frac{h_x h_y}{6}\left\{w_q(z_{i,j}) + w_q(z_{i-1,j}) + w_q(z_{i,j-1})\right\}.$$

Therefore, the finite element approximation to the quadratic (6.53) leads to a quadratic programming problem of the following form:

$$\min\{q(v) : v \in \Omega\}, \tag{6.54}$$

where q is the quadratic function:

$$q(v) = \frac{1}{2}\sum\left(q_{i,j}^L(v) + q_{i,j}^U(v)\right) - h_x h_y \sum w_l(z_{i,j}) v_{i,j}. \tag{6.55}$$

Observe that in this formulation the quadratic $q_{i,j}^L$ is defined only when $0 \le i \le n_x$ and $0 \le j \le n_y$, while $q_{i,j}^U$ is defined when $1 \le i \le n_x + 1$ and $1 \le j \le n_y + 1$. Besides, note that for the torsion problem, $w_q = 1$ and $w_l = c$, and the feasible set Ω is $\Omega = \{v \in R^{n_x n_y} : |v_{i,j}| \le d_{i,j}\}$, where $d_{i,j}$ is the value of $dist(., \partial D)$ at vertices $z_{i,j}$.

Considering $D = (0, 1) \times (0, 1)$, $c = 5$, and $nx = 200$, $ny = 200$, then a minimization problem with 40000 variables is obtained. The solution of this application without simple bounds is illustrated in Figure 6.6. By introducing the simple bounds $0 \le v \le 0.01$, the algorithm SPG gives the solution presented in Figure 6.7.

Table 6.4 contains the performances of the SPG algorithm with quadratic or cubic interpolation in line search for different values of the simple bounds: $0 \le v \le 1$, $0 \le v \le 0.1$, and $0 \le v \le 0.01$. In this table, *#iter* represents the number of iterations to get the solution and *#f* and *#g* are the number of evaluations of the function and the number of evaluations of its gradient, respectively. *cpu* is the running CPU time in seconds. The number of values of the objective functions retained in nonmonotone linear search is $M = 10$. Observe that the cubic interpolation is more benefic versus the quadratic interpolation.

Table 6.5 presents the performances of the algorithm L-BFGS-B for solving this application. In this table, *#sg* represents the total number of segments explored during searching the generalized Cauchy point, while *#act* is the number of active constraints in the final generalized Cauchy point. Here $m = 5$ is the number of the pairs $\{s_i, y_i\}$, $i = 1, \ldots, m$, stored for the memoryless BFGS updating.

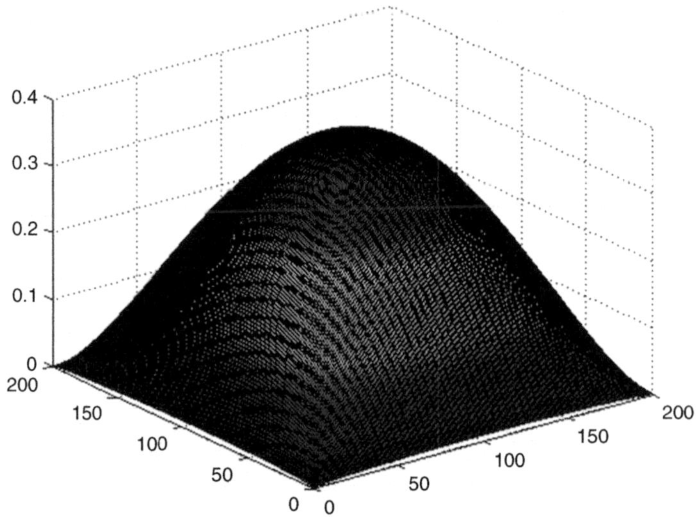

Figure 6.6 Solution of the application A1 without simple bounds $nx = 200$, $ny = 200$

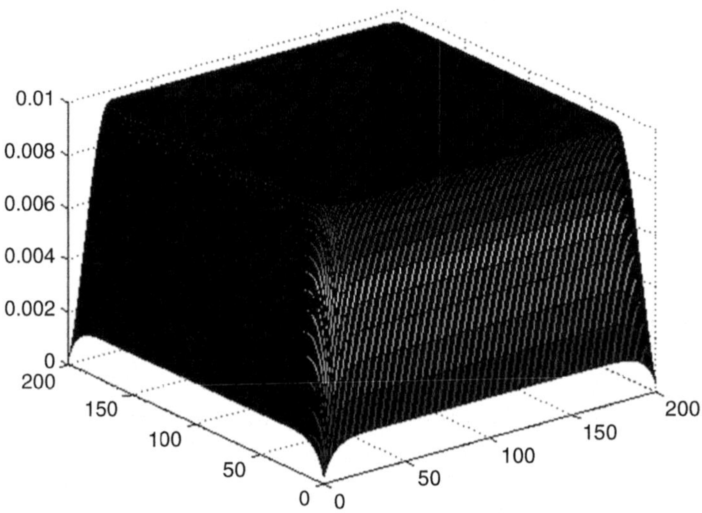

Figure 6.7 Solution of the application A1 with simple bounds $0 \le v \le 0.01$ $nx = 200$, $ny = 200$

Table 6.4 Elastic-plastic torsion. SPG. 40000 variables. $M = 10$

	Quadratic interpolation				Cubic interpolation			
	#iter	*#f*	*#g*	*cpu*	*#iter*	*#f*	*#g*	*cpu*
$0 \le v \le 1$	3311	5433	3312	113.25	2663	3600	2664	83.21
$0 \le v \le 0.1$	981	1504	982	59.49	719	940	720	40.59
$0 \le v \le 0.01$	152	190	153	7.63	118	132	119	5.66

Table 6.5 Elastic-plastic torsion. L-BFGS-B. 40000 variables. $m=5$

	#iter	#f	#sg	#act	cpu
$0 \le v \le 1$	173	184	177	0	1.89
$0 \le v \le 0.1$	117	122	1774	13036	1.082
$0 \le v \le 0.01$	41	45	62	30920	2.74

Application A2 (Pressure Distribution in a Journal Bearing)
This problem consists in determining the pressure distribution in a thin film of lubricant between two circular cylinders. The infinite-dimensional version of this problem is of the following form:

$$\min\{q(v) : v \in K\}, \tag{6.56}$$

$$q(v) = \frac{1}{2}\int_D w_q(x)\|\nabla v(x)\|^2 dx - \int_D w_l(x)v(x)dx$$

with

$$w_q(z_1, z_2) = (1 + \varepsilon\cos z_1)^3, \quad w_l(z_1, z_2) = \varepsilon \sin z_1$$

for some constant $\varepsilon \in (0,1)$ and $D = (0, 2\pi) \times (0, 2b)$, where $b > 0$ is again an arbitrary constant. The convex set K is defined as:

$$K = \left\{v \in H_0^1(D) : v \in D, \ v \ge 0\right\}.$$

A finite element approximation to the journal bearing problem is obtained as in the torsion problem. Thus, a quadratic programming problem of the form (6.54) where q is the quadratic defined by (6.55) is obtained. In the case of the journal bearing problem, $w_q(\xi_1, \xi_2) = (1 + \varepsilon\cos\xi_1)^3$ and $w_l(\xi_1, \xi_2) = \varepsilon\sin\xi_1$. The feasible set Ω is given by $\Omega = \left\{v \in \mathbb{R}^{n_x n_y} : v_{i,j} \ge 0\right\}$. Considering $b = 10$ and $\varepsilon = 0.1$, as well as a discretization $n_x \times n_y$ of the domain $D = (0, 2\pi) \times (0, 2b)$, where $nx = 200$ and $ny = 200$, then the solution of this application without simple bounds is represented in Figure 6.8. Introducing simple bounds as $0 \le v \le 1$, then the SPG algorithm gives the solution from Figure 6.9.

Tables 6.6 and 6.7 present the performances of SPG and L-BFGS-B for different simple bounds, respectively.

Numerical results for this problem are given, for example, by Lin and Cryer (1985), Cimatti and Menchi (1978), and Moré and Toraldo (1991).

Application A3 (Optimal Design with Composite Materials)
This application requires determining the placement of two elastic materials in the cross-section of a rod with maximal torsional rigidity. The formulation of this problem is given in Averick, Carter, Moré, and Xue (1992) and follows the presentation from Goodman, Kohn, and Reyna (1986).

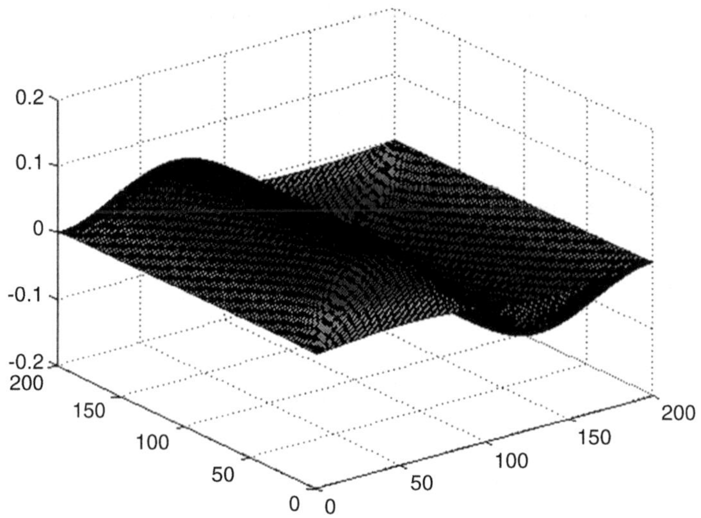

Figure 6.8 Solution of application A2 without simple bounds $nx = 200$, $ny = 200$

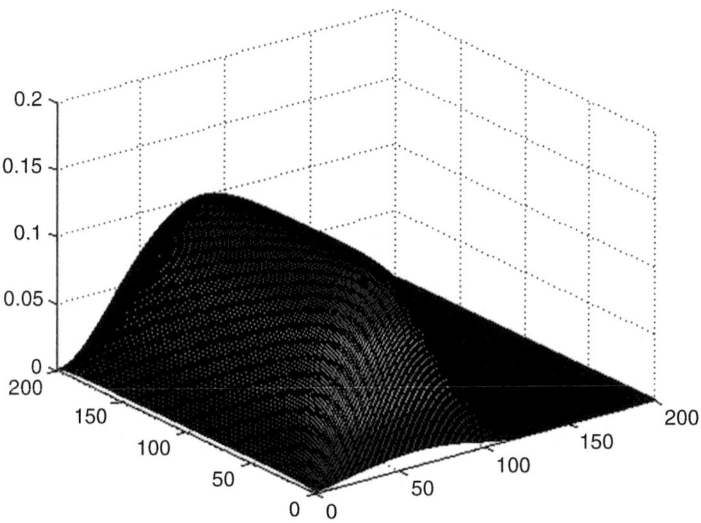

Figure 6.9 Solution of application A2 with simple bound $0 \leq v \leq 1$ $nx = 200$, $ny = 200$

Table 6.6 Pressure distribution in a journal bearing. SPG. 40000 variables. $M = 10$

	Quadratic interpolation				Cubic interpolation			
	#iter	*#f*	*#g*	*cpu*	*#iter*	*#f*	*#g*	*cpu*
$-1 \leq v \leq 1$	4179	7071	4180	140.73	2663	3553	2664	83.96
$0 \leq v \leq 1$	4964	7796	4965	152.02	4808	6301	4809	133.78
$-1 \leq v \leq 0$	3881	6179	3882	116.35	3154	4170	3155	86.99
$-0.5 \leq v \leq 0.5$	4054	6820	4055	143.09	2969	4105	2970	94.84

Table 6.7 Pressure distribution in a journal bearing. L-BFGS-B. 40000 variables. $m = 5$

	#iter	#f	#sg	#act	cpu
$-1 \le v \le 1$	623	648	623	0	7.05
$0 \le v \le 1$	2419	2429	25995	12862	2.62
$-1 \le v \le 0$	377	391	796	12904	3.84

Let $D \subset \mathbb{R}^2$ be a bounded domain and let $w < |D|$, where $|D|$ is the area of D. The solution of the optimal design problem is a subset Ω of D that solves the problem:

$$\min\{F(v,\Omega) : v \in H_0^1(D), |\Omega| = w\}, \tag{6.57}$$

where

$$F(v,\Omega) = \int_D \left\{\frac{1}{2}\mu(x)\|\nabla v(x)\|^2 + v(x)\right\} dx,$$

and $\mu(x) = \mu_1$ for $x \in \Omega$, and $\mu(x) = \mu_2$ for $x \notin \Omega$. The reciprocals of the constants μ_1 and μ_2 are the shear moduli of the elastic materials in the rod. It is assumed that $\mu_1 < \mu_2$.

Goodman, Kohn, and Reyna (1986) formulate the optimal design problem in terms of a family of problems of the form:

$$\min\{f_\lambda(v) : v \in H_0^1(D)\},$$

where $f_\lambda : H_0^1(D) \to \mathbb{R}$ is the functional

$$f_\lambda(v) = \int_D \{\psi_\lambda(\|\nabla v(x)\|) + v(x)\} dx$$

and $\psi_\lambda : \mathbb{R} \to \mathbb{R}$ is a piecewise quadratic function. In this formulation, λ is a Lagrange multiplier associated with the optimal design problem. The piecewise quadratic $\psi_\lambda : \mathbb{R} \to \mathbb{R}$ is of the form:

$$\psi_\lambda(t) = \begin{cases} \frac{1}{2}\mu_2 t^2, & 0 \le t \le t_1, \\ \mu_2 t_1\left(t - \frac{1}{2}t_1\right), & t_1 \le t \le t_2, \\ \frac{1}{2}\mu_1(t^2 - t_2^2) + \mu_2 t_1\left(t_2 - \frac{1}{2}t_1\right), & t_2 \le t, \end{cases}$$

with the breakpoints t_1 and t_2 defined by:

$$t_1 = \left(2\lambda\frac{\mu_1}{\mu_2}\right)^{1/2} \quad \text{and} \quad t_2 = \left(2\lambda\frac{\mu_2}{\mu_1}\right)^{1/2}.$$

The definition of these breakpoints implies that $\mu_1 t_2 = \mu_2 t_1$ and thus ψ is continuously differentiable. The solution of the optimum design problem considered by Averick, Carter, Moré, and Xue (1992) is the minimizing f_λ for a fixed value of λ, where $\mu_1 = 1$ and $\mu_2 = 2$, such that $t_1^2 = \lambda$ and $t_2^2 = 2\lambda$.

A finite element approximation to this problem is obtained by minimizing f_λ over the space of piecewise linear functions v with values v_{ij} at z_{ij}, where $z_{ij} \in \mathbb{R}^2$ are the vertices of a triangulation of D with grid spacings h_x and h_y. The values v_{ij} are obtained by solving the minimization problem:

$$\min\left\{\sum \left(f_{i,j}^L(v) + f_{i,j}^U(v) + h_x h_y v_{i,j}\right) : v \in \mathbb{R}^n\right\},$$

where the functions $f_{i,j}^L$ and $f_{i,j}^U$ are defined by

$$f_{i,j}^L(v) = \frac{h_x h_y}{2}\psi_\lambda\left(d_{i,j}^+(v)\right), \quad f_{i,j}^U(v) = \frac{h_x h_y}{2}\psi_\lambda\left(d_{i,j}^-(v)\right),$$

with

$$d_{i,j}^{\pm}(v) = \left\{\left(\frac{v_{i\pm1,j} - v_{i,j}}{h_x}\right)^2 + \left(\frac{v_{i,j\pm1} - v_{i,j}}{h_y}\right)^2\right\}^{1/2}.$$

Observe that in this formulation $f_{i,j}^L$ is defined only for $0 \le i \le n_x$ and $0 \le j \le n_y$, while $f_{i,j}^U$ is defined for $1 \le i \le n_x + 1$ and $1 \le j \le n_y + 1$.

Considering $\mu_1 = 1$ and $\mu_2 = 2$, then Tables 6.8 and 6.9 show the performances of SPG and L-BFGS-B, respectively. Figure 6.10 presents the solution of this application for $nx = 200$ and $ny = 200$. Figure 6.11 shows the solution of this application with simple bounds $-0.02 \le v \le 0$.

Table 6.8 Optimal design with composite materials. SPG. 40000 variables. $M = 10$

	Quadratic interpolation				Cubic interpolation			
	#iter	*#f*	*#g*	*cpu*	*#iter*	*#f*	*#g*	*cpu*
$-1 \le v \le 0$	22515	34686	22516	1259.82	20969	27964	20970	1083.20
$-0.02 \le v \le 0$	1855	2769	1856	86.21	2228	2890	2229	95.74

Table 6.9 Optimal design with composite materials. L-BFGS-B. 40000 variables. $m = 5$

	#iter	*#f*	*#sg*	*#act*	*cpu*
$-1 \le v \le 0$	646	649	744	0	7.94
$-0.02 \le v \le 0$	124	127	734	7420	1.38

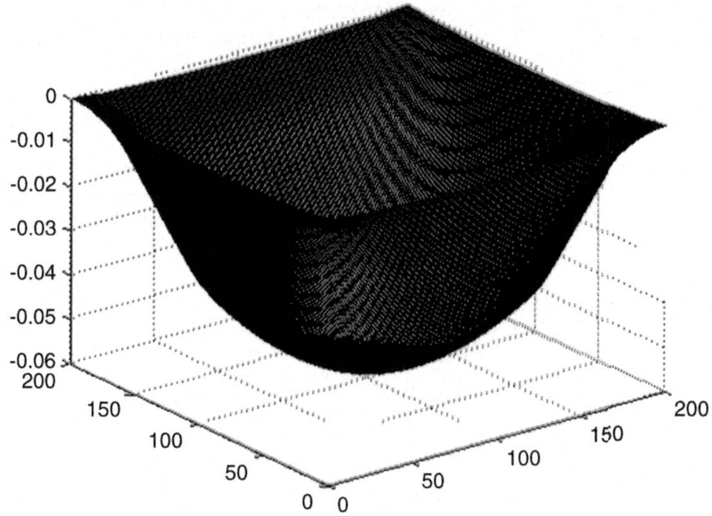

Figure 6.10 Solution of application A3 without simple bounds $nx = 200$, $ny = 200$

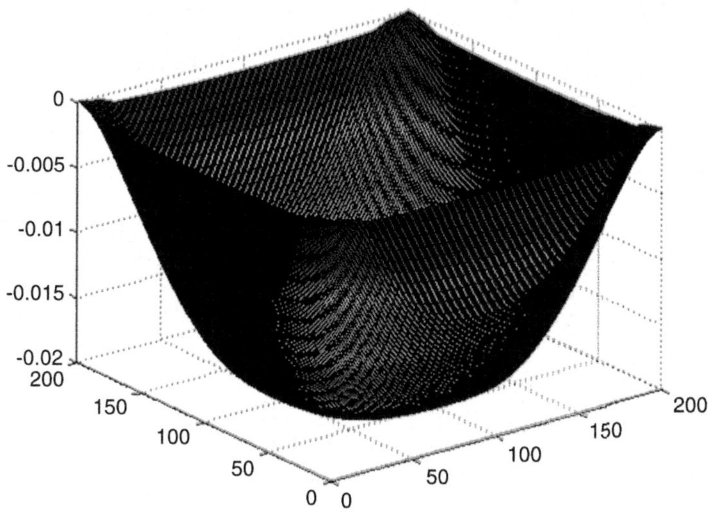

Figure 6.11 Solution of application A3 with simple bounds $-0.02 \le v \le 0$ $nx = 200$, $ny = 200$

Application A4 (Inhomogeneous Superconductors: 1-D Ginzburg-Landau)
This problem arises in the solution of the Ginzburg-Landau equations for inhomogeneous superconductors in the absence of a magnetic field. The formulation of this application is based on the work of Garner and Benedek (1990) and is presented in Averick, Carter, Moré, and Xue (1992).

The optimization problem is to minimize the Gibbs free energy as a function of the temperature. The infinite-dimensional version of this problem is of the form:

$$\min\{f(v) : v(-d) = v(d), v \in C^1[-d, d]\}, \tag{6.58}$$

where $2d$ is the width of the material and f is the Gibbs free energy function. In this problem

$$f(v) = \frac{1}{2d}\int_{-d}^{d}\left\{\alpha(\xi)|v(\xi)|^2 + \frac{1}{2}\beta(\xi)|v(\xi)|^4 + \frac{\hbar^2}{4m}|v'(\xi)|^2\right\}d\xi,$$

the functions α and β are piecewise constants for a fixed value of the temperature, $\hbar$ is Planck's constant (1.05459e-27 erg-sec), and m is the mass of the electron (9.11e-28 grams).

The functions α and β are constant in the intervals that correspond to the lead and the tin. Since in this problem the lead in the material corresponds to the interval $[-d_S, d_S]$ and the tin in the remaining part of the interval $[-d, d]$, the functions α and β are defined by:

$$\alpha(\xi) = \begin{cases} \alpha_N, & -d \le \xi \le -d_S, \\ \alpha_S, & -d_S < \xi \le d_S, \\ \alpha_N, & d_S < \xi \le d, \end{cases} \qquad \beta(\xi) = \begin{cases} \beta_N, & -d \le \xi \le -d_S, \\ \beta_S, & -d_S < \xi \le d_S, \\ \beta_N, & d_S < \xi \le d. \end{cases}$$

The constants α_S and α_N are negative, but β_S and β_N are positive.

A finite element approximation to the superconductivity problem is obtained by minimizing f over the space of the piecewise linear functions v with values v_i at t_i, where $-d = t_1 < t_2 < \cdots < t_n < t_{n+1} = d$. It is assumed that there are indices n_1 and n_2 such that $t_{n_1} = -d_S$ and $t_{n_2} = d_S$, where $1 < n_1 < n_2 < n$. This guarantees that the t_i does not straddle a point of discontinuity of the functions α and β. The values v_i are obtained by solving the following minimization problem:

$$\min\left\{\frac{1}{2d}\sum_{i=1}^{n} f_i(v) : v \in \mathbb{R}^n\right\},$$

where

$$f_i(v) = h_i\left\{\frac{\alpha_i}{3}\frac{v_{i+1}^3 - v_i^3}{v_{i+1} - v_i} + \frac{\beta_i}{10}\frac{v_{i+1}^5 - v_i^5}{v_{i+1} - v_i} + \frac{\hbar^2}{4m}\left(\frac{v_{i+1} - v_i}{h_i}\right)^2\right\},$$

with $h_i = t_{i+1} - t_i$ the length of the i-th interval and the constants α_i and β_i as the values of the functions α and β in the interval $[t_i, t_{i+1}]$. The constraint that $v(-d) = v(d)$ is taken into account by requiring that $v_{n+1} = v_1$.

Considering $d = 3.2$ Å and the temperature $T = 5$, for $n = 1000$, the solution of this application without simple bounds is presented in Figure 6.12. Introducing simple bounds on variables $0 \le v \le 0.001$, the corresponding solution given by SPG is given in Figure 6.13.

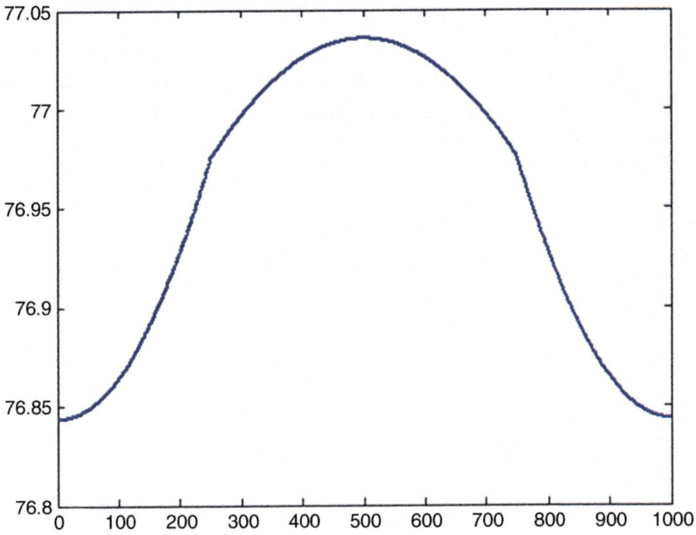

Figure 6.12 Solution of application A4 without simple bounds. $n = 1000$.

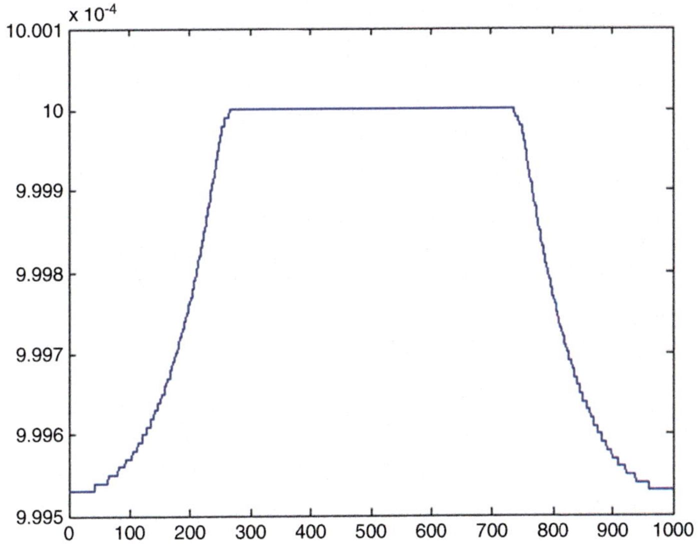

Figure 6.13 Solution of application A4 with simple bounds $0 \leq v \leq 0.001$ $n = 1000$.

Tables 6.10 and 6.11 present the performances of SPG and L-BFGS-B for solving this application with different simple bounds, respectively.

Application A5 (Steady-State Combustion)
This application is taken from Averick, Carter, Moré, and Xue (1992). The infinite-dimensional optimization problem is as follows:

Table 6.10 1-D Ginzburg-Landau problem. SPG. 1000 variables. $M=10$

	Quadratic interpolation				Cubic interpolation			
	#iter	*#f*	*#g*	*cpu*	*#iter*	*#f*	*#g*	*cpu*
$0\le v\le 10^8$	3465	6001	3466	0.81	4373	6001	4373	0.87
$0\le v\le 10$	3466	6001	3468	1.92	4374	6001	4375	2.20

Table 6.11 1-D Ginzburg-Landau problem. L-BFGS-B. 1000 variables. $m=5$

	#iter	*#f*	*#sg*	*#act*	*cpu*
$0\le v\le 10^8$	34895	36076	34895	0	6.30
$0\le v\le 10$	10278	10643	11438	391	1.74

$$\min\left\{f_\lambda(v) : v\in H_0^1(D)\right\}, \tag{6.59}$$

where $f_\lambda : H_0^1(D) \to \mathbb{R}$ is the functional

$$f_\lambda(v) = \int_D \left\{\frac{1}{2}\|\nabla v(x)\|^2 - \lambda\exp[v(x)]\right\}\mathrm{d}x,$$

and $\lambda\ge 0$ is a known parameter. This problem is the variational formulation of the boundary value problem:

$$-\Delta v(x) = \lambda\exp[v(x)],\ x\in D,$$
$$v(x) = 0,\ x\in\partial D,$$

where Δ is the Laplacian operator. Aris (1975) and Bebernes and Eberly (1989) discuss this application in the context of combustion problems.

An interesting property of the variational Bratu problem is that f_λ is unbounded below for any $\lambda>0$. This can be seen by noting that if v is any positive constant function, then $f_\lambda(\alpha v)\to -\infty$ as $\alpha\to\infty$. Another interesting property of the variational Bratu problem is that if $\lambda_{FK}>0$ is the Frank-Kamenetskii parameter, then f_λ has a unique minimizer for $\lambda\in[0,\lambda_{FK}]$, but no minimizers for $\lambda>\lambda_{FK}$. If D is the unit square, then $\lambda_{FK}\approx 6.81$, known as the Frank-Kamenetskii parameter.

A finite element approximation to this problem is obtained by minimizing f over the space of piecewise linear functions v with values v_{ij} at z_{ij}, where $z_{ij}\in\mathbb{R}^2$ are the vertices of a triangulation of D with grid spacings h_x and h_y. The values of v_{ij} are computed by solving the following minimization problem:

$$\min\left\{\sum\left(f_{i,j}^L(v)+f_{i,j}^U(v)\right) : v\in\mathbb{R}^n\right\},$$

where

$$f_{i,j}^L(v) = \frac{h_xh_y}{4}\left\{\left(\frac{v_{i+1,j}-v_{i,j}}{h_x}\right)^2 + \left(\frac{v_{i,j+1}-v_{i,j}}{h_y}\right)^2 - \lambda\mu_{i,j}^L\right\},$$

$$\mu_{i,j}^L = \frac{2}{3}\left\{\exp(v_{i,j})+\exp(v_{i+1,j})+\exp(v_{i,j+1})\right\},$$

and

$$f_{i,j}^U(v) = \frac{h_x h_y}{4}\left\{\left(\frac{v_{i-1,j} - v_{i,j}}{h_x}\right)^2 + \left(\frac{v_{i,j-1} - v_{i,j}}{h_y}\right)^2 - \lambda\mu_{i,j}^U\right\},$$

$$\mu_{i,j}^U = \frac{2}{3}\left\{\exp(v_{i,j}) + \exp(v_{i-1,j}) + \exp(v_{i,j-1})\right\}.$$

In this formulation $f_{i,j}^L$ is defined only for $0 \le i \le n_x$ and $0 \le j \le n_y$, while $f_{i,j}^U$ is defined when $1 \le i \le n_x + 1$ and $1 \le j \le n_y + 1$.

Considering $\lambda = 5$, Figure 6.14 shows the solution of this application where $nx = 200$ and $ny = 200$. Figure 6.15 presents the solution of the application with simple bounds $0 \le v \le 0.05$, where $\lambda = 5$, $nx = 200$, and $ny = 200$.

Tables 6.12 and 6.13 give the performances of SPG and of L-BFGS-B for solving this application with different simple bounds on variables, respectively.

From these tables we see that subject to the CPU computing time, L-BFGS-B, is the best algorithm, followed by SPG with cubic interpolation and then by SPG with quadratic interpolation.

Notes and References

Recent advances in simple bound constrained optimization are presented in Hager and Zhang (2006b) and Tröltzsch (2007). The algorithms for solving this class of problems use either a line-search- or a trust-region-oriented framework to guarantee global convergence to a local solution. The algorithmic components of line-search methods are subspace minimization, inexact line search, the active-set method, the gradient projection method, the projected Newton method, and the gradient projection – reduced Newton method and the gradient projection – conjugate gradient

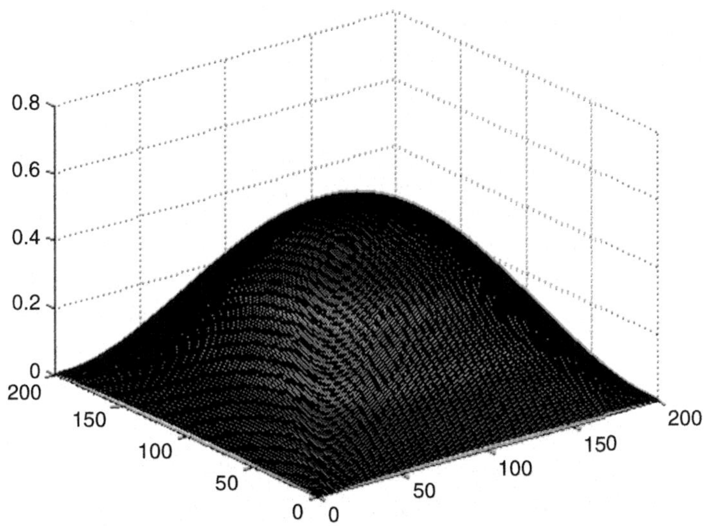

Figure 6.14 Solution of application A5 without simple bounds $nx = 200$, $ny = 200$

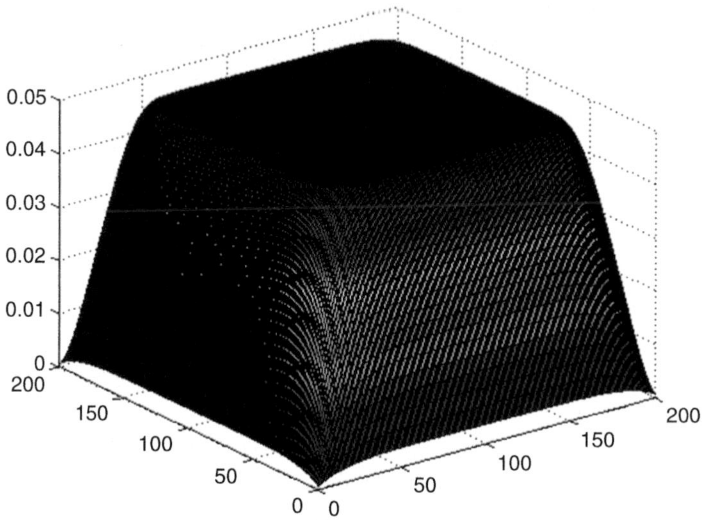

Figure 6.15 Solution of application A5 with simple bounds $0 \le v \le 0.05$ $nx = 200$, $ny = 200$

Table 6.12 Steady-state combustion. SPG. 40000 variables. $M = 10$

	Quadratic interpolation				Cubic interpolation			
	#iter	*#f*	*#g*	*cpu*	*#iter*	*#f*	*#g*	*cpu*
$0 \le v \le 1$	5134	8418	5135	421.92	5352	7236	5353	395.13
$0 \le v \le 0.2$	1843	2795	1844	137.32	1764	2286	1765	121.01
$0 \le v \le 0.1$	1140	1724	1141	81.52	665	854	666	43.45

Table 6.13 Steady-state combustion. L-BFGS-B. 40000 variables. $m = 5$

	#iter	*#f*	*#sg*	*#act*	*cpu*
$0 \le v \le 1$	360	372	424	0	5.48
$0 \le v \le 0.2$	185	196	1012	7960	2.76
$0 \le v \le 0.1$	128	130	909	14796	1.83

method. On the other hand, the algorithmic components of trust-region methods are the generalized Cauchy point and the trust-region method. For simple bound constrained optimization, there are some specialized solvers (software). LANCELOT-B (Conn, Gould, & Toint, 1992b) and TRON (Lin & Moré, 1999) use trust-region method. The other solvers like DOT (Vanderplaats, 1995), L-BFGS-B, TN-BC, and IPOPT (Wächter & Biegler, 2006) use line-search methods. IPOPT uses a primal-dual interior point filter line-search method (see Chapter 20). Another approach completely different from the above categories of solvers is based on the penalty functions method of Facchinei, Lucidi, and Palagi (2002).

Our development follows the papers of Kelley (1999), Byrd, Lu, Nocedal, and Zhu (1995) (L-BFGS-B), Birgin, Martínez, and Raydan (2000, 2001) (SPG), and Nash (1984a, 1984b, 1985) (TNBC).

Chapter 7
Penalty and Augmented Lagrangian Methods

This chapter introduces two very important concepts in constrained nonlinear optimization. These are *penalty* and *augmented Lagrangian* concepts. The idea is that both these concepts replace the original problem by a sequence of sub-problems in which the constraints are expressed by terms added to the objective function. The penalty concept is implemented in two different methods. The *quadratic penalty method* adds to the objective function a multiple of the square of the violation of each constraint and solves a sequence of unconstrained optimization sub-problems. Simple and enough intuitive, this approach has some important deficiencies. The *nonsmooth exact penalty method*, on the other hand, solves a single unconstrained optimization problem. In this approach, a popular function is the l_1 penalty function. The problem with this method is that the nonsmoothness may create complications in numerical implementations. Finally, the second concept is the *multiplier method* or the *augmented Lagrangian method*, which explicitly uses Lagrange multiplier estimates in order to avoid the ill-conditioning of the quadratic penalty method.

The best known methods for solving nonlinear optimization problems combine the penalty concept with the augmented Lagrangian in a penalty barrier with quadratic approximation of the inequality constraints (SPENBAR), the minimization of a modified augmented Lagrangian subject to linearized constraints (MINOS), or the minimization of the augmented Lagrangian subject to simple bounds on variables (LANCELOT). In this chapter we shall present the theory behind SPENBAR (Andrei, 1996a, 1996b, 1996c, 1996d, 1996e, 1998a, 2015) and MINOS (Murtagh & Saunders, 1978, 1980, 1982, 1987, 1995) algorithms described in this book.

N. Andrei, *Continuous Nonlinear Optimization for Engineering Applications in GAMS Technology*, Springer Optimization and Its Applications 121,
DOI 10.1007/978-3-319-58356-3_7

7.1 The Quadratic Penalty Method

Let us consider the equality constrained optimization problem:

$$\min\{f(x) : h_i(x) = 0, \quad i \in E\} \tag{7.1}$$

where $f : \mathbb{R}^n \to \mathbb{R}$ and $h_i : \mathbb{R}^n \to \mathbb{R}$, $i \in E \triangleq \{1, \ldots, p\}$, are continuously differentiable functions defined on $\mathbb{R}^n$. The *quadratic penalty function* $Q(x, \sigma)$ for problem (7.1) is

$$Q(x, \sigma) \triangleq f(x) + \frac{\sigma}{2} \sum_{i \in E} h_i^2(x), \tag{7.2}$$

where $\sigma > 0$ is the *penalty parameter*. Observe that if σ is increased to ∞, the constraint violations are penalized. Therefore, the idea is to consider a sequence of $\{\sigma_k\}$ with $\sigma_k \to \infty$ as $k \to \infty$, and to seek the approximate minimum x_k of $Q(x, \sigma_k)$ for each k. Since the penalty term in (7.2) is smooth, any unconstrained optimization method can be used for x_k computation. Therefore, a sequence of unconstrained minimization sub-problems $\min_{x \in \mathbb{R}^n} Q(x, \sigma_k)$ are solved for each value of the penalty parameter σ_k. Clearly, in searching for x_k, the approximate minimum x_{k-1} of the previous sub-problem may be used.

Now, let us consider the general nonlinear optimization problem:

$$\min\{f(x) : c_i(x) \geq 0, i \in I, h_i(x) = 0, i \in E\}, \tag{7.3}$$

where $f : \mathbb{R}^n \to \mathbb{R}$, $c_i : \mathbb{R}^n \to \mathbb{R}$, $i \in I \triangleq \{1, \ldots, m\}$, and $h_i : \mathbb{R}^n \to \mathbb{R}$, $i \in E \triangleq \{1, \ldots, p\}$, which contains both inequality and equality constraints. For this problem, the following quadratic penalty function may be defined:

$$Q(x, \sigma) \triangleq f(x) + \frac{\sigma}{2} \sum_{i \in E} h_i^2(x) + \frac{\sigma}{2} \sum_{i \in I} \left([c_i(x)]^-\right)^2, \tag{7.4}$$

where $[y]^- = \max\{-y, 0\}$. In this case, the quadratic penalty function may be less smooth than the original objective and constraints functions of problem (7.1).

The quadratic penalty method is based on the minimization of the quadratic penalty function in the following framework.

Algorithm 7.1 Quadratic penalty method

1.	Consider an initial starting point x_0^s. Select an initial value $\sigma_0 > 0$ for the penalty parameter and a nonnegative sequence $\{\tau_k\}$ with $\tau_k \to 0$. Set $k = 0$.
2.	Find an approximate minimum x_k of the sub-problem $\min_{x \in \mathbb{R}^n} Q(x, \sigma_k)$ starting at x_k^s, and terminate the iterations when $\|\nabla_x Q(x, \sigma_k)\| \leq \tau_k$.
3.	If a test of convergence of the method is satisfied, then stop with x_k as an approximate solution of the problem.

(continued)

4.	Choose a new penalty parameter $\sigma_{k+1} > \sigma_k$.
5.	Choose a new starting point x_{k+1}^s, set $k = k+1$ and go to step 2. ♦

Some remarks are in order:

1. The sequence of the penalty parameters $\{\sigma_k\}$ may be adaptively selected with respect to the difficulty of minimizing the penalty function at each iteration k. When minimization of $Q(x, \sigma_k)$is too expensive, then σ_{k+1} can be chosen near σ_k, for example, $\sigma_{k+1} = 1.5\sigma_k$. On the other hand, if the minimization of $Q(x, \sigma_k)$ is relatively easy to be done, then σ_{k+1} can be selected larger than σ_k, for example, $\sigma_{k+1} = 10\sigma_k$. Of course, some other schemes for enlarging $\sigma_k$may be adopted.
2. In case the problem has only equality constraints, then $Q(x, \sigma_k)$ is smooth. Therefore, the algorithms for unconstrained optimization (in step 2) can be used to find an approximate solution x_k. However, the minimization of $Q(x, \sigma_k)$ becomes more difficult as σ_k becomes larger. Near the minimum, the Hessian $\nabla^2 Q(x, \sigma_k)$ becomes ill-conditioned, and in this case the quasi-Newton or the conjugate gradient method performs very poorly.

Theorem 7.1 *Suppose that* x_k, *for any* $k \geq 0$, *is the exact global minimum of* $Q(x, \sigma_k)$ *defined in* (7.2), *determined by the* Algorithm 7.1. *Suppose that* $\sigma_k \to \infty$. *Then, every limit point* x^* *of the sequence* $\{x_k\}$ *is a global solution of the problem* (7.1).

Proof. Let $\bar{x}$ be a global solution of (7.1), that is, $f(\bar{x}) \leq f(x)$, for all x satisfying $h_i(x) = 0, i \in E$. Since x_k is a minimum of $Q(x, \sigma_k)$ for each k, we have that $Q(x_k, \sigma_k) \leq Q(\bar{x}, \sigma_k)$, which leads to the inequality:

$$f(x_k) + \frac{\sigma_k}{2} \sum_{i \in E} h_i^2(x_k) \leq f(\bar{x}) + \frac{\sigma_k}{2} \sum_{i \in E} h_i^2(\bar{x}) = f(\bar{x}). \tag{7.5}$$

By rearranging the terms of (7.5), we get:

$$\sum_{i \in E} h_i^2(x_k) \leq \frac{2}{\sigma_k} [f(\bar{x}) - f(x_k)]. \tag{7.6}$$

Suppose that x^* is a limit point of $\{x_k\}$ so that there is an infinite sequence K such that $\lim_{k \in K} x_k = x^*$. By taking the limit as $k \to \infty$, $k \in K$, on both sides of (7.6), we obtain

$$\sum_{i \in E} h_i^2(x^*) = \lim_{k \in K} \sum_{i \in E} h_i^2(x_k) \leq \lim_{k \in K} \frac{2}{\sigma_k} [f(\bar{x}) - f(x_k)] = 0,$$

where the last equality follows from the fact that $\sigma_k \to \infty$. Therefore, we have that $h_i(x^*) = 0$ for all $i \in E$, so that x^* is feasible. Additionally, by taking the limit as $k \to \infty$ for $k \in K$ in (7.5), by nonnegativity of σ_k and of each $h_i^2(x_k)$, we have that:

$$f(x^*) \le f(x^*) + \lim_{k\in K} \frac{\sigma_k}{2} \sum_{i\in E} h_i^2(x_k) \le f(\bar{x}).$$

Since x^* is a feasible point whose objective function is no larger than that of the global solution $\bar{x}$, it follows that x^* is a global solution too, as claimed. ♦

Observe that this result requires finding the global minimum of each sub-problem. This property of convergence to the global solution of (7.1) cannot be attained in general. Nocedal and Wright (2006) prove the convergence of the sequence $\{x_k\}$ when inexact minimization of $Q(x, \sigma_k)$ is allowed, as in the following theorem.

Theorem 7.2 *Suppose that the tolerances and the penalty parameters satisfy* $\tau_k \to 0$ *and* $\sigma_k \to \infty$. *Then, if a limit point* x^* *of the sequence* $\{x_k\}$ *is infeasible, then it is a stationary point of the function* $\|h(x)\|^2$. *On the other hand, if a limit point* x^* *is feasible and the constraint gradients* $\nabla h_i(x^*)$, $i \in E$, *are linearly independent, then* x^* *is a KKT point for the problem* (7.1). *For such points, for any infinite sequence K such that* $\lim_{k\in K} x_k = x^*$, *we have that*

$$\lim_{k\in K} - \sigma_k h_i(x_k) = \lambda_i^*, \tag{7.7}$$

for all $i \in E$, *where* $\lambda^* = \left[\lambda_1^*, \ldots, \lambda_p^*\right]$ *is the multiplier vector that satisfies the KKT optimality conditions* (5.21) *for the equality constrained problem* (7.1).

Proof. From (7.2) we have:

$$\nabla Q(x_k, \sigma_k) = \nabla f(x_k) + \sum_{i\in E} \sigma_k h_i(x_k) \nabla h_i(x_k). \tag{7.8}$$

Therefore, from the termination criterion in step 2 of Algorithm 7.1, it follows that:

$$\left\| \nabla f(x_k) + \sum_{i\in E} \sigma_k h_i(x_k) \nabla h_i(x_k) \right\| \le \tau_k. \tag{7.9}$$

Having in view that $\|a\| - \|b\| \le \|a + b\|$, from (7.9) we get:

$$\left\| \sum_{i\in E} h_i(x_k) \nabla h_i(x_k) \right\| \le \frac{1}{\sigma_k} [\tau_k + \|\nabla f(x_k)\|]. \tag{7.10}$$

Let x^* be a limit point of the sequence $\{x_k\}$. Then, there is a subsequence K such that $\lim_{k\in K} x_k = x^*$. When $k \to \infty$ for $k \in K$, $\tau_k + \|\nabla f(x_k)\| \to \|\nabla f(x_k)\|$. Since $\sigma_k \to \infty$, the right-hand side of (7.10) approaches zero. Therefore, from the corresponding limit of the left-hand side of (7.10), we have:

$$\sum_{i \in E} h_i(x^*) \nabla h_i(x^*) = 0. \tag{7.11}$$

Now, if the constraints gradients $\nabla h_i(x^*)$ are linearly dependent, then from (7.11) $h_i(x^*) \neq 0$. In this case x^* is a stationary point of the function $\|h(x)\|^2$. On the other hand, if the constraint gradients $\nabla h_i(x^*)$ are linearly independent at the limit point x^*, from (7.11) it follows that $h_i(x^*) = 0$ for all $i \in E$, i.e., x^* is feasible. Hence, the KKT condition (5.13b) is satisfied. We need to check (5.13a) and to show that (7.7) holds.

Let $A(x)^{\mathrm{T}} = [\nabla h_i(x)]_{i \in E}$ be the matrix of the constraint gradients and $\lambda_k = -\sigma_k h(x_k)$. From (7.9) it follows that:

$$A(x_k)^{\mathrm{T}} \lambda_k = \nabla f(x_k) - \nabla Q(x_k, \sigma_k), \quad \|\nabla Q(x_k, \sigma_k)\| \leq \tau_k. \tag{7.12}$$

For all $k \in K$ sufficiently large, the matrix $A(x_k)$ is of full rank, so $A(x_k)A(x_k)^{\mathrm{T}}$ is nonsingular. From the first relation of (7.12) we get:

$$\lambda_k = \left[A(x_k)A(x_k)^{\mathrm{T}}\right]^{-1} A(x_k)[\nabla f(x_k) - \nabla Q(x_k, \sigma_k)]. \tag{7.13}$$

Now, taking the limit as $k \in K$goes to ∞, we obtain that:

$$\lim_{k \to \infty, k \in K} \lambda_k = \lambda^* = \left[A(x^*)A(x^*)^{\mathrm{T}}\right]^{-1} A(x^*) \nabla f(x^*). \tag{7.14}$$

Therefore, taking the limit in (7.9), we get

$$\nabla f(x^*) - A(x^*)^{\mathrm{T}} \lambda^* = 0, \tag{7.15}$$

i.e., λ^* satisfies the first KKT condition (5.13a) for problem (7.1). In conclusion, x^* is a KKT point for (7.1) with unique Lagrange multiplier vector λ^*. ♦

Observe that the quantities $\sigma_k h_i(x_k)$ may be used as estimates of the Lagrange multipliers λ_i^*, $i \in E$. This is important for the analysis of the augmented Lagrangian method.

It is worth seeing the ill-conditioning in the Hessian $\nabla^2 Q(x, \sigma_k)$.This analysis is extremely important in choosing effective algorithms for the minimization in step 2 of Algorithm 7.1 as well as the linear algebra calculations at each iteration k. From (7.2) we get:

$$\nabla Q(x, \sigma_k) = \nabla f(x) + \sum_{i \in E} \sigma_k h_i(x) \nabla h_i(x). \tag{7.16}$$

$$\nabla^2 Q(x, \sigma_k) = \nabla^2 f(x) + \sum_{i \in E} \sigma_k h_i(x) \nabla^2 h_i(x) + \sigma_k A(x)^{\mathrm{T}} A(x), \tag{7.17}$$

where $A(x)^{\mathrm{T}} = [\nabla h_i(x)]_{i \in E}$. In the conditions of Theorem 7.2, from (7.7) it follows that the sum of the first two terms on the right-hand side of (7.17) is approximately equal to the Hessian of the Lagrangian function. Therefore, when x is close to the minimum point of $Q(x, \sigma_k)$,it follows that:

$$\nabla^2 Q(x, \sigma_k) \approx \nabla^2 L(x, \lambda^*) + \sigma_k A(x)^{\mathrm{T}} A(x). \tag{7.18}$$

Observe that $\nabla^2 Q(x, \sigma_k)$ is the sum of two matrices: a matrix whose elements are independent of σ_k and a matrix of rank $card(E)$ whose nonzero eigenvalues are of order σ_k. Usually, the number of constraints $card(E)$ is smaller than the number of variables n. In this case, the last term in (7.18) is singular. Therefore, the whole matrix $\nabla^2 Q(x, \sigma_k)$ has some eigenvalues approaching a constant, while the others are of order σ_k. Since $\sigma_k \to \infty$, it follows that the ill-conditioning of the matrix $\nabla^2 Q(x, \sigma_k)$ is increasing as k increases. *This is the major difficulty associated to the quadratic penalty method.* We see that this ill-conditioning is responsible for possible inaccuracy in computations of the Newton step p for the sub-problem $\min_{x \in \mathbb{R}^n} Q(x, \sigma_k)$, which is obtained by solving the linear algebraic system:

$$\nabla^2 Q(x, \sigma_k) p = -\nabla Q(x, \sigma_k). \tag{7.19}$$

Clearly, the poor conditioning of this system will lead to significant errors in the value of p, regardless of the computational technique used to solve (7.19).

7.2 Nonsmooth Penalty Method

As already seen, the quadratic penalty function is not exact. Its minimum point is generally not the same as the solution of the nonlinear optimization problem for any positive value of the penalty parameter σ. This penalty function is called *inexact*. In this context, the *exact penalty function* is introduced, in which for certain choices of its penalty parameter, a single minimization with respect to x can yield to the exact solution of the nonlinear optimization problem (Han & Mangasarian 1979). In this section we shall consider the *nonsmooth exact penalty functions* (Nocedal & Wright 2006).

For the general nonlinear optimization problem (7.3), the l_1 penalty function is defined by

$$\varphi_1(x, \sigma) = f(x) + \sigma \sum_{i \in E} |h_i(x)| + \sigma \sum_{i \in I} [c_i(x)]^-, \tag{7.20}$$

where $[y]^- = \max\{0, -y\}$. The following result is proved by Han and Mangasarian (1979).

Theorem 7.3 *Suppose that x^* is a strict local solution of the problem* (7.3) *at which the first-order necessary optimality conditions* (5.21) *hold with the Lagrange multipliers $\mu_i^*, i \in I$ and $\lambda_i^*, i \in E$. Then x^* is a local minimum of $\varphi_1(x, \sigma)$ for all $\sigma > \sigma^*$, where*:

$$\sigma^* = \max\{\max_{i\in I}\{|\mu_i^*|\}, \max_{i\in E}\{|\lambda_i^*|\}\}. \tag{7.21}$$

If, in addition, the second-order sufficient conditions (5.22) *hold and* $\sigma > \sigma^*$, *then* x^* *is a strict local minimum of* $\varphi_1(x,\sigma)$. ♦

The following definition characterizes the stationary points of φ_1, even if φ_1 is not differentiable.

Definition 7.1 *A point* $\bar{x} \in \mathbb{R}^n$ *is a stationary point of the penalty function* $\varphi_1(x,\sigma)$ *if*

$$D(\varphi_1(\bar{x},\sigma);p) \geq 0, \tag{7.22}$$

for all $p \in \mathbb{R}^n$. *Similarly,* $\bar{x}$ *is a stationary point of the measure of infeasibility*

$$g(x) = \sum_{i\in E} |h_i(x)| + \sum_{i\in I} [c_i(x)]^- \tag{7.23}$$

if $D(g(\bar{x}),p) \geq 0$ *for all* $p \in \mathbb{R}^n$.

If a point is infeasible for (7.3) *but stationary with respect to the infeasibility measure* g, *then it is an infeasible stationary point.* ♦

The following theorem shows that under some mild assumptions, the stationary points of $\varphi_1(x,\sigma)$ correspond to the KKT points of the constrained nonlinear optimization problem (7.3).

Theorem 7.4 *Suppose that* $\bar{x}$ *is a stationary point of the penalty function* $\varphi_1(x,\sigma)$ *for all* $\sigma > \bar{\sigma} > 0$, *where* $\bar{\sigma}$ *is a certain threshold. Then, if* $\bar{x}$ *is feasible for the nonlinear optimization problem* (7.3), *it satisfies the KKT optimality conditions* (5.21) *for* (7.3). *If* $\bar{x}$ *is not feasible for* (7.3), *then it is an infeasible stationary point.*

Proof. Suppose that $\bar{x}$ is feasible. Therefore:

$$D(\varphi_1(\bar{x},\sigma);p) = \nabla f(\bar{x})^{\mathrm{T}} p + \sigma \sum_{i\in E} \left|\nabla h_i(\bar{x})^{\mathrm{T}} p\right| + \sigma \sum_{i\in I\cap A(\bar{x})} \left[\nabla c_i(\bar{x})^{\mathrm{T}} p\right]^-,$$

where $A(\bar{x})$ is the set of active constraints at $\bar{x}$. Consider any direction p that satisfies: $\nabla h_i(\bar{x})^{\mathrm{T}} p = 0$ for all $i \in E$ and $\nabla c_i(\bar{x})^{\mathrm{T}} p \geq 0$ for all $i \in I \cap A(\bar{x})$. Therefore:

$$\sum_{i\in E} \left|\nabla h_i(\bar{x})^{\mathrm{T}} p\right| + \sum_{i\in I\cap A(\bar{x})} \left[\nabla c_i(\bar{x})^{\mathrm{T}} p\right]^- = 0. \tag{7.24}$$

By the stationarity assumption on $\varphi_1(\bar{x})$, it follows that

$$D(\varphi_1(\bar{x},\sigma);p) = \nabla f(\bar{x})^{\mathrm{T}} p \geq 0,$$

for all p satisfying (7.24). By Theorem A4.1 (Farkas) it follows that

$$\nabla f(\bar{x}) = \sum_{i\in A(x^*)} \bar{\lambda}_i \nabla c_i(\bar{x}),$$

for some coefficients $\overline{\lambda}_i$ with $\overline{\lambda}_i \geq 0$, $i \in I \cap A(\overline{x})$. But, this expression implies that the KKT optimality conditions (5.21) hold. The second part of the proof is very simple and is omitted in here. ♦

With these results, the following algorithm based on the classical l_1 penalty function may be presented.

Algorithm 7.2 l_1 penalty method

1.	Consider an initial starting point x_0^s. Select an initial value $\sigma_0 > 0$ for the penalty parameter and a tolerance $\tau > 0$. Set $k = 0$.
2.	Find an approximate minimum x_k of the $\varphi_1(x, \sigma_k)$ starting at x_k^s.
3.	If $g(x_k) \leq \tau$, then stop with x_k as an approximate solution of the problem.
4.	Choose a new penalty parameter $\sigma_{k+1} > \sigma_k$.
5.	Choose a new starting point x_{k+1}^s, set $k = k + 1$, and go to step 2. ♦

Due to nonsmoothness, the minimization of $\varphi_1(x, \sigma_k)$ in step 2 above is the most difficult operation in Algorithm 7.2. A practical l_1 penalty method will be presented below.

The scheme for updating the penalty parameter σ_k in step 4 is to increase it by an arbitrary constant factor, for example, $\sigma_{k+1} = 5\sigma_k$. If σ_k is too small, then many cycles of the Algorithm 7.2 may be required. On the other hand, if σ_k is too large, then the penalty function will be difficult to minimize.

As already seen, $\varphi_1(x, \sigma_k)$ is nonsmooth. The gradient of $\varphi_1(x, \sigma_k)$ is not defined at any x for which $c_i(x) = 0$, for some $i \in I$, and $h_i(x) = 0$, for some $i \in E$. In order to avoid the difficulties for minimizing $\varphi_1(x, \sigma_k)$, the strategy is to consider a *simplified model* of this function and to try to find the minimum of this model. The most direct idea is to define the model by linearizing the constraints and by replacing the objective function with a quadratic function as follows:

$$q(p; \sigma) = f(x) + \nabla f(x)^{\mathrm{T}} p + \frac{1}{2} p^{\mathrm{T}} W p + \sigma \sum_{i \in E} \left| h_i(x) + \nabla h_i(x)^{\mathrm{T}} p \right|$$
$$+ \sigma \sum_{i \in I} \left[c_i(x) + \nabla c_i(x)^{\mathrm{T}} p \right]^{-}, \tag{7.25}$$

where $W$is a symmetric matrix which contains second derivatives information about the functions of the problem (7.3). As we can see, the model (7.25) is not smooth, but it can be reformulated as a *smooth quadratic programming problem* by introducing the artificial variables r_i, s_i,$i \in E$, and t_i,$i \in I$, as follows:

$$\min_{p,r,s,t} f(x) + \nabla f(x)^{\mathrm{T}} p + \frac{1}{2} p^{\mathrm{T}} W p + \sigma \sum_{i \in E} (r_i + s_i) + \sigma \sum_{i \in I} t_i$$

subject to: (7.26)

$$h_i(x) + \nabla h_i(x)^{\mathrm{T}} p = r_i - s_i, \quad i \in E,$$
$$c_i(x) + \nabla c_i(x)^{\mathrm{T}} p \geq -t_i, \quad i \in I,$$
$$r, s, t \geq 0.$$

This problem can be solved by any standard quadratic programming solver.

In this algorithm, by using the quadratic programming (7.26), a crucial problem remains the updating of the penalty parameter σ_k. Some discussions about this aspect of the algorithm are presented by Nocedal and Wright (2006). An idea is to choose an initial value of the penalty parameter and to increase it until the feasibility has been attained. Another variant is to update the penalty parameter so that σ_k to be greater than an estimate of the Lagrange multipliers computed at x_k.

The development of the filter method (Fletcher & Leyffer, 2002), which does not require any penalty parameter, placed the nonsmooth penalty methods in a cone of penumbra. However, these methods have not been completely abandoned, partly because of their ability to handle degenerate problems (see Remark 5.8).

7.3 Augmented Lagrangian Method

Following Nocedal and Wright (2006), this section discusses the augmented Lagrangian method, known as the *method of multipliers*. Hestenes (1969) and Powell (1969) independently proposed these methods as a possibility of reducing the ill-conditioning by introducing explicit Lagrange multiplier estimates into the function to be minimized, known as the *augmented Lagrange function*. In the following, we consider the problems with equalities constraints: $\min\{f(x) : h_i(x) = 0, \ i \in E\}$. As seen in Theorem 7.2, the approximate minimum x_k of $Q(x, \sigma_k)$ does not satisfy the feasibility constraints $h_i(x) = 0, i \in E$. Instead, the constraints are perturbed (see (7.7)) so that

$$h_i(x_k) \approx -\lambda_i^* / \sigma_k, \tag{7.27}$$

for all $i \in E$. Indeed, as $\sigma_k \to \infty$, $h_i(x_k) \to 0$. However, is it possible to avoid this systematic perturbation of $Q(x, \sigma_k)$, that is to make the approximate minimizers satisfy more nearly the constraints $h_i(x) = 0$, even for moderate values of σ_k?

The augmented Lagrangian function

$$L_A(x, \lambda, \sigma) \triangleq f(x) - \sum_{i \in E} \lambda_i h_i(x) + \frac{\sigma}{2} \sum_{i \in E} h_i^2(x), \tag{7.28}$$

achieves this goal by including the explicit estimate of the Lagrange multipliers λ_i. Observe that the augmented Lagrangian (7.28) is a combination of the Lagrangian function and of the quadratic penalty function. An algorithm based on the augmented Lagrangian function at iteration k fixes the penalty parameter σ to a value $\sigma_k > 0$ and fixes λ at the current estimate λ^k. Thus, a minimization of the augmented Lagrangian, with respect to x, is performed. Let x_k be the approximate minimum of $L_A(x, \lambda^k, \sigma_k)$. Therefore, by the optimality conditions for unconstrained optimization (see Theorem 5.3), we have:

$$\nabla L_A(x_k, \lambda^k, \sigma_k) = \nabla f(x_k) - \sum_{i \in E} \left[\lambda_i^k - \sigma_k h_i(x_k)\right] \nabla h_i(x_k) = 0. \tag{7.29}$$

Now, comparing this with the optimality condition (7.15) for (7.1), it follows that

$$\lambda_i^* \approx \lambda_i^k - \sigma_k h_i(x_k), \tag{7.30}$$

for all $i \in E$. From (7.30) we get

$$h_i(x_k) \approx -\frac{1}{\sigma_k}\left(\lambda_i^* - \lambda_i^k\right),$$

for all $i \in E$. Therefore, if λ^k is close to the optimal multiplier vector λ^*, the infeasibility at x_k will be much smaller than $(1/\sigma_k)$, rather than being proportional to $(1/\sigma_k)$ as in (7.27). The relation (7.30) suggests a formula for updating the current estimate λ^k of the Lagrange multiplier vector by using the approximate minimum x_k just calculated as:

$$\lambda_i^{k+1} = \lambda_i^k - \sigma_k h_i(x_k), \quad i \in E. \tag{7.31}$$

With these theoretical developments, the augmented Lagrangian algorithm for nonlinear optimization problems with equality constraints can be presented as follows.

Algorithm 7.3 Augmented Lagrangian method: Equality constraints

1.	Consider the initial starting points x_0^s and λ^0. Select an initial value $\sigma_0 > 0$ for the penalty parameter and a tolerance $\tau_0 > 0$. Set $k = 0$.
2.	Find an approximate minimum x_k of the $L_A(x, \lambda^k, \sigma_k)$ starting at x_k^s. If $\|\nabla L_A(x_k, \lambda^k, \sigma_k)\| \leq \tau_k$, stop the iterations.
3.	If a convergence test for (7.1) is satisfied, then stop with x_k as an approximate solution of the problem.
4.	Update the Lagrange multipliers as in (7.31) to get λ^{k+1}.
5.	Choose a new penalty parameter $\sigma_{k+1} \geq \sigma_k$.
6.	Set the starting point for the new iteration as $x_{k+1}^s = x_k$.
7.	Choose the tolerance τ_{k+1}, set $k = k + 1$ and go to step 2. ♦

It was proved that the convergence of this method is assured without increasing σ indefinitely (Nocedal & Wright, 2006). In other words, the ill-conditioning of this method is reduced, thus the choice of the starting point x^s_{k+1} is less critical than in the previous algorithms based on quadratic or nonsmooth penalty methods. The tolerance τ_k could be chosen by using the infeasibility measure $\sum_{i \in E} |h_i(x_k|$ and the penalty parameter σ may be updated (increased) if the reduction in this infeasibility measure is insufficient at the current iteration (Nocedal & Wright, 2006).

Theorem 7.5 *Let x^* be a local solution of* (7.1) *at which the gradients $\nabla h_i(x^*)$, $i \in E$, are linearly independent and the second-order sufficient conditions given in* Theorem 5.13 *are satisfied for $\lambda = \lambda^*$. Then, there is a threshold $\overline{\sigma}$ such that for all $\sigma \geq \overline{\sigma}$, x^* is a strict local minimum of the augmented Lagrangian $L_A(x, \lambda^*, \sigma)$.*

Proof. We show that x^* satisfies the second-order sufficient conditions for a strict local minimum of $L_A(x, \lambda^*, \sigma)$ given in Theorem 5.6, for all σ sufficiently large, i.e.:

$$\nabla L_A(x^*, \lambda^*, \sigma) = 0, \tag{7.32}$$

$$\nabla^2 L_A(x^*, \lambda^*, \sigma) \text{ is positive definite.} \tag{7.33}$$

Now, as x^* is a local solution of (7.1) at which LICQ is satisfied, from Theorem 5.11 (see Remark 5.3) it follows that $\nabla L(x^*, \lambda^*) = 0$ and $h_i(x^*) = 0$, $i \in E$. Therefore:

$$\begin{aligned} \nabla L_A(x^*, \lambda^*, \sigma) &= \nabla f(x^*) - \sum_{i \in E} \left[\lambda_i^* - \sigma h_i(x^*)\right] \nabla h_i(x^*) \\ &= \nabla f(x^*) - \sum_{i \in E} \lambda_i^* \nabla h_i(x^*) = \nabla L(x^*, \lambda^*) = 0, \end{aligned}$$

verifies (7.32), independently of σ. Further on, to verify (7.33), consider

$$\nabla^2 L_A(x^*, \lambda^*, \sigma) = \nabla^2 L(x^*, \lambda^*) + \sigma A(x^*)^{\mathrm{T}} A(x^*),$$

where $A(x)^{\mathrm{T}} = [\nabla h_i(x)]_{i \in E}$.

If $\nabla^2 L_A(x^*, \lambda^*, \sigma)$ is not positive definite, then for each integer $k \geq 1$, we could choose a vector w_k with $\|w_k\| = 1$, such that

$$0 \geq w_k^{\mathrm{T}} \nabla^2 L_A(x^*, \lambda^*, k) w_k = w_k^{\mathrm{T}} \nabla^2 L(x^*, \lambda^*) w_k + k \|A(x^*) w_k\|_2^2, \tag{7.34}$$

and therefore:

$$\|A(x^*) w_k\|_2^2 \leq -\frac{1}{k} w_k^{\mathrm{T}} \nabla^2 L(x^*, \lambda^*) w_k \to 0, \tag{7.35}$$

as $k \to \infty$. Since the vectors $\{w_k\}$ lie in a compact set (the surface of the unit sphere), they have an accumulation point w. From (7.35) it follows that $A(x^*) w = 0$. Now, from (7.34) we have:

$$w_k^{\mathrm{T}}\nabla^2 L(x^*,\lambda^*)w_k \leq -k\|A(x^*)w_k\|_2^2 \leq 0,$$

So, by taking the limits, we have $w^{\mathrm{T}}\nabla^2 L(x^*,\lambda^*)w \leq 0$. But this inequality contradicts the second-order condition in Theorem 5.13, which states that we must have $w^{\mathrm{T}}\nabla^2 L(x^*,\lambda^*)w > 0$ for all nonzero vectors w with $A(x^*)w = 0$. Therefore, (7.33) holds for all σ sufficiently large. ♦

The following theorem, proved by Bertsekas (1999), gives the conditions under which there is a minimum of the augmented Lagrangian $L_A(x, \lambda, \sigma)$ that lies close to x^*, conditions that specify some error bounds on both x_k and λ^{k+1} solutions of the sub-problem at iteration k.

Theorem 7.6 *Suppose that the assumptions of* Theorem 7.5 *are satisfied at* x^* *and* λ^*. *Let* $\overline{\sigma}$ *be chosen as in the theorem. Then, there exist positive scalars* δ, ε, *and* M *such that the following claims hold:*

(a) *For all* λ^k *and* σ_k *satisfying*

$$\|\lambda^k - \lambda^*\| \leq \sigma_k\delta, \quad \sigma_k \geq \overline{\sigma}, \tag{7.36}$$

the problem $\min_{x\in\mathbb{R}^n} L_A\left(x, \lambda^k, \sigma_k\right)$ *subject to* $\|x - x^*\| \leq \varepsilon$ *has a unique solution* x_k. *Moreover:*

$$\|x_k - x^*\| \leq M\|\lambda^k - \lambda^*\|/\sigma_k. \tag{7.37}$$

(b) *For all* λ^k *and* σ_k *satisfying* (7.36), *we have*

$$\|\lambda^{k+1} - \lambda^*\| \leq M\|\lambda^k - \lambda^*\|/\sigma_k, \tag{7.38}$$

where λ^{k+1} *is given by* (7.31).

(c) *For all* λ^k *and* σ_k *satisfying* (7.36), *the matrix* $\nabla^2 L_A(x_k, \lambda^k, \sigma_k)$ *is positive definite and the constraint gradients* $\nabla h_i(x_k), i \in E$, *are linearly independent.* ♦

The theorem of Bertsekas (1999) is important. It illustrates some salient properties of the augmented Lagrangian approach. The bound (7.37) shows that x_k will be close to x^* if λ^k is accurate *or* if the penalty parameter σ_k is large. Therefore, in this approach of the augmented Lagrangian method, there are *two* ways of improving the accuracy of x_k, whereas in the quadratic penalty approach, there is only one: increase σ_k. The bound (7.37) shows that the accuracy of the multipliers can be increased by choosing a sufficiently large value of σ_k. Finally, the last observation of the theorem shows that under the given conditions, the second-order sufficient optimality conditions for the unconstrained minimization are satisfied for the kth sub-problem. Therefore, very good numerical performances can be expected by applying any standard unconstrained minimization technique.

Example 7.1 Let us solve the following optimization problem by using the augmented Lagrangian method:

$$\min f(x) \equiv x_1^2 + 8x_2^2$$

subject to:

$$h_1(x) \equiv x_1 + x_2 - 4 = 0.$$

Observe that the augmented Lagrange function is:

$$L_A(x, \lambda, \sigma) = x_1^2 + 8x_2^2 - \lambda(x_1 + x_2 - 4) + \sigma(x_1 + x_2 - 4)^2.$$

Consider the initial point $x_0 = [0 \;\; 4]^T$, the initial values for the penalty parameter $\sigma_0 = 1$ and the Lagrange multiplier $\lambda^0 = 0$. By using the Newton method for minimizing the augmented Lagrange function in step 2 of the Algorithm 7.3, we get the results from Table 7.1.

As a comparison, in Table 7.2 we present the optimization process for solving the problem by using the quadratic penalty method.

This example shows that the inclusion of the Lagrange multiplier term in the augmented Lagrangian $L_A(x, \lambda, \sigma)$ represents a significant improvement over the quadratic penalty method.

Table 7.1 Augmented Lagrangian method

k	σ	λ	x_1	x_2	$f(x_k)$	$h_1(x_k)$
0	1	0	0	4	128	0
1	10	3.767400	1.882353	0.235294	3.986159	−1.88235
2	100	6.837930	3.418960	0.427370	13.15051	−0.153661
3	1000	7.108704	3.554350	0.444290	14.21260	−0.00135
4	10,000	7.1111	3.555555	0.444444	14.22222	−0.120e-5

Table 7.2 Quadratic penalty method

k	σ	x_1	x_2	$f(x_k)$	$h_1(x_k)$
0	1	0	4	128	0
1	10	1.882353	0.235294	3.986151	−1.882350
2	100	3.265306	0.408163	11.99500	−0.326530
3	1000	3.524229	0.440528	13.97271	−0.035242
4	10,000	3.552390	0.444049	14.19697	−0.003552
5	100,000	3.555240	0.444404	14.21969	−0.000355
6	1,000,000	3.555524	0.444440	14.22197	−0.355e-4
7	10,000,000	3.555555	0.444444	14.22222	−0.355e-5

7.4 Criticism of the Penalty and Augmented Lagrangian Methods

Let us see the fundamentals of the penalty and of the augmented Lagrangian methods. For this, we shall consider the nonlinear optimization problem with equality constraints:

$$\min\{f(x) : h_i(x) = 0, \quad i \in E\}, \tag{7.39}$$

where $f : \mathbb{R}^n \to \mathbb{R}$ and $h_i : \mathbb{R}^n \to \mathbb{R}$, $i \in E \triangleq \{1, \ldots, p\}$, are continuously differentiable on $\mathbb{R}^n$. Define $h(x) = [h_1(x), \ldots, h_p(x)]$. A local solution (x^*, λ^*) satisfies the first-order KKT necessary optimality conditions:

$$\nabla h(x)^{\mathrm{T}}\lambda = \nabla f(x), \tag{7.40a}$$

$$h(x) = 0, \tag{7.40b}$$

where $\lambda \in \mathbb{R}^p$ is the vector of the Lagrange multipliers.

It is common knowledge that in everyday life, we always try to find a balance (an equilibrium) between what is desirable (to minimize the objective function) and what is legally achievable (to satisfy the constraints). This multiple objective point of view represents the essence of the penalty methods. For example, the quadratic penalty method combines these two requirements and solves a sequence of unconstrained optimization sub-problems:

$$\min_{x \in \mathbb{R}^n} Q(x, \sigma) = f(x) + \frac{1}{2}\sigma \|h(x)\|^2, \tag{7.41}$$

parameterized by the penalty parameter $\sigma > 0$. Thus, a trajectory of points $x^*(\sigma)$ is obtained as solutions of (7.41), as well as an increasing sequence of penalty parameters. We must let σ become large in order to achieve near feasibility, but at least the penalty function is *smooth*. Therefore, we may apply the Newton method or the quasi-Newton methods for solving (7.41). Let us introduce the Lagrange function:

$$L(x, \lambda) = f(x) - \lambda^{\mathrm{T}} h(x). \tag{7.42}$$

The derivatives of the quadratic penalty function are

$$\nabla Q(x, \sigma) = \nabla f(x) + \nabla h(x)^{\mathrm{T}}\lambda, \tag{7.43}$$

$$\nabla^2 Q(x, \sigma) = \nabla^2 f(x) + \sum_{i \in E} \lambda_i \nabla^2 h_i(x) + \sigma \nabla h(x)^{\mathrm{T}} \nabla h(x), \tag{7.44}$$

where $\lambda_i = \sigma h_i(x), i \in E$, and $\lambda = [\lambda_1, \ldots, \lambda_p]$.

Observe that for $x^*(\sigma)$, $\nabla Q(x^*(\sigma),\sigma)=0$. Defining $\lambda^*(\sigma)=\sigma h(x^*(\sigma))$, then we see that $(x^*(\sigma),\lambda^*(\sigma))$ is the *exact solution* of a perturbed form of problem (7.39):

$$\min\{f(x) : h(x) = h(x^*(\sigma))\}. \tag{7.45}$$

If the Jacobian matrix $\nabla h(\bar{x})$ has full row rank and $\bar{x}$ is a unique local minimizer for (7.39), we can show that the full Hessian $\nabla^2 Q(x,\sigma)$ is positive definite at $(x^*(\sigma),\lambda^*(\sigma))$ for sufficiently large σ. Thus, the penalty function is *convex* for large σ, and the minimum $x^*(\sigma)$ exists. Therefore, the problem (7.41) can be very well solved by the Newton method.

We see that the Newton method for minimizing (7.41) would generate a search direction d by solving the linear system $\nabla^2 Q(x,\sigma)d= -\nabla Q(x,\sigma)$, i.e.,

$$\left(\nabla^2 L(x,\lambda) + \sigma\nabla h(x)^{\mathrm{T}}\nabla h(x)\right)d = -\left(\nabla f(x) + \sigma\nabla h(x)^{\mathrm{T}} h(x)\right), \tag{7.46}$$

where $\nabla^2 L(x,\lambda)$ is defined with $\lambda=\sigma h(x)$. The difficulty with this approach is that the system (7.46) is ill-conditioned for large σ (assuming $p<n$). This is one reason why the quadratic penalty approach (enough convenient as smoothness) proved unsuccessful for solving nonlinear optimization problems. A solution for eliminating this deficiency was given by Gould (1986), who introduced the vector $q=\sigma(h(x)+\nabla h(x)d)$ at the current point x. With this, the Newton system (7.46) is equivalent to

$$\begin{bmatrix} \nabla^2 L(x,\lambda) & \nabla h(x)^{\mathrm{T}} \\ \nabla h(x) & -\frac{1}{\sigma}I \end{bmatrix}\begin{bmatrix} d \\ q \end{bmatrix} = -\begin{bmatrix} \nabla f(x) \\ h(x) \end{bmatrix}, \tag{7.47}$$

which contains no large numbers and may be preferable for computing the search direction d. If (x,λ) is close to a local optimum (x^*,λ^*) and σ is large, any ill-conditioning in (7.47) reflects the sensitivity of (x^*,λ^*) to perturbations in the data of the problem.

Unfortunately, although d can be reliably computed from (7.47) when σ is large, this does not save the quadratic penalty method. When $h(x)$ is not very small, then d leads away from the linearization of the constraints $h(x)=0$ at the current point x, and therefore the Newton method is likely to be too slow. It is obvious that the algebraic trick used does not eliminate the deficiency of the quadratic Newton approach.

Let us now continue to study the problem (7.39) with equality constraints, bearing in mind the difficulties encountered with the quadratic penalty method when σ becomes very large. Let (x^*,λ^*) be a local minimizer and assume that the Jacobian matrix $\nabla h(x)$ has full row rank at x^*, i.e., $\mathrm{rank}(\nabla h(x^*))=p$. In this minimum point, the gradient of the Lagrange function must be zero. Therefore, the required solution (x^*,λ^*) is a *stationary point* of the Lagrangian. In general, we cannot find x^* by minimizing $L(x,\lambda)$ as a function of x, even if we set $\lambda=\lambda^*$.

The second-order optimality condition for (x^*, λ^*) to be an isolated local minimizer is that the Hessian of the Lagrange function $\nabla^2 L(x^*, \lambda^*)$ should be positive definite within the null space of $\nabla h(x^*)$. That is, the Hessian should satisfy the condition $z^T \nabla^2 L(x^*, \lambda^*) z > 0$ for all nonzero vectors z satisfying $\nabla h(x^*)^T z = 0$. The following result, proved by Debreu (1952) on quadratic forms, is relevant.

Theorem 7.7 *(Debreu). Let H be an $n \times n$ symmetric matrix and J an $m \times n$ matrix with $m \leq n$. If $z^T H z > 0$ for every nonzero z satisfying $Jz = 0$, then for all σ sufficiently large $H + \sigma J^T J$ is positive definite.* ♦

This result suggests that we should add to the Lagrangian function a term whose Hessian is $\sigma \nabla h(x)^T \nabla h(x)$. It is exactly what we already did when we introduced the augmented Lagrangian:

$$L_A(x, \lambda, \sigma) = f(x) - \lambda^T h(x) + \frac{1}{2}\sigma \|h(x)\|^2. \tag{7.48}$$

This may be looked upon as a modification to the Lagrangian *or* as a shifted quadratic penalty function. For a given λ and σ, the derivatives of $L_A(.)$ are

$$\nabla L_A(x, \lambda, \sigma) = \nabla f(x) - \nabla h(x)^T \widehat{\lambda}, \tag{7.49a}$$

$$\nabla^2 L_A(x, \lambda, \sigma) = \nabla^2 f(x) - \sum_{i \in E} \widehat{\lambda}_i \nabla^2 h_i(x) + \sigma \nabla h(x)^T \nabla h(x), \tag{7.49b}$$

where $\widehat{\lambda} = \lambda + \sigma h(x)$. Observe that (7.49b) illustrates Debreu's theorem.

As already seen, the augmented Lagrangian method for solving (7.39) proceeds by choosing λ and σ as *judiciously* as possible, and then by minimizing $L_A(x, \lambda, \sigma)$ with respect to x. The resulting x, solution of this minimizing problem, is used to choose a new λ and σ, and the process repeats itself. The auxiliary vector $\widehat{\lambda}$ simplifies the above process and proves to be useful in its own right.

Observe that if σ is reasonably large, minimizing L_A will tend to make $\|h(x)\|$ small, *for any values* of λ. Also, the Hessian $\nabla^2 L_A(x, \lambda, \sigma)$ will tend to have positive curvature in the null space of $\nabla h(x)$. Therefore, the problem $\min L_A(x, \lambda, \sigma)$ has a solution. On the other hand, if λ is close to λ^*, since the minimization of L_A makes $\|\nabla L_A\|$ small, we see that $(x, \lambda) \approx (x, \lambda^*)$ almost satisfies (7.40a). Moreover, if it also happens that $\|h(x)\|$ is small (because σ is large enough), then (x, λ) will almost satisfy (7.40b) as well. Therefore, the strategy is to check that $\|h(x)\|$ is suitably small after each approximate minimization of L_A. If so, λ is updated as $\lambda = \lambda - \sigma h(x)$. If not, σ is *judiciously* increased and λ remains the same. Under *favorable conditions*, we hope that $(x, \lambda) \to (x^*, \lambda^*)$ before σ becomes too large (Saunders, 2015a).

This weakness of the augmented Lagrangian method (judicious selection of σ, favorable conditions, etc.) determined some modifications of it. Two algorithms based on the modifications of the augmented Lagrangian are presented in this book. One is SPENBAR (Breitfeld & Shanno, 1994a, 1994b, 1994c; Andrei, 1996a,

1996b, 1996c, 1996d, 1996e, 1998a, 2015). In it, the inequality constraints are imbedded into a modified logarithmic penalty function. The other one is MINOS (Murtagh & Saunders, 1978, 1980, 1982, 1987, 1995). It uses a combination of successive linear programming with a modification of the augmented Lagrangian.

Notes and References

The penalty and the augmented Lagrangian are two very important concepts in nonlinear optimization. The content of this chapter is based on the books: (Nocedal & Wright, 2006; Bertsekas, 1982b). The criticism of penalty and of the augmented Lagrangian methods are taken from (Saunders, 2015a; Andrei, 2015). The quadratic penalty function was first proposed by Courant (1943). The augmented Lagrangian was proposed independently by Hestenes (1969) and Powell (1969). The l_1 penalty method was developed by Fletcher. All these concepts are implemented in different combinations in very efficient nonlinear optimization packages. For example, the l_1 penalty method that uses linear programming sub-problems has been implemented as part of KNITRO software package (see Chapter 14). The best implementation of the augmented Lagrangian methods is MINOS (Murtagh & Saunders, 1995) (see Chapter 9). We also mention LANCELOT (Conn, Gould, & Toint, 1992b). Both MINOS and KNITRO are imbedded in the GAMS technology. In LANCELOT, simple bounds are treated explicitly, and all the other constraints are converted to equations and incorporated into an augmented Lagrangian function. Although this approach proves to be robust (Conn, Gould, & Toint, 1992b, 1996), it has a number of drawbacks. One of these is that the augmentation may not be the ideal way to treat linear constraints. A more attractive approach is to handle all linear constraints explicitly (Conn, Gould, Sartenaer, & Toint 1996). We mention that the methods that use the linearly constrained sub-problems at their heart are described, for example, by Rosen and Kreuser (1972), Robinson (1972, 1974), Murtagh and Saunders (1978).

Chapter 8
A Penalty-Barrier Algorithm: SPENBAR

Let us consider the following general nonlinear optimization problem:

$$\min f(x)$$

subject to: (8.1)

$$\begin{aligned} & c_i(x) \geq 0, \quad i \in I_c, \\ & c_i(x) = 0, \quad i \in E, \\ & x_j \geq l_j, \quad j \in I_l, \\ & x_j \leq u_j, \quad j \in I_u, \end{aligned}$$

where $x \in \mathbb{R}^n, E \triangleq \{1, \ldots, m_e\}$ and $I_c \triangleq \{1, \ldots, m\}$. The functions $c_i : \mathbb{R}^n \to \mathbb{R}$, $i \in I_c \cup E$, are assumed to be twice continuously differentiable on $\mathbb{R}^n$. I_l, $I_u \subseteq \{1, \ldots, n\}$. To simplify the presentation of the algorithm, the simple bounds on the variables are also denoted $c_i(x)$. Define I_{sb} as the set of indices such that for all $j \in I_l \cup I_u$ there is an $i \in I_{sb}$ with the property:

$$c_i(x) = x_j - l_j \geq 0, \text{ for } j \in I_l, \text{or} \tag{8.2a}$$

$$c_i(x) = u_j - x_j \geq 0, \text{ for } j \in I_u. \tag{8.2b}$$

The set of all inequality constraints including the simple bounds is denoted as $I = I_c \cup I_{sb}$. Let $|I| = p$, i.e., I has p elements. Suppose that:

(*i*) The domain $X_B = \{x \in \mathbb{R}^n : l_j \leq x_j, j \in I_l, \ x_j \leq u_j, j \in I_u\}$, which is a compact set, has a nonempty interior.
(*ii*) The functions $f(x)$, $c_i(x)$, $i \in E \cup I_c$ are twice continuously differentiable on X_B.

N. Andrei, *Continuous Nonlinear Optimization for Engineering Applications in GAMS Technology*, Springer Optimization and Its Applications 121,
DOI 10.1007/978-3-319-58356-3_8

The main idea of the penalty-barrier method to find a solution for problem (8.1) is to solve a sequence of unconstrained minimization sub-problems where the solution x_k of the sub-problem at iteration k is considered as an initial point for the sub-problem at the next iteration. The suggested algorithm *combines* the augmented Lagrangian with a penalty function (a composite function which includes the augmented Lagrangian and the log-barrier function) in which the parameters are updated in such a way as to obtain a KKT point for problem (8.1). The constraints are treated separately as follows.

The equality constraints, $c_i(x) = 0$, $i \in E$, are imbedded into the *augmented Lagrangian*. Therefore, to minimize $f(x)$ subject to $c_i(x) = 0$, $i \in E$, the augmented Lagrangian is defined as

$$L(x, \sigma, \lambda) \triangleq f(x) + \sum_{i=1}^{m_e} \lambda_i c_i(x) + \frac{1}{2\sigma} \sum_{i=1}^{m_e} c(x)^2, \tag{8.3}$$

where λ_i, $i = 1, \ldots, m_e$, are estimates of the Lagrange multipliers and $\sigma > 0$ is the penalty parameter. The reason for introducing the augmented Lagrangian function is to make sure that the penalty parameter σ does not need to be enlarged too much as to avoid the ill-conditioning of the augmented Lagrangian. The augmented Lagrangian method, also known as the multiplier method, was proposed independently by Hestenes (1969) and Powell (1969).

The inequality constraints $c_i(x) \geq 0$, $i \in I_c$, are imbedded in a *logarithmic penalty function*. Therefore, to minimize $f(x)$ subject to $c_i(x) \geq 0$, $i \in I_c$, we consider the log-barrier function:

$$B(x, \sigma) \triangleq f(x) - \sigma \sum_{i=1}^{m} \log(c_i(x)), \tag{8.4}$$

where $\sigma > 0$ is the penalty parameter. For solving the problem $\min\{f(x) : c_i(x) \geq 0, i = 1, \ldots, m\}$, the following algorithm can be presented.

Algorithm 8.1 General barrier

1.	Select an initial feasible point x_0 and a value for the penalty parameter $\sigma_0 > 0$. Set $k = 0$.
2.	Solve the sub-problem $\min_{x \in \mathbb{R}^n} B(x, \sigma_k)$ obtaining a local solution x_{k+1}.
3.	If σ_k is small enough, then stop. Otherwise, select $\sigma_{k+1} < \sigma_k$, set $k = k + 1$, and go to step 2. ♦

Observe that when $\sigma_k \to 0$, the Hessian of $B(x_k, \sigma_k)$ is ill-conditioned. On the other hand, selection of a good value for σ_{k+1} is difficult. The first-order minimizing condition for $B(x, \sigma_k)$ is:

$$\nabla f(x) - \sum_{i=1}^{m} \frac{\sigma_k}{c_i(x)} \nabla c_i(x) = 0. \tag{8.5}$$

If $\sigma_k \to 0$, then $\sigma_k/c_i(x) \to \lambda_i$, which is the Lagrange multiplier associated to the ith inequality constraint. Therefore, the estimates of the Lagrange multipliers associated to the active constraints are computed as the ratio of two quantities, both of them tending to zero. More exactly, observe that when $1 > c_i(x) > 0$, $i \in I_c$, then $\log(c_i(x)) < 0$. Hence, the second term on the right-hand side of (8.4) implies $B(x,\sigma) \gg f(x)$ when any of the constraint functions is small and positive. Suppose that the inequality constraint problem $\min\{f(x) : c_i(x) \geq 0, i \in I_c\}$ has a solution (x^*, λ^*). Suppose also that $\overline{\sigma}$ is a positive constant such that for all $\sigma_k < \overline{\sigma}$, the Hessian matrix $\nabla^2 B(x, \sigma_k)$ of the barrier function (8.4) is positive definite for all feasible x. If x_k denotes the solution of the unconstrained problem $\min\{B(x, \sigma_k)\}$, then $x_k \to x^*$ as $\sigma_k \to 0$. Moreover, from (8.5) $\sigma_k/c_i(x_k) \to \lambda_i^*$, $i \in I_c$, as $\sigma_k \to 0$.

Jittorntrum and Osborne (1980) modified the log-barrier function (8.4) by introducing the estimates of the Lagrange multipliers as:

$$J(x,\sigma,\lambda) \triangleq f(x) - \sigma \sum_{i=1}^{m} \lambda_i \log(c_i(x)). \tag{8.6}$$

However, this modified log-barrier function presents the same numerical difficulties as those of function $B(x, \sigma)$.

Polyak (1992) suggested the following *modified log-barrier function*:

$$M(x,\sigma,\lambda) \triangleq f(x) - \sigma \sum_{i \in I} \lambda_i \log\left(1 + \frac{c_i(x)}{\sigma}\right), \tag{8.7}$$

where $\lambda_i, i = 1, \ldots, p$, are nonnegative estimates of the Lagrange multipliers associated to the inequality constraints and $\sigma > 0$ is the barrier parameter. For this modified log-barrier function, Polyak (1992) established the convergence properties similar to those given by Bertsekas (1982b) for the multiplier method. Under reasonably mild conditions, Polyak showed that there exists a threshold value $\overline{\sigma} > 0$ of the barrier parameter such that for any fixed $\sigma < \overline{\sigma}$, i.e., by only updating the Lagrange multipliers $\lambda_i, i = 1, \ldots, p$, the sequence (x^k, λ^k) converges, at least at a linear rate, to a local solution (x^*, λ^*) of the problem (8.1). Very encouraging computational results based on the log-barrier function (8.7) are reported in Breitfeld and Shanno (1994a), Ben-Tall, Yuzefovich, and Zibulevsky (1992), and Nash, Polyak, and Sofer (1994).

The simple bounds $x_j \geq l_j, j \in I_l, x_j \leq u_j, j \in I_u$, are handled by means of the classical logarithmic barrier terms.

Therefore, the classical log-barrier function for inequality constraints is defined by

$$C(x,\sigma) \triangleq f(x) - \sigma \sum_{i \in I} \log(c_i(x)), \tag{8.8}$$

where $\sigma > 0$ is the barrier parameter. The log-barrier function (8.8) was introduced by Frish (1955) and developed by Fiacco and McCormick (1968).

In our approach we consider a combination of all the above functions which define a *general penalty-barrier function*. This is imbedded into the general scheme of the penalty method for which the global convergence was proved. Mainly, the algorithm considers a sequence of simple bounded minimization sub-problems which are solved by means of the simple bound minimization techniques (see Chapter 6).

8.1 The Penalty-Barrier Method

For the general nonlinear optimization problem (8.1), the following penalty-barrier function is defined:

$$F(x, \sigma_k, \lambda^k, s^k, \beta^k) = f(x) - \sigma_k \sum_{i \in I} \lambda_i^k \Psi^k(c_i(x)) + \sum_{i=1}^{m_e} \lambda_i^k c_i(x) + \frac{1}{2\sigma_k} \sum_{i=1}^{m_e} c_i(x)^2, \tag{8.9}$$

where $k \geq 0$ and

$$\Psi^k(c_i(x)) = \log\left(s_i^k + \frac{c_i(x)}{\sigma_k}\right), \quad \text{for} \quad c_i(x) \geq -\beta^k \sigma_k s_i^k \text{ and } i \in I_c, \tag{8.10a}$$

$$\Psi^k(c_i(x)) = Q^k(c_i(x)), \quad \text{for} \quad c_i(x) < -\beta^k \sigma_k s_i^k \text{ and } i \in I_c, \tag{8.10b}$$

$$\Psi^k(c_i(x)) = \log\left(\frac{c_i(x)}{\sigma_k}\right), \quad \text{for } i \in I_{sb}, \tag{8.10c}$$

where $Q^k(c_i(x))$ is a quadratic extrapolation function defined as:

$$Q(c_i(x)) = \frac{1}{2} q_i^a (c_i(x))^2 + q_i^b c_i(x) + q_i^c. \tag{8.11}$$

The coefficients q_i^a, q_i^b, and q_i^c are determined in such a way that the function value as well as the first and the second derivatives of $Q(c_i(x))$ match the corresponding values of the logarithmic terms (8.10a) at $c_i(\widehat{x}) = -\beta^k \sigma_k s_i^k$. A simple computation shows that:

$$q_i^a = \frac{-1}{\left(s_i^k \sigma_k (1 - \beta^k)\right)^2}, \tag{8.12}$$

$$q_i^b = \frac{1 - 2\beta^k}{s_i^k \sigma_k (1 - \beta^k)^2}, \tag{8.13}$$

$$q_i^c = \frac{\beta^k(2-3\beta^k)}{2(1-\beta^k)^2} + \log\left[s_i^k(1-\beta^k)\right]. \tag{8.14}$$

The values of the penalty parameter σ_k are positive scalars which tend to zero as k goes to infinity. For the shifts s_i^k and the extrapolation parameter β^k, we impose

$$0 \le \beta^k \le \beta_u < 1, \tag{8.15a}$$

$$1 \le s_i^k \le s_u < +\infty, \quad i \in I_c, \tag{8.15b}$$

where $\beta_u \ge 0$ and $s_u \ge 1$ for all $i \in I_c$ and $k \ge 0$.

The algorithm finds an approximate local minimum of (8.1) by repeatedly solving the kth sub-problem:

$$\min_{x \in X_B} F\left(x, \sigma_k, \lambda^k, s^k, \beta^k\right), \tag{8.16}$$

such that

$$\left\| \nabla_x F\left(x_{k+1}, \sigma_k, \lambda^k, s^k, \beta^k\right) \right\|_2 \le \varepsilon_k, \tag{8.17}$$

where ε_k is the convergence tolerance corresponding to the kth sub-problem and $\varepsilon_k \to 0$ as $k \to +\infty$. Observe that (8.16) is a simple bounded constrained optimization.

Extrapolating the modified log-barrier terms was suggested by Ben-Tall, Yuzefovich, and Zibulevsky (1992). The quadratic extrapolating terms are defined beyond the singularities of the logarithmic terms (8.10a). This allows an efficient line search for unconstrained minimization and reducing the penalty-barrier parameter σ without restrictions. To ensure that the logarithmic terms (8.10a) are well defined in case the quadratic extrapolation was not used, the shifts s^k are selected such that

$$\frac{c_i(x)}{\sigma_k} - s^k > 0 \tag{8.18}$$

for all $i \in I$, $k \ge 0$, and $x = x_k$ is the starting point for the kth sub-problem.

Updating the multiplier estimates λ^k and the penalty-barrier parameters σ_k is made in such a way that the iterates are convergent to a stationary point x^* of the problem (8.1), i.e., there exists λ^* such that (x^*, λ^*) satisfies the first-order optimality conditions:

$$\nabla f(x^*) - \sum_{i \in E \cup I} \lambda_i^* \nabla c_i(x^*) = 0, \tag{8.19a}$$

$$\sum_{i \in E \cup I} \lambda_i^* c_i(x^*) = 0, \tag{8.19b}$$

$$c_i(x^*) \geq 0 \quad \text{and} \quad \lambda_i^* \geq 0, \quad i \in I, \tag{8.19c}$$

$$c_i(x^*) = 0, \quad i \in E. \tag{8.19d}$$

The initialization of the estimates of the Lagrange multipliers is given by a vector λ^0 computed as solution of the problem:

$$\min_{\lambda_I \geq 0} \left\| \nabla f(x_0) - \sum_{i \in E \cup I} \lambda_i \nabla c_i(x_0) \right\|_2, \tag{8.20}$$

where $\lambda_I = (\lambda_i)_{i \in I}$ and x_0 is the initial estimation of x. Having in view the combinatorial nature of solving (8.20), a procedure is to compute a solution $\widehat{\lambda}$ of (8.20) by neglecting the restrictions $\lambda_I \geq 0$ and then to set

$$\lambda_i^0 = \begin{cases} \min\left\{\max\left\{\lambda_l, \widehat{\lambda}_i\right\}, \lambda_u\right\}, & \text{if} \quad i \in I \quad \text{or} \quad i = 1, \ldots, m_e, \quad \widehat{\lambda}_i \geq 0, \\ \max\left\{\min\left\{-\lambda_l, \widehat{\lambda}_i\right\}, -\lambda_u\right\}, & \text{if} \quad i = 1, \ldots, m_e, \quad \widehat{\lambda}_i < 0, \end{cases} \tag{8.21}$$

where λ_l and λ_u are lower and upper bounds for the initial estimates of the multipliers, which are imposed by numerical reasons (Breitfeld & Shanno, 1994c). Usually, $\lambda_l = 1$ and $\lambda_u = 100$.

The first derivative of $F(x, \sigma_k, \lambda^k, s^k, \beta^k)$ is

$$\nabla_x F\left(x, \sigma_k, \lambda^k, s^k, \beta^k\right) = \nabla f(x) - \sigma_k \sum_{i \in I} \lambda_i^k \Psi'(c_i) \nabla c_i(x) - \sum_{i=1}^{m_e} \left[\lambda_i^k - \frac{c_i(x)}{\sigma_k}\right] \nabla c_i(x), \tag{8.22}$$

where $\Psi'(c_i)$ is the first derivative of Ψ^k with respect to c_i, i.e.:

$$\Psi'(c_i(x)) = \begin{cases} \dfrac{1}{\sigma_k s_i^k + c_i(x)}, & c_i(x) \geq -\beta^k \sigma_k s_i^k, \quad i \in I_c, \\ q_i^a c_i(x) + q_i^b, & c_i(x) < -\beta^k \sigma_k s_i^k, \quad i \in I_c, \\ \dfrac{1}{c_i(x)}, & i \in I_{sb}, \end{cases} \tag{8.23}$$

where q_i^a and q_i^b are given by (8.12) and (8.13), respectively. The first-order optimality conditions (8.19) and (8.22) suggest the following scheme for updating the Lagrange multiplier estimates:

$$\lambda_i^{k+1} = \begin{cases} \lambda_i^k - \dfrac{c_i(x_{k+1})}{\sigma_k}, & i \in E, \\ \sigma_k \lambda_i^k \Psi'(c_i(x_{k+1})), & i \in I, \end{cases} \tag{8.24}$$

where x_{k+1} is the solution of the kth sub-problem (8.16) satisfying (8.17). Observe that (8.24) requires the components λ_i^0, $i \in I$, to be positive, otherwise, if there is a $j \in I$ with $\lambda_j^0 = 0$, then for all $k \geq 1$, $\lambda_j^k = 0$.

The penalty-barrier parameter σ is initialized by a positive scalar σ_0 and then updated by

$$\sigma_{k+1} = \frac{\sigma_k}{\gamma}, \tag{8.25}$$

for $k \geq 0$, where $\gamma > 1$ ensures that $\sigma_k \to 0$ as $k \to +\infty$. Breitfeld and Shanno (1994c) suggest $\sigma_0 = 10^{-1}$ and $\gamma = 10$.

The parameter β defines how close to singularities the logarithmic terms (8.10a) are extrapolated. Breitfeld and Shanno (1994c) show that if β is too small ($0 \leq \beta^k < 0.5$), then the minimum of an individual penalty term (8.10a) or (8.10b) might be very far from the feasibility domain of (8.1), and therefore this term does not impose a penalty for being infeasible. On the other hand, if β is too close to one ($\beta = 0.99$), then big function values of the logarithmic terms are no longer avoided, and therefore ill-conditioning is introduced into the penalty-barrier function. Hence, β is initialized by β^0 and updated as

$$\beta^{k+1} = \beta^k \theta^k, \tag{8.26}$$

for $k \geq 0$, where β^0 and θ^k are scalars such that $0 \leq \beta^k \leq \beta_u$ for all $k \geq 0$ with $\beta_u < 1$. Breitfeld and Shanno (1994c) suggest $\beta^0 = 0.9$ and $\theta^k = 1$, for all $k \geq 0$. Some other sequences $\{\theta^k\}$ are considered in Ben-Tal, Yuzefovich, and Zibulevsky (1992).

As described in Breitfeld and Shanno (1994c), the shifting parameters s allow scaling of the constraints in order to avoid ill-conditioning of the terms (8.10b). We can show it by introducing the general shifts s into the modified barrier terms. Scaling of the inequality constraints such that $\|c(x)\| = 1$ can be achieved without reformulating problem (8.1). To see this, consider x for which $\|c(x)\| > 0$ and let $s = \|c(x)\| > 0$. Then, the scaled constraint is defined as $\widehat{c}(x) = c(x)/s \geq 0$. Clearly, $\|\widehat{c}(x)\| = 1$. With these, Polyak's modified log-barrier term (8.7) for $\widehat{c}(x)$, is:

$$\log\left(1 + \frac{\widehat{c}(x)}{\sigma}\right) = \log\left(1 + \frac{c(x)}{s\sigma}\right) = \log\left(s + \frac{c(x)}{\sigma}\right) - \log(s).$$

The last term, $\log(s)$, is constant with respect to x and therefore can be ignored in the minimization of the penalty-barrier function. Therefore, scaling the inequality constraints can be readily achieved by simply choosing the appropriate shifts s.

On the other hand, in order to avoid ill-conditioning of the penalty-barrier function introduced by the quadratic extrapolation, we set

$$s_i^k = \min\{\max\{1, -c_i(x_k)\}, s_u\}, \tag{8.27}$$

for all $i \in I_c$ and $k \geq 0$, which basically scales the quadratic terms with respect to the constraint values. Usually, $s_u = 10^3$.

Concerning the stopping criteria, we take the advantage that we estimate the Lagrange multipliers along the iterations for solving (8.16). Therefore, a KKT point can be identified and we terminate the penalty-barrier method for that $k \geq 0$ for which

$$v_1^k \leq \tau, \tag{8.28}$$

or

$$v_2^k \leq \tau \text{ and } v_3^k \leq \tau_f, \tag{8.29}$$

where $\tau > 0$ is the convergence tolerance, $\tau_f = 10^{-2}\tau$ and

$$v_1^k = \max\left\{\max_{i=1,\ldots,m_e}\{|c_i(x_k)|\}, \quad -\min_{i\in I}\{c_i(x_k)\}, \right.$$
$$\left. \frac{\sum_{i\in E\cup I} |\lambda_i^k c_i(x_k)|}{1 + \|x_k\|_2}, \quad \frac{\left\|\nabla f(x_k) - \sum_{i\in E\cup I} \lambda_i^k \nabla c_i(x_k)\right\|_\infty}{1 + \|x_k\|_2} \right\},$$

$$v_2^k = \max\{\max_{i\in E}|c_i(x_k)|, \quad -\min_{i\in I} c_i(x_k)\},$$
$$v_3^k = \frac{|f(x_k) - f(x_{k-1})|}{1 + |f(x_k)|}.$$

Observe that if $v_1^k \leq \tau$, then the feasibility, the scaled complementarity, and the scaled first-order optimality conditions are satisfied, i.e., the current point is a KKT one with accuracy τ.

Finally, in order to have the penalty-barrier algorithm, we must specify the optimization method used for solving (8.16) for each set of parameters $(\sigma_k, \lambda^k, s^k, \beta^k)$. Since (8.16) is a simple bounded constrained optimization problem, we can apply any method for solving this type of problems (spectral projected gradient algorithm by Birgin, Martínez, and Raydan (2000, 2001), limited memory BFGS with gradient projection by Byrd, Lu, and Nocedal (1995), or truncated Newton with simple bounds (Nash, 1984a, 1984b, 1985). In our implementation of the algorithm, we have used the truncated Newton with simple bounds (TNBC) (Nash, 1984a, 1984b, 1985). The approximate solution x_k is used as the starting point for the next sub-problem. Now, the following penalty-barrier algorithm with quadratic extrapolation of the inequality constraints can be presented (Andrei, 1996a, 1996b, 1996c, 1996d, 1996e, 1998a).

Algorithm 8.2 Penalty-barrier: SPENBAR (Andrei)

1.	Choose: $x_0 \in \text{int}(X_B)$, $\tau > 0$, and the sequences $\{\varepsilon_k\}$ and $\{\theta_k\}$. Determine λ^0 as in (8.21) and $s_i^0 = \min\{\max\{1, \ -c_i(x_0)\}, \ s_u\}$, $i \in I$. Set $k=0$.
2.	Test of convergence. If $v_1^k \leq \tau$ or if $v_2^k \leq \tau$ and $v_3^k \leq \tau_f$, stop. Otherwise go to step 3.
3.	Using x_k as the initial point, determine $x_{k+1} \in \text{int}(X_B)$ for which: $$\frac{\left\|\nabla_x F(x, \sigma_k, \lambda^k, s^k, \beta^k)\right\|_2}{1 + \|x\|_2} \leq \varepsilon_k.$$
4.	Update the parameters: $$\lambda_i^{k+1} = \begin{cases} \lambda_i^k - \dfrac{c_i(x_{k+1})}{\sigma_k}, & i \in E, \\ \sigma_k \lambda_i^k \Psi'(c_i(x_k)), & i \in I, \end{cases}$$ $$\sigma_{k+1} = \sigma_k/\gamma,$$ $$\beta^{k+1} = \beta^k \theta^k,$$ $$s_i^k = \min\{\max\{1, -c_i(x_{k+1})\}, \ s_u\}, \quad i \in I.$$ Set $k = k+1$ and go to step 2. ♦

We notice that SPENBAR has two types of iterates. The first one is the so-called the *major* (*external*) *iteration* in which the test of convergence in step 2 is checked. The second is the *minor* (*internal*) *iteration* in which the simple bounded optimization sub-problem (8.16) in step 3 is solved.

8.2 Global Convergence

Let $\{x_k\}$ be the sequence generated by the penalty-barrier algorithm. The global convergence is proved under the assumption that the sequence $\{x_k\}$ converges. To prove the global convergence of $\{x_k\}$, suppose that the following three additional assumptions hold:

(*iii*) The set X_B is bounded.
(*iv*) The gradients of the active constraints at the limit point of the sequence $\{x_k\}$ generated by the algorithm are linearly independent.
(*v*) The initial estimates of the Lagrange multipliers corresponding to the inequalities and simple bounds are positive.

The following two propositions proved by Breitfeld and Shanno (1994c) refer to the estimates of the Lagrange multipliers corresponding to the inequality constraints including the simple bounds under the updating formula (8.24).

Proposition 8.1 *(Positivity of Lagrange multipliers). Suppose that (i), (ii), and (v) hold. Under the updating scheme* (8.24), *the estimates of the Lagrange multipliers λ_i^k are positive for all $i \in I$ and $k \geq 0$.*

Proof. For all $i \in I$, by assumption $\lambda_i^0 > 0$. The proposition is proved by induction and suppose that $\lambda_i^k > 0$ for all $i \in I$ and some $k \geq 0$. For case (8.10a) and all $i \in I_c$ and $k \geq 0$, from (8.24) we get:

$$\lambda_i^{k+1} = \frac{\lambda_i^k}{s_i^k + c_i(x_{k+1})/\sigma_k} > 0.$$

The positivity follows from $c_i(x_{k+1}) \geq -\beta^k \sigma_k s_i^k$ and

$$s_i^k + \frac{c_i(x_{k+1})}{\sigma_k} \geq s_i^k(1-\beta^k) \geq 1 - \beta_u > 0,$$

where $\beta_u < 1$ is the upper bound of β. For case (8.10b), from (8.12), since $c_i(x_{k+1}) < -\beta^k \sigma_k s_i^k$, it follows that

$$\lambda_i^{k+1} = \sigma_k \lambda_i^k \left(\frac{-c_i(x_{k+1})}{\left(s_i^k \sigma_k (1-\beta^k)\right)^2} + \frac{1-2\beta^k}{s_i^k \sigma_k (1-\beta^k)^2} \right)$$
$$> \lambda_i^k \left(\frac{\beta^k}{s_i^k (1-\beta^k)^2} + \frac{1-2\beta^k}{s_i^k (1-\beta^k)^2} \right) = \frac{\lambda_i^k}{s_i^k(1-\beta^k)} \geq \frac{\lambda_i^k}{s_u} > 0$$

for all $i \in I_c$ and $k \geq 0$, where s_u is the upper bound of the shifts s. For simple bounds, since $\sigma_k > 0$ and x_{k+1} is strictly feasible with respect to the simple bounds, it follows that

$$\lambda_i^{k+1} = \lambda_i^k \frac{\sigma_k}{c_i(x_{k+1})} > 0,$$

for all $i \in I_{sb}$ and $k \geq 0$. ♦

Proposition 8.2 *(Complementarity slackness). Suppose that (i), (ii), and (v) hold, and the sequence $\{x_k\}$ generated by the algorithm converges to the point $\bar{x}$. Let $I_{in}(\bar{x}) = \{i : \ c_i(\bar{x}) > 0, \ \ i \in I\}$ be the set of indices of the inactive inequality constraints at $\bar{x}$. Then*

$$\{\lambda_i^k\}_{k \geq 0} \to \bar{\lambda}_i = 0$$

for all $i \in I_{in}(\bar{x})$.

Proof. For every $i \in I_{in}(\bar{x})$ there is a $\delta > 0$ such that $c_i(\bar{x}) \geq \delta$. Therefore, for all $k \geq \bar{k}$ and $i \in I_{in}(\bar{x})$, there exists a $\bar{k} \geq 0$ such that:

$$c_i(x_k) \geq \frac{\delta}{2} > 0. \tag{8.30}$$

Observe that for the positive general constraints, the logarithmic terms (8.10a) are utilized. For estimating the multipliers of the general constraints, from (8.24), (8.30), and $s_i^k \geq 1$, it follows that

$$\lambda_i^{k+1} = \frac{\lambda_i^k}{s_i^k + c_i(x_{k+1})/\sigma_k} \leq \lambda_i^k \frac{\sigma_k}{\sigma_k + c_i(x_{k+1})} \leq \lambda_i^k \frac{\sigma_k}{\sigma_k + \delta/2}, \tag{8.31}$$

for all $i \in I_{in}(\bar{x}) \cap I_c$ and $k \geq \bar{k}$. From (8.31) we see that $\lambda_i^{k+1} < \lambda_i^k$ for $k \geq \bar{k}$. Therefore,

$$\lambda_i^{k+1} \leq \lambda_i^k \frac{\sigma_k}{\sigma_k + \delta/2}, \tag{8.32}$$

for all $i \in I_{in}(\bar{x}) \cap I_c$ and $k \geq \bar{k}$. On the other hand, for simple bounds:

$$\lambda_i^{k+1} = \lambda_i^k \frac{\sigma_k}{c_i(x_{k+1})} \leq \lambda_i^k \frac{\sigma_k}{\delta/2} \tag{8.33}$$

for all $i \in I_{in}(\bar{x}) \cap I_{sb}$ and $k \geq \bar{k}$. Since $\sigma_k \to 0$, there is a $\widehat{k} \geq \bar{k}$ such that $\sigma_k \leq \delta/2$ for all $k \geq \widehat{k}$, and therefore (8.33) implies $\lambda_i^{k+1} < \lambda_i^k$ for $k \geq \widehat{k}$. Using this together with (8.32), we get

$$\lambda_i^{k+1} \leq \lambda_i^{\hat{k}} \frac{\sigma_k}{\delta/2} \tag{8.34}$$

for all $k \geq \widehat{k}$ and $i \in I_{in}(\bar{x})$. Since $\sigma_k \to 0$ as $k \to +\infty$, λ_i^k is positive from Proposition 8.1 and $\lambda_i^{\hat{k}}$ is finite, it follows that $\{\lambda_i^k\}_{k \geq 0} \to 0$ for all $i \in I_{in}(\bar{x})$, which proves the proposition. ♦

The following result establishes the convergence of the sequence $\{x_k\}$ and of the estimates sequence of the Lagrange multipliers to a KKT point which satisfies the first-order optimality conditions (8.19). We follow the proof given by Breitfeld and Shanno (1994c).

Theorem 8.1 *(KKT theorem). Suppose that $\{x_k\}$ converges to a point $\bar{x}$, and the assumptions (i)–(v) hold. Then, the sequence of the multiplier estimates $\{\lambda^k\}$ converges to a limit point $\bar{\lambda}$ such that $(\bar{x}, \bar{\lambda})$ satisfies the first-order optimality conditions* (8.19) i.e., *it is a KKT point of the problem* (8.1).

Proof. As we can see, the assumptions *(i)* and *(ii)* guarantee that $F(x, \sigma_k, \lambda^k, s^k, \beta^k)$ is well defined for all $x \in \text{int}(X_B)$, $k \geq 0$, and there is an $\widehat{x} \in \text{int}(X_B)$ for which:

$$F(\widehat{x}, \sigma_k, \lambda^k, s^k, \beta^k) < \infty. \tag{8.35}$$

Now, let us define:

$$F\left(x, \sigma_k, \lambda^k, s^k, \beta^k\right) = \infty \quad \text{for} \quad x \in X_B \backslash \operatorname{int}(X_B). \tag{8.36}$$

Observe that $F(x, \sigma_k, \lambda^k, s^k, \beta^k)$ is continuous and X_B is a compact set because it is bounded and closed. Therefore, by Theorem 1.1 (Weierstrass), $F(x, \sigma_k, \lambda^k, s^k, \beta^k)$ attains its minimum in X_B, which together with (8.35) and (8.36) implies that the approximate minimum $x_{k+1} \in \operatorname{int}(X_B)$, for all $k \geq 0$. Therefore $\bar{x} \in X_B$.

Now, let us show that the estimates of the Lagrange multipliers corresponding to the active constraints are convergent. For simplicity, let us denote $F(x_k, \sigma_{k-1}, \lambda^{k-1}, s^{k-1}, \beta^{k-1})$ by F^k.Therefore, from (8.22) and (8.24) we have

$$\nabla_x F^k = \nabla f(x_k) - \sum_{i \in I} \lambda_i^k \nabla c_i(x_k) - \sum_{i \in E} \lambda_i^k \nabla c_i(x_k), \tag{8.37}$$

for all $k \geq 1$. In the following, we consider the multiplier estimates of the inactive constraints at $\bar{x}$ separately from the others. The vectors or matrices corresponding to the constraints that are inactive at $\bar{x}$ are denoted by the subscript *in*. For example, the vector of multiplier estimates of the inactive constraints at $\bar{x}$ is denoted by λ_{in} with $\lambda_{in} = (\lambda_i)_{i \in I_{in}(x)}$. On the other hand, the vectors and matrices corresponding to the remaining constraints are denoted by the subscript *ac*. These are all equalities and the inequalities that are active at $\bar{x}$, i.e., all the indices $i \in E \cup I \backslash I_{in}(\bar{x})$. With these, (8.37) can be rewritten as

$$\nabla_x F^k = \nabla f(x_k) - \nabla C_{ac}(x_k)^{\mathrm{T}} \lambda_{ac}^k - \nabla C_{in}(x_k)^{\mathrm{T}} \lambda_{in}^k, \tag{8.38}$$

where $\nabla C_{ac}(.)$ is the Jacobian matrix whose rows are the gradients of the constraints c_i, $i \in E \cup I \backslash I_{in}(\bar{x})$ and $\nabla C_{in}(.)$ is the Jacobian matrix corresponding to the inactive constraints at $\bar{x}$. Observe that *(iv)* implies that $\nabla C_{ac}(\bar{x})$ has full rank. Therefore, there is an $\bar{k} \geq 0$ such that $\nabla C_{ac}(x_k)$ has full rank for all $k \geq \bar{k}$. Thus, (8.38) can be rewritten as

$$\lambda_{ac}^k = \left[\nabla C_{ac}(x_k) \nabla C_{ac}(x_k)^{\mathrm{T}}\right]^{-1} \nabla C_{ac}(x_k) \left[\nabla f(x_k) - \nabla C_{in}(x_k)^{\mathrm{T}} \lambda_{in}^k - \nabla_x F^k\right]$$

for all $k \geq \bar{k}$. Since $x \in \operatorname{int}(X_B)$, $\nabla C_{in}(x)$ is a bounded matrix (by assumption *(ii)*), $\lambda_i^k \to 0$ for $i \in I_{in}(\bar{x})$ (by Proposition 8.2) and $\|\nabla_x F^k\| \to 0$, it follows that if $k \to +\infty$, then:

$$\lambda_{ac}^k \to \left[\nabla C_{ac}(\bar{x}) \nabla C_{ac}(\bar{x})^{\mathrm{T}}\right]^{-1} \nabla C_{ac}(\bar{x}) \nabla f(\bar{x}). \tag{8.39}$$

Now, let us define:

$$\bar{\lambda}_{ac} = \left[\nabla C_{ac}(\bar{x}) \nabla C_{ac}(\bar{x})^{\mathrm{T}}\right]^{-1} \nabla C_{ac}(\bar{x}) \nabla f(\bar{x}). \tag{8.40}$$

Observe that by *(ii)*, the components of $\bar{\lambda}_{ac}$ are finite. From (8.38) and (8.40), since $\|\nabla_x F^k\| \to 0$ and $\lambda_i^k = 0$ for all $i \in I_{in}(\bar{x})$, we get:

$$\nabla f(\bar{x}) - \sum_{i \in E \cup I} \bar{\lambda}_i \nabla c_i(\bar{x}) = 0.$$

From Proposition 8.2 and from (8.39), we see that there exist $\delta > 0$ and $\bar{k} \geq 0$ with $\bar{\lambda}_i - \delta \leq \lambda_i^k \leq \bar{\lambda}_i + \delta$ for all $i \in E \cup I$ and $k \geq \bar{k}$, which means that the sequence $\{\lambda^k\}_{k \geq \bar{k}}$ is bounded. From (8.24) we get

$$\lambda_i^{k+1} = \lambda_i^k - \frac{c_i(x_{k+1})}{\sigma_k}, \tag{8.41}$$

for all $i \in E$ and $k \geq 0$. This, together with the boundedness of λ^k, imply that the sequence $\{c_i(x_{k+1})/\sigma_k\}_{k \geq \bar{k}}$ is bounded for all $i \in E$. Since $\sigma_k \to 0$, we must have $c_i(\bar{x}) = 0$ for all $i \in E$. Therefore, the feasibility of $\bar{x}$ with respect to the equality constraints is established.

Now, let us show that $\bar{x}$ is feasible with respect to the inequality constraints, i.e., $c_i(\bar{x}) \geq 0$ for all $i \in I$. Observe that it is necessary to consider the general constraints c_i, $i \in I_c$, because the simple bounds are always strictly feasible. Let us assume that there is an index $j \in I_c$ with $c_j(\bar{x}) = -\delta < 0$. Then, there exists a $\bar{k} \geq 0$ such that

$$c_j(x_k) \leq -\frac{\delta}{2} \tag{8.42}$$

for all $k \geq \bar{k}$. Now, from (8.15) it follows that $-\beta^k \sigma_k s_j^k \to 0$ when $\sigma_k \to 0$. Thus, there is a $\widehat{k} \geq \bar{k}$ with

$$c_j(x_k) \leq -\frac{\delta}{2} < -\beta^k \sigma_k s_j^k \tag{8.43}$$

for all $k \geq \widehat{k}$, and hence, the quadratic extrapolation is used. Consequently, (8.15) and (8.42) imply that

$$\begin{aligned}
\lambda_j^{k+1} &= \sigma_k \lambda_j^k \left(\frac{-c_j(x_{k+1})}{\left(s_j^k \sigma_k (1 - \beta^k)\right)^2} + \frac{1 - 2\beta^k}{s_j^k \sigma_k (1 - \beta^k)^2} \right) \\
&\geq \lambda_j^k \left(\frac{\delta/2}{\sigma_k \left(s_j^k (1 - \beta^k)\right)^2} - \frac{1}{s_j^k (1 - \beta^k)^2} \right) \geq \lambda_j^k \left(\frac{\delta/2}{\sigma_k s_u^2} - \frac{1}{(1 - \beta_u)^2} \right)
\end{aligned} \tag{8.44}$$

for all $k \geq \widehat{k}$, where $\widehat{k}$ is chosen as in (8.43) with the additional requirement that:

$$\sigma_{\hat{k}} \leq \frac{\delta}{2s_u^2\left(1 + 1/(1-\beta_u)^2\right)}.$$

Then, from (8.44) it follows that $\lambda_j^{k+1} \geq \lambda_j^k$ for $k \geq \widehat{k}$, and therefore:

$$\lambda_j^{k+1} \geq \lambda_j^{\hat{k}}\left(\frac{\delta/2}{\sigma_k s_u^2} - \frac{1}{(1-\beta_u)^2}\right) \tag{8.45}$$

for $k \geq \widehat{k}$. As $\sigma_k \to 0$, (8.45) implies that $\lambda_j^k \to +\infty$, which contradicts the finiteness of the multipliers $\overline{\lambda}_j$, $j \in I_c$ (see (8.40) and the Proposition 8.2). Therefore $\bar{x}$ is feasible.

Finally, like in Breitfeld and Shanno (1994c), let us argue that the complementary slackness holds at $(\bar{x}, \overline{\lambda})$, i.e.:

$$\overline{\lambda}_i c_i(\bar{x}) = 0 \tag{8.46}$$

for all $i \in E \cup I$. If $c_i(\bar{x}) = 0$ for some $i \in E \cup I$,it is clear that (8.46) is satisfied. Otherwise, $c_i(\bar{x}) = \delta > 0$, from Proposition 8.2 it follows that $\overline{\lambda}_i = 0$. Thus (8.46) holds. Furthermore, from Proposition 8.1 it follows that $\overline{\lambda}_i \geq 0$ for all $i \in I$.

In conclusion, we have proved that (x_k, λ^k) converges to a point $(\bar{x}, \overline{\lambda})$ satisfying (8.19). Hence, the limit point is a KKT point of the problem (8.1), which completes the proof of the theorem. ♦

The convergence of the penalty-barrier algorithm without assuming that the sequence of iterates $\{x_k\}$ is convergent is proved in (Breitfeld & Shanno, 1994c). The convergence of the algorithm when $\{x_k\}$ is not convergent is proved under the assumption that the gradients of the active constraints at a limit point of $\{x_k\}_{k \in K}$ are linearly independent, where $K$is a subset of $\{k : k \geq 0\}$ and the sequence of the Lagrange multiplier estimates $\{\lambda^k\}_{k \geq 0}$ stays bounded. The proof is quite technical, and it is not presented here.

For convex programming, the convergence results under mild conditions are given by Ben-Tall and Zibulevsky (1993). For the modified log-barrier method, under the convexity assumption, the convergence is given in Jensen and Polyak (1992). Moreover, Polyak (1992) proves the convergence for the modified log-barrier method applied to nonconvex nonlinear programs under nondegeneracy and second-order sufficiency assumptions. The above theoretical results assume the exactness of the solutions to the unconstrained sub-problems. For an augmented Lagrangian algorithm, Conn, Gould, and Toint (1991) establish the global convergence allowing inexact minimization of the augmented Lagrangian without the assumption that the sequence of the Lagrange multiplier estimates stays bounded. A similar approach used in SPENBAR algorithm was considered by Conn, Gould, and Toint (1992a). The Lagrangian barrier function is comprised of similar modified barrier terms as (8.10a). Conn, Gould, and Toint (1992a) remark that the convergence theory extends to a composite Lagrangian barrier, i.e., the augmented Lagrangian algorithm. In this algorithm the update strategy also allows to show that the penalty parameter is bounded away from zero, and therefore the potential ill-conditioning of the augmented Lagrangian function can be avoided. Also, Powell (1969) suggests updating the parameters in a similar way, depending on the reduction in the infeasibility of the constrained problem.

In the following let us present some examples illustrating the numerical performances of SPENBAR, in which the sub-problems (8.16) are solved by truncated Newton with simple bounds (TNBC).

Example 8.1 (Schittkowski 1987, pp. 186)

$$\min\Big[-5(x_1+x_2)-4x_3-x_1x_3-6x_4-\frac{5x_5}{1+x_5}-\frac{8x_6}{1+x_6}$$
$$-10\big(1-2\exp(-x_7)+\exp(-2x_7)\big)\Big]$$

subject to:

$$2x_4+x_5+0.8x_6+x_7-5=0,$$
$$x_2^2+x_3^2+x_5^2+x_6^2-5=0,$$
$$10-x_1-x_2-x_3-x_4-x_5-x_6-x_7\geq 0,$$
$$5-x_1-x_2-x_3-x_4\geq 0,$$
$$5-x_1-x_3-x_5-x_6^2-x_7^2\geq 0,$$
$$x_i\geq 0,\ i=1,\ldots,7.$$

Consider the initial point $x_0=[0.1, 0.1, \ldots, 0.1]$, for which $f(x_0)=-3.282377$. Table 8.1 presents the results given by SPENBAR.

Table 8.1 Iterations generated by SPENBAR. $\sigma_0=0.1$, $\beta=0.9$, $\tau=10^{-8}$

k	1	2	3	4	5
σ_k	0.1	0.01	0.001	0.0001	0.00001
itin	40	41	39	8	2
nf	200	162	175	42	9
x_1	1.460992	1.46663	1.468803	1.468808	1.468808
x_2	1.940633	1.983322	1.983970	1.983971	1.983971
x_3	0.464774	0.354626	0.351885	0.351878	0.351878
x_4	1.202136	1.195411	1.195342	1.195342	1.195342
x_5	0.577817	0.570095	0.569399	0.569396	0.569396
x_6	0.774228	0.784720	0.784745	0.784746	0.784746
x_7	1.379950	1.411455	1.412121	1.412122	1.412122
$f(x)$	−37.68229	−37.41242	−37.41296	−37.41296	−37.41296

In Table 8.1 *itin* and *nf* represent the number of iterations and the number of function evaluations, respectively, in the truncated Newton method for solving the sub-problem (8.16).

Note that toward the end of the solving process, the number of iterations *itin* is smaller and smaller. This is a characteristic of SPENBAR (Andrei, 1996a, 1996b, 1996c, 1996d, 1996e). We see that SPENBAR takes five *major iterations*, at each of them one sub-problem (8.16) being solved, and a total of 130 *minor iterations* for solving (8.16) by truncated Newton with simple bounds. The *total number of function evaluations* is 588. Observe that along the iterations, the values of the penalty parameter σ_k are very reasonable. In SPENBAR the Jacobian matrices $\nabla c(x)$ are considered as sparse matrices. Therefore, it is able to solve large-scale nonlinear optimization problems. ♦

Remark 8.1 Observe that the numerical results for solving this problem are obtained by using the truncated Newton method with simple bounds (TNBC). In general, the best method for solving the unconstrained minimization is the Newton method. For nonconvex problems the Newton direction is not necessarily a descent one. To overcome this deficiency, Gill and Murray (1974b) suggested a modified Cholesky factorization which finds a search direction corresponding to a positive definite approximation to the Hessian. Another approach was given by Breitfeld and Shanno (1994b). Their idea of modifying the Newton method was to use the Newton direction obtained with the original Hessian as long as it is a direction of descent, even in the case of an indefinite Hessian, and only if the search direction is not a descent one is the modification of the Cholesky factorization of Gill and Murray (1974b) used.

However, due to its high storage ($O(n^2)$) and arithmetic costs ($O(n^3)$), the (modified) Newton method can only be applied efficiently to problems of small or moderate dimensions or to large-scale problems for which the sparsity in the Hessian can by utilized. If it is not the case, then it is better to use the truncated Newton or the limited memory BFGS method. The truncated Newton method was introduced by Dembo and Steihaug (1983). It uses a preconditioned linear conjugate gradient algorithm to find an approximate solution to the Newton system. Details on this method are given in (Nash, 1985). On the other hand, the limited memory BFGS method is an adaptation of the BFGS method to large-scale problems. Instead of explicitly forming the approximations to the Hessian, a fixed number of vectors are stored, which define the approximation matrices implicitly, as described in (Liu & Nocedal, 1989). A numerical comparison and a brief outline of the L-BFGS and of the truncated Newton algorithms are given in (Nash & Nocedal, 1991). Both methods roughly require the same amount of storage. The arithmetic cost of one typical truncated Newton iteration is about eight times the cost for one typical L-BFGS iteration, but characteristically, the truncated Newton method needs fewer iterations than the L-BFGS algorithm. This is the reason we selected TNBC algorithm to solve (8.16). ♦

Example 8.2 (*Optimization of a heavy body*). (Brown & Bartholomew-Biggs, 1987; Andrei, 2003, Application A10, pp. 356).

$$\min(x_1 + 0.5x_3 + x_4)$$

subject to:

$$\begin{aligned}
&x_1 - x_6^2 - 0.1 = 0,\\
&x_3 - x_7^2 = 0,\\
&x_4 - x_8^2 = 0,\\
&z^2 + y^2 - 4 = 0,\\
&zu + vy = 0,\\
&uy - vz - 0.7 = 0,
\end{aligned}$$

where:

$$\begin{aligned}
&z = 0.1(x_1 + x_3 + x_4) + 0.01\left(x_1^2 + 2x_1x_3 + 2x_1x_4\right)\cos x_2 + 0.01x_4^2 \cos x_5,\\
&y = 1 + 0.1(x_1 + x_3 + x_4) + 0.01\left(x_1^2 + 2x_1x_3 + 2x_1x_4\right)\sin x_2 + 0.01x_4^2 \sin x_5,\\
&u = 0.1 + 0.02(x_1 \cos x_2 + x_4 \cos x_5),\\
&v = 0.1 + 0.02(x_1 \sin x_2 + x_4 \sin x_5).
\end{aligned}$$

Considering the initial point $x_0 = [5.67, 5.23, 11.96, 23.88, -2.61, 2.34, 3.45, 4.89]$, for which $f(x_0) = 35.53$, Table 8.2 presents the results of the optimization process, as well as the solution.

For solving this problem, SPENBAR needs seven major iterations and a total of 651 minor iterations used by the truncated Newton method. The total number

Table 8.2 Iterations generated by SPENBAR. $\sigma_0 = 0.1$, $\beta = 0.9$, $\tau = 10^{-8}$

k	1	2	3	4	5	6	7
σ_k	0.1	0.01	0.001	0.0001	1.e-5	1.e-6	1.e-7
itin	44	117	190	290	5	4	1
nf	264	665	1018	1412	20	17	6
x_1	0.203143	0.007214	0.305438	0.39424	0.39417	0.39417	0.39417
x_2	7.237269	5.48815	7.430341	7.42649	7.4266	7.4266	7.4266
x_3	7.695956	2.232681	13.48405	13.70247	13.703	13.703	13.703
x_4	0.081001	9.94434	23.58007	24.03176	24.0325	24.0325	24.0325
x_5	−0.90282	−1.23978	−1.91034	−1.91831	−1.9183	−1.9183	−1.9183
x_6	0.305193	−0.18e-8	0.543286	0.619871	0.61981	0.61981	0.61981
x_7	−2.75607	1.49421	3.67206	3.701685	3.70175	3.70175	3.70175
x_8	−0.32e-9	−3.15316	−4.85593	−4.90222	−4.9023	−4.9023	−4.9023
$f(x)$	4.13212	11.06789	30.62753	31.27724	31.2782	31.2782	31.2782

of function evaluations is 3402. The final value of the objective at local optimum is $f(x^*) = 31.27823435$. ♦

Application DES (*Distribution of electrons on a sphere*). (Dolan, Moré, & Munson, 2004; Andrei, 2015). This application, known as the Thomson problem, is described in Chapter 4. In Table 8.3 we present the performances of SPENBAR for solving this application for different numbers of electrons on the sphere.

Table 8.3 Performances of SPENBAR. $\sigma_0 = 0.1$, $\beta = 0.9$, $\tau = 10^{-8}$. Application DES. Thomson problem

n	*me*	*mc*	*itM*	*itm*	*nf*	*qe*	*cpu*	*vfo*
150	50	0	6	402	4095	0	2.23	1055.182314
225	75	0	6	638	6600	0	7.47	2454.369689
300	100	0	6	838	9663	0	18.78	4448.350634
450	150	0	6	1138	14,508	0	61.30	10236.26400
525	175	0	7	1356	18,206	0	103.73	14034.95857
600	200	0	7	1680	25,854	0	190.41	18438.975151
900	300	0	7	2679	37,903	0	618.82	42131.673280
1200	400	0	7	3876	52,568	0	1006.89	75583.037847

In Table 8.3 we have: *n* = the number of variables, *me* = the number of equality constraints, *mc* = the number of inequality constraints, *itM* = the number of the major iterations, *itm* = the number of the minor iterations in truncated Newton method for solving (8.16), *nf* = the total number evaluations of functions in truncated Newton method for solving (8.16), *qe* = the total number of quadratic extrapolations, *cpu* = the cpu time (seconds) to get a local solution to the problem, *vfo* = the optimal value of the objective function.

Application HANG (*Hanging chain*). (Dolan, Moré, & Munson, 2004; Andrei, 2015). This application is presented in Chapter 4. Table 8.4 contains the performances of SPENBAR for solving this application for different number of discretization points, with $a = 1$, $b = 3$, and $L = 4$.

Table 8.4 Performances of SPENBAR. $\sigma_0 = 0.1$, $\beta = 0.9$, $\tau = 10^{-8}$. Application HANG. Dog curve

n	*me*	*mc*	*itM*	*itm*	*nf*	*qe*	*cpu*	*vfo*
50	1	0	3	1338	11,341	0	0.50	5.068577
100	1	0	3	3191	29,213	0	2.42	5.068505
200	1	0	3	3539	32,715	0	12.37	5.068486
400	1	0	4	4873	43,951	0	13.06	5.068481
800	1	0	3	15,896	147,351	0	86.43	5.068480
1000	1	0	3	21,465	196,275	0	144.29	5.068480
2000	1	0	5	102,285	780,770	0	1084.85	5.088175

Numerical study In Chapter 4 we have presented a number of 18 real nonlinear optimization applications. In Table 8.5 of this numerical study, we see the numerical performances of SPENBAR for solving nine applications.

Table 8.5 Performances of SPENBAR. $\sigma_0 = 0.1$, $\beta = 0.9$, $\tau = 10^{-8}$

	n	*me*	*mc*	*itM*	*itm*	*nf*	*qe*	*cpu*	*vfo*
ELCH	10	3	0	6	215	1109	0	0.04	−47.761090
ALKI	10	3	8	8	419	2064	1290	0.06	−1768.8070
PREC	8	0	6	5	87	403	39	0.01	3.9511635
PPSE	9	6	0	9	137	643	0	0.03	5055.0118
MSP3	13	0	15	9	1434	9687	8128	0.33	97.587531
MSP5	16	0	21	9	1256	9189	8413	0.5	174.787136
POOL	34	20	0	8	1282	11,620	0	1.0	2785.8000
TRAFO	6	0	2	6	171	744	173	0.02	135.075955
LATHE	10	1	14	10	781	5079	2296	0.19	−4430.5793

Notes and References

The SPENBAR algorithm presented in Andrei (1996a, 1996b, 1996c, 1996d, 1996e, 1998a) combines the augmented Lagrangian with a log-barrier function (a composite function which includes the augmented Lagrangian and the classical log-barrier function) in which the parameters are updated in such a way as to obtain a KKT point for the considered nonlinear optimization problem. The first development of this algorithm based on the penalty-barrier function was given by Breitfeld and Shanno (1994a, 1994b, 1994c). The equality and inequality constraints are treated separately. In this variant, the penalty-barrier function includes only the equality and inequality constraints. Thus, the algorithm reduces to minimizing a sequence of simple-bound optimization sub-problems. The simple bounds on variables can also be introduced in the composite function, thus obtaining another variant of the penalty-barrier algorithm, in which a sequence of unconstrained minimization sub-problems is used. For both variants the convergence theory is the same (Ben-Tal & Zybulevsky, 1993; Conn, Gould, & Toint, 1991, 1992a, 1996). However, the numerical performances could be different. *SPENBAR illustrates the importance of modifying the classical augmented Lagrangian in order to avoid the ill-conditioning*. In the next chapter we will see a totally different and more sophisticated modification of the augmented Lagrangian, based on the linearization of constraints and on their departure from linearity: the MINOS algorithm elaborated by Murtagh and Saunders (1978, 1980, 1982, 1995).

Chapter 9
Linearly Constrained Augmented Lagrangian: MINOS

In this chapter we present one of the most respectable algorithms and softwares for solving general nonlinear optimization problems given by Murtagh and Saunders (1978, 1980, 1982, 1995). The main idea behind this method is to generate a step by minimizing the Lagrangian or the augmented Lagrangian subject to the linearizations of the constraints.

The origins of the linearly constrained augmented Lagrangian method can be found in the papers of Robinson (1972) and Rosen and Kreuser (1972). Let us consider the general nonlinear optimization problem with equality constraints:

$$\min\{f(x): \ h(x) = 0, \ l \leq x \leq u\} \tag{9.1}$$

For a given point (x_k, λ_k), in Robinson's method we define:

- *Linear approximation* to $h(x)$: $\bar{h}_k(x) = h(x_k) + \nabla h(x_k)(x - x_k)$,
- *Departure from linearity*: $d_k(x) = h(x) - \bar{h}_k(x)$,
- *Modified Lagrangian*: $M_k(x) = f(x) - \lambda_k^T d_k(x)$.

With these, the next iteration of Robinson's method is obtained as a solution of the following sub-problem:

$$\min_{x \in \mathbb{R}^n} \{M_k(x): \ \bar{h}_k(x) = 0, \ l \leq x \leq u\}. \tag{9.2}$$

Under suitable conditions, Robinson (1972) proves that the sequence of sub-problem solutions $\{(x_k, \lambda_k)\}$ converges *quadratically* to a solution of (9.1).

On the other hand, in MINOS, as described in Murtagh and Saunders (1982), the *penalty* term of the *augmented* Lagrangian is included in the sub-problem objective in an attempt to improve the convergence from arbitrary starting points. Therefore, a *modified augmented* Lagrangian sub-problem is obtained:

N. Andrei, *Continuous Nonlinear Optimization for Engineering Applications in GAMS Technology*, Springer Optimization and Its Applications 121,
DOI 10.1007/978-3-319-58356-3_9

$$\min_{x\in\mathbb{R}^n}\left\{f(x) - \lambda_k^{\mathrm{T}} d_k(x) + \frac{1}{2}\sigma_k \|d_k(x)^2\|\right\}$$

subject to (9.3)

$$\bar{h}_k(x) = 0, \quad l \le x \le u.$$

Observe that this is equivalent to the normal augmented Lagrangian because $d_k(x)$ and $h(x)$ are the same when the linearized constraints are satisfied $(\bar{h}_k(x) = 0)$. An important benefit is that if $h(x)$ involves only some of the variables nonlinearly, i.e., it is not hard nonlinear, then $M_k(x)$ has the same property, whereas $\|h(x)\|^2$ appears to be more nonlinear.

MINOS uses the simplex or the reduced-gradient iterations to satisfy the linearized constraints for each sub-problem. It limits the number of "minor iterations" performed on (9.3) in a heuristic way in order to avoid excessive optimization within a wrong subspace. MINOS also monitors $\|x_{k+1} - x_k\|$ and $\|\lambda_{k+1} - \lambda_k\|$, and if they seem to be large, then the step toward (x_{k+1}, λ_{k+1}) is heuristically shortened. Only if $\|h(x)\|$ has increased substantially should σ_k be increased.

9.1 MINOS for Linear Constraints

As described in Murtagh and Saunders (1982), MINOS is a particular implementation of Wolfe's reduced-gradient algorithm (1967). It is designed to solve large-scale problems with nonlinear objective functions, expressed in the following standard format:

$$\min f(x_N) + c^T x, \tag{9.4a}$$

subject to:

$$Ax = b, \tag{9.4b}$$

$$l \le x \le u, \tag{9.4c}$$

where A is a $m \times n$ matrix with $m \le n$. The vector x is partitioned into the linear portion x_L and the nonlinear portion x_N:

$$x = \begin{bmatrix} x_N \\ x_L \end{bmatrix}.$$

The components of x_N are called the *nonlinear variables*. On the other hand, the components of x_L are the *linear variables*. Observe that in the objective function, the nonlinear variables are incorporated in a nonlinear part of it. Possibly, the part of $c^{\mathrm{T}}x$ involving x_N may be included into $f(x_N)$. It is quite clear that if $f(x_N) = 0$, then (9.4) is a linear programming problem. MINOS is an extension of the revised simplex method (Dantzig, 1963; Vanderbei, 2001) and is able to deal with nonlinear terms by using quasi-Newton procedures.

As known, in the simplex method, a basic solution is characterized by having at most m *basic variables* lying between their bounds, while the rest of $n - m$ *nonbasic variables* are equal to one bound or another. An associated square basis matrix B is extracted from the columns of the constraint matrix A, and along the iterations of the simplex algorithm, the columns of B are replaced one at a time.

In case of nonlinear optimization problems, we cannot expect an optimal point to be a basic solution. However, if the number of nonlinear variables is small, then it seems reasonable to suppose that an optimal solution will be *nearly* basic. Thus, as described in Murtagh and Saunders (1978), as a generalization of the simplex algorithm, the concept of *superbasic variables* is introduced. Therefore, partitioning the vector x as $x = [x^B \; x^S \; x^N]^{\mathrm{T}}$, then the linear constraints (9.4b) may be written as:

$$Ax = [B \quad S \quad N]\begin{bmatrix} x^B \\ x^S \\ x^N \end{bmatrix} = b. \tag{9.5}$$

The matrix B is square and nonsingular, as in the simplex algorithm. Therefore, it is invertible. The matrix S is $m \times s$ with $0 \le s \le n - m$, and N is the remaining columns of A. Clearly, the associated variables x^B, x^S, and x^N are called the *basics*, *superbasics*, and *nonbasics*, respectively. As in the simplex algorithm, both basics and superbasics are free to vary between their bounds. The superbasic variables may be moved in any direction in order to improve the objective value. On the other hand, the basic variables are obliged to change in a definite way to maintain the feasibility of the constraints $Ax = b$.

The following theorem shows that the optimal solutions of a nonlinear optimization problem are often near basic (Murtagh & Saunders, 1978).

Theorem 9.1 *Let us suppose that a nonlinear optimization problem has t variables occurring nonlinearly in either the objective function or constraints. Then, an optimal solution exists, in which the number of superbasic variables s satisfies $s \le t$.*

Proof. Let us assume that the nonlinear variables are fixed at their optimal values. The remaining problem is a linear programming one for which a basic solution exists ($s = 0$). The conclusion of the theorem follows immediately if the nonlinear variables are now regarded as superbasics in the original problem. Observe that at the very beginning $s = t$, but if any of the nonlinear variables are actually on a bound, then we can label them as nonbasic, i.e., $s < t$. ♦

Let us now detail the method of MINOS for linear constraints. Suppose that f is twice continuously differentiable. Therefore, it can be expanded in Taylor's series with remainder of second order:

$$f(x + \Delta x) = f(x) + g(x)^{\mathrm{T}}\Delta x + \frac{1}{2}\Delta x^T H(x + \gamma\Delta x)\Delta x, \tag{9.6}$$

where $0 \le \gamma \le 1$, $g(x) = \nabla f(x)$, and $H(x + \gamma\Delta x)$ is the Hessian of f evaluated at some point between x and $x + \Delta x$. Observe that if $f(x)$ is a quadratic function, then H is a constant matrix. Now, let us partition Δx and $g(x)$ corresponding to the partition of A: $\Delta x = [\Delta x^B \quad \Delta x^S \quad \Delta x^N]^{\mathrm{T}}$ and $g = [g_B \quad g_S \quad g_N]^{\mathrm{T}}$.

If $f(x)$ is quadratic, then a constrained stationary point at $x + \Delta x$ is obtained by requiring the following two properties of the step Δx to hold:

Property 1 (Feasibility)

$$\begin{bmatrix} B & S & N \\ & & I \end{bmatrix} \begin{bmatrix} \Delta x^B \\ \Delta x^S \\ \Delta x^N \end{bmatrix} = 0. \tag{9.7}$$

This property says that the step remains on the surface given by the intersection of the active constraints, i.e., the new point $x + \Delta x$ is feasible. From (9.7) observe that $\Delta x^N = 0$.

Property 2 (Optimality)

$$\begin{bmatrix} g_B \\ g_S \\ g_N \end{bmatrix} + H \begin{bmatrix} \Delta x^B \\ \Delta x^S \\ \Delta x^N \end{bmatrix} = \begin{bmatrix} B^{\mathrm{T}} & 0 \\ S^{\mathrm{T}} & 0 \\ N^{\mathrm{T}} & I \end{bmatrix} \begin{bmatrix} \mu \\ \lambda \end{bmatrix}, \tag{9.8}$$

i.e., the gradient at $x + \Delta x$, given by the left-hand side of (9.8), is orthogonal to the surface of active constraints and thus is expressible as a linear combination of the active constraint normals. Moreover, for the optimality of $x + \Delta x$, the negative gradient must be orthogonal outside of the feasibility domain. For (9.4) we impose that $\lambda_j \le 0$ if $x_j^N = u_j$, or $\lambda_j \ge 0$ if $x_j^N = l_j$, $j = 1, \ldots, n - m - s$. Here μ and λ are the Lagrange multipliers.

For a general function $f(x)$, the step Δx may not lead directly to a stationary point, but we may use Properties 1 and 2 to get a feasible descent direction. Now, from (9.7) we get:

$$\begin{aligned} B\Delta x^B + S\Delta x^S &= 0, \\ \Delta x^N &= 0. \end{aligned} \tag{9.9}$$

Therefore,

$$\Delta x^B = -W\Delta x^S, \tag{9.10}$$

where:

$$W = B^{-1}S. \tag{9.11}$$

Thus:

$$\Delta x = \begin{bmatrix} -W \\ I \\ 0 \end{bmatrix} \Delta x^S.$$

So, we can work only with Δx^S. The matrix W is not explicitly computed because B^{-1} is represented as the product form of inverse or the elimination form of inverse (Andrei, 2011a).

Observe that (9.8) may be simplified when it is multiplied to the left by the matrix:

$$\begin{bmatrix} \mathrm{I} & 0 & 0 \\ -W^T & \mathrm{I} & 0 \\ 0 & 0 & \mathrm{I} \end{bmatrix}. \tag{9.12}$$

With this, from the first row of (9.8) multiplied by (9.12), we get an expression for the estimates of the Lagrange multipliers for the general constraints:

$$g_B + [I \quad 0 \quad 0]H \begin{bmatrix} -W \\ I \\ 0 \end{bmatrix} \Delta x^S = B^{\mathrm{T}}\mu. \tag{9.13}$$

Observe that if $\Delta x^S = 0$, which means that x is stationary, we have:

$$B^{\mathrm{T}}\mu = g_B. \tag{9.14}$$

Let u be the solution of (9.14), which, as we can see, is analogous to the *pricing vector* in the revised simplex algorithm.

Considering now the third row of (9.8) multiplied by (9.12), we get:

$$g_N + [0 \quad 0 \quad I]H \begin{bmatrix} -W \\ I \\ 0 \end{bmatrix} \Delta x^S = N^{\mathrm{T}}\mu + \lambda. \tag{9.15}$$

Again, if $\Delta x^S = 0$, then this equation reduces to

$$\lambda = g_N - N^T\mu, \tag{9.16}$$

which is analogous to the vector of *reduced costs* in linear programming.

Finally, considering the second row of (9.8) multiplied by (9.12), we obtain:

$$[-W^T \quad I \quad 0]H \begin{bmatrix} -W \\ I \\ 0 \end{bmatrix} \Delta x^S = -h, \tag{9.17}$$

where:

$$h = \begin{bmatrix} -W^{\mathrm{T}} & I & 0 \end{bmatrix} g = g_S - S^{\mathrm{T}} u. \tag{9.18}$$

The form of the equation (9.17) suggests that

$$\begin{bmatrix} -W^{\mathrm{T}} & I & 0 \end{bmatrix} H \begin{bmatrix} -W \\ I \\ 0 \end{bmatrix} \tag{9.19}$$

can be regarded as the *reduced Hessian* and $h = \begin{bmatrix} -W^{\mathrm{T}} & I & 0 \end{bmatrix} g$ as a *reduced gradient*. Observe that (9.17) gives a Newton step in the independent variables Δx^S. Therefore, $\|h\| = 0$ becomes a necessary condition for a stationary point on the current set of active constraints, which, if the reduced Hessian is nonsingular, implies that $\|\Delta x^S\| = 0$.

Practically, an $R^{\mathrm{T}}R$ factorization of the reduced Hessian (9.19) is computed, where R is an upper triangular matrix such that the step Δx^S is obtained from the system:

$$\left(R^{\mathrm{T}} R\right) \Delta x^S = -h \tag{9.20}$$

by forward and backward substitution, like in the simplex algorithm with LU factorization of basis.

As described in Nocedal and Wright (2006); Gill and Murray (1974a) have considered a class of algorithms in which the search direction along the surface of the active constraints is characterized as being in the range of a matrix Z which is orthogonal to the matrix of constraint normals. Thus, if $\widehat{A}x = \widehat{b}$ is the current set of $n - s$ active constraints, then Z is an $n \times s$ matrix such that:

$$\widehat{A} Z = 0. \tag{9.21}$$

Using this idea, the main steps to be executed at each iteration, to get a feasible direction p, are as follows:

(*i*) Compute the reduced gradient $\bar{g} = Z^{\mathrm{T}} g$.
(*ii*) Compute an approximation to the reduced Hessian $\bar{H} = Z^{\mathrm{T}} H Z$.
(*iii*) Compute an approximate solution to the system $\bar{H}\bar{p} = -\bar{g}$.
(*iv*) Compute the search direction $p = Z\bar{p}$.
(*v*) Perform a line search to find an approximation to the stepsize α^*

$$f(x + \alpha^* p) = \min_{\alpha} \{ f(x + \alpha p) : x + \alpha p \ \text{feasible} \}.$$

This algorithm is very general. Two conditions are imposed on the matrix Z: to be of full rank on columns and to verify (9.21). Therefore Z may have any form which verifies the above conditions. In particular, in MINOS, Z has the following form:

$$Z = \begin{bmatrix} -W \\ I \\ 0 \end{bmatrix} = \begin{bmatrix} -B^{-1}S \\ I \\ 0 \end{bmatrix}. \tag{9.22}$$

This is a very convenient representation which will be used both in MINOS with linear constraints and in MINOS with nonlinear ones. Note that this matrix is not computed. The algorithm works only with S and triangular (LU) factorization of B.

Now, let us present the MINOS algorithm with linear constraints. For this, suppose that we have the following elements:

1. A feasible vector x satisfying $[B \quad S \quad N]x = b$, and $l \leq x \leq u$.
2. The function value $f(x)$, and the gradient $g(x) = [g_B \quad g_S \quad g_N]^T$.
3. The number of superbasic variables s, where $0 \leq s \leq n - m$.
4. A LU factorization of the $m \times m$-basis matrix B.
5. An R^TR factorization of the quasi-Newton approximation to the matrix Z^THZ.
6. A vector u, solution of the system $B^Tu = g_B$.
7. The reduced gradient $h = g_S - S^Tu$.
8. Small positive convergence tolerances ε_h and ε_λ.

With these developments, the MINOS algorithm for solving nonlinear optimization problems with linear constraints is as follows.

Algorithm 9.1 MINOS: linear constraints (Murtagh and Saunders)

1.	Initialization of the algorithm with the above described elements. Set $k = 1$, $x_k = x$, $f(x_k) = f_k$, and $g(x_k) = g_k$.
2.	Test for convergence. If $\lVert h \rVert > \varepsilon_h$, then go to step 4, otherwise continue with step 3.
3.	Compute an estimation of the Lagrange multipliers. Modify the number of superbasics: (a) Compute $\lambda = g_N - N^Tu$. (b) Since for the verification of the KKT conditions, all the variables fixed at the upper (lower) bound must have the Lagrange multipliers negative (positive), determine: $\lambda_1 = \max\{\lambda_j : \lambda_j > 0\}$ for x_j^N fixed at its upper bound u_j, $\lambda_2 = \min\{\lambda_j : \lambda_j < 0\}$ for x_j^N fixed at its lower bound l_j. If $\lambda_1 \leq \varepsilon_\lambda$ and $\lambda_2 \leq \varepsilon_\lambda$, stop, the current solution is optimal. (c) Otherwise, choose x_i^N corresponding to $\lvert\lambda_i\rvert = \max\{\lambda_1, \lvert\lambda_2\rvert\}$ as a new superbasic variable. (1) Augment the matrix S with the column corresponding to variable x_i^N. (2) Augment the reduced gradient h with λ_i. (3) Add a suitable new column to R. (d) Update $s = s + 1$.

(continued)

4.	Compute the search direction p: (a) Solve the system $(R^{\mathrm{T}}R)p_S = -h$. (b) Solve the system $(LU)p_B = -Sp_S$. (c) Set $p = [p_B \quad p_S \quad 0]^{\mathrm{T}}$.
5.	Determine the maximum value of the stepsize: (a) Find $\alpha_{max} \geq 0$ as the greatest positive value of α for which one component of the vector $\left[x^B + \alpha p_B \quad x^S + \alpha p_S\right]^{\mathrm{T}}$ has reached one of its bounds. (b) If $\alpha_{max} = 0$, then go to step 10, otherwise continue with step 6.
6.	Line search. Find α^* such that: $f(x_k + \alpha^* p) = \min\{f(x_k + \alpha p) : 0 < \alpha \leq \alpha_{max}\}$.
7.	Update the solution as $x_{k+1} = x_k + \alpha^* p$ and set $k = k + 1$.
8.	Compute $f(x_k) = f_k$ and $g(x_k) = \nabla f(x_k) = g_k$.
9.	Compute the reduced gradient: (a) Solve the system $(U^{\mathrm{T}}L^{\mathrm{T}})u = g_B$. (b) Compute the new reduced gradient $\bar{h} = g_S - S^{\mathrm{T}}u$. (c) Using α, p, and the change in reduced gradient $\bar{h} - h$, modify R corresponding to a quasi-Newton update of $R^{\mathrm{T}}R$. (d) Set $h = \bar{h}$. (e) If $\alpha = \alpha_{max}$, that is one component of x^B or x^S has reached one of their bounds, then continue with step 10, otherwise ($\alpha < \alpha_{max}$) go to step 2.
10.	Here $\alpha_{max} = 0$. At this step of the algorithm, a basic variable or a superbasic variable, say with index i, has reached one of its bounds. (a) If the variable i is basic, then: (1) Interchange the ith basic variable x_i^B (i.e., the column corresponding to x_i^B from B) with the jth superbasic variable x_j^S (i.e., the column corresponding to x_j^S from$S$). The index j is chosen to keep B nonsingular. (2) Update L and U factors, the matrix R, and the vector u. (3) Compute the new reduced gradient $h = g_S - S^{\mathrm{T}}u$ and go to step 10.c). (b) If the i variable is superbasic, then define $j = i - m$. (c) Make the jth variable in S nonbasic at the appropriate bound: (1) Delete column j from S and place it in N. (2) Delete column j from R. (3) Delete the jth component from the vectors x^S and h. (4) Restore R to triangular form. (d) Set $s = s - 1$ and go to step 2. ♦

Table 9.1 MINOS for solving some large-scale linear programs

	Characteristics			Optimization		
	m	*n*	*nz*	*iter*	*z*	*cpu*
cq9	9451	13,778	157,598	104,150	0.5055445e6	449.47
cre-b	9649	72,447	328,542	183,473	0.2312964e8	678.23
cre-d	8927	69,980	312,626	240,460	0.2445497e8	829.64
ge	10,339	11,098	53,763	15,524	0.5581281e7	51.39
ken-11	14,694	21,349	70,354	15,307	−0.69723e10	71.21
ken-13	28,633	42,659	139,834	45,461	−0.10257e10	440.56
pds-10	16,559	48,763	140,063	74,814	0.267270e11	404.45
stocfor3	16,675	15,695	74,004	14,082	−0.399767e5	72.55
osa-07	1118	23,949	167,643	1915	0.5357225e6	2.02
osa-14	2338	52,460	367,220	4116	0.1106462e7	7.91

Table extracted from Andrei (2011a, pp. 325)
In Table 9.1 we have: $m =$ the number of linear constraints, $n =$ the number of variables, $nz =$ the number of nonzeros in the matrix of linear constraints, $iter =$ the number of iterations necessary to get the optimal solution, $z =$ the value of the objective, and $cpu =$ cpu computing time (seconds) for solving the problem. The numerical experiments were run on a Workstation Intel Pentium 4 with 1.8 GHz.

An iteration of the MINOS algorithm for linear constraints is roughly equivalent to an iteration of the revised simplex algorithm on a $m \times n$ linear programming problem, plus an iteration of a quasi-Newton algorithm on an unconstrained optimization problem with s variables. MINOS with linear constraints is an advanced software, written in Fortran, which takes the sparsity of A and uses an efficient LU factorization of the basis matrix B based on the "*bump and spike*" algorithm (Hellerman & Rarick, 1971, 1972). Murtagh and Saunders (1978) present plenty of details on updating the sparse matrix factorizations, quasi-Newton updates, basis change, removal of one superbasic variable, addition of one superbasic variable, convergence tests, use of the first and of the second derivatives, forward and backward transformations of vectors, quadratic programs, etc.

Numerical Study (MINOS for Linear Programming)
To see the performance of MINOS for solving linear programming problems, in Table 9.1 we present the results of solving some large-scale linear programming problems (the terms $f(x_N)$ in (9.4a) is zero). These are taken from Netlib collection (Gay, 1985) or collection of linear programs described in Carolan, Hill, Kennington, Niemi, and Wichmann (1990) (see Andrei (2011a, pp.325)).

9.2 MINOS for Nonlinear Constraints

In the following, we shall extend MINOS for solving general nonlinear optimization problems of the form:

$$\min f^0(x) + c^{\mathrm{T}}x + d^{\mathrm{T}}y \tag{9.23a}$$

subject to:

$$f(x) + A_1 y = b_1, \tag{9.23b}$$

$$A_2 x + A_3 y = b_2, \tag{9.23c}$$

$$l \leq \begin{bmatrix} x \\ y \end{bmatrix} \leq u, \tag{9.23d}$$

where $f(x) = \left[f^1(x) \quad \cdots \quad f^{m_1}(x)\right]$, $b_1 \in \mathbb{R}^{m_1}$, $b_2 \in \mathbb{R}^{m_2}$, which emphasize the *linear part* as well as the *nonlinear part* of the problem. Define $m = m_1 + m_2$. The first n_1 variables x are called the *nonlinear variables*. They occur nonlinearly in either the objective function or in the first m_1 constraints. The variables y represent the *linear variables*. They include the slack variables. The problem may have purely linear constraints (9.23c). Suppose that the functions $f^i(x)$, $i = 0, \ldots, m_1$, are twice continuously differentiable with the gradients $g^i(x)$, $i = 0, \ldots, m_1$, and there exists a local minimum x^* with the corresponding Lagrange multipliers λ^* such that the first- and second-order KKT conditions hold.

As described by Murtagh and Saunders (1982), the solution process of this problem consists of a sequence of *major iterations*, each one involving a linearization of the nonlinear constraints at some current point x_k corresponding to a first-order Taylor's series approximation:

$$f^i(x) = f^i(x_k) + g^i(x_k)^{\mathrm{T}}(x - x_k) + O\left(\|x - x_k\|^2\right).$$

Define

$$\tilde{f}(x, x_k) \triangleq f(x_k) + J(x_k)(x - x_k), \tag{9.24}$$

or $\tilde{f} = f_k + J_k(x - x_k)$, where $J(x)$ is $m_1 \times n_1$-Jacobian matrix of function $f(x)$. Observe that

$$f - \tilde{f} = (f - f_k) - J_k(x - x_k) \tag{9.25}$$

contains the high-order nonlinear terms in Taylor's expansion of $f(x)$ about the current point x_k.

At the kth major iteration, the following *linearly constrained sub-problem* is formed:

$$\min_{x,y} L(x, y, x_k, \lambda_k, \rho) =$$
$$f^0(x) + c^T x + d^T y - \lambda_k^T (f - \tilde{f}) + \frac{1}{2}\rho (f - \tilde{f})^T (f - \tilde{f}) \quad (9.26a)$$

subject to:

$$\tilde{f} + A_1 y = b_1, \quad (9.26b)$$
$$A_2 x + A_3 y = b_2, \quad (9.26c)$$
$$l \leq \begin{bmatrix} x \\ y \end{bmatrix} \leq u. \quad (9.26d)$$

The objective function $L(.)$ given by (9.26a) is a *modified augmented Lagrangian* in which $f - \tilde{f}$ is used instead of the conventional constraint violation $f + A_1 y - b_1$. The partial derivatives of $L(.)$ are:

$$\begin{aligned} \frac{\partial L(x,y)}{\partial x} &= g^0(x) + c - (J - J_k)^T \left[\lambda_k - \rho(f - \tilde{f})\right], \\ \frac{\partial L(x,y)}{\partial y} &= d. \end{aligned} \quad (9.27)$$

Observe that the nonlinearities in $L(.)$ involve x but not y, which means that the sub-problem has the same nonlinear variables as the original one. The modified Lagrangian was used by Robinson (1972) with $\rho = 0$. The use of a penalty term to ensure that the augmented Lagrangian maintains a positive definite Hessian in the appropriate subspace was suggested by Arrow and Solow (1958) and later considered by Hestenes (1969), Powell (1969), and Sargent and Murtagh (1973).

In the following, in order to specify the MINOS algorithm for nonlinear constraints, it is necessary to define procedures for the choice of Lagrange multipliers λ_k and of the penalty parameter ρ.

Choice of λ_k. The best choice is $\lambda_k = \lambda^*$, but the optimal value of multipliers is not known. Therefore, the simplest choice is $\lambda_k = \widehat{\lambda}$, where $\widehat{\lambda}$ is the multiplier corresponding to the linearized constraints at the solution of the previous sub-problem. Let us assume that the problem does not have linear constraints. Then $\widehat{\lambda}$ is the solution of the system $B^T \widehat{\lambda} = g_B$ from the end of the previous iteration. Moreover, $\widehat{\lambda}$ also verifies the system $S^T \widehat{\lambda} = g_S$. As we know, g is zero for all slack variables, and it follows that $\widehat{\lambda}_i = 0$ if the ith linearized constraint is inactive. Therefore, the choice $\lambda_k = \widehat{\lambda}$ ensures that an inactive nonlinear constraint will be excluded from the Lagrangian term $\lambda_k^T (f - \tilde{f})$ in the next sub-problem.

Choice of ρ. As known, x^* need not be a local minimum of the Lagrangian function. If we assume that $J(x^*)$ is of full rank, then λ^* exists, so that

$$L(x, \lambda) = f^0(x) + c^T x + d^T y - \lambda^T (f + A_1 y - b_1),$$

is stationary at (x^*, λ^*), but $L(x^*, \lambda^*)$ may have a negative curvature in x at x^*. If we consider that the constraints are satisfied at x^* as equalities and ignore the inactive ones, then a necessary (sufficient) condition for x^* to be a local minimum is

$$Z(x^*)^{\mathrm{T}} \frac{\partial L(x^*, \lambda^*)}{\partial x} = 0$$

and

$$Z(x^*)^{\mathrm{T}} \frac{\partial^2 L(x^*, \lambda^*)}{\partial x^2} Z(x^*)$$

to be positive semidefinite (positive definite), where $Z(x^*)$ is defined in (9.22). Therefore, if the search is restricted to the linearly constrained subspace defined by $Z(x^*)$, then we do seek a minimum of the Lagrangian. We may expect that when x_k is sufficiently close to x^* for $J(x_k)$ to be close to $J(x^*)$, we may minimize (9.26a) with $\rho = 0$. As it is discussed in Murtagh and Saunders (1982), the difficulty arises when x_k is far away from x^*, since in this case the linearized constraints may define a subspace where perhaps a saddle point would be closer to x^* than a minimum would be. Successive minima of (9.26) with $\rho = 0$ may therefore fail to converge to x^*. The addition of a penalty term $\rho\left(f - \tilde{f}\right)^{\mathrm{T}}\left(f - \tilde{f}\right)$ imposes the correct curvature properties on (9.26a) for a sufficiently large $\rho > 0$.

To illustrate the importance of the penalty term in (9.26), let us consider the following problem with equality constraints:

$$\min f^0(x)$$

subject to

$$f(x) = 0, \tag{9.28}$$

where the functions are twice continuously differentiable with bounded Hessians. Assume that at some point x^*, the Jacobian $J(x^*)$ is of full rank that there is a λ^* such that $\partial f^0 / \partial x = J(x^*)^{\mathrm{T}} \lambda^*$, and that the reduced Hessian

$$Z(x^*)^{\mathrm{T}} \left[\partial^2 L(x^*, \lambda^*) / \partial x^2\right] Z(x^*)$$

is positive definite, i.e., the sufficient condition of optimality is satisfied for x^* to be a local minimum. As in Murtagh and Saunders (1982), the following theorems can be presented.

Theorem 9.2 *Let* (x_k, λ_k) *be an approximate solution to* (9.28) *and let* $\left(\widehat{x}, \widehat{\lambda}\right)$ *be a solution to the linearized sub-problem:*

$$\min f^0(x) - \lambda_k^{\mathrm{T}}(f - \tilde{f}) + \frac{1}{2}\rho(f - \tilde{f})^{\mathrm{T}}(f - \tilde{f})$$

subject to (9.29)

$$\tilde{f}(x, x_k) = 0,$$

where $\tilde{f}(x, x_k)$ *is given by (9.24). If* $\widehat{\lambda} - \lambda_k = \varepsilon_1$ *and* $f(\widehat{x}) = \varepsilon_2$, *then* $\left(\widehat{x}, \widehat{\lambda}\right)$ *is also a solution to the perturbed problem*:

$$\min f^0(x) + (\varepsilon_1 + \rho\varepsilon_2)^{\mathrm{T}}(f - \tilde{f})$$

subject to (9.30)

$$f(x) = \varepsilon_2,$$

for ε_1 *and* ε_2 *sufficiently small.*

Proof. If $\left(\widehat{x}, \widehat{\lambda}\right)$ is a solution of (9.29), then we must have $\tilde{f} = 0$ and

$$g^0(\widehat{x}) - \left(\widehat{J} - J_k\right)^{\mathrm{T}}\lambda_k + \rho\left(\widehat{J} - J_k\right)^{\mathrm{T}}(f - \tilde{f}) = J_k^{\mathrm{T}}\widehat{\lambda},$$

where J_k is the Jacobian at x_k, but $\widehat{J}$, f, and $\tilde{f}$ are evaluated at $\widehat{x}$. Now, adding $\left(\widehat{J} - J_k\right)^{\mathrm{T}}\widehat{\lambda}$ to both sides of the above equality and inserting the expression for ε_1 and $\varepsilon_2$we get

$$g^0(\widehat{x}) - \left(\widehat{J} - J_k\right)^{\mathrm{T}}\varepsilon_1 + \rho\left(\widehat{J} - J_k\right)^{\mathrm{T}}\varepsilon_2 = \widehat{J}^{\mathrm{T}}\widehat{\lambda},$$

which shows that $\left(\widehat{x}, \widehat{\lambda}\right)$ also satisfies the conditions for a stationary point of (9.30). Observe that the Hessians for the Lagrangian function of (9.29) and (9.30) differ only by the amount $\rho\left(\widehat{J} - J_k\right)^{\mathrm{T}}\left(\widehat{J} - J_k\right)$ at the solution of (9.30), which is of order $\rho\|\Delta x_k\|$, where $\Delta x_k = \widehat{x} - x_k$. Hence, for sufficiently small ε_1, ε_2, and Δx_k, if the reduced Hessian of (9.29) is positive definite at $\widehat{x}$, then by continuity the reduced Hessian of (9.30) will also be positive definite, thus satisfying the sufficient conditions for a local minimum of (9.30) at $\widehat{x}$. ♦

Theorem 9.3 *Let* (x_k, λ_k) *be an approximate solution of* (9.28) *and let* $\left(\widehat{x}, \widehat{\lambda}\right)$ *be a solution of the linearized sub-problem:*

$$\min f^0(x) - \lambda_k^{\mathrm{T}}(f - \tilde{f}) + \frac{1}{2}\rho f^{\mathrm{T}} f$$

subject to: (9.31)

$$\tilde{f}(x, x_k) = 0.$$

If $\widehat{\lambda} - \lambda_k = \varepsilon_1$ *and* $f(\widehat{x}) = \varepsilon_2$, *then* $\left(\widehat{x}, \widehat{\lambda}\right)$ *is also a solution to the perturbed problem:*

$$\min f^0(x) + \varepsilon_1^T (f - \tilde{f}) + \rho \varepsilon_2^T f$$

subject to: (9.32)

$$f(x) = \varepsilon_2.$$

Proof. Analogous to the proof of Theorem 9.2. ♦

As remarked in Murtagh and Saunders (1982), if ε_1 and ε_2 are sufficiently small, then ρ can safely be reduced to zero. Apparently, the problem (9.30) is less sensitive than (9.32) to deviations from its optimum. Thus, let Δx be an arbitrary small change to the solution $\widehat{x}$ of (9.30). The objective function for (9.30) then differs from the true objective $f^0(x)$ by an amount $\delta_1 = (\varepsilon_1 + \rho\varepsilon_2)^T (f - \tilde{f})$. Clearly, $|\delta_1| \le (\|\varepsilon_1\| + \rho\|\varepsilon_2\|) O(\|\Delta x\|^2)$. On the other hand, for (9.32) the analogous deviation is:

$$\delta_2 = \varepsilon_1^T (f - \tilde{f}) + \rho \varepsilon_2^T f = \varepsilon_1^T (f - \tilde{f}) + \rho \varepsilon_2^T \left(\widehat{f} + \widehat{J} \Delta x + O\left(\|\Delta x\|^2\right)\right).$$

In this case $|\delta_2| \le (\|\varepsilon_1\| + \rho\|\varepsilon_2\|) O\left(\|\Delta x\|^2\right) + \rho\|\varepsilon_2\|^2 + \rho\left\|\widehat{J}^T \varepsilon_2\right\| \|\Delta x\|$. Since δ_1 is of order $\|\Delta x\|^2$ while δ_2 is of order $\|\Delta x\|$, it follows that the modified penalty term in (9.29) has a theoretical advantage over the conventional penalty term of (9.31).

Having in view the above theoretical developments, the MINOS algorithm for solving the general nonlinear optimization problems is as follows.

Algorithm 9.2 MINOS: nonlinear constraints (Murtagh and Saunders)

1.	Consider $k = 0$. Choose the initial estimates x_k, y_k, and λ_k. Select a value for the penalty parameter $\rho > 0$ as well as a convergence tolerance $\varepsilon_c > 0$.
2.	Solve the linear sub-problem: (a) Given x_k, y_k, λ_k, and ρ, using the Algorithm 9.1. (MINOS for linear constraints), solve the linear constrained sub-problem (9.26) obtaining x_{k+1}, y_{k+1}, and u, where u is the vector of the Lagrange multipliers associated to the constraints from (9.26). (b) Form the vector λ_{k+1} where the first m_1 components of it are those of u.
3.	Test for convergence. If (x_k, y_k) satisfies the KKT optimality conditions, then stop.

(continued)

4.	If $\dfrac{\left\lVert f(x_{k+1}) + A_1 y_{k+1} - b_1 \right\rVert}{1 + \left\lVert [x_{k+1} \quad y_{k+1}] \right\rVert} \le \varepsilon_c$, and $\dfrac{\lVert \lambda_{k+1} - \lambda_k \rVert}{1 + \lVert \lambda_{k+1} \rVert} \le \varepsilon_c$, then set $\rho = 0$; otherwise set $\rho = 10\rho$.
5.	Consider a re-linearization of the constraints in x_{k+1}, set $k = k + 1$, and go to step 2. ♦

In Murtagh and Saunders (1982) we find some details on the algorithm concerning sparse matrices techniques for solving linear algebraic systems, infeasible sub-problems, user options, verification of the gradients option, Jacobian options (dense or sparse), evaluation options (the constraints and the gradients are evaluated only once per major iteration, or they are evaluated as often as the objective), convergence conditions (major iterations, minor iterations, radius of convergence, row tolerance), etc. Let us present some numerical examples.

Example 9.1 Consider the nonlinear optimization problem (Wright (No.4), 1976), (Murtagh & Saunders, 1982).

$$\min (x_1 - 1)^2 + (x_1 - x_2)^2 + (x_2 - x_3)^3 + (x_3 - x_4)^4 + (x_4 - x_5)^4$$

subject to:

$$x_1 + x_2^2 + x_3^3 = 2 + 3\sqrt{2},$$
$$x_2 - x_3^2 + x_4 = -2 + 2\sqrt{2},$$
$$x_1 x_5 = 2.$$

Considering the initial points

$$A = [1 \ 1 \ 1 \ 1 \ 1]^T, \ B = [2 \ 2 \ 2 \ 2 \ 2]^T, \ C = [-1 \ 3 \ -0.5 \ -2 \ -3]^T,$$
$$D = [-1 \ 2 \ 1 \ -2 \ -2]^T, \ E = [-2 \ -2 \ -2 \ -2 \ -2]^T,$$

the Algorithm 9.2 (MINOS with nonlinear constraints) gives the following results:

$$A, B \to x_1^* = [1.11663 \quad 1.22043 \quad 1.53779 \quad 1.97277 \quad 1.79110]^T,$$
$$f(x_1^*) = 0.0029307,$$
$$C \to x_2^* = [-0.703393 \quad 2.63570 \quad -0.0963618 \quad -1.79799 \quad -2.84336]^T,$$
$$f(x_2^*) = 44.022089,$$
$$D \to x_3^* = [-1.27305 \quad 2.41035 \quad 1.19486 \quad -0.154239 \quad -1.57103]^T,$$
$$f(x_3^*) = 27.8718714,$$
$$E \to x_4^* = [-2.79087 \quad -3.00414 \quad 0.20538 \quad 3.87474 \quad -0.71662]^T,$$
$$f(x_4^*) = 607.03036.$$

Table 9.2 MINOS with different initializations. Example 9.1

	A	B	C	D	E
itm	8	9	7	6	8
it	30	21	14	17	21
evf	64	49	35	35	55

In Table 9.2 we see the characteristics of the optimization process, where *itm* is the number of the major iterations, *it* is the number of minor iterations necessary for solving the sub-problems with linear constraints, and *evf* is the number of evaluations of the objective and of the constraints including their gradients.

Example 9.2 Consider the problem (Wright (No.9), 1976), (Murtagh & Saunders, 1982):

$$\min 10x_1x_4 - 6x_2^2x_3 + x_1^3x_2 + 9\sin(x_5 - x_3) + x_2^3x_4^2x_5^4$$

subject to:

$$\begin{aligned}
&x_1^2 + x_2^2 + x_3^2 + x_4^2 + x_5^2 \le 20,\\
&x_1^2x_3 + x_4x_5 \ge -2,\\
&x_2^2x_4 + 10x_1x_5 \ge 5.
\end{aligned}$$

For the initial points

$$A = [1\ \ 1\ \ 1\ \ 1\ \ 1]^T,\ B = [1\ \ -3\ \ 1\ \ -1\ \ 2]^T,\ C = [5\ \ -5\ \ 1\ \ -3\ \ 1]^T,$$
$$D = [10\ \ -1\ \ 1\ \ 10\ \ 2]^T,$$

the Algorithm 9.2 gives the following results:

$$\begin{aligned}
A \to x_1^* &= [-0.08145\quad 3.69238\quad 2.48741\quad 0.37713\quad 0.17398]^T,\\
f(x_1^*) &= -210.407817,\\
B, C \to x_2^* &= [1.47963\quad -2.63661\quad 1.05467\quad -1.61151\quad 2.67388]^T,\\
f(x_2^*) &= -2500.584488,\\
D \to x_3^* &= [-0.0774\quad -2.58139\quad 0.01143\quad 2.10766\quad 2.98229]^T,\\
f(x_3^*) &= -6043.539113.
\end{aligned}$$

Table 9.3 contains the characteristics of the optimization process initialized in different points.

Practical Hints in GAMS

Two types of options may be used by MINOS in GAMS: options specified through the option statement and options specified through model suffixes. Each option has a default value that is appropriate for most problems. Suppose we have an

Table 9.3 MINOS with different initializations. Example 9.2

	A	*B*	*C*	*D*
itm	10	8	20	9
it	51	25	105	46
evf	136	64	195	97

optimization problem specified in the GAMS source file as `model mymodel/all/;`. All these options can be entered in the option file `minos.opt`, after setting the `mymodel.OPTFILE = 1;` in GAMS source file.

MINOS is able to solve a large diversity of optimization problems: linear programming (LP), nonlinear optimization (NLP), optimization problems with discontinuous or non-differentiable functions (DNLP), linear mixed integer optimization problems (RMIP), and nonlinear mixed integer optimization problems (RMINLP). These nonlinear optimization problems may be solved by MINOS through options specified by the option statement. For example, if we have to solve a linear programming problem, then in the GAMS source file, the following line `option LP = MINOS;` instructs MINOS that it has to solve a linear programming problem, etc.

The performances of MINOS can be controlled by a number of parameters and options which are described as option statements or model suffixes in *GAMS – The Solver Manuals* (2005, pp.219–250). The options are classified as output-related options, options affecting the tolerances, options affecting the iteration limits, other algorithmic options, etc.

Numerical Study

In Chap. 4 we have presented a number of 18 real nonlinear optimization applications. In Tables 9.4 and 9.5 of this numerical study, we see the numerical performances of MINOS for solving 12 small-scale nonlinear optimization applications and 6 large-scale nonlinear optimization applications of different dimensions, respectively.

Notes and References

MINOS is one of the most respectable software for nonlinear optimization. Our description of MINOS is based on the papers of Murtagh and Saunders (1978, 1980, 1982, 1995). It uses a projected augmented Lagrangian for solving general nonlinear optimization problems. At each major iteration, a linearly constrained nonlinear sub-problem is solved, where the linear constraints constitute all the linear constraints and the linearizations of the nonlinear constraints of the problem. The objective function of this sub-problem is an augmented Lagrangian which contains the departure from linearity of the nonlinear constraints. The sub-problems are solved by using a reduced gradient algorithm along with a quasi-Newton algorithm. The quasi-Newton algorithm provides a search direction along which a line search is performed to improve the value of the objective function and to reduce the infeasibilities. MINOS integrates a lot of computational ingredients. The most important is LUSOL, a set of procedures for computing and updating LU

Table 9.4 Performances of MINOS small-scale nonlinear application

	n	*me*	*mc*	*itM*	*itm*	*nf*	*nr*	*s*	*cpu*	*vfo*
ELCH	10	3	0	1	29	84	1	7	0.02	−47.761090
ALKI	10	3	8	12	29	74	74	1	0.02	−1768.8070
PREC	8	0	6	17	71	171	171	3	0.02	4.286758
PPSE	9	6	0	6	13	17	17	1	0.01	5055.0118
MSP3	13	0	15	7	60	95	95	1	0.01	97.587531
MSP5	16	0	21	10	89	139	139	0	0.01	174.787136
POOL	34	20	0	5	60	133	133	3	0.01	2569.800
TRAFO	6	0	2	8	41	83	83	4	0.00	135.075962
LATHE	10	1	14	23	150	327	327	0	0.02	−4430.08793
DES	150	50	0	66	2344	5838	5838	100	2.75	1055.182314
CSTC	303	200	0	6	160	320	320	100	0.19	−3.480074
DIFF	396	324	0	1	48	0	0	0	0.02	0.00

Table 9.5 Performances of MINOS large-scale nonlinear applications

	n	*me*	*mc*	*itM*	*itm*	*nf*	*s*	*cpu*	*vfo*
HANG	1002	501	0	94	3681	8642	496	4.46	5.06855
	2002	1001	0	209	8252	19,712	987	27.49	5.06851
	4002	2001	0	447	17,791	42,955	1950	205.64	5.06849
	8002	4001	0	704	28,153	68,784	3385	1024.63	5.06893
FLOW	1182	734	0	1	997	14	0	0.19	0.00
FLOWO	1556	1005	0	1	709	2	0	0.25	0.00
POL	4004	3000	0	28	1002	3173	0	5.29	14.14484
	6004	4500	0	40	1502	4821	0	12.87	14.14341
	8004	6000	0	53	2000	6524	0	22.45	14.12985
	10,004	7500	0	66	2499	8489	0	36.52	14.14179
CAT	3003	2000	0	11	272	1183	0	2.18	−0.04805
	6003	4000	0	18	536	2430	0	9.06	−0.04805
	9003	6000	0	23	796	3703	0	20.73	−0.04804
CONT	2505	2000	0	1	506	663	1	0.09	1.013239
	5005	4000	0	1	1005	1314	2	0.20	1.005922
	7500	6000	0	1	1509	1978	1	0.69	1.004561
	10,005	8000	0	1	2005	2627	0	1.33	1.004072

In Tables 9.4 and 9.5 we have: n = the number of variables, me = the number of equality constraints, mc = the number of inequality constraints, itM = the number of major iterations, itm = the number of minor iterations, nf = the number of evaluations of objective function, nr = the number of evaluations of constraints, s = the number of superbasic variables, cpu = cpu computing time for solving the problem (seconds), vfo = the value of the objective function.

factors of a general sparse matrix A. The original LUSOL procedures are described in Gill, Murray, Saunders, and Wright (1987). The main factorization uses a traditional Markowitz (1957) strategy with threshold partial pivoting. It is designed to allow A, being square or rectangular with arbitrary rank, to factorize $A = LU$

directly by finding suitable row and column orderings, to replace a column or a row of A, to add or delete a column or a row of A, etc. LUSOL continues to evolve with the addition of rank-revealing LU factorizations for sparse matrices, using either threshold rook pivoting or threshold complete pivoting (Saunders, 2015a, 2015b). MINOS is imbedded in GAMS technology.

Chapter 10
Quadratic Programming

One of the most important nonlinear optimization problems is quadratic programming, in which a quadratic objective function is minimized with respect to linear equality and inequality constraints. These kinds of problems are present in many methods as sub-problems and in real applications from different areas of activity as mathematical models of these applications. At the very beginning, we consider the equality-constrained quadratic programming, after which the inequality-constrained programming will be presented.

10.1 Equality-Constrained Quadratic Programming

The expression of an *equality-constrained quadratic program* is as follows

$$\min q(x) \equiv \frac{1}{2}x^{\mathrm{T}}Gx + c^{\mathrm{T}}x$$

subject to (10.1)

$$a_i^{\mathrm{T}}x + b_i = 0, \quad i \in E \triangleq \{1, \ldots, m\},$$

where G is an $n \times n$ symmetric matrix and $a_i, i \in E$, are vectors in $\mathbb{R}^n$. All the vectors $a_i, i \in E$, can be assembled into matrix A such that the constraints from (10.1) can be compactly written as $Ax + b = 0$, where A is an $m \times n$ matrix. Suppose that $m \leq n$. If the Hessian matrix G is positive semidefinite, we say that (10.1) is a *convex quadratic program*. In this case the problem is similar in difficulty to a linear program. *Strictly convex quadratic programs* are those in which G is positive definite. *Nonconvex quadratic programs* have G an indefinite matrix.

The first-order optimality conditions for (10.1) are

N. Andrei, *Continuous Nonlinear Optimization for Engineering Applications in GAMS Technology*, Springer Optimization and Its Applications 121,
DOI 10.1007/978-3-319-58356-3_10

$$Gx^* + c - A^{\mathrm{T}}\lambda^* = 0, \tag{10.2a}$$

$$Ax^* + b = 0, \tag{10.2b}$$

where x^* is the solution and λ^* is the vector of the Lagrange multipliers. Rearranging (10.2), these equations become a system of $n+m$ linear equations:

$$\begin{bmatrix} G & -A^{\mathrm{T}} \\ -A & 0 \end{bmatrix} \begin{bmatrix} x^* \\ \lambda^* \end{bmatrix} = \begin{bmatrix} -c \\ b \end{bmatrix}. \tag{10.3}$$

In the following, we consider Z to denote the $n \times (n-m)$ matrix whose columns are the bases for the null space of A. That is, Z has full rank and satisfies $AZ = 0$.

Factorization of the Full KKT System For solving (10.1), the direct way, known as the factorization of the full KKT system, is to form the linear system (10.3) and to find x^* and λ^* by solving it, let us say, by the Gaussian elimination (see Appendix B6). Observe that although the matrix from (10.3) is symmetric, the zeros on the main diagonal of this matrix imply that it is not positive definite, that is the Cholesky factorization is not suitable. If G is positive definite, then the feasible stationary point obtained from (10.3) is the minimum of (10.1). Otherwise, to confirm that x^* is the minimum point, the second-order condition $z^{\mathrm{T}}Gz > 0$ for all vectors z such that $Az = 0$,must be checked.

The system (10.3) can be rearranged in a very suitable form for computation by expressing x^* as $x^* = x + p$, where x is some estimation of the solution and p is the desired step. Using this notation, from the system (10.3) we obtain

$$\begin{bmatrix} G & A^{\mathrm{T}} \\ A & 0 \end{bmatrix} \begin{bmatrix} -p \\ \lambda^* \end{bmatrix} = \begin{bmatrix} g \\ h \end{bmatrix}, \tag{10.4}$$

where $g = c + Gx$ and $h = Ax + b$.The matrix in (10.4) is called the *KKT matrix*.

Theorem 10.1 *Let A be of full rank and assume that $Z^{\mathrm{T}}GZ$ is positive definite. Then the KKT matrix*

$$K = \begin{bmatrix} G & A^{\mathrm{T}} \\ A & 0 \end{bmatrix} \tag{10.5}$$

is nonsingular and therefore there is a unique vector pair (x^*, λ^*) satisfying (10.3).

Proof. Suppose that there are vectors w and v such that:

$$\begin{bmatrix} G & A^{\mathrm{T}} \\ A & 0 \end{bmatrix} \begin{bmatrix} w \\ v \end{bmatrix} = 0. \tag{10.6}$$

Since $Aw = 0$, from (10.6) it follows that:

$$0 = \begin{bmatrix} w^{\mathrm{T}} & v^{\mathrm{T}} \end{bmatrix} \begin{bmatrix} G & A^{\mathrm{T}} \\ A & 0 \end{bmatrix} \begin{bmatrix} w \\ v \end{bmatrix} = w^{\mathrm{T}} G w.$$

But w lies in the null space of A. Therefore, there is a vector $u \in \mathbb{R}^{n-m}$, such that $w = Zu$. With these, we have

$$0 = w^{\mathrm{T}} G w = u^{\mathrm{T}} Z^{\mathrm{T}} G Z u,$$

which by positive definiteness of $Z^{\mathrm{T}}GZ$ implies $u = 0$. Therefore, $w = 0$ and by (10.6), $A^{\mathrm{T}}v = 0$. Now, we suppose that A is of full rank on rows. Therefore, $v = 0$. The conclusion is that the equation (10.6) is satisfied only if $w = 0$ and $v = 0$, i.e., the matrix K is nonsingular. ♦

Theorem 10.2 *Let A be of full rank and assume that $Z^{\mathrm{T}}GZ$ is positive definite. Then x^* satisfying* (10.3) *is the unique global solution of* (10.1).

Proof. Let x be any other feasible point satisfying $Ax = -b$. Let $p = x^* - x$. Since $Ax^* = -b$, we have $Ap = 0$. Substituting $x = x^* - p$ in $q(x)$ from (10.1), we get:

$$\begin{aligned} q(x) &= \frac{1}{2}(x^* - p)^{\mathrm{T}} G (x^* - p) + c^{\mathrm{T}}(x^* - p) \\ &= \frac{1}{2} p^{\mathrm{T}} G p - p^{\mathrm{T}} G x^* - c^{\mathrm{T}} p + q(x^*). \end{aligned}$$

But from (10.3) we have $Gx^* = -c + A^{\mathrm{T}}\lambda^*$. Therefore, from $Ap = 0$ we have:

$$p^{\mathrm{T}} G x^* = p^{\mathrm{T}}(-c + A^{\mathrm{T}}\lambda^*) = -p^{\mathrm{T}} c.$$

With this,

$$q(x) = \frac{1}{2} p^{\mathrm{T}} G p + q(x^*).$$

Since p lies in the null space of A, there is a vector $u \in \mathbb{R}^{n-m}$, such that $p = Zu$. Hence:

$$q(x) = \frac{1}{2} u^{\mathrm{T}} Z^{\mathrm{T}} G Z u + q(x^*).$$

But $Z^{\mathrm{T}}GZ$ is positive definite, hence $q(x) > q(x^*)$, except the situation in which $u = 0$, that is when $x = x^*$. Therefore, x^* is the unique global solution of (10.1). ♦

Schur-Complement Method The use of (10.3) to get a solution for (10.1) may be efficient only when the matrices G and A are sparse. Otherwise, another method known as the Schur complement may be used, in which x^* and λ^* are separately determined. For example, if we multiply (10.2a) by AG^{-1} and then use (10.2b) to eliminate x^*, we can obtain λ^* as solution of the following linear system:

$$\left(AG^{-1}A^{\mathrm{T}}\right)\lambda^* = AG^{-1}c - b. \tag{10.7}$$

Clearly, x^* is computed from:

$$Gx^* = A^{\mathrm{T}}\lambda^* - c. \tag{10.8}$$

A crude analysis of the computational effort shows that solving (10.7) and (10.8) requires $O(m^3) + O(n^3)$ multiplications. This is less than $O((n+m)^3)$ multiplication needed to solve (10.3). However, this approach requires to perform operations with G^{-1} and to compute factorization of the $m \times m$ matrix $AG^{-1}A^{\mathrm{T}}$.

The name "Schur complement" derives from the fact that by applying block Gaussian elimination to (10.3) using G as the pivot, the following block upper triangular system

$$\begin{bmatrix} G & -A^{\mathrm{T}} \\ 0 & -AG^{-1}A^{\mathrm{T}} \end{bmatrix} \begin{bmatrix} x^* \\ \lambda^* \end{bmatrix} = \begin{bmatrix} -c \\ b - AG^{-1}c \end{bmatrix} \tag{10.9}$$

is obtained. In linear algebra terminology, the matrix $AG^{-1}A^{\mathrm{T}}$ is called the Schur complement of G.

Null-Space Method This method does not require the matrix G to be nonsingular. It assumes only that A has full row rank and that the matrix $Z^{\mathrm{T}}GZ$ is positive definite, where Z is an $n \times (n-m)$ matrix whose columns span the null space of A. The matrix $Z^{\mathrm{T}}GZ$ is called the *reduced Hessian*. Clearly, this method requires knowledge of the null-space basis matrix Z.

The method considers a solution in terms of its components in two subspaces which are *normal* and *tangential* to the constraints. As we know (see Chapter 5), the optimality conditions involve feasible directions which lie in the tangent space of the constraints. If an n-variable optimization problem involves m linear constraints, $Ax=b$, then a feasible direction can be any member of the $(n-m)$-dimensional subspace of vectors z which satisfy $Az=0$. As we said before, let Z be an $n \times (n-m)$ matrix whose columns span the null space of A. This means that Zw is a feasible direction for any $(n-m)$ vector w. Also, let Y be an $n \times m$ dimensional arbitrary matrix, such that the matrix $[Y \quad Z]$ is nonsingular. Suppose that p in (10.4) is partitioned so that

$$p = Yp_Y + Zp_Z, \tag{10.10}$$

where $p_Y \in \mathbb{R}^m$ and $p_Z \in \mathbb{R}^{n-m}$. In this context, the components Yp_Y and Zp_Z are called the *vertical step* and the *horizontal step*, respectively.Substituting p given by (10.10) into the second equation from (10.4), since $AZ=0$, we get:

$$(AY)p_Y = -h. \tag{10.11}$$

But A is a matrix of full rank m and $[Y \quad Z]$ is an $n \times n$ nonsingular matrix. Therefore, the matrix

$$A[Y \quad Z] = [AY \quad 0] \tag{10.12}$$

is of rank m. Hence, AY is an $m \times m$ nonsingular matrix. p_Y is well defined as solution of the system (10.11).Now, substituting p given by (10.10) into the first equation from (10.4), we get:

$$-GYp_Y - GZp_Z + A^{\mathrm{T}}\lambda^* = g. \tag{10.13}$$

Premultiplying (10.13) with Z^{T}, we obtain:

$$\left(Z^{\mathrm{T}}GZ\right)p_Z = -Z^{\mathrm{T}}GYp_Y - Z^{\mathrm{T}}g. \tag{10.14}$$

Since $Z^{\mathrm{T}}GZ$ is a symmetric positive definite matrix, it follows that p_Z is well defined as solution of the system (10.14). Since the reduced Hessian $Z^{\mathrm{T}}GZ$ is positive definite, then the Cholesky factorization can be used to solve (10.14). In conclusion, the total step p can be computed as $p = Yp_Y + Zp_Z$, where p_Y is solution of (10.11) and p_Z is solution of (10.14), computed in this order.

To obtain the Lagrange multipliers, we multiply the first line of (10.4) with Y^{T} to get the linear system:

$$(AY)^{\mathrm{T}}\lambda^* = Y^{\mathrm{T}}g + Y^{\mathrm{T}}Gp. \tag{10.15}$$

In this method, solving the linear algebraic systems (10.11) and (10.14) requires $O(m^2) + O((n-m)^3)$ multiplications, which can be appreciable less than $O((n+m)^3)$ multiplications needed to solve the KKT system (10.3), especially when $m \approx n$.

Example 10.1 Let us consider the null-space method for solving the following quadratic programming problem

$$\min Q(x) = \frac{1}{2}\underbrace{\left(x_1^2 + x_2^2 + 2x_3^2 + x_4^2 + x_5^2\right)}_{x^{\mathrm{T}}Gx} + \underbrace{(x_1 + x_3 + x_5)}_{c^T x}$$

subject to the linear constraints $Ax = b$, where:

$$\underbrace{\begin{bmatrix} 1 & 0 & 2 & 3 & 1 \\ 2 & 1 & 2 & 1 & 1 \\ 4 & 3 & 1 & 2 & 2 \end{bmatrix}}_{A} \begin{bmatrix} x_1 \\ x_2 \\ x_3 \\ x_4 \\ x_5 \end{bmatrix} = \underbrace{\begin{bmatrix} 7 \\ 7 \\ 12 \end{bmatrix}}_{b}.$$

In this problem $n = 5$ and $m = 3$. Using the reduced echelon technique, a basis Z for the null space of A, as well as the matrix Y chosen such that $[Y \quad Z] \in \mathbb{R}^{n\times n}$ is nonsingular are:

$$Z = \begin{bmatrix} -13 & -3 \\ 15 & 3 \\ 5 & 1 \\ 1 & 0 \\ 0 & 1 \end{bmatrix} \in \mathbb{R}^{n\times(n-m)}, \quad Y = \begin{bmatrix} 1 & 0 & 0 \\ 0 & 1 & 0 \\ 0 & 0 & 1 \\ 0 & 0 & 0 \\ 0 & 0 & 0 \end{bmatrix} \in \mathbb{R}^{n\times m}.$$

Consider $x = [0 \quad 1 \quad 0 \quad 1 \quad 1]^{\mathrm{T}}$ an initial estimation of the solution with which we compute:

$$h = Ax - b = [-3 \quad -4 \quad -5]^{\mathrm{T}}, \; g = Gx + c = [1 \quad 1 \quad 1 \quad 1 \quad 2]^{\mathrm{T}}.$$

From (10.11) we get:

$$p_Y = -(AY)^{-1}h = -\begin{bmatrix} 5 & -6 & 2 \\ -6 & 7 & -2 \\ -2 & 3 & -1 \end{bmatrix} \begin{bmatrix} -3 \\ -4 \\ -5 \end{bmatrix} = \begin{bmatrix} 1 \\ 0 \\ 1 \end{bmatrix}.$$

Similarly, from (10.14) we obtain:

$$p_Z = -\left(Z^{\mathrm{T}}GZ\right)^{-1}\left\{Z^{\mathrm{T}}GYp_Y + Z^{\mathrm{T}}g\right\}$$
$$= -\frac{1}{509}\begin{bmatrix} 21 & -94 \\ -94 & 445 \end{bmatrix}\left\{\begin{bmatrix} -3 \\ -1 \end{bmatrix} + \begin{bmatrix} 8 \\ 3 \end{bmatrix}\right\} = \begin{bmatrix} 0.163064 \\ -0.825147 \end{bmatrix}.$$

Therefore, from (10.10) we get:

$$p = Yp_Y + Zp_Z = \begin{bmatrix} 1.355609 \\ -0.029481 \\ 0.990173 \\ 0.163064 \\ -0.825147 \end{bmatrix}.$$

Using the known initial estimation x, the local optimal solution of the problem is:

$$x^* = x + p = \begin{bmatrix} 1.355609 \\ 0.970519 \\ 0.990173 \\ 1.163064 \\ 0.174853 \end{bmatrix}.$$

The most complicated steps of the algorithm are determination of a basis of the null space of A and solving the linear algebraic systems (10.11) and (10.14). ♦

Remark 10.1 (Bartholomew-Biggs, 2008) The null-space method for quadratic programming with equality constraints can be very easily extended to the nonlinear optimization with linear equality constraints

$$\min\{f(x):\ \ Ax+b=0\}, \tag{10.16}$$

where $f:\mathbb{R}^n \to \mathbb{R}$ is continuously differentiable and $A \in \mathbb{R}^{m \times n}$ is a constant matrix. As in the unconstrained optimization, the idea is to use an iterative scheme based on the local quadratic approximation of f. That is, in a neighborhood of a solution estimate x, we suppose

$$f(x+p) \approx f(x) + \nabla f(x)^{\mathrm{T}}p + \frac{1}{2}p^{\mathrm{T}}Bp, \tag{10.17}$$

where either $B=\nabla^2 f(x)$ or $B \approx \nabla^2 f(x)$ is a quasi-Newton approximation of the Hessian. The following algorithm solves the problem (10.16) by using the quadratic approximation (10.17) at each iteration. The algorithm uses an update approximation to the Hessian matrix rather that the exact Hessian $\nabla^2 f(x)$ and a line search to improve the values of the nonquadratic function f.

Therefore, the problem is now to minimize (10.17) subject to $Ap+(Ax+b)=0$. The optimality conditions for this problem are:

$$\begin{aligned} Bp + \nabla f(x) - A^{\mathrm{T}}\lambda &= 0, \\ Ap + (Ax+b) &= 0. \end{aligned}$$

Next, we apply the technique of the null-space method. Let Z be an $n \times (n-m)$ matrix whose columns span the null space of A. The choice of Z is not unique, but one way of getting it is by *orthogonal factorization* of the constraint Jacobian A. This factorization yields an orthogonal $n \times n$ matrix Q and an $m \times m$ lower triangular matrix L such that $AQ=R=[L \quad 0]$ and $Q^{\mathrm{T}}Q=I$. Now, let us define Y as the matrix composed of the first m columns of Q and Z the matrix consisting of the remaining $(n-m)$ columns. It can be shown that $AZ=0$, $AY=L$, and $Y^{\mathrm{T}}Z=0$.

If p solves the quadratic approximation of (10.16), then its components in Y-subspace and Z-subspace can be determined as follows. Suppose that $\bar{y} \in \mathbb{R}^m$ and $\bar{z} \in \mathbb{R}^{n-m}$ are vectors such that:

$$p = Y\bar{y} + Z\bar{z}.$$

Since $AZ=0$, the optimality condition $Ap+(Ax+b)=0$ implies:

$$AY\bar{y} + (Ax+b) = 0.$$

But $AY = L$. Therefore, $\bar{y}$ can be found by solving a lower triangular system of equations:

$$AY\bar{y} = -(Ax + b).$$

Now, premultiplying by Z^T the optimality condition $Bp + \nabla f(x) - A^T\lambda = 0$ and having in view that $AZ = 0$, we get $Z^T BZ\bar{z} + Z^T\nabla f(x) + Z^T BY\bar{y} = 0$. Therefore, $\bar{z}$ can be found by solving the system

$$Z^T BZ\bar{z} = -Z^T\nabla f(x) - Z^T BY\bar{y},$$

where $Z^T BZ$ is a symmetric positive definite matrix. Knowing $\bar{y}$ and $\bar{z}$, we can compute $p = Y\bar{y} + Z\bar{z}$ as a correction of the estimate x.

As in the quadratic case, in this context $Z^T\nabla f(x)$ is called the *reduced gradient* of f and $Z^T BZ$ is the *reduced Hessian* of f. Observe that both Y and Z, basis matrices for the range and null spaces of the constraints matrix A, are kept constant along the iterations of the algorithm. The corresponding algorithm is as follows.

Algorithm 10.1 Reduced gradient for linear equality constraints

1.	Choose an initial feasible point x_0. Choose B_0 as a positive definite approximation of $\nabla^2 f(x_0)$. Compute Y and Z as basis matrices for the range and the null spaces of the constraints matrix A. Choose $\varepsilon > 0$ sufficiently small. Set $k = 0$ and compute $\nabla f(x_0)$
2.	Test for convergence. If $\|Z^T\nabla f(x_k)\| \leq \varepsilon$, then stop
3.	Determine $\bar{y}$ as solution of the system $AY\bar{y} = -(Ax_k + b)$
4.	Determine $\bar{z}$ as solution of the system $(Z^T B_k Z)\bar{z} = -Z^T\nabla f(x_k) - Z^T B_k Y\bar{y}$
5.	Set $p_k = Y\bar{y} + Z\bar{z}$
6.	Compute the estimate of Lagrange multipliers λ_{k+1} as solution of the linear system $(AY)^T\lambda = Y^T\nabla f(x_k) + Y^T B_k p_k$
7.	Perform a line search to determine s such that $f(x_k + sp_k) < f(x_k)$. Set $x_{k+1} = x_k + sp_k$ as a new estimate of the solution
8.	Using a quasi-Newton update, from B_k compute a new estimate B_{k+1}. Set $k = k + 1$ and continue with step 2 ♦

Since the objective is not quadratic, it follows that $x_k + p_k$ is not guaranteed to be a better point than x_k. Therefore, the Algorithm 10.1 is equipped with a line search (in step 7), which may be exact or based on the Wolfe line-search conditions.

The algorithm uses a quasi-Newton update of B_k to get B_{k+1} as a new estimate of the Hessian. The BFGS update has the advantage that it generates positive definite matrices B_{k+1}. This will ensure that the approximation of the reduced Hessian $Z^T B_k Z$ is also positive definite, and hence the p_k is a descent direction.

It is worth saying that in constrained optimization, the true Hessian $\nabla^2 f(x^*)$ is often indefinite. This suggests that a positive definite updating scheme, based for example on BFGS formula, may be inconsistent with making B_k a good approximation. However, this is not a problem because the second-order optimality conditions only relate to the null space of the binding constraint normals. In this subspace the optimality conditions require the Hessian to be positive definite.

Under reasonable assumptions about the function and constraints and about the Hessian, it can be shown that the search directions and the step lengths in the feasible subspace will satisfy the Wolfe conditions. Therefore, the iterations will converge to a point where the reduced gradient is close to zero. ♦

10.2 Inequality-Constrained Quadratic Programming

An inequality-constrained quadratic program has the following form

$$\min q(x) \equiv \frac{1}{2} x^T G x + c^T x \tag{10.18a}$$

subject to

$$a_i^T x = b_i, \quad i \in E, \tag{10.18b}$$

$$a_i^T x \geq b_i, \quad i \in I, \tag{10.18c}$$

where G is an $n \times n$ symmetric matrix, c, x, and $\{a_i\}, i \in E \cup I$, are vectors in $\mathbb{R}^n$, and E and I are finite set of indices. The Lagrangian for this problem is:

$$L(x, \lambda) = \frac{1}{2} x^T G x + c^T x - \sum_{i \in E \cup I} \lambda_i \left(a_i^T x - b_i\right).$$

As in Definition 5.12, the active-set $A(x^*)$ consists of the indices of the constraints from (10.18) for which equality holds at x^*:

$$A(x^*) = \left\{ i \in E \cup I : \ a_i^T x = b_i \right\}. \tag{10.19}$$

Supposing that the constraint qualification LICQ is satisfied, specializing the general KKT conditions (5.21) to this problem, we find that any solution x^* of (10.18) satisfies the following first-order optimality conditions for some Lagrange multipliers $\lambda_i^*, i \in A(x^*)$:

$$Gx^* + c - \sum_{i \in A(x^*)} \lambda_i^* a_i = 0, \quad (10.20a)$$

$$a_i^T x^* = b_i, \ \text{for all} \ i \in A(x^*), \quad (10.20b)$$

$$a_i^T x^* \geq b_i, \ \text{for all} \ i \in I \backslash A(x^*), \quad (10.20c)$$

$$\lambda_i^* \geq 0, \ \text{for all} \ i \in I \cap A(x^*). \quad (10.20d)$$

If G is positive definite, then (10.18) is called "*convex quadratic programming.*" For convex quadratic programming, the conditions (10.20) are sufficient for x^* to be a global solution, as proved in the next theorem.

Theorem 10.3 *If* x^* *satisfies the conditions* (10.20) *for some* $\lambda_i^*, i \in A(x^*)$ *and* G *is positive semidefinite, then* x^* *is a global solution of* (10.18).

Proof. If x is any other feasible solution for (10.18) then we have $a_i^T x = b_i$, for all $i \in E$, and $a_i^T x \geq b_i$, for all $i \in A(x^*) \cap I$. Hence, $a_i^T(x - x^*) = 0$ for all $i \in E$ and $a_i^T(x - x^*) \geq 0$ for all $i \in A(x^*) \cap I$. Now, using (10.20) we see that:

$$(x - x^*)^T(Gx^* + c) = \sum_{i \in E} \lambda_i^* a_i^T(x - x^*) + \sum_{i \in A(x^*) \cap I} \lambda_i^* a_i^T(x - x^*) \geq 0. \quad (10.21)$$

Now, having in view (10.21), after some algebra we get

$$q(x) = q(x^*) + (x - x^*)^T(Gx^* + c) + \frac{1}{2}(x - x^*)^T G(x - x^*)$$
$$\geq q(x^*) + \frac{1}{2}(x - x^*)^T G(x - x^*) \geq q(x^*),$$

since G is positive semidefinite. Therefore, we have shown that $q(x) \geq q(x^*)$ for any feasible x, i.e., x^* is a global solution. ♦

If the reduced Hessian $Z^T GZ$ is positive definite, where Z is a null-space basis matrix for the active constraint Jacobian matrix, which is the matrix whose rows are a_i^T for all $i \in A(x^*)$, then the second-order sufficient conditions for x^* to be a local minimum hold. If G is not positive definite, then the general problem (10.18) may have more than one strict local solution. Such problems are called "*nonconvex quadratic programming*" or "*indefinite quadratic programming.*"

For solving the inequality-constrained quadratic programming problems (10.18) we can use the *active-set methods* or *interior point methods*. In case of convex quadratic programming, there are three varieties of active-set methods: *primal* (Beale, 1959; Fletcher, 1971; Bunch & Kaufman, 1977), *dual* (Lemke, 1962), and *primal-dual* (Gonçalves, 1972; Goldfarb & Idnani, 1983). The active-set methods find a step from one iteration to the next one by solving a quadratic sub-problem in which some of the inequality constraints (10.18c) and all the equality constraints (10.18b) are imposed as equalities. This subset of constraints is referred to as the *working set*, which at iteration k is denoted as W_k. The case in which G is an indefinite matrix raises some complications in the algorithms which are outside the purposes of this book. For nonconvex quadratic programming, see Gould, Orban, and Toint (2005). In the following, we shall discuss the primal active-set method and the primal-dual active-set method by Goldfarb and Idnani (1983). The interior point methods for solving (10.18) are a particularization of the interior point methods for general nonlinear optimization (see Chapter 17), and we do not present them here.

The Primal Active-set Method

The idea of the primal active-set method is as follows. Given the iterate x_k and the working set W_k, test whether x_k minimizes the quadratic q in the subspace defined by the working set. If not, a step p is computed by solving an equality-constrained quadratic programming sub-problem in which the constraints corresponding to the working set W_k are regarded as equalities and the rest of constraints are temporarily ignored. Let us define:

$$p = x - x_k, \quad g_k = Gx_k + c.$$

With these, the objective function (10.18a) becomes

$$q(x) = q(x_k + p) = \frac{1}{2}p^{\mathrm{T}}Gp + g_k^{\mathrm{T}}p + \delta_k,$$

where $\delta_k = \frac{1}{2}x_k^{\mathrm{T}}Gx_k + c^{\mathrm{T}}x_k$ is independent of p.

Since δ_k can be dropped from the expression of $q(x)$, the quadratic sub-problem which must be solved at the kth iteration is

$$\min \frac{1}{2}p^{\mathrm{T}}Gp + g_k^{\mathrm{T}}p \tag{10.22a}$$

subject to:

$$a_i^{\mathrm{T}}p = 0, \quad i \in W_k. \tag{10.22b}$$

Let p_k be the solution of (10.22). Observe that for each $i \in W_k$, since $a_i^{\mathrm{T}}(x_k + \alpha p_k) = a_i^{\mathrm{T}}x_k = b_i$ for all α, it follows that the value of $a_i^{\mathrm{T}}x$ does not change as we move along p_k. Therefore, since the constraints in W_k are satisfied at x_k, they are also satisfied at $x_k + \alpha p_k$, for any value of α. It is clear that, since G is positive definite, the solution of (10.22) can be computed by any method for solving equality-constrained quadratic programming.

Supposing that the optimal p_k is nonzero, the problem is how far to move along this direction to conserve the feasibility. If $x_k + p_k$ is feasible with respect to all the constraints, we set $x_{k+1} = x_k + p_k$. Otherwise set

$$x_{k+1} = x_k + \alpha_k p_k, \tag{10.23}$$

where the stepsize α_k is computed to be the largest value in the range $[0, 1]$ for which all the constraints are satisfied. We must consider the definition of α_k for constraints $i \notin W_k$, since, as we said before, the constraints in W_k are satisfied at $x_k + \alpha p_k$, for any value of α. If $a_i^T p_k \geq 0$ for some $i \notin W_k$, then for all $\alpha_k \geq 0$ we have $a_i^T(x_k + \alpha_k p_k) \geq a_i^T x_k \geq b_i$. Therefore, the constraint i will be satisfied for all nonnegative α_k. On the other hand, if $a_i^T p_k < 0$ for some $i \notin W_k$, then $a_i^T(x_k + \alpha_k p_k) \geq b_i$ only if:

$$\alpha_k \leq \frac{b_i - a_i^T x_k}{a_i^T p_k}. \tag{10.24}$$

To maximize the decrease in q, we must take α_k to be as large as possible in $[0, 1]$ subject to satisfying feasibility. Hence, α_k is determined as:

$$\alpha_k \triangleq \min\left\{1,\ \min\left\{\frac{b_i - a_i^T x_k}{a_i^T p_k},\ i \notin W_k,\ a_i^T p_k < 0\right\}\right\}. \tag{10.25}$$

The constraint i for which the minimum in (10.25) is achieved is called *blocking constraint*, like in linear programming (Andrei, 2011a).

If $\alpha_k < 1$, that is the step along p_k was blocked by some constraint not in W_k, then a new working set W_{k+1} is generated by adding one of the blocking constraints to W_k.

The iterations are continued in this way by adding constraints to the working set until a point $\widehat{x}$ that minimizes the quadratic objective function over its current working set $\widehat{W}$ has been attained. Such a point is easy to be recognized because the sub-problem (10.22) has solution $p = 0$. Since $p = 0$ satisfies the optimality conditions (10.3) for (10.22), we have

$$\sum_{i \in \widehat{W}} a_i \widehat{\lambda}_i = g = G\widehat{x} + c, \tag{10.26}$$

for some Lagrange multipliers $\widehat{\lambda}_i$, $i \in \widehat{W}$. It follows that $\widehat{x}$ and $\widehat{\lambda}$ satisfy the first KKT optimality condition (10.20a) if we define the multipliers corresponding to the inequality constraints that are not in the working set to be zero. On the other hand, because $\widehat{x}$ is also feasible with respect to all the constraints, it follows that (10.20) are satisfied at this point.

Algorithm 10.2 Active-set method for convex quadratic programming

1.	Compute a feasible initial point x_0. Set W_0 to be a subset of the active constraints at x_0. Set $k=0$
2.	Test a criterion for stopping the iterations
3.	Solve the equality quadratic programming sub-problem (10.22) to find p_k
4.	If $p_k=0$, then compute the Lagrange multipliers $\widehat{\lambda}_i$ that satisfy (10.26) with $\widehat{W}=W_k$. If $\widehat{\lambda}_i \geq 0$, for all $i \in W_k \cap I$, then stop with the solution $x^* = x_k$; otherwise determine $j=\text{argmin}\left\{\widehat{\lambda}_j : \; j \in W_k \cap I\right\}$, and set $x_{k+1}=x_k$ and $W_{k+1}=W_k \backslash \{j\}$ If $p_k \neq 0$, then from (10.25) compute α_k and set $x_{k+1}=x_k+\alpha_k p_k$. If there are blocking constraints, then obtain W_{k+1} by adding one of the blocking constraints to W_k; otherwise set $W_{k+1}=W_k$ and go to step 2 ♦

Plenty and important details on the primal active-set method are presented in Nocedal and Wright (2006). They concern the signs of the multipliers corresponding to the inequality constraints in the working set (the indices $i \in \widehat{W} \cap I$), various techniques used to determine an initial feasible point, the strategy of removing the constraint corresponding to the most negative Lagrange multiplier, the strategy of adding or deleting at most one constraint at each iteration, the finite termination of the active-set algorithm on strictly convex quadratic programming, the updating of factorizations, etc.

An Algorithm for Positive Definite Hessian

This active-set algorithm is based on a repeated use of the optimality conditions for an equality quadratic programming problem. In the following, we consider quadratic programming which has only inequality constraints. The extension to mixed equality and inequality quadratic problems is straightforward. Therefore, let us consider the problem

$$\min \frac{1}{2}x^{\mathrm{T}}Gx + c^{\mathrm{T}}x \tag{10.27a}$$

subject to

$$\widehat{A}x + \widehat{b} \geq 0, \tag{10.27b}$$

where we assume that G is positive definite. The method begins by identifying active set of constraints. This is an estimate of the ones which are binding at the solution. Now, let A be the matrix and b the vector formed from the rows of $\widehat{A}$ and $\widehat{b}$ corresponding to active constraints. With these, let $\left(\tilde{x}, \widetilde{\lambda}\right)$ be a trial solution by minimizing (10.27a) subject to

$$Ax + b = 0,$$

obtained by using an algorithm for the equality quadratic programming problem. If

$$\widehat{A}\tilde{x} + \widehat{b} \geq 0, \tag{10.28a}$$

$$\tilde{\lambda}^{\mathrm{T}}\left(\widehat{A}\tilde{x} + \widehat{b}\right) = 0, \ \ and \ \tilde{\lambda} \geq 0, \tag{10.28b}$$

then the optimality conditions (10.20) are all satisfied, and the problem (10.27) is solved by $x^* = \tilde{x}$ and $\lambda^* = \tilde{\lambda}$. On the other hand, if $\left(\tilde{x}, \tilde{\lambda}\right)$ is not optimal, we must change the active set and solve another equality-constrained quadratic programming sub-problem. This process is repeated until the active set has become the binding set for the problem (10.27). The choice of a new active set is based on (10.28). Two ideas are followed here. The first one is that any new constraints which are violated at $\tilde{x}$ can be regarded as candidates for being added to the current active set. Secondly, any active constraints which correspond to a negative component of $\widehat{\lambda}$ are candidates for deletion from the current active set (Bartholomew-Biggs, 2008). The following algorithm formalizes the ideas just outlined, where $\widehat{a}_i$ denote the ith row of $\widehat{A}$.

Algorithm 10.3 Active-set method with positive definite Hessian

1.	Choose an initial point x and set $\lambda_i = 0, i = 1, \ldots, m$
2.	Test the optimality conditions. If $\widehat{A}x + \widehat{b} \geq 0,\ Gx + c - \widehat{A}^{\mathrm{T}}\lambda = 0$ and $\lambda \geq 0$, then stop
3.	Identify the active constraints as those for which: $$\widehat{a}_i^{\mathrm{T}}x + \widehat{b}_i < 0 \text{ or } (\widehat{a}_i^{\mathrm{T}}x + \widehat{b}_i = 0 \text{ and } \lambda_i \geq 0).$$ Renumber these constraints so that the active set is $i = 1, \ldots, t$
4.	Set $g = Gx + c$ and $b_i = \widehat{a}_i^{\mathrm{T}}x + \widehat{b}_i$, for $i = 1, \ldots, t$
5.	Find p and μ as solution for the equality quadratic programming problem $\min \frac{1}{2}p^{\mathrm{T}}Gp + g^{\mathrm{T}}p$ subject to $\widehat{a}_i^{\mathrm{T}}p + b_i = 0, i = 1, \ldots, t$
6.	Set $(s = 1, \lambda_i = \mu_i, i = 1, \ldots, t)$ and $(\lambda_i = 0, i = t+1, \ldots, m)$
7.	For $i = t+1, \ldots, m$, i.e., for all *inactive* constraints if $\widehat{a}_i^{\mathrm{T}}p < 0$ set $$s = \min\left\{s, -\frac{\widehat{a}_i^{\mathrm{T}}x + \widehat{b}_i}{\widehat{a}_i^{\mathrm{T}}p}\right\}$$
8.	Set $x = x + sp$ and go to step 2 ♦

The algorithm may have different variants according to the methods for solving the problems in step 5 and the rules regarding how many constraints may be added to or dropped from the active set. The stepsize computed in step 7 checks all the inactive constraints that might be violated by a step along the search direction p and ensures that no more than one constraint can be added to the active set on the current iteration. Details can be found in Fletcher (1971).

Reduced Gradient for Inequality Constraints

The active-set strategy combined with the null-space approach can be very easily extended to solving problems with nonquadratic objective function $f(x)$ and linear inequality constraints $\widehat{a}_i^T x + \widehat{b}_i \geq 0, i = 1, \ldots, m$. In this case, the Y and Z matrices used in the null-space method must be computed whenever the active set is changed (Bartholomew-Biggs, 2008).

Algorithm 10.4 Reduced gradient for linear inequality constraints

1.	Choose an initial point x_0 and set $\lambda^0 = 0$. Choose B_0 as a positive definite estimate of $\nabla^2 f(x_0)$. Select a tolerance $\varepsilon > 0$. Set $k = 0$
2.	Compute $g_k = \nabla f(x_k)$
3.	Select active constraints as those with $\widehat{a}_i^T x_k + \widehat{b}_i = 0$ and $\lambda_i^k \geq 0$
4.	Consider A_k as the matrix of active constraint normals at x_k
5.	Compute Y_k and Z_k as basis matrices for the range and the null spaces of matrix A_k
6.	Determine z as solution of the linear system $Z_k^T B_k Z_k z = -Z_k^T g_k$. Set $p_k = Z_k z$
7.	Find λ^{k+1} as solution of the linear system $Y_k^T A_k^T \lambda = Y_k^T g_k + Y_k^T B_k p_k$
8.	Step length computation. Perform a line search to get $x_{k+1} = x_k + s p_k$ so that $f(x_{k+1}) < f(x_k)$
9.	Compute a quasi-Newton update of B_k to get B_{k+1}
10.	Test of convergence. If $\|Z_k^T g_k\| \leq \varepsilon$, then stop; otherwise set $k = k+1$ then go to step 2 ♦

Observe that the computations in step 8 of the step length must ensure that no new constraints are violated. Therefore the stepsize s is subject to an upper limit which allows at most one new constraint to become binding at x_{k+1}. This can be computed as in step 7 of the Algorithm 10.3. The difficulty with this algorithm is that the matrices Y and Z used in the null-space method must be computed whenever the active set is changed in step 4 of the algorithm.

Reduced Gradient for Simple Bounds

An important situation in which Z can be very easily computed and therefore makes the reduced-gradient approach very attractive is when the problem has only simple bounds on variables: $l_i \le x_i \le u_i$, $i = 1, \ldots, n$. Suppose that $l_i < u_i$, $i = 1, \ldots, n$. In this case, at the start of each iteration, the variables are classified in two classes: those which are *fixed* (i.e., on their bounds) and those which are *free*. If x_k is the current estimate at the start of iteration k, then the bound on the ith variable is active if x_k is fixed, which means that:

$$(x_{k_i} = l_i \text{ and } g_{k_i} > 0) \text{ or } (x_{k_i} = u_i \text{ and } g_{k_i} < 0).$$

In this case the matrix Z whose columns span the space of the free variables can be simply taken as a partition of the identity matrix. When the iterate x_{k+1} is computed from x_k along a search direction p_k, the stepsize must ensure that no new bounds are violated. A maximum stepsize to force each variable x_{k_i} to stay within its bounds can be computed as:

$$s_i^{\max} = \begin{cases} (u_i - x_{ki})/p_{k_i} & \text{if } p_{k_i} > 0, \\ (l_i - x_{ki})/p_{k_i} & \text{if } p_{k_i} < 0. \end{cases}$$

Algorithm 10.5 Reduced gradient for simple bounds constraints

1.	Choose an initial point x_0 and B_0 as a positive definite estimate of $\nabla^2 f(x_0)$. Select a tolerance $\varepsilon > 0$. Set $k = 0$
2.	Compute $g_k = \nabla f(x_k)$
3.	Set Z_k as the $n \times n$ identity matrix
4.	For $i = 1, \ldots, n$ repeat: if x_{k_i} satisfies $(x_{k_i} = l_i$ and $g_{k_i} > 0)$ or $(x_{k_i} = u_i$ and $g_{k_i} < 0)$, then delete the $i$thcolumn of Z_k
5.	Determine z as solution of the linear system $Z_k^T B_k Z_k z = -Z_k^T g_k$. Set $p_k = Z_k z$
6.	Use a line search along p_k to find s so that $f(x_k + s p_k) < f(x_k)$
7.	For each free variable x_{k_i} compute $s_i^{\max}$ and set $s = \min\{s, s_i^{\max}\}$
8.	Set $x_{k+1} = x_k + s p_k$
9.	Compute a quasi-Newton update of B_k to get B_{k+1}
10.	Test of convergence. If $\lVert Z_k^T g_k \rVert \le \varepsilon$,then stop; otherwise set $k = k + 1$ then go to step 2 ♦

Details on this algorithm can be found in Bartholomew-Biggs (2008).

The Primal-Dual Active-Set Method

In the following we present the algorithm by Goldfarb and Idnani (1983) for solving inequality-constrained quadratic programming problems

$$\min Q(x) \triangleq \frac{1}{2} x^{\mathrm{T}} G x + c^{\mathrm{T}} x \tag{10.29a}$$

subject to

$$s_i(x) = a_i^{\mathrm{T}} x + b_i = 0, \;\; i = 1, \ldots, m_e, \tag{10.29b}$$

$$s_i(x) = a_i^{\mathrm{T}} x + b_i \geq 0, \;\; i = m_e + 1, \ldots, m, \tag{10.29c}$$

$$l \leq x \leq u, \tag{10.29d}$$

where G is an $n \times n$ symmetric and positive definite matrix and $a_i \in \mathbb{R}^n$ for all $i = 1, \ldots, m$. In general, the algorithms for quadratic programming have two phases. In the first one, a feasible point is obtained, and then in the second phase, the optimality is achieved while maintaining feasibility.

The algorithm of Goldfarb and Idnani, is of the active-set type. By an active set, we mean a subset of the m constraints in (10.29) that are satisfied as equalities by the current estimate x of the solution to the quadratic programming problem (10.29). If we denote the set of indices of the constraints (without simple bounds) by $I = \{1, \ldots, m\}$, then the active set is denoted by $A \subseteq I$. Let $P(J)$ be the sub-problem with the objective function (10.29a) subject only to the subset of the constraints indexed by $J \subset I$. For example, $P(\emptyset)$ is the problem of finding the unconstrained minimum of (10.29). If the solution x of a sub-problem $P(J)$ lies on some linearly independent active set of constraints indexed by $A \subseteq J$, then the pair (x, A) is called a solution. Clearly, if (x, A) is a solution for the sub-problem $P(J)$, then it is also a solution for the sub-problem $P(A)$. By linear independence of a set of constraints, we mean that the normals corresponding to these constraints are linearly independent. In the following we outline the algorithm given by Goldfarm and Idnani (1983):

1. Assume that we know an arbitrary solution (x, A).
2. Choose a violated constraint $p \in I \backslash A$.
3. If $P(A \cup \{p\})$ is infeasible, then stop; the problem (10.29) is infeasible.
4. Or else, obtain a new pair $(\bar{x}, \bar{A} \cup \{p\})$, where $\bar{A} \subseteq A$ and $Q(\bar{x}) > Q(x)$, and set $(x, A) \leftarrow (\bar{x}, \bar{A} \cup \{p\})$.
5. Continue with step 2 until all the constraints have been satisfied.

Since G is positive definite, it follows that the unconstrained minimum of function $Q(x)$ defined in (10.29a) is $x^0 = -G^{-1} c$. Therefore, the above procedure can start with the pair $(x^0, \emptyset)$.

To describe the algorithm, it is necessary to introduce some notation like in Goldfarb and Idnani (1983). The matrix of the normal vectors of the constraints in the active-set A is denoted by N, and the cardinality of A is denoted by q. Denote $A^+ = A \cup \{p\}$, where $p \in I \backslash A$, and A^- is a proper subset of A which contains one fewer element than A. N^+ and N^- represent the matrices of normals corresponding to A^+ and A^-, respectively. e_i represents the i–th column of the unity matrix, and I_k is the $k \times k$ identity matrix. n^+ represents the normal vector n_p added to N to obtain N^+; n^- represents the column deleted from N to get N^-.

If the columns from N are linearly independent, then we may introduce the following operators:

$$N^* = \left(N^T G^{-1} N\right)^{-1} N^T G^{-1} \tag{10.30}$$

and

$$H = G^{-1}(I - NN^*) = G^{-1} - G^{-1}N\left(N^T G^{-1} N\right)^{-1} N^T G^{-1}. \tag{10.31}$$

We see that N^* is the pseudo-inverse, or generalized Moore-Penrose, of N in the space of variables obtained through the transformation $y = G^{1/2}x$. On the other hand, H is the inverse of the reduced Hessian of function $Q(x)$ with respect to the active set of constraints. If $\widehat{x}$ is a point in the $(n-q)$ dimensional manifold $M = \left\{x \in \mathbb{R}^n : n_i^T x = b_i,\ i \in A\right\}$ and $g(\widehat{x}) \equiv \nabla Q(\widehat{x}) = G\widehat{x} + c$ is the gradient of the objective function $Q(x)$ at $\widehat{x}$, then the minimum of $Q(x)$ over M is attained in point $\bar{x} = \widehat{x} - Hg(\widehat{x})$. For $\bar{x}$ to be the optimal solution of the sub-problem $P(A)$, we must have

$$g(\bar{x}) = Nu(\bar{x}), \tag{10.32}$$

where the vector of Lagrange multipliers $u(\bar{x}) \geq 0$. From the definitions of N^* and H, it follows that at such a point:

$$u(\bar{x}) = N^* g(\bar{x}) \geq 0 \tag{10.33}$$

and

$$Hg(\bar{x}) = 0. \tag{10.34}$$

Observe that these conditions are necessary as well as sufficient for x to be the optimal solution of $P(A)$. Let us define $r = N^* n^+$ as the Lagrange multipliers associated to the infeasible constraints. Define H^+ as in (10.31) with N replaced by N^+. A similar notation is used for $(N^+)^*$ and u^+.

The algorithm presented below follows the dual approach described above and makes use of the operators N^* and H defined in (10.30) and (10.31), respectively (Goldfarb & Idnani, 1983). An efficient implementation does not explicitly compute and does not store these operators. Instead, this implementation stores and updates the matrices $J = Q^T L^{-1}$ and R obtained from the Cholesky and QR factorization of $G = LL^T$ and $L^{-1}N = Q\begin{bmatrix} R \\ 0 \end{bmatrix}$.

Algorithm 10.6 Dual algorithm for quadratic programming

1.	Find the unconstrained minimum of function $Q(x)$ from (10.29a). Compute $x=-G^{-1}c, H=G^{-1}, A=\varnothing$, and $q=0$
2.	Choose a violated constraint, if any, and compute $s_j(x)$ for all $j \in \mathcal{M}$. If $V=\{j \in \mathcal{M}: \ s_j(x)<0\}=\varnothing$, then stop; the current solution is optimal. Otherwise choose $p \in V$, and set $n^+=n_p$ and $u^+=[u \quad 0]^{\mathrm{T}}$. If $q=0$, then set $u=0$. $(A^+=A \cup \{p\})$
3.	Determine the search direction. Compute $z=Hn^+$ as the direction in primal space. If $q>0$, then set $r=N^*n^+$ the negative of the direction in the dual space
4.	Compute the step length: (i) Compute t_1 the maximum step in the dual space without violation of the dual feasibility. If $r \le 0$ or $q=0$, then set $t_1=\infty$; otherwise set: $$t_1=\min\left\{\frac{u_j^+(x)}{r_j}:\ r_j>0,\ j=1,\ldots,q\right\}=\frac{u_l^+(x)}{r_l}$$ In step 5 below, the element $k \in I$ corresponds to the lth element form A. (ii) Compute t_2 the minimum step in the primal space such that the pth constraint becomes feasible. If $\lvert z \rvert=0$, then set $t_2=\infty$; otherwise set: $$t_2=-\frac{s_p(x)}{z^{\mathrm{T}}n^+}$$ (iii) Step length computation: $t=\min(t_1,t_2)$
5.	Determine the pair and take the step: (i) If $t=\infty$,then no step in primal or dual space is executed, stop; the sub-problem $P(A^+)$ is infeasible, and hence the problem (10.29) is infeasible (ii) If $t_2=\infty$, then set $u^+=u^++t\begin{bmatrix}-r\\1\end{bmatrix}$ and delete the constraint k, i.e., set $A=A\backslash\{k\}$, $q=q-1$; update H and N^* and go to step 3 (iii) Step in the primal and dual space. Set $x=x+tz$, $$u^+=u^++t\begin{bmatrix}-r\\1\end{bmatrix}$$ If $t=t_2$,then set $u=u^+$ and add the constraint p to$A$, that is $A=A \cup \{p\}$; set $q=q+1$; update H and N^* and go to step 2 If $t=t_1$, then drop the constraint k, i.e., set $A=A\backslash\{k\}$, $q=q-1$; update H and N^* and go to step 3 ♦

Goldfarb and Idnani (1983) prove that the algorithm will solve the quadratic programming problem (10.29) or indicates that it has no feasible solution in a finite number of steps.

The implementation of the above algorithm, in a stable manner, is based on the Cholesky factorization

$$G = LL^{\mathrm{T}} \tag{10.35}$$

of the symmetric and positive definite matrix G and the QR factorization

$$B = Q\begin{bmatrix} R \\ 0 \end{bmatrix} = [Q_1 \quad Q_2]\begin{bmatrix} R \\ 0 \end{bmatrix} \tag{10.36}$$

of the $n \times q$ matrix:

$$B = L^{-1}N. \tag{10.37}$$

In these relations, L is an $n \times n$ lower triangular matrix, R is a $q \times q$ upper triangular matrix, and $[Q_1 \quad Q_2]$ is an $n \times n$ orthogonal matrix partitioned so that Q_1 has q columns. By substituting (10.35), (10.36) and (10.37) in (10.30) and (10.31), the operators H and N^* can be expressed as

$$H = J_2 J_2^{\mathrm{T}} \tag{10.38}$$

and

$$N^* = R^{-1} J_1^{\mathrm{T}}, \tag{10.39}$$

where:

$$[J_1 \quad J_2] = \left[L^{-\mathrm{T}}Q_1 \quad L^{-\mathrm{T}}Q_2\right] = L^{-\mathrm{T}}Q = J. \tag{10.40}$$

Although the algorithm is based on QR factorization, the orthogonal matrix Q from (10.36) is not stored; rather, the matrix $J = L^{-\mathrm{T}}Q$ is stored and updated, since whenever Q is called by the algorithm, it appears in conjunction with $L^{-\mathrm{T}}$.

In Algorithm 10.6, it is necessary to compute the vectors $z = Hn^+$ and $r = N^* n^+$. If we compute

$$d = J^{\mathrm{T}} n^+ = \begin{bmatrix} J_1^{\mathrm{T}} \\ J_2^{\mathrm{T}} \end{bmatrix} n^+ = \begin{bmatrix} d_1 \\ d_2 \end{bmatrix}, \tag{10.41}$$

then from (10.38) and (10.39), we get:

$$z = J_2 d_2, \quad r = R^{-1} d_1. \tag{10.42}$$

In the following, like in Goldfarb and Idnani (1983), let us present some details concerning the updating of the factors J and R when a constraint is added to or deleted from the active set.

Updating the Factors J and R Whenever a constraint is added to or deleted from the set of active constraints, the factors J and R must be updated. Numerically stable methods for updating the QR factorization of a matrix when it is modified by the addition or deletion of a single column are very well known (Daniel, Graggs, Kaufman, & Stewart, 1976; Gill, Golub, Murray, & Saunders, 1974; Goldfarb, 1975). In the updating scheme described by Goldfarb and Idnani (1983), the Givens matrices Q_{ij} which are equal to the identity matrix, with elements (i,i), (i,j), (j,i), and (j,j) replaced by t,s, s, and $-t$, respectively, where $t=\cos\theta$ and $s=\sin\theta$ for some value of θ, are used. Having in view the special structure, computations involving Givens matrices can be illustrated by considering the 2×2 matrix:

$$\widehat{Q}=\begin{bmatrix} t & s \\ s & -t \end{bmatrix}.$$

Since $t^2+s^2=1$, it follows that $\widehat{Q}$ is orthogonal. In all computations $\widehat{Q}$ is chosen in such a way that the vector $w=\begin{bmatrix} w_1 & w_2 \end{bmatrix}^{\mathrm{T}}$ is transformed into the vector $\begin{bmatrix} \omega & 0 \end{bmatrix}^{\mathrm{T}}$, where $\omega=\pm\left(w_1^2+w_2^2\right)^{1/2}$. To accomplish this, the following elements are computed:

$$\mu=\max\{|w_1|,|w_2|\},\ \omega=\operatorname{sign}(w_1)\mu\left[\left(\frac{w_1}{\mu}\right)^2+\left(\frac{w_2}{\mu}\right)^2\right]^{1/2},$$

$$t=\frac{w_1}{\omega}$$

and

$$s=\frac{w_2}{\omega}.$$

To compute $\widehat{y}=\widehat{Q}y=\widehat{Q}^{\mathrm{T}}y$, where $y=\begin{bmatrix} y_1 & y_2 \end{bmatrix}^{\mathrm{T}}$, the algorithm follows the scheme given in Gill, Golub, Murray and Saunders (1974) and determines:

$$v=\frac{w_2}{\omega+w_1}=\frac{s}{1+t},$$

and

$$\widehat{y}_1=ty_1+sy_2,\ \widehat{y}_2=v(y_1+\widehat{y}_1)-y_2. \tag{10.43}$$

Observe that if $\widehat{Q}$ (or $\widehat{Q}^{\mathrm{T}}$) is applied to a $2\times n$ matrix, then the above computing scheme saves $n-1$ multiplications but introduces $n+1$ additions over the usual procedure for matrix multiplications. Also, observe that the sign of ω is chosen so that there is no cancelation in the computation of ν.

Adding a Constraint When the constraint p with normal n^+ is added to the active-set A, then the factorization (10.36) is replaced by:

$$B^+ = \begin{bmatrix} B & L^{-1}n^+ \end{bmatrix} = Q^+ \begin{bmatrix} R^+ \\ 0 \end{bmatrix}. \tag{10.44}$$

Therefore, from (10.36), (10.40), and (10.41), it follows that:

$$Q^{\mathrm{T}}B^+ = \begin{bmatrix} R & d_1 \\ 0 & d_2 \end{bmatrix}.$$

Thus the factorization (10.44) can be obtained as.

$$Q^+ = Q\begin{bmatrix} I_q & 0 \\ 0 & \bar{Q}^{\mathrm{T}} \end{bmatrix} \quad \text{and} \quad R^+ = \begin{bmatrix} R & d_1 \\ 0 & \delta \end{bmatrix},$$

where $\delta = \|d_2\|$ and $\bar{Q} = Q_{1,2}Q_{2,3}\cdots Q_{n-q-1,n-q}$ is the product of the Givens matrices chosen so that:

$$\bar{Q}d_2 = \pm\delta e_1. \tag{10.45}$$

Moreover, we see that

$$J^+ = L^{-\mathrm{T}}Q^+ = \begin{bmatrix} J_1 & J_2\bar{Q}^{\mathrm{T}} \end{bmatrix} = \begin{bmatrix} J_1^+ & J_2^+ \end{bmatrix}, \tag{10.46}$$

where J_1 has q columns, $J_2\bar{Q}^{\mathrm{T}}$ has $n-q$ columns, J_1^+ has $q+1$ columns, and J_2^+ has $n-q-1$ columns.

Dropping a Constraint When the lth constraint is deleted from the active set, that is when the lth column of N is deleted to form N^+, then

$$Q_1^TB^- = Q_1^TL^{-1}N^- = \begin{bmatrix} R_1 & S \\ 0 & T \end{bmatrix},$$

where the partitioned matrix on the right is equal to R but with its lth column eliminated. If $l \neq q$, then T is a $(q-l+1)\times(q-l)$ upper-Hessenberg matrix. Again, a sequence of $q-l$ Givens matrices can be chosen so that the product of these matrices $\bar{Q} = Q_{q-l,q-l+1}\cdots Q_{2,3}Q_{1,2}$ reduces T to an upper triangular matrix R_2, i.e.,

$$\bar{Q}T = R_2. \tag{10.47}$$

Thus, the new factors are obtained as:

$$R^{-} = \begin{bmatrix} R_1 & S \\ 0 & R_2 \end{bmatrix} \quad and \ J^{-} = J \begin{bmatrix} I_{l-1} & 0 & 0 \\ 0 & \bar{Q}^{\mathrm{T}} & 0 \\ 0 & 0 & I_{n-q} \end{bmatrix}. \tag{10.48}$$

The matrices $\bar{Q}$ from (10.45) and (10.47) are not computed. Instead, the Givens matrices from which they are formed, which successively introduce zeros into d_2 and T, are applied directly to the rows of J^{T} using the computational scheme (10.44).

When the step directions z for the primal variables and $-r$ for the dual variables given by (10.42) are computed, we need the vector $d = J^{\mathrm{T}} n^{+}$. When a constraint is deleted from the active set, then the same orthogonal transformations that are applied to update J^{T} can be applied to d. If these operations are executed, then the updated d can be used to compute the new vectors z and r after the bases change and to determine the appropriate orthogonal transformation $\bar{Q}$ in (10.45) needed when the pth constraint is finally added to the active set.

The Algorithm 10.6 (dual for quadratic programming) is one of the most advanced algorithms for solving quadratic programming problems with equality and inequality constraints. Its efficiency is coming from two sources: the dual method and the implementation of an advanced computational scheme based on QR factorization which is updated by Givens matrices. One of the first implementation of the Algorithm 10.6 was given by Powell (1983) (Fortran package ZQPCVX). A professional extension was given by Schittkowski (2005) (Fortran package QLD).

Example 10.2 Let us consider the following quadratic programming problem (Andrei, 1999, pp. 792; Andrei, 2003, pp. 329)

$$\min Q(x) \equiv \sum_{i=1}^{15} x_i^2 + \sum_{i=1}^{14} (x_i + x_{i+1})$$

subject to:

$$\begin{aligned}
&0.1x_1 + 0.1x_7 + 0.3x_8 + 0.2x_9 + 0.2x_{11} - 1 = 0,\\
&0.1x_2 + 0.2x_8 + 0.3x_9 + 0.4x_{10} + x_{11} - 2 = 0,\\
&0.1x_3 + 0.2x_8 + 0.3x_9 + 0.4x_{10} + 2x_{11} - 3 = 0,\\
&x_4 + x_8 + 0.5x_9 + 0.5x_{10} + x_{11} - x_{12} - 3 = 0,\\
&2x_5 + x_6 + 0.5x_7 + 0.5x_8 + 0.25x_9 + 0.25x_{10} + 0.5x_{11} - x_{13} - 4 = 0,\\
&x_4 + x_6 + x_8 + x_9 + x_{10} + x_{11} - x_{14} - 5 = 0,\\
&0.1x_1 + 1.2x_7 + 1.2x_8 + 1.4x_9 + 1.1x_{10} + 2x_{11} - x_{15} - 6 = 0,\\
&0 \le x_i \le 2, \quad i = 1, \ldots, 15.
\end{aligned}$$

Table 10.1 contains information on the optimization process given by the Algorithm 10.6 in QLD implementation by Schittkowski (2005). To get a solution, the algorithm needs 8 iterations.

In Table 10.1 we see the evolution of the values of the objective function along the iterations. Notice also the evolution of the sum of constraints violation, which tends to zero, as well as the evolution of the norm of the KKT conditions. The number of active constraints is seven; the problem has only equality constraints. The stepsize is equal to one along the iterations, like in the Newton method from unconstrained optimization. In Table 10.2 we see the initial point, the bounds on variables, and the solution given by QLD. Observe that the initial point is not feasible.

Table 10.1 Optimization process by QLD

k	$Q(x_k)$	svr	nra	t	kkt
1	97.000000	27.800000	7	0	68.760324
2	46.471283	0.266453e-14	7	1	17.341127
3	31.302619	0.155431e-14	7	1	10.560131
4	22.807444	0	7	1	0.740935
5	22.179315	0.199840e-14	7	1	0.168103
6	22.037532	0.666133e-15	7	1	0.232076
7	21.921296	0.111022e-15	7	1	0.670174e-06
8	21.921296	0.888178e-15	7	1	0.303832e-14

In Table 10.1 we have: $k =$ the number of iterations, $Q(x_k) =$ the value of the objective function, $svr =$ the sum of constraints violation, $nra =$ the number of active constraints, $t =$ the stepsize, $kkt =$ the value of the KKT conditions at x_k.

Table 10.2 Initial point, solution, and bounds on variables

	l	x_0	x^*	u
1	0	2.1	0.400273	2
2	0	2.1	0.130566	2
3	0	2.1	0	2
4	0	2.1	0	2
5	0	2.1	0.903379	2
6	0	2.1	0.416858	2
7	0	2.1	0	2
8	0	2.1	1.509334	2
9	0	2.1	1.522805	2
10	0	2.1	0.537945	2
11	0	2.1	1.013057	2
12	0	1	0.552766	2
13	0	1	0	2
14	0	1	0	2
15	0	1	0.601009	2

Application Dynamic (Andrei, 2013b, pp.145) This application is a representation of the problem concerning the scheduling of three generators to meet the demand for power over a given period of time. The variables x_{3k+i} denote the output from the generator i at time t_k. The constraints of the problem are upper and lower limits on the power available from each generator, bounds on the amount by which the output from a generator can change from time t_k to t_{k+1}, and the condition that at each time t_k the power generated must at least satisfy the demand. The mathematical model is described in Pant, Thangaraj and Singh (2009) and is as follows

$$\min \sum_{k=0}^{4} (2.3x_{3k+1} + 0.0001x_{3k+1}^2 + 1.7x_{3k+2} + 0.0001x_{3k+2}^2 + \\ 2.2x_{3k+3} + 0.00015x_{3k+3}^2)$$

subject to:

$$\begin{aligned}
&x_1 + x_2 + x_3 \geq 60, \\
&x_4 + x_5 + x_6 \geq 50, \\
&x_7 + x_8 + x_9 \geq 70, \\
&x_{10} + x_{11} + x_{12} \geq 85, \\
&x_{13} + x_{14} + x_{15} \geq 100, \\
&-7 \leq x_{3k+1} - x_{3k-2} \leq 6, \quad k = 1, \ldots, 4, \\
&-7 \leq x_{3k+2} - x_{3k-1} \leq 7, \quad k = 1, \ldots, 4, \\
&-7 \leq x_{3k+3} - x_{3k} \leq 6, \quad k = 1, \ldots, 4.
\end{aligned}$$

The variables are bounded as follows:

$$\begin{array}{lll}
8 \leq x_1 \leq 21, & 0 \leq x_{3k+1} \leq 90, & k = 1, \ldots, 4 \\
43 \leq x_2 \leq 57, & 0 \leq x_{3k+2} \leq 120, & k = 1, \ldots, 4, \\
3 \leq x_3 \leq 16, & 0 \leq x_{3k+3} \leq 60, & k = 1, \ldots, 4.
\end{array}$$

Considering the initial point $x_i^0 = (l_i + u_i)/2$, $i = 1, \ldots, 15$, the Algorithm 10.6 gives the following solution:

$$\begin{array}{lllll}
x_1^* = 8, & x_2^* = 49, & x_3^* = 3, & x_4^* = 1, & x_5^* = 52, \\
x_6^* = 0, & x_7^* = 5, & x_8^* = 59, & x_9^* = 6, & x_{10}^* = 7, \\
x_{11}^* = 66, & x_{12}^* = 12, & x_{13}^* = 0, & x_{14}^* = 95, & x_{15}^* = 5.
\end{array}$$

The value of the objective function at this point is $Q(x^*) = 653.4427$.

Table 10.3 gives some information about the optimization process with QLD (Schittkowski, 2005). To get the above solution, the algorithm needs 18 iterations.

Table 10.3 Dynamic. Optimization process by QLD

k	$Q(x_k)$	svr	nra	t	kkt
1	1228.3246	0.42e + 02	29	0	0.12e + 04
2	1115.8453	0	6	1	0.11e + 03
3	1006.9820	0	4	1	0.32e + 03
4	762.61922	0.11e-13	4	1	0.76e + 02
5	708.25997	0.89e-14	8	1	0.26e + 02
6	692.67975	0.71e-14	9	1	0.15e + 01
7	691.18494	0.36e-14	8	1	0.70e + 01
8	684.96921	0	8	1	0.11e + 02
9	674.81722	0.18e-13	8	1	0.48e + 01
10	670.99302	0.67e-15	9	1	0.28e + 01
11	668.92085	0.75e-14	8	1	0.21e + 01
12	667.49688	0.53e-14	8	1	0.64e + 00
13	666.86138	0.12e-13	7	1	0.32e + 01
14	663.69246	0.18e-14	7	1	0.12e + 02
15	655.26305	0.18e-14	10	1	0.48e + 00
16	654.78703	0.53e-14	10	1	0.78e + 00
17	654.12603	0	10	1	0.12e + 01
18	653.44270	0	11	1	0.20e-30

Notes and References

The quadratic programming problem has its importance in the nonlinear optimization area; many optimization problems are being modeled as quadratic programs. The most famous is the portfolio optimization formulated by Markowitz (1952). This chapter is based on Nocedal and Wright (2006) and Sun and Yuan (2006). Methods for general quadratic programming are presented in Gill, Murray, Saunders and Wright (1984) and Gould (1991). A quadratic programming bibliography is presented by Gould and Toint (2012). Quadratic programming is intensively used in sequential quadratic programming, one of the most efficient methods in nonlinear optimization. We have considered here only the active-set methods for quadratic programming. The interior point methods for quadratic programming can be very easily adapted from the general theory (see Chapter 17). Interior point methods for convex quadratic programming can be found in Wright (1997) and Vanderbei (2001). The numerical comparison of active set and interior point methods for convex quadratic programming shows that interior point methods are generally much faster on large problems (Gould & Toint, 2002a). However, if a warm start is available, the active-set methods may be generally preferable. Active-set methods for convex quadratic programming are implemented in many codes: QPOPT (Gill & Murray, 1978), BQPD (Fletcher, 2000), and QPA (Gould & Toint, 2002b). The interior point solvers for quadratic programming are CPLEX (ILOG CPLEX, 2002), XPRESS-MP (Guéret, Prins, & Sevaux, 2002), MOSEK (Andresen and Andresen 2000), etc. The problem of determining whether a feasible point for a nonconvex quadratic program is a NP-hard problem was proven in Murty and Kabadi (1987). Also, the problem of determining whether a given point is a local minimizer is NP-hard (Vavasis, 1990).

Chapter 11
Sequential Quadratic Programming (SQP)

SQP is an active-set method. In this chapter we consider both the equality-constrained and the inequality-constrained sequential quadratic programming.

Equality-Constrained Problems Let us consider the equality-constrained nonlinear optimization problem

$$\min f(x) \tag{11.1a}$$

subject to

$$h(x) = 0, \tag{11.1b}$$

where $f: \mathbb{R}^n \to \mathbb{R}$ and $h: \mathbb{R}^n \to \mathbb{R}^m$ are smooth functions. The idea of the SQP method is to model (11.1) at the current point x_k by a quadratic programming sub-problem and then to use the solution of this sub-problem to define a new iterate x_{k+1}. The most challenging is how to design the quadratic sub-problem so that it yields a good approximation of (11.1). Perhaps the simplest derivation of the SQP methods is to apply Newton method to the KKT optimality conditions for the nonlinear problem (11.1).

The Lagrange function for (11.1) is

$$L(x, \lambda) = f(x) - \lambda^T h(x), \tag{11.2}$$

where $\lambda \in \mathbb{R}^m$ is the vector of the Lagrange multipliers associated to the constraints of the problem. Let us define $A(x)$ as the Jacobian matrix of the constraints, that is,

$$A(x)^T = [\nabla h_1(x), \ldots, \nabla h_m(x)] \in \mathbb{R}^{n \times m}, \tag{11.3}$$

where $h_i(x)$, $i = 1, \ldots, m$, are the components of the vector $h(x)$. The first-order optimality conditions of the equality-constrained problem can be written as a system of $n + m$ equations with $n + m$ unknowns x and λ:

N. Andrei, *Continuous Nonlinear Optimization for Engineering Applications in GAMS Technology*, Springer Optimization and Its Applications 121,
DOI 10.1007/978-3-319-58356-3_11

$$F(x,\lambda) = \begin{bmatrix} \nabla f(x) - A(x)^{\mathrm{T}}\lambda \\ h(x) \end{bmatrix} = 0. \tag{11.4}$$

Observe that any solution (x^*, λ^*) of (11.1) for which $A(x^*)$ has full rank satisfies (11.4). To find such a point, solution of the nonlinear system (11.4), the Newton method is the most suitable. The Jacobian of (11.4) with respect to x and λ is given by:

$$F'(x,\lambda) = \begin{bmatrix} \nabla_{xx}^2 L(x,\lambda) & -A(x)^{\mathrm{T}} \\ A(x) & 0 \end{bmatrix}. \tag{11.5}$$

The Newton step from the iterate (x_k, λ_k) is given by

$$\begin{bmatrix} x_{k+1} \\ \lambda_{k+1} \end{bmatrix} = \begin{bmatrix} x_k \\ \lambda_k \end{bmatrix} + \begin{bmatrix} p_x \\ p_\lambda \end{bmatrix}, \tag{11.6}$$

where p_x and p_λ are solutions of the Newton system:

$$\begin{bmatrix} \nabla_{xx}^2 L(x_k,\lambda_k) & -A(x_k)^{\mathrm{T}} \\ A(x_k) & 0 \end{bmatrix} \begin{bmatrix} p_x \\ p_\lambda \end{bmatrix} = \begin{bmatrix} -\nabla f(x_k) + A(x_k)^{\mathrm{T}}\lambda_k \\ -h(x_k) \end{bmatrix}. \tag{11.7}$$

The Newton iteration (11.6) is well defined when the KKT matrix in (11.7) is nonsingular. As we know (see Chapter 5), this matrix is nonsingular if the following two assumptions hold:

C1) *The Jacobian $A(x)$ of the constraints has full row rank.*
C2) *The matrix $\nabla_{xx}^2 L(x,\lambda)$ is positive definite on the tangent space of the constraints, that is, $p^{\mathrm{T}}\nabla_{xx}^2 L(x,\lambda)p > 0$ for all $p \neq 0$ for which $A(x)p = 0$.*

Observe that the first assumption is exactly the linear independence constraint qualification (LICQ) discussed in Chapter 5 (see Remark 5.2). The second assumption holds whenever (x,λ) is close to the optimum point (x^*,λ^*), and the second-order sufficient condition is satisfied at the solution (see Theorem 5.13).

This approach, which theoretically is very well justified, has an alternative that illustrates the motivation of the sequential quadratic programming. Let us focus on the Newton iteration given by (11.6) and (11.7), and let us assume that at the iteration (x_k, λ_k), the problem (11.1) is modeled as a quadratic program

$$\min_{p\in\mathbb{R}^n} f(x_k) + \nabla f(x_k)^{\mathrm{T}} p + \frac{1}{2} p^{\mathrm{T}} \nabla_{xx}^2 L(x_k,\lambda_k) p \tag{11.8a}$$

subject to:

$$A(x_k)p + h(x_k) = 0. \tag{11.8b}$$

If the assumptions C1 and C2 hold, then this problem has a unique solution (p_k, l_k) that satisfies:

$$\nabla_{xx}^2 L(x_k, \lambda_k) p_k + \nabla f(x_k) - A(x_k)^T l_k = 0, \quad (11.9a)$$

$$A(x_k) p_k + h(x_k) = 0. \quad (11.9b)$$

But we see that the vectors p_k and l_k can be immediately identified as the solution of the Newton system (11.7). Now, if we subtract $A(x_k)^T\lambda_k$ from both sides of the first equation in (11.7), we get:

$$\begin{bmatrix} \nabla_{xx}^2 L(x_k, \lambda_k) & -A(x_k)^T \\ A(x_k) & 0 \end{bmatrix} \begin{bmatrix} p_k \\ \lambda_{k+1} \end{bmatrix} = \begin{bmatrix} -\nabla f(x_k) \\ -h(x_k) \end{bmatrix}. \quad (11.10)$$

Therefore, having in view the nonsingularity of the coefficient matrix of (11.10), we see that $\lambda_{k+1} = l_k$ and p_k represent a solution of the quadratic program (11.8a) and (11.8b) of the system (11.7).

In conclusion, *the new iterate* (x_{k+1}, λ_{k+1}) *can be defined either as the solution of the quadratic programming problem* (11.8) *or as the iterate generated by the Newton method* (11.6) *and* (11.7) *applied to the optimality conditions of the problem* (11.1). Both these approaches are useful. The Newton point of view is suitable for analysis, whereas the sequential quadratic programming enables us to derive practical algorithms and to extend the technique to the inequality-constrained case (Nocedal & Wright, 2006).

Based on these theoretical developments, the following sequential quadratic programming algorithm for equality constraints can be presented.

Algorithm 11.1 Sequential quadratic programming – equality constraints

1.	Choose an initial point (x_0, λ_0) and set $k=0$
2.	If a test for stopping the iterations is satisfied, stop; otherwise go to step 3
3.	Evaluate: $f(x_k)$, $\nabla f(x_k)$, $\nabla_{xx}^2 L(x_k, \lambda_k)$, $h(x_k)$, and $A(x_k)$
4.	Solve the quadratic sub-problem (11.8) to obtain a solution (p_k, l_k)
5.	Set $x_{k+1} = x_k + p_k$ and $\lambda_{k+1} = l_k$. Set $k = k+1$ and go to step 2 ♦

It is obvious that in the objective function (11.8a), the linear term $\nabla f(x_k)^T p$ can be replaced by $\nabla_x L(x_k, \lambda_k)^T p$, since the constraints (11.8b) make the two choices equivalent. In this case, (11.8a) is exactly a quadratic approximation of the Lagrangian function. This is the main motivation for the choice of the quadratic model (11.8): *first replace the nonlinear optimization problem* (11.1) *by the problem of minimizing the Lagrangian subject to the equality constraints* (11.1b), *and then make a quadratic approximation to the Lagrangian and a linear approximation to the constraints to obtain the quadratic program* (11.8).

Now, let us present the conditions that guarantee the local convergence of the Algorithm 11.1. Consider that the algorithm uses exact second derivatives.

Theorem 11.1 *Assume that the point* x^* *is a local solution of the problem* (11.1) *at which the following conditions hold:*

(i) The functions f and h are twice continuously differentiable in a neighborhood of x^* *with Lipschitz continuous second derivatives.*

(ii) The linear independence constraint qualification (see Remark 5.2) *holds at* x^*. *This condition implies that the KKT conditions* (5.21) *are satisfied for some vector of multipliers* λ^*.

(iii) The second-order sufficient conditions (see Theorem 5.16) *hold at* (x^*, λ^*).

Then, if (x_0, λ_0) *is sufficiently close to* (x^*, λ^*), *the pairs* (x_k, λ_k) *generated by the Algorithm 11.1 quadratically converge to* (x^*, λ^*).

Proof. This result follows directly from the convergence of the Newton method applied to the nonlinear system (11.4). ♦

The Algorithm 11.1 can be specified in some variants. A version of the algorithm uses a quasi-Newton estimate of $\nabla^2_{xx}L(x_k, \lambda_k)$ rather than calculating it from the second derivatives see Wilson (1963), Han (1976), and Powell (1978a). In step 5 of the algorithm, we can use the line search to determine $x_{k+1} = x_k + sp_k$ ensuring $P(x_{k+1}) < P(x_k)$, where P denotes some penalty function. Different penalty functions can be tried. The line search is important because it forces a reduction in a composite function involving both the objective f and the constraints h. It ensures that the new point x_{k+1} is, in a measurable way, an improvement on x_k and therefore provides a basis for a proof of convergence.

Inequality-Constrained Problems The above-described sequential quadratic programming approach can be extended to the general nonlinear optimization problem

$$\min f(x) \tag{11.11a}$$

subject to

$$c_i(x) = 0, \quad i \in E, \tag{11.11b}$$

$$c_i(x) \geq 0, \quad i \in I, \tag{11.11c}$$

where $E \triangleq \{1, \ldots, m_e\}$ and $I \triangleq \{m_e + 1, \ldots, m\}$ are sets on indices for the equality and inequality constraints, respectively. Assume that all the functions of the problem (11.11) are twice continuously differentiable.

As in the case of equality-constrained problems, at the current point x_k, the problem (11.11) is modeled as a quadratic program by linearizing both the equality and the inequality constraints to obtain

$$\min_{p \in \mathbb{R}^n} f(x_k) + \nabla f(x_k)^{\mathrm{T}} p + \frac{1}{2} p^{\mathrm{T}} \nabla^2_{xx} L(x_k, \lambda_k) p \tag{11.12a}$$

subject to:

$$\nabla c_i(x_k)^{\mathrm{T}} p + c_i(x_k) = 0, \quad i \in E, \tag{11.12b}$$

$$\nabla c_i(x_k)^{\mathrm{T}} p + c_i(x_k) \geq 0, \quad i \in I. \tag{11.12c}$$

For solving the sub-problem (11.12), we can use the algorithms for quadratic programming described in Chapter 10. The new iterate is given by (x_k+p_k, λ_{k+1}), where p_k and λ_{k+1} are the solution and the corresponding Lagrange multiplier of (11.12). The algorithm is as follows.

Algorithm 11.2 Sequential quadratic programming – inequality constraints

1.	Choose an initial point (x_0, λ_0) and set $k=0$
2.	If a test for stopping the iterations is satisfied, stop; otherwise go to step 3
3.	Evaluate: $f(x_k)$, $\nabla f(x_k)$, $\nabla^2_{xx}L(x_k, \lambda_k)$, $c_i(x_k)$, and $\nabla c_i(x_k), i \in E \cup I$
4.	Solve the quadratic problem (11.12) to obtain a solution (p_k, l_k)
5.	Set $x_{k+1}=x_k+p_k$ and $\lambda_{k+1}=l_k$. Set $k=k+1$ and go to step 2 ♦

In this approach the set of active constraints at the solution of (11.12) is the guess of the active set at the solution of the nonlinear problem (11.11). If the SQP method is able to correctly identify the optimal active set (i.e., the active set at the solution of (11.11)), then the method works like a Newton method for equality-constrained optimization. In this case the convergence will be rapid. The following result, given by Robinson (1974), presents the conditions under which this behavior of the algorithm holds.

Theorem 11.2 *Assume that* x^* *is a local solution of* (11.11) *at which the KKT conditions are satisfied for some* λ^*. *Also, assume that the linear independence constraint qualification* (Remark 5.2), *the strict complementarity* (Definition 5.22), *and the second-order sufficient conditions* (Theorem 5.13) *hold at* (x^*, λ^*). *Then, if* (x_k, λ_k) *is sufficiently close to* (x^*, λ^*), *there is a local solution of the sub-problem* (11.12) *whose active set is the same as the active set of the nonlinear program* (11.11) *at* x^*. ♦

Remark 11.1 In designing SQP algorithms for solving the general nonlinear optimization problem (11.11), two methods are known. The first, which has been described above, solves the quadratic sub-problem (11.12) at each iteration, taking the active set at the solution of this sub-problem as a guess of the optimal active set. This approach is called IQP (*inequality-constrained QP*). It has proved to be quite effective and successful in practice. Its main disadvantage is the expense of solving the general quadratic sub-problem (11.12), which can be high when the problem is large. However, an improvement of this situation could be: as the iterates converge to the solution, solving the quadratic sub-problem (11.12) becomes economical if the information from the previous iteration is used to make a good guess of the optimal solution to the current sub-problem. This strategy is called *warm start*. The second approach selects a subset of constraints at each iteration to be the so-called *working set* and solves only equality-constrained sub-problems of the form (11.8), where the constraints in the working set are imposed as equalities and all other constraints are ignored. This approach is called EQP (*equality-constrained QP*). In it the working set is updated at every iteration by rules based on the Lagrange multiplier estimates or by solving an auxiliary sub-problem. The EQP is more effective because the equality-constrained quadratic sub-problems are less expensive to solve than (11.12), when the problem is large. An example of an EQP method is the sequential linear-quadratic programming (SLQP) method, which is to be discussed in Section 11.6. Another successful EQP is the gradient projection method in the context of bound constrained quadratic programming. ♦

Numerical experience has shown that the SQP method often converges to a solution from remote starting points. Along the years there has been a considerable interest in understanding and improving the design and implementation of algorithms based on the idea of sequential quadratic programming. Nocedal and Wright (2006) present some details about these improvements referring to enforcing convergence from remote starting points, handling inconsistent linearizations, full quasi-Newton approximations, damped BFGS updating, reduced Hessian quasi-Newton approximations, merit function in sequential quadratic programming using nonsmooth penalty functions or augmented Lagrangians, second-order correction to overcome the Maratos effect, etc. In the following we describe some ingredients needed to produce *practical* SQP algorithms able to solve large-scale nonlinear optimization problems (Nocedal and Wright, 2006).

11.1 Reduced Hessian Quasi-Newton Approximations

Let us consider the equality-constrained problem (11.1). By examining the KKT system (11.10), we see that the part of step p_k in the range space of $A(x_k)^T$ is completely determined by the line $A(x_k)p_k = -h(x_k)$. The Lagrangian $\nabla^2_{xx}L(x_k, \lambda_k)$ affects only the part of p_k in the orthogonal subspace, i.e., the null-space of A_k. Therefore, it seems reasonable to consider quasi-Newton methods that find approximations to only that part of $\nabla^2_{xx}L(x_k, \lambda_k)$ which affects the component of p_k in the null-space of A_k. This is the idea of reduced Hessian quasi-Newton approximations which we are to detail in this section. For this, we consider the solution of the step equations (11.10) by means of the null-space approach presented in Section 10.1.

Let us define the matrices Y_k and Z_k whose columns span the range space of $A(x_k)^T$ and the null-space of $A(x_k)$, respectively. Using these matrices, p_k can be written as

$$p_k = Y_k p_Y + Z_k p_Z, \tag{11.13}$$

where p_Y is the *normal component* and p_Z is the *tangential component* of p_k. By substituting p_k from (11.13) in (11.10), the following system is obtained

$$[A(x_k)Y_k]p_Y = -h(x_k), \tag{11.14a}$$

$$\left[Z_k^T \nabla^2_{xx}L(x_k, \lambda_k)Z_k\right]p_Z = -Z_k^T \nabla^2_{xx}L(x_k, \lambda_k)Y_k p_Y - Z_k^T \nabla f(x_k), \tag{11.14b}$$

to get the components p_Y and p_Z, in this order. Observe that from the first line of (11.10), the Lagrange multipliers λ_{k+1} can be computed by solving the linear system:

$$[A(x_k)Y_k]^T\lambda_{k+1} = Y_k^T\left[\nabla f(x_k) + \nabla_{xx}^2 L(x_k\lambda_k)p_k\right]. \tag{11.15}$$

In practical implementations some simplifications are often considered. The idea is to avoid the computation of $\nabla_{xx}^2 L(x_k, \lambda_k)$ by introducing several approximations in the null-space approach.

One simplification in the null-space approach is to remove the cross term $Z_k^T \nabla_{xx}^2 L(x_k, \lambda_k) Y_k p_Y$, in (11.14b), thus obtaining the simpler system:

$$\left[Z_k^T \nabla_{xx}^2 L(x_k, \lambda_k) Z_k\right] p_Z = -Z_k^T \nabla f(x_k). \tag{11.16}$$

Dropping the cross term is motivated when $Z_k^T \nabla_{xx}^2 L(x_k, \lambda_k) Z_k$ is replaced by a quasi-Newton approximation because the normal component p_Y usually converges to zero faster than the tangential component p_Z, thereby making (11.16) a good approximation of (11.14b).

Another simplification consists in deleting the term involving p_k from the right-hand side of (11.15), thus decoupling the computations of p_k and λ_{k+1}. A motivation of this simplification is that p_k converges to zero as we approach the solution, whereas $\{\nabla f(x_k)\}$ does not. If we choose $Y_k = A(x_k)^T$ (when $A(x_k)$ has full row rank), we obtain:

$$\widehat{\lambda}_{k+1} = \left[A(x_k)A(x_k)^T\right]^{-1} A(x_k)\nabla f(x_k). \tag{11.17}$$

11.2 Merit Functions

Often, SQP methods use a merit function to see whether a trial step should be accepted. In line-search methods, the merit function controls the size of the step. In trust-region methods, it determines whether the step is accepted or rejected and whether the trust-region radius should be adjusted (Nocedal & Wright, 2006). A variety of merit functions has been proposed in the literature, including nonsmooth penalty functions (l_1 merit function) and augmented Lagrangians. For the step computation and evaluation of a merit function, the inequality constraints are converted to equalities by introducing the slack variables. Therefore, in the following we consider only the nonlinear equality-constrained problem (11.1).

The l_1 merit function for (11.1) is defined by:

$$P(x, \sigma) = f(x) + \sigma \|h(x)\|_1. \tag{11.18}$$

Another merit function was proposed by Wright (1976) and Schittkowski (1981, 1983) as augmented Lagrangian

$$L(x,\sigma,\lambda)=f(x)-\lambda^{\mathrm{T}}h(x)+\frac{1}{2}\sigma h(x)^{\mathrm{T}}h(x), \tag{11.19}$$

where λ is an estimation of the Lagrange multipliers and $\sigma>0$ is the penalty parameter.

In a line-search method, a step $\alpha_k p_k$ is accepted if the following sufficient decrease condition holds

$$P(x_k+\alpha_k p_k,\sigma_k)\le P(x_k,\sigma_k)+\eta\alpha_k P'_{p_k}(x_k,\sigma_k), \tag{11.20}$$

where $P'_{p_k}(x_k,\sigma_k)$ is the directional derivative of $P(.)$ in the direction p_k and η is a parameter sufficiently small, $\eta\in(0,1)$. If p_k is a descent direction, that is, $P'_{p_k}(x_k,\sigma_k)<0$, then the reduction (11.20) is resembling the rule of Armijo used in unconstrained optimization (Andrei, 2009). The following theorem shows that if σ is chosen sufficiently large, then the descent condition holds.

Theorem 11.3 *Let* p_k *and* λ_{k+1} *be generated by the system* (11.10). *Then the directional derivative of* $P(.)$ *in the direction* p_k*, satisfies:*

$$P'_{p_k}(x_k,\sigma_k)=\nabla f(x_k)^{\mathrm{T}}p_k-\sigma\|h(x_k)\|_1. \tag{11.21}$$

Moreover, we have:

$$P'_{p_k}(x_k,\sigma_k)\le -p_k^{\mathrm{T}}\nabla_{xx}^2L(x_k,\lambda_k)p_k-\left(\sigma-\|\lambda_{k+1}\|_\infty\right)\|h(x_k)\|_1. \tag{11.22}$$

Proof. By applying Taylor's theorem, we get

$$\begin{aligned}P(x_k+\alpha p,\sigma)-P(x_k,\sigma)&=f(x_k+\alpha p)-f(x_k)+\sigma\|h(x_k+\alpha p)\|_1-\sigma\|h(x_k)\|_1\\&\le\alpha\nabla f(x_k)^{\mathrm{T}}p+\gamma\alpha^2\|p\|^2+\sigma\|h(x_k)+\alpha A(x_k)p\|_1-\sigma\|h(x_k)\|_1,\end{aligned}$$

where γ is a positive constant which bounds the second derivative terms. If $p=p_k$ is given by (11.10), then $A(x_k)p_k=-h(x_k)$, so for $\alpha\le 1$ it follows that:

$$P(x_k+\alpha p_k,\sigma)-P(x_k,\sigma)\le\alpha\left[\nabla f(x_k)^{\mathrm{T}}p_k-\sigma\|h(x_k)\|_1\right]+\alpha^2\gamma\|p_k\|^2.$$

Similarly, we obtain the following lower bound:

$$P(x_k+\alpha p_k,\sigma)-P(x_k,\sigma)\ge\alpha\left[\nabla f(x_k)^{\mathrm{T}}p_k-\sigma\|h(x_k)\|_1\right]-\alpha^2\gamma\|p_k\|^2.$$

Taking limits, we conclude that

$$P'_{p_k}(x_k, \sigma_k) = \nabla f(x_k)^{\mathrm{T}} p_k - \sigma \|h(x_k)\|_1,$$

which proves (11.21).

Now, since p_k satisfies the first equation from the system (11.10), then:

$$P'_{p_k}(x_k, \sigma_k) = -p_k^{\mathrm{T}} \nabla_{xx}^2 L(x_k, \lambda_k) p_k + p_k^{\mathrm{T}} A(x_k)^{\mathrm{T}} \lambda_{k+1} - \sigma \|h(x_k)\|_1. \tag{11.23}$$

But from the second equation of the system (11.10), the term $p_k^{\mathrm{T}} A(x_k)^{\mathrm{T}} \lambda_{k+1}$ from (11.23) can be replaced by $-h(x_k)^{\mathrm{T}} \lambda_{k+1}$. Having in view the Hölder inequality (see Appendix A), we have $-h(x_k)^{\mathrm{T}} \lambda_{k+1} \leq \|h(x_k)\|_1 \|\lambda_{k+1}\|_\infty$; then from (11.23) we obtain (11.22). ♦

A strategy for selecting σ, which is appropriate both for line-search and trust-region algorithms, has in view the effect of the step on a model of the merit function. In Nocedal and Wright (2006), a quadratic model for function $P(.)$ is defined as

$$q_\sigma(p) = f(x_k) + \nabla f(x_k)^{\mathrm{T}} p + \frac{\mu}{2} p^{\mathrm{T}} \nabla_{xx}^2 L(x_k, \lambda_k) p + \sigma m(p), \tag{11.24}$$

where $m(p) = \|h(x_k) + A(x_k) p\|_1$, and μ is a parameter to be defined. The strategy is as follows. After computing a step p_k, the penalty parameter σ is chosen large enough such that

$$q_\sigma(0) - q_\sigma(p_k) \geq \rho \sigma [m(0) - m(p_k)], \tag{11.25}$$

for some value of parameter $\rho \in (0, 1)$. Now, from (11.24) and (11.8b), we see that the inequality (11.25) is satisfied for:

$$\sigma \geq \frac{\nabla f(x_k)^{\mathrm{T}} p_k + (\mu/2) p_k^{\mathrm{T}} \nabla_{xx}^2 L(x_k, \lambda_k) p_k}{(1 - \rho) \|h(x_k)\|_1}. \tag{11.26}$$

If the value of σ from the previous iteration of the SQP method satisfies (11.26), then it is left unchanged. Otherwise, σ is increased so that it satisfies this inequality with some margins. The constant μ is used to handle the situations in which the Hessian $\nabla_{xx}^2 L(x_k, \lambda_k)$ is not positive definite. A simple strategy for μ is:

$$\mu = \begin{cases} 1, & \text{if } \; p_k^{\mathrm{T}} \nabla_{xx}^2 L(x_k, \lambda_k) p_k > 0, \\ 0, & \text{otherwise.} \end{cases} \tag{11.27}$$

11.3 Second-Order Correction (Maratos Effect)

In constrained nonlinear optimization, we practically face two purposes. One is to minimize the objective function, and another one is to satisfy the constraints. A way to balance these competing goals is using merit functions or filters. The idea is that a step generated by an algorithm will be accepted only if it leads to a sufficient reduction in the merit function or if it is acceptable to the filter. However, the following situation may appear. Suppose that an algorithm for solving nonlinear optimization problems generates a step that reduces the objective function value but increases the violation of the constraints, a phenomenon known as the *Maratos effect* (1978). The following example due to Powell (1986) illustrates this situation.

Example 11.1 Let us consider the problem

$$\min f(x_1, x_2) = 2\left(x_1^2 + x_2^2 - 1\right) - x_1$$

subject to (11.28)

$$h(x_1, x_2) = x_1^2 + x_2^2 - 1 = 0.$$

The solution of the problem is $x_1^* = 1$, $x_2^* = 0$, and the Lagrange multiplier is $\lambda^* = 3/2$. Moreover, $\nabla_{xx}^2 L(x^*, \lambda^*) = I$.

Consider an iterate x_k as $x_k = [\cos\theta \;\; \sin\theta]^T$. We immediately see that x_k is feasible for any value of θ. Suppose that our algorithm computes the following step

$$p_k = \begin{bmatrix} \sin^2\theta \\ -\sin\theta\cos\theta \end{bmatrix}, \tag{11.29}$$

which determines the next point (iterate):

$$x_k + p_k = \begin{bmatrix} \cos\theta + \sin^2\theta \\ (1 - \cos\theta)\sin\theta \end{bmatrix}.$$

Now, we see that:

$$\|x_k + p_k - x^*\|_2 = 2\sin^2(\theta/2) \quad \text{and} \quad \|x_k - x^*\|_2 = 2|\sin(\theta/2)|.$$

Hence,

$$\frac{\|x_k + p_k - x^*\|_2}{\|x_k - x^*\|_2^2} = \frac{1}{2},$$

i.e., this step tends Q-quadratically to the solution of the problem. However, if we insist and compute, we find that:

$$f(x_k + p_k) = \sin^2\theta - \cos\theta > -\cos\theta = f(x_k),$$
$$h(x_k + p_k) = \sin^2\theta > h(x_k) = 0.$$

Therefore, even if the considered step determines a Q-quadratic convergence, the values of the objective function increase and the constraints are more and more unsatisfied. This behavior occurs for any value of θ even if the initial point is arbitrarily close to the solution.

For this example, let us consider the sequential quadratic programming algorithm in which p_k is determined as solution of the quadratic program (11.8) in which $\nabla^2_{xx}L(x_k, \lambda_k)$ is replaced by $\nabla^2_{xx}L(x^*, \lambda^*) = I$. Hence, consider the current point $x_k = [\cos\theta \quad \sin\theta]^T$. Since

$$f(x_k) = -\cos\theta, \quad \nabla f(x_k) = \begin{bmatrix} 4\cos\theta - 1 \\ 4\sin\theta \end{bmatrix}, \quad A(x_k)^T = \begin{bmatrix} 2\cos\theta \\ 2\sin\theta \end{bmatrix},$$

the quadratic sub-problem (11.8) is

$$\min_p (4\cos\theta - 1)p_1 + 4p_2\sin\theta + \frac{1}{2}p_1^2 + \frac{1}{2}p_2^2$$

subject to:

$$p_2 + p_1 \mathrm{ctg}\theta = 0.$$

By solving this sub-problem, we get the direction

$$p_k = \begin{bmatrix} \sin^2\theta \\ -\sin\theta\cos\theta \end{bmatrix},$$

which coincides with (11.29), showing the failure of the sequential quadratic programming algorithm. ♦

The Maratos effect shows that for many merit functions, a superlinear convergent step may not be accepted, thus preventing the algorithm from fast convergence. There are three ways to overcome the Maratos effect. The first one is to relax the line-search conditions. The second one is to use a second-order correction step $\widehat{p}_k$ which satisfies $P(x_k + p_k + \widehat{p}_k) < P(x_k)$. In this way, $p_k + \widehat{p}_k$ is an acceptable step, and it is still a superlinear convergent step. The third one is to use smooth exact penalty functions as merit functions.

As we said, the technique for avoiding the Maratos effect is to introduce the second-order corrections, as it was suggested by Coleman and Conn (1982a, 1982b), Fletcher (1982), Mayne and Polak (1982), Fukushima (1986), etc.

Suppose that the SQP algorithm has computed a step p_k as solution of the sub-problem (11.12). If this step yields an increase in the merit function $P(x_k, \sigma)$, then a possible cause is that the linear approximation of the constraints given by (11.12) is not accurate enough. To overcome this situation, a natural solution is to re-solve (11.12) with the linear terms $c_i(x_k) + \nabla c_i(x_k)^{\mathrm{T}} p$ replaced by quadratic approximations:

$$c_i(x_k) + \nabla c_i(x_k)^{\mathrm{T}} p + \frac{1}{2} p^{\mathrm{T}} \nabla^2 c_i(x_k) p. \tag{11.30}$$

However, the resulting sub-problem with quadratic constraints is too difficult to solve. Instead, the following procedure is used. The constraints are evaluated in the new point $x_k + p_k$, and then we make use of their approximation by Taylor's theorem:

$$c_i(x_k + p_k) \cong c_i(x_k) + \nabla c_i(x_k)^{\mathrm{T}} p_k + \frac{1}{2} p_k^{\mathrm{T}} \nabla^2 c_i(x_k) p_k. \tag{11.31}$$

Assuming that the second-order step p, which is not known, will not be too different from p_k, then the last term in (11.30) can be approximated as:

$$p^{\mathrm{T}} \nabla^2 c_i(x_k) p = p_k^{\mathrm{T}} \nabla^2 c_i(x_k) p_k. \tag{11.32}$$

Therefore, by making this substitution in (11.30) and by using the approximation (11.31), the following *second-order correction sub-problem* is obtained

$$\min_{p \in \mathbb{R}^n} f(x_k) + \nabla f(x_k)^{\mathrm{T}} p + \frac{1}{2} p^{\mathrm{T}} \nabla_{xx}^2 L(x_k, \lambda_k) p \tag{11.33a}$$

subject to:

$$\nabla c_i(x_k)^{\mathrm{T}} p + \left(c_i(x_k + p_k) - \nabla c_i(x_k)^{\mathrm{T}} p_k \right) = 0, \quad i \in E, \tag{11.33b}$$

$$\nabla c_i(x_k)^{\mathrm{T}} p + \left(c_i(x_k + p_k) - \nabla c_i(x_k)^{\mathrm{T}} p_k \right) \geq 0, \quad i \in I. \tag{11.33c}$$

Let $\widehat{p}_k$ be the solution of (11.33). Observe that the second-order correction step implies the evaluation of the constraints in point $x_k + p_k$, i.e., $c_i(x_k + p_k)$, $i \in E \cup I$, which can be a laborious activity. Therefore, a strategy is to use this correction step only if the increase in the merit function is concomitant with an increase in the constraint norm. When the step p_k is generated by the SQP algorithm based on (11.12) then, near a solution satisfying the second-order sufficient conditions, the algorithm based on (11.33) takes either the full step p_k or the corrected step $p_k + \widehat{p}_k$.

11.4 Line-Search SQP Algorithm

Taking into consideration the above developments, we see that there is a large variety of line-search SQP algorithms that differ in many respects, such as the way in which the Hessian approximation is computed (limited-memory BFGS that is suitable for large-scale problems), the step acceptance mechanism using different techniques to avoid the Maratos effect, the use of the merit functions or of the filters, etc. In the following we present a practical line-search SQP algorithm for solving the general nonlinear optimization problem (11.11). To keep the description as simple as possible, the algorithm below does not include details to ensure the feasibility of the sub-problem or the second-order correction mechanisms. Rather, the algorithm is simply obtained by solving the sub-problem (11.12). Assume that the quadratic sub-problem (11.12) is convex. Therefore, it can be solved by means of the active-set methods for quadratic programming described in Chapter 10.

Algorithm 11.3 Line-search SQP algorithm

1.	Choose an initial pair (x_0, λ_0), as well as the parameters $\eta \in (0, 1/2)$ and $\tau \in (0, 1)$. Evaluate $f(x_0)$, $\nabla f(x_0)$, $c(x_0)$, and $A(x_0)$. If a quasi-Newton approximation of the Hessian of the Lagrange function is used, then choose an initial $n \times n$ symmetric and positive definite approximation B_0; otherwise compute $\nabla^2_{xx} L(x_0, \lambda_0)$. Set $k = 0$.
2.	Test a criterion for stopping the iterations.
3.	Compute the search direction p_k as solution of the quadratic sub-problem (11.12). Let $\widehat{\lambda}$ be the corresponding Lagrange multipliers.
4.	Set $p_\lambda = \widehat{\lambda} - \lambda_k$.
5.	Choose σ_k to satisfy (11.26) with $\mu = 1$. Set $\alpha_k = 1$.
6.	Inner while loop. Select $\eta \in (0, 1)$ and $\tau_\alpha \in (0, \tau]$. While $$P(x_k + \alpha_k p_k, \sigma_k) > P(x_k, \sigma_k) + \eta \alpha_k P'_{p_k}(x_k, \sigma_k),$$ reset $\alpha_k = \tau_\alpha \alpha_k$; otherwise go to step 7.
7.	Set $x_{k+1} = x_k + \alpha_k p_k$ and $\lambda_{k+1} = \lambda_k + \alpha_k p_\lambda$.
8.	Evaluate $f(x_{k+1})$, $\nabla f(x_{k+1})$, $c(x_{k+1})$, and $A(x_{k+1})$. If it is the case, evaluate $\nabla^2_{xx} L(x_{k+1}, \lambda_{k+1})$.
9.	If a quasi-Newton approximation method is used, then set $$s_k = \alpha_k p_k \quad \text{and} \quad y_k = \nabla_x L(x_{k+1}, \lambda_{k+1}) - \nabla_x L(x_k, \lambda_{k+1}),$$ and obtain B_{k+1} by updating B_k using a quasi-Newton method (BFGS).
10.	Set $k = k + 1$ and go to step 2. ♦

An efficient implementation of the Algorithm 11.3 is to use the *warm start* procedure. The working set for each quadratic programming sub-problem is initialized with the working set of the previous SQP iteration. In step 6 of the algorithm where the inner while loop is implemented instead of a merit function, a filter may be used. In step 9 the limited memory BFGS updating approach can be used, especially for large-scale problems.

11.5 Trust-Region SQP Method

The simplest way to formulate a trust-region SQP method is to add a trust-region constraint to the sub-problem (11.12), as

$$\min_{p\in\mathbb{R}^n} f(x_k) + \nabla f(x_k)^{\mathrm{T}} p + \frac{1}{2} p^{\mathrm{T}} \nabla^2_{xx} L(x_k, \lambda_k) p \tag{11.34a}$$

subject to:

$$\nabla c_i(x_k)^{\mathrm{T}} p + c_i(x_k) = 0, \quad i \in E, \tag{11.34b}$$

$$\nabla c_i(x_k)^{\mathrm{T}} p + c_i(x_k) \geq 0, \quad i \in I. \tag{11.34c}$$

$$\|p\| \leq \Delta_k. \tag{11.34d}$$

The problem with this formulation is that even if the constraints (11.34) are compatible, it is often possible that the sub-problem (11.34) may not always have a solution because of the trust-region constraint (11.34d). To solve this possible conflict between the linear constraints (11.34), it is not appropriate to increase Δ_k until the set of steps p satisfying the linear constraints intersects the trust region. A more appropriate viewpoint is not to satisfy the linearized constraints exactly at every step; rather, we try to improve the feasibility of these constraints at each step and to satisfy them exactly only if the trust-region constraint permits it. This point of view can be implemented in three ways: by relaxation methods, penalty methods, and filter methods.

A Relaxation Method for Equality-Constrained Optimization Consider the equality-constrained nonlinear optimization problem (11.1). At iterate x_k, the SQP step is computed by solving the sub-problem

$$\min_{p\in\mathbb{R}^n} f(x_k) + \nabla f(x_k)^{\mathrm{T}} p + \frac{1}{2} p^{\mathrm{T}} \nabla^2_{xx} L(x_k, \lambda_k) p \tag{11.35a}$$

subject to

$$A(x_k)p + h(x_k) = r_k, \tag{11.35b}$$

$$\|p\|_2 \leq \Delta_k, \tag{11.35c}$$

where $r_k \in \mathbb{R}^m$ is the *relaxation vector*. The relaxation vector is selected as the smallest vector such that (11.35) are consistent for some reduced value of trust-region radius Δ_k. To achieve this, firstly solve the sub-problem

$$\min_y \|A(x_k)y + h(x_k)\|_2^2 \tag{11.36a}$$

subject to:

$$\|y\|_2 \leq 0.8\Delta_k. \tag{11.36b}$$

Let y_k be the solution of (11.36). Define the relaxation vector as:

$$r_k = A(x_k)y_k + h(x_k). \tag{11.37}$$

In the following step, p_k is computed by (11.35), and define the new iterate $x_{k+1} = x_k + p_k$. The new multiplier estimates λ_{k+1} is computed as:

$$\lambda_{k+1} = \left[A(x_k)A(x_k)^{\mathrm{T}}\right]^{-1} A(x_k)\nabla f(x_k). \tag{11.38}$$

Now, for computing an approximate solution p_k of (11.35), one method is the projected conjugate gradient. The idea is to apply this algorithm to (11.35a) and (11.35b), monitoring the satisfaction of the trust-region constraint (11.35c) and stopping the algorithm if the boundary of this region is reached or if a negative curvature is detected. A merit function that takes into consideration all these ingredients is the nonsmooth l_2 function $P_2(x,\sigma) = f(x) + \sigma\|h(x)\|_2$. This merit function is modeled by

$$q_\sigma(p) = f(x_k) + \nabla f(x_k)^{\mathrm{T}} p + \frac{1}{2} p^T \nabla_{xx}^2 L(x_k, \lambda_k) p + \sigma m(p), \tag{11.39}$$

where:

$$m(p) = \|h(x_k) + A(x_k)p\|_2. \tag{11.40}$$

The penalty parameter σ is selected large enough as to satisfy (11.25). The acceptability of step p_k is monitored by the ratio:

$$\rho_k = \frac{ared_k}{pred_k} = \frac{P_2(x_k,\sigma) - P_2(x_k + p_k,\sigma)}{q_\sigma(0) - q_\sigma(p_k)}. \tag{11.41}$$

Now, a description of the trust-region SQP algorithm for solving equality-constrained nonlinear optimization problem (11.1), $\min\{f(x) : h(x) = 0\}$, is as follows Omojokun (1989), Lalee, Nocedal, and Plantenga (1998), and Nocedal and Wright (2006).

Algorithm 11.4 Trust-region SQP algorithm

1.	Choose the initial point x_0 and the initial trust-region radius $\Delta_0 > 0$. Choose the constants $\varepsilon > 0$ and $\eta,\ \gamma \in (0, 1)$. Set $k = 0$.
2.	Evaluate $f(x_k)$, $\nabla f(x_k)$, $h(x_k)$, and $A(x_k)$.
3.	Compute the Lagrange multipliers: $\lambda_{k+1} = [A(x_k)A(x_k)^T]^{-1}A(x_k)\nabla f(x_k)$.
4.	If $\|\nabla f(x_k) - A(x_k)^T\lambda_k\|_\infty < \varepsilon$ and $\|h(x_k)\|_\infty < \varepsilon$, then stop. x_k is the solution of the problem.
5.	Solve the sub-problem (11.36) with respect to y_k.
6.	Compute the relaxation parameter $r_k = A(x_k)y_k + h(x_k)$.
7.	Compute $\nabla^2_{xx}L(x_k, \lambda_k)$, or a quasi-Newton approximation of it.
8.	Compute p_k as solution of the quadratic programming sub-problem (11.35).
9.	Choose σ, large enough to satisfy (11.26).
10.	Compute the ratio $\rho_k = ared_k/pred_k$.
11.	If $\rho_k > \eta$, then set $x_{k+1} = x_k + p_k$; choose Δ_{k+1} to satisfy $\Delta_{k+1} \geq \Delta_k$ and go to step 12. If $\rho_k \leq \eta$, then set $x_{k+1} = x_k$; compute $\Delta_{k+1} = \gamma\|p_k\|$ and go to step 12.
12.	Set $k = k + 1$ and continue with step 2. ♦

It is quite simple to see that a second-order correction can be introduced in this algorithm to avoid the Maratos effect. The main computational cost of this algorithm is given by the projected conjugate gradient iteration for solving the quadratic programming sub-problem (11.35).

Sequential l_1 *Quadratic Programming for Equality and Inequality Constraints* In this method the linearized constraints (11.34b) and (11.34c) are placed into the objective of the quadratic program as an l_1 penalty term. Thus, the following sub-problem is obtained

$$\min_p q_\sigma(p) \triangleq f(x_k) + \nabla f(x_k)^T p + \frac{1}{2}p^T\nabla^2_{xx}L(x_k,\lambda_k)p$$
$$+\sigma\sum_{i\in E}\left|c_i(x_k) + \nabla c_i(x_k)^T p\right| + \sigma\sum_{i\in I}\left[c_i(x_k) + \nabla c_i(x_k)^T p\right]^-$$

subject to (11.42)

$$\|p\|_\infty \leq \Delta_k,$$

where σ is the penalty parameter and $[y]^- = \max\{0, \ -y\}$. When introducing slack variables, this problem is reformulated as

$$\min_{p,v,w,t} f(x_k) + \nabla f(x_k)^{\mathrm{T}} p + \frac{1}{2} p^{\mathrm{T}} \nabla^2_{xx} L(x_k, \lambda_k) p + \sigma \sum_{i \in E} (v_i + w_i) + \sigma \sum_{i \in I} t_i$$

subject to (11.43)

$$\begin{aligned} &\nabla c_i(x_k)^{\mathrm{T}} p + c_i(x_k) = v_i - w_i, \quad i \in E, \\ &\nabla c_i(x_k)^{\mathrm{T}} p + c_i(x_k) \geq -t_i, \quad i \in I, \\ &v, w, \ t \geq 0, \\ &\|p\|_\infty \leq \Delta_k. \end{aligned}$$

The constraints of this problem are always consistent. Moreover, since the trust region has been defined using the l_∞ norm, (11.43) is a smooth quadratic program which can be solved by means of a quadratic programming algorithm.

To determine the acceptance of the step, the algorithm uses the l_1 merit function:

$$P(x, \sigma) = f(x) + \sigma \sum_{i \in E} |c_i(x)| + \sigma \sum_{i \in I} [c_i(x)]^-. \tag{11.44}$$

Observe that q_σ defined in (11.42) can be viewed as a model of the merit function (11.44) at x_k, in which we approximate each constraint function c_i by its linearization and in which f is replaced by a quadratic function whose curvature term includes information from both the objective and constraints.

After computing step p_k as solution of (11.43), from (11.41) the ratio ρ_k is computed by using this time the above merit function and defining q_σ by (11.42). The step is accepted or rejected according to standard trust-region rules, as described in Algorithm 11.4. Clearly, a second-order correction step can be introduced in order to prevent the Maratos effect.

11.6 Sequential Linear-Quadratic Programming (SLQP)

As we have already seen, the SQP methods presented above require the solution of a general quadratic programming problem with equality and inequality constraints at each iteration. However, this is a difficult task, especially when the second derivative information is incorporated into the algorithm. This approach imposes a limit on the size of the problem that can be solved in practice by these methods.

The sequential linear-quadratic programming method, which we briefly describe here, tries to overcome this situation by computing the step in two phases. In the first one, a linear programming (LP) problem is formed and solved to identify a working set W. The second one consists of an equality-constrained quadratic programming phase in which the constraints in the working set W are imposed as equalities. The total step is a combination of the steps obtained in the linear programming and equality-constrained phases.

In LP phase the following problem is solved

$$\min_{p\in\mathbb{R}^n} \; f(x_k) + \nabla f(x_k)^T p \tag{11.45a}$$

subject to:

$$\nabla c_i(x_k)^T p + c_i(x_k) = 0, \quad i \in E, \tag{11.45b}$$

$$\nabla c_i(x_k)^T p + c_i(x_k) \geq 0, \quad i \in I, \tag{11.45c}$$

$$\|p\|_\infty \leq \Delta_k^{LP}. \tag{11.45d}$$

Observe that this linear programming problem differs from the standard SQP sub-problem (11.34) only in that the second-order term $p^T \nabla_{xx}^2 L(x_k, \lambda_k) p$ has been omitted and that a l_∞ norm has been used to define the trust region. Since the constraints of (11.45) may be inconsistent, instead of (11.45), a l_1 penalty reformulation is defined

$$\begin{aligned} \min_p l_\sigma(p) \triangleq\, & f(x_k) + \nabla f(x_k)^T p \\ & + \sigma \sum_{i\in E} \left| c_i(x_k) + \nabla c_i(x_k)^T p \right| + \sigma \sum_{i\in I} \left[c_i(x_k) + \nabla c_i(x_k)^T p \right]^- \end{aligned}$$

subject to: (11.46)

$$\|p\|_\infty \leq \Delta_k^{LP}.$$

Now, introducing the slack variables, (11.46) can be reformulated as a linear programming problem. Let p^{LP} be the solution of (11.46). From this solution the following explicit estimate of the optimal active set can be obtained:

$$\begin{aligned} A_k\left(p^{LP}\right) = & \left\{ i \in E : \; c_i(x_k) + \nabla c_i(x_k)^T p^{LP} = 0 \right\} \\ & \cup \left\{ i \in I : \; c_i(x_k) + \nabla c_i(x_k)^T p^{LP} = 0 \right\}. \end{aligned}$$

Similarly, the set V_k of the violated constrained is defined as:

$$\begin{aligned} V_k\left(p^{LP}\right) = & \left\{ i \in E : \; c_i(x_k) + \nabla c_i(x_k)^T p^{LP} \neq 0 \right\} \\ & \cup \left\{ i \in I : \; c_i(x_k) + \nabla c_i(x_k)^T p^{LP} < 0 \right\}. \end{aligned}$$

The working set W_k is defined as some linearly independent subset of the active-set $A_k(p^{LP})$. To ensure the progress of the algorithm on the penalty function (11.44), the Cauchy step is defined as

$$p^C = \alpha^{LP} p^{LP}, \tag{11.47}$$

where $\alpha^{LP} \in (0\,,1]$ is the stepsize that provides sufficient decrease in the model q_σ defined in (11.42). With the working set W_k, an equality-constrained quadratic programming is solved where the constraints in W_k are considered as equalities, ignoring all the others. Thus, the following sub-problem is obtained

$$\min_p f(x_k) + \frac{1}{2} p^{\mathrm{T}} \nabla^2_{xx} L(x_k, \lambda_k) p + \left(\nabla f(x_k) + \sigma_k \sum_{i \in V_k} \gamma_i \nabla c_i(x_k) \right)^{\mathrm{T}} p$$

subject to: (11.48)

$$\begin{aligned} &\nabla c_i(x_k)^{\mathrm{T}} p + c_i(x_k) = 0, \quad i \in E \cap W_k, \\ &\nabla c_i(x_k)^{\mathrm{T}} p + c_i(x_k) = 0, \quad i \in I \cap W_k, \\ &\|p\|_2 \leq \Delta_k, \end{aligned}$$

where the scalars γ_i represent the algebraic sign of the ith violated constraint. Observe that the trust-region constraint in (11.48) is spherical and Δ_k is different from the trust-region radius Δ_k^{LP} used in (11.46). Let p^Q be the solution of (11.48). The total step p_k of the SLQP method is computed as

$$p_k = p^C + \alpha^Q \left(p^Q - p^C\right),$$

where $\alpha^Q \in [0, 1]$ is the stepsize that approximately minimizes the model q_σ defined in (11.42). Byrd, Gould, Nocedal, and Waltz (2004) argue that the choice of radius Δ_{k+1}^{LP} for the LP phase is more delicate, since it influences our guess of the optimal active set. The value of Δ_{k+1}^{LP} should be set to be a little larger than the total step p_k, subject to some other restrictions. The Lagrange multiplier estimates λ_k used in the evaluation of the Hessian $\nabla^2_{xx} L(x_k, \lambda_k)$ are computed as in (11.38) using the working set W_k and modified so that $\lambda_i \geq 0$, $i \in I$. The advantage of using SLQP is that well-established algorithms for solving large-scale versions of linear programming and equality quadratic sub-problems are readily available.

Notes and References

The content of this chapter is based on the theoretical developments from Nocedal and Wright (2006). Sequential quadratic programming (SQP) is one of the most effective methods for solving nonlinear optimization problems of different structure and complexity. SQP is a very important ingredient in many algorithms for nonlinear optimization. The motivation for which this approach is suitable for solving nonlinear optimization problems is based on the fact that locally, any optimization problem can be very well approximated by a quadratic model. SQP can be used both in the line-search and trust-region frameworks. The algorithms DONLP (see Chapter 12), NLPQLP (see Chapter 13), KNITRO (see Chapters 14 and 19), SNOPT (see Chapter 15), and CONOPT (see Chapter 16) described in this book use SQP as a crucial ingredient in different computational structures. If the penalty and augmented Lagrangian methods (SPENBAR and MINOS) are effective when most of the constraints are linear, SQP methods (DONLP, NLPQLP,

KNITRO, SNOPT, and CONOPT) show their strength when the problems have significant nonlinearities in constraints. A competing approach for handling nonlinear optimization problems is the interior point methods, implemented for example in KNITRO and IPOPT.

There are two types of SQP methods. In the inequality quadratic programming approach, at each iteration a general inequality-constrained quadratic programming sub-problem is solved in order to compute a step and to generate an estimate of the optimal active set. On the other hand, equality quadratic programming methods decouple these computations and operate in two phases: In the first phase, an estimate of the optimal active set is determined. In the second one, an equality-constrained quadratic programming sub-problem is solved to find the step.

There is a vast literature on SQP. These methods were proposed by Wilson (1963) in his Ph.D. thesis and developed inter alia by Han (1976, 1977), Powell (1977, 1978a, 1978b) Byrd, Schnabel, and Shultz (1987), Byrd, Tapia, and Zhang (1990), etc. For literature surveys see the papers by Boggs and Tolle (1989, 1995) and Gould, Orban, and Toint (2005).

The SLQP approach was proposed by Fletcher and Sainz de la Maza (1989) and was further developed by Byrd, Gould, Nocedal, and Waltz (2004). The second-order corrections were suggested by Coleman and Conn (1982a) Mayne and Polak (1982), and Gabay (1982). The filter SQP was developed by Fletcher and Leyffer (2002) and Fletcher, Leyffer, and Toint (1999).

Chapter 12
A SQP Method Using Only Equality-Constrained Sub-problems: DONLP

Let us consider the general nonlinear optimization problem:

$$
\begin{aligned}
& \min f(x) \\
\text{subject to:} \quad & \\
& e(x) = 0, \\
& c(x) \geq 0,
\end{aligned}
\tag{12.1}
$$

where the functions $f: \mathbb{R}^n \to \mathbb{R}$, $e : \mathbb{R}^n \to \mathbb{R}^{m_e}$, and $c : \mathbb{R}^n \to \mathbb{R}^m$ are supposed to be twice continuously differentiable. Also, suppose that:

$$X = \{x \in \mathbb{R}^n : e(x) = 0, \ \ c(x) \geq 0\} \neq \varnothing.$$

For solving this problem, we present the quadratic programming algorithm DONLP elaborated by Peter Spellucci (1985, 1993, 1994, 1995, 1998).

The sequential quadratic programming is one of the most efficient methods for solving nonlinear optimization problems of small to medium size. However, if $m \gg n$ and $c_j(x) \gg 0$, for some j, then these methods may require a lot of evaluations of $\nabla c_j(x)$. Besides, the computational complexity of a quadratic programming problem is in general larger than that of a system of linear equations. Therefore, a good idea is to consider only the "near active" or violated inequalities in the quadratic programming sub-problem and to treat them as equalities. Given the current point x_k as an estimate of the local solution x^* and a symmetric positive semidefinite matrix B_k, then a correction p_k of x_k is computed as solution of the equality-constrained sub-problem:

N. Andrei, *Continuous Nonlinear Optimization for Engineering Applications in GAMS Technology*, Springer Optimization and Its Applications 121,
DOI 10.1007/978-3-319-58356-3_12

$$\min \nabla f(x_k)^{\mathrm{T}} p + \frac{1}{2} p^{\mathrm{T}} B_k p$$

subject to: (12.2)

$$\nabla e(x_k)^{\mathrm{T}} p + e(x_k) = 0,$$
$$\nabla c_{A_k}(x_k)^{\mathrm{T}} p + c_{A_k}(x_k) = 0,$$

where $A_k \supseteq \{i : i \in \{1, \ldots, m\}$ and $c_i(x_k) \leq 0\}$ is the so-called *working set*, suitably determined.

In order to have p_k well defined by (12.2), it is sufficient to require B_k satisfy the condition $y^{\mathrm{T}} B_k y > 0$ for all $y \in N(A_k, x_k) \triangleq \{z : N_{A_k}^{\mathrm{T}} z = 0\}$, $y \neq 0$, where

$$N_{A_k} = [\nabla e(x_k), \nabla c_{A_k}(x_k)] \tag{12.3}$$

is of full rank. Suppose that the strict complementarity slackness conditions (see Definition 5.22) hold at x_k. Therefore, these conditions correspond to the well-known regularity (see Definition 5.20) and to the second-order sufficient conditions (see Theorem 5.13).

Given x_k and p_k, the algorithm computes the next estimate of the solution x^* as

$$x_{k+1} = x_k + \alpha_k p_k, \tag{12.4}$$

where α_k is the stepsize determined by minimizing the l_1 penalty function

$$P(x) = \|e(x)\|_1 + \|c(x)^-\|_1, \tag{12.5}$$

and for any vector y, y^- is its negative part, i.e., $y_i^- = y_i$ if $y_i < 0$ and $y_i^- = 0$ otherwise. This l_1 penalty function gives reliable global convergence of the algorithm and can be cheaply evaluated. The Maratos effect (see Section 11.3) can be avoided by a second-order correction of infeasibility.

Let us consider

$$L(x, \lambda, \mu) = f(x) - \lambda^{\mathrm{T}} c(x) - \mu^{\mathrm{T}} e(x) \tag{12.6}$$

as the Lagrange function associated to (12.1). In order to obtain the global convergence to a feasible and locally optimal solution, suppose that the following hypotheses hold:

H0: Functions f, e, c are twice continuously differentiable on $\mathbb{R}^n$.

H1: The extension of the feasible set $X(\tau_0) = \{x \in \mathbb{R}^n : P(x) \leq \tau_0\}$ is a compact for some $\tau_0 > 0$.

H2: For all $x \in X(\tau_0)$ the matrix

$$N_{A_0}(x) = [\nabla e(x) \;\; \nabla c_{A_0}(x)] \tag{12.7}$$

is of full rank, where:

$$A_0(x) = \{i \in \{1,\ldots,m\} : c_i(x) \le 0\}. \tag{12.8}$$

H3: A local solution x^* of the problem (12.1) satisfies the strict complementarity slackness and the second-order sufficient condition, namely:

$$c(x^*) + \lambda^* > 0, \tag{12.9}$$

$$\nabla_x L(x^*, \lambda^*, \mu^*) = 0, \lambda^* \ge 0, c(x^*)^T \lambda^* = 0, \tag{12.10}$$

$$z^T \nabla_{xx}^2 L(x^*, \lambda^*, \mu^*) z > 0, \quad \text{for all } z \in N(A_0(x^*), x^*) \backslash \{0\}. \tag{12.11}$$

If hypotheses *H1* and *H2* are satisfied, then there exists the penalty weights $u > 0$, $u \in \mathbb{R}^m$ and $v > 0$, $v \in \mathbb{R}^{m_e}$, such that

$$\Phi(x, u, v) \triangleq f(x) + \sum_{i=1}^{m_e} v_i |e_i(x)| - \sum_{i=1}^{m} u_i \min\{0, c_i(x)\} \tag{12.12}$$

is an exact penalty function for (12.1) on $X(\tau_0)$. The hypothesis *H3* is needed in order to identify the correct working set (Spellucci, 1995).

If *H2* is satisfied, then for all $x \in X(\tau_0)$, the left inverse of $N_{A_0}(x)$

$$N_{A_0}(x)^{(-)} \triangleq \left[N_{A_0}(x)^T N_{A_0}(x) \right]^{-1} N_{A_0}(x)^T \tag{12.13}$$

exists and is uniformly bounded with respect to x. Moreover, by the compactness of $X(\tau_0)$, continuity, and *H2*, it follows that there exists some $\delta_0 > 0$, such that these statements also hold for the possible larger matrix:

$$N_{A(x,\delta)}(x) = \left[\nabla e(x) \quad \nabla c_{A(x,\delta)}(x) \right], \tag{12.14}$$

where:

$$A(x, \delta) = \{i \in \{1,\ldots,m\} : c_i(x) \le \delta\}, \quad \delta_0 \ge \delta \ge 0. \tag{12.15}$$

Note that $\tau_0 > 0$ is known. It is specified by the user in the initialization step of the algorithm. However, δ_0 is not known. A procedure for computation of δ_0 is given by Spellucci (1993), as will be described later.

Every step of Spellucci's algorithm consists of four basic parts: (1) estimation of the working set, (2) computation of the correction of the primal and the estimates for the dual variables, (3) adaptive determination of the penalty weights, and (4) determination of the stepsize.

As we have already mentioned, the algorithm computes the directions of descent p for Φ at $x \in X(\tau_0)$ and the estimates of the multipliers μ and λ_A as solution of an equality-constrained quadratic programming problem. This consists in solving the KKT system:

$$\begin{bmatrix} B & N_A \\ N_A^T & 0 \end{bmatrix} \begin{bmatrix} p \\ \begin{bmatrix} -\mu \\ -\lambda_A \end{bmatrix} \end{bmatrix} = -\begin{bmatrix} \nabla f(x) \\ h \end{bmatrix}, \tag{12.16}$$

where $A = A(x, \delta)$. The values of δ are computed from the local information available at point x. This requires the use of two choices for the component h of the right-hand side of (12.16), requiring two solutions of the system by using one decomposition of its matrix. If δ is enough small, then N_A is of full rank (see hypothesis *H2*). Therefore, we check this condition first and reduce δ if necessary. Then the system (12.16) has a unique solution provided that $Z^T BZ$ is regular, where Z is an orthonormal basis of $N(A, x)$. Assume that $Z^T BZ$ is positive definite. We can consider $Z^T BZ$ as an approximation of the projected Hessian, which is positive definite if *H3* holds.

In these conditions, for suitable values of the penalty weights u^* and v^*, it is possible that:

$$\Phi(x_{k-1}, u^*, v^*) - \Phi(x_k, u^*, v^*) \geq O\left(\|x_{k-1} - x^*\|^2\right) \geq O\left(\|x_k - x^*\|^2\right).$$

Therefore, we can consider

$$\delta_0 = \left(\Phi\left(x_{k-1}, u^{k-1}, v^{k-1}\right) - \Phi\left(x_k, u^{k-1}, v^{k-1}\right)\right)^{1/\sigma} \tag{12.17}$$

with $\sigma > 2$ as the first proposal for δ_k. u^{k-1} and v^{k-1} are the current penalty weights. The value of σ can be modified as it will be seen in the algorithm. In fact the second proposal for δ_k is computed by using an estimate of the violation of the KKT conditions at the current point. This is obtained from (12.16) using $h = 0$. The maximum of these two proposals for δ_k is finally taken (Spellucci, 1995). If we are far from the solution or if δ is too large, then negative multipliers might appear for some inequality constraints in the working set, thus indicating a wrong choice of $A(x, \delta)$. Spellucci recommends the modification of p by a specific choice of h in (12.16). The stepsize is computed using the Armijo's procedure.

Algorithm 12.1 DONLP (Spellucci)

1.	Initialization of parameters: $0 < \delta_{\min} \ll 1$, $r \in N$, $\sigma > 2$, $\tau_1 > 1$, $\rho > 0$, $0 < \beta < 1$, $0 < \gamma < 1/2$, $0 < \eta_0 \ll 1$. ($\delta_{\min} = 1000\varepsilon$, $r = 3$, $\sigma = 7/3$, $\tau_1 = 1.1$, $\rho = 1$, $\beta = 1/2$, $\gamma = 0.01$, and $\eta_0 = 10^{-4}$.) Consider $h_0 = 1$ and $\delta_0 \gg \delta_{\min}$, ($\delta_0 = 1$). Set$k = 0$
2.	Set $j = 0$
3.	Compute the first proposal for δ_k: If $k > 0$, set $\delta_k^{(0)} = \min\left\{\delta_0, \left(\Phi\left(x_{k-1}, u^{k-1}, v^{k-1}\right) - \Phi\left(x_k, u^{k-1}, v^{k-1}\right)\right)^{1/\sigma}\right\}$, otherwise $\delta_0^{(0)} = \delta_0$

(continued)

4.	Let $A_k = A\left(x_k, \delta_k^{(j)}\right)$
5.	Check the rank condition. If N_{A_k} is not of full rank and $\delta_0 > \delta_{\min}$, then set $\delta_0 = \delta_0\beta$ and go to step 3. Otherwise stop, there is a $O(\delta_{\min})$ perturbation of problem (12.1) which does not satisfy *H2*
6.	Choose B_k positive definite on $N(A_k, x_k)$
7.	Compute the estimate of multipliers and the error in the KKT conditions. Solve the system: $$\begin{bmatrix} B_k & N_{A_k} \\ N_{A_k}^{\mathrm{T}} & 0 \end{bmatrix} \begin{bmatrix} p_k^{[0]} \\ -\begin{bmatrix} \mu^{[0]k} \\ \lambda_{A_k}^{[0]k} \end{bmatrix} \end{bmatrix} = \begin{bmatrix} -\nabla f(x_k) \\ 0 \end{bmatrix}$$
8.	Based on the errors in KKT conditions, compute the second proposal for δ_k: If $j = 0$, then set $$\delta_k^{(1)} = \delta_0 \min\left\{1,\ \left(\left\|\left(\lambda_{A_k}^{[0]k}\right)^{-}\right\| + P(x_k) + \left\|B_k p_k^{[0]}\right\|\right)^{2/\sigma}\right\};$$ otherwise go to step 9
9.	Check whether the second proposal for δ_k can be used: If $j = 0$, $\delta_k^{(1)} > \delta_k^{(0)}$ and $A\left(x_k, \delta_k^{(1)}\right)$ is not included in $A_k = A\left(x_k, \delta_k^{(0)}\right)$, then set $j = 1$ and continue with step 4; otherwise go to step 10
10.	Define δ_k and compute p_k as solution of an equality-constrained quadratic programming problem with modified constraint values: $$\delta_k = \max\left\{\delta_k^{(0)}, \delta_k^{(1)}\right\}$$ Solve the linear system: $$\begin{bmatrix} B_k & N_{A_k} \\ N_{A_k}^{\mathrm{T}} & 0 \end{bmatrix} \begin{bmatrix} p_k \\ -\begin{bmatrix} \mu^k \\ \lambda_{A_k}^k \end{bmatrix} \end{bmatrix} = -\begin{bmatrix} \nabla f(x_k) \\ \begin{bmatrix} e(x_k) \\ \Delta_{A_k} c_{A_k}(x_k) + \varphi_{A_k} \end{bmatrix} \end{bmatrix},$$ where, for $i \in A_k$: $$\Delta_i \triangleq \begin{cases} -1, & \text{if} \quad c_i(x_k) \geq 0 \quad \text{and} \quad \lambda_i^{[0]k} < 0, \\ 1, & \text{otherwise,} \end{cases}$$ $$\varphi_i \triangleq \begin{cases} \lambda_i^{[0]k}, & \text{if} \quad \lambda_i^{[0]k} < 0, \\ 0, & \text{otherwise.} \end{cases}$$ Here, $\Delta_{A_k} \triangleq \mathrm{diag}(\Delta_i : \quad i \in A_k)$
11.	Test for termination: If $p_k = 0$ and $\lambda_{A_k}^k \geq 0$, then stop. $x_k = x^*$ is a first-order KKT point

(continued)

12.	Compute the new tentative penalty weights: Let $$\tilde{u}_i^k = \begin{cases} \rho, & \text{if } i \notin A_k \text{ and } k = 0, \\ u_i^{k-1}, & \text{if } i \notin A_k \text{ and } k > 0, \\ \left\|\lambda_i^{[0]k}\right\| + \rho, & \text{otherwise,} \end{cases}$$ $$\tilde{v}_i^k = \left\|\mu_i^{[0]k}\right\| + \rho, \ i = 1, \ldots, m_e$$
13.	Final choice of the penalty weights: If $k=0$, then set $u^k = \tilde{u}^k$, $v^k = \tilde{v}^k$. If $k>0$ and $\tilde{u}_i^k < u_i^{k-1}$ for some $i \in A_k$, or $\tilde{v}_i^k < v_i^{k-1}$ for some $i \in \{1, \ldots, m_e\}$ and $$\Phi\left(x_0, \tilde{u}^k, \tilde{v}^k\right) - \Phi\left(x_k, \tilde{u}^k, \tilde{v}^k\right) \geq \eta_k h_k,$$ then accept the new tentative multipliers: $$u^k = \tilde{u}^k, \ v^k = \tilde{v}^k,$$ $$\eta_{k+1} = \begin{cases} \eta_k, & \text{if } h_k \leq k/r, \\ \tau_1 \eta_k, & \text{otherwise,} \end{cases} \quad h_{k+1} = h_k + 1;$$ otherwise, reject them and increase the penalty weights as far as it is necessary: $$u_i^k = \begin{cases} u_i^{k-1}, & \text{if } i \notin A_k, \\ \left\|\lambda_i^{[0]k}\right\| + 2\rho, & \text{if } i \in A_k \text{ and } u_i^{k-1} \leq \left\|\lambda_i^{[0]k}\right\| \leq u_i^{k-1} + \rho, \\ \max\left\{\tilde{u}_i^k, u_i^{k-1}\right\}, & \text{otherwise,} \end{cases}$$ $$v_i^k = \begin{cases} \left\|\mu_i^{[0]k}\right\| + 2\rho, & \text{if } v_i^{k-1} \leq \left\|\mu_i^{[0]k}\right\| \leq v_i^{k-1} + \rho, \\ \max\left\{\tilde{v}_i^k, v_i^{k-1}\right\}, & \text{otherwise.} \end{cases}$$ $\eta_{k+1} = \eta_k$, $h_{k+1} = h_k$
14.	Compute the stepsize: Compute α_k as the largest number in the sequence $1, \beta, (\beta)^2, \ldots$ which satisfies the following three conditions: (a) $x_k + \alpha_k p_k \in X(\tau_0)$, (b) $\Phi(x_k, u^k, v^k) - \Phi(x_k + \alpha_k p_k, u^k, v^k) \geq -\alpha_k \gamma D\Phi(x_k, p_k, u^k, v^k)$, where $$-D\Phi\left(x_k, p_k, u^k, v^k\right) = p_k^T B_k p_k + \left(\mu^k - \mu^{[0]k}\right)^T e(x_k) + \rho P(x_k) + \left\|\left(\lambda_{A_k}^{[0]k}\right)^-\right\|^2 + \left(\lambda_{A_k} - \lambda_{A_k}^{[0]k}\right)^T \left(\Delta_{A_k} c_{A_k}(x_k) + \varphi_{A_k}\right) + \left\|\left(\lambda_{A_{3,k}}^{[0]k}\right)^T\right\| c_{A_{3,k}}(x_k),$$ where $A_{3,k} = \{i \in A_k : \ c_i(x_k) > 0\}$. (c) If $P(x_k) > \tau_0/2$, then: $$P(x_k) - P(x_k + \alpha_k p_k) \geq \alpha_k \gamma P(x_k)$$
15.	Set $x_{k+1} = x_k + \alpha_k p_k$ and go to step 2 ♦

Some comments on the algorithm are necessary.

1. Concerning the choice of δ_k, observe that the algorithm in steps 3–9 computes the sequences $\{\delta_k^{(0)}\}$ and $\{\delta_k^{(1)}\}$ for identifying the active set. The quantity $\left\|\left(\lambda_{A_k}^{[0]k}\right)^{-}\right\| + P(x_k) + \left\|B_k p_k^{[0]}\right\|$ (computed in step 8) is a measure of the error in the KKT conditions where the estimate of the active set is $A\left(x_k, \delta_k^{(0)}\right)$. If hypothesis *H3* is satisfied, then this term is of order $\|x_k - x^*\|$, and therefore from the expression of $\delta_k^{(1)}$, it follows that $\|x_k - x^*\| = o\left(\max\left\{\delta_k^{(0)}, \delta_k^{(1)}\right\}\right)$. Moreover, the computation of $\delta_k^{(1)}$ is based on local information and might correct a too small $\delta_k^{(0)}$. Therefore, the steps 4–7 are repeated if $\delta_k^{(1)} > \delta_k^{(0)}$, and $A\left(x_k, \delta_k^{(0)}\right) \neq A\left(x_k, \delta_k^{(1)}\right)$. If $\delta_{\min}$ is sufficiently small and *H1* and *H2* are satisfied, then unsuccessful termination in step 5 will never occur.
2. The rank condition on N_{A_k} from step 5 can be verified, for example, by QR or LU factorization of this matrix. These factorizations can be used for solving the systems from steps 7 and 10. Step 7 can be considered as a test on the condition $\nabla_x L = 0$. If $p_k^{[0]} = 0$, then x_k is a stationary point of the Lagrangian $L\left(., \lambda_{A_k}^{[0]k}, \mu^{[0]k}\right)$, in which $\lambda_i^{[0]k} = 0$ for $i \notin A_k$. If some multipliers $\lambda_i^{[0]k}$ are negative, then the active set is wrongly identified and some indices from A_k must be deleted. This is made in step 10.
3. In step 12, at each iteration the algorithm computes new values for the penalty parameters $\tilde{u}^k$ and $\tilde{v}^k$. Since $\tilde{u}^k > \lambda^*$ and $\tilde{v}^k > |\mu^*|$, it follows that the estimates $\tilde{u}^k$ and $\tilde{v}^k$ of the Lagrange multipliers can be increased with a small constant ρ to define $\tilde{u}^k$ and $\tilde{v}^k$, respectively.
4. A technique for updating B_k in step 6 of the algorithm is not specified. This is dependent on the quantity of information available at every step. For small- and medium-scale problems the update technique of Pantoja-Mayne (1991), with the modifications given by Heinz and Spellucci (1994), are useful. For large-scale problems, the BFGS and the limited memory BFGS updates are the most efficient.
5. The convergence of the algorithm is proved by Spellucci (1998). If the functions of the problems are twice continuously differentiable, the feasible set is a compact, the strict complementarity conditions (Definition 5.22) hold and the matrix $N_{A_k} = [\nabla e(x_k), \nabla c_{A_k}(x_k)]$ is of full rank, then it is proved that if for any k the matrices B_k are positive definite and the sequences $\{B_k\}$ and $\left\{\left(Z_k^T B_k Z_k\right)^{-1}\right\}$ are bounded, then the algorithm is superlinear convergent. In these conditions it follows that the sequences $\{\|\lambda^{[0]k}\|\}$ and $\{\|\mu^{[0]k}\|\}$ remain bounded, and therefore the sequences $\left\{\|\tilde{u}^k\|\right\}$ and $\left\{\|\tilde{v}^k\|\right\}$ also remain bounded. Therefore, $\Phi\left(., \tilde{u}^k, \tilde{v}^k\right)$ is bounded from below on $X(\tau_0)$ for any k. It follows that a decrease of the penalty weights in step 13 is only possible a finite number of times.

DONLP is a much elaborated algorithm for solving general nonlinear optimization problems using the sequential quadratic programming. Some details on linear algebra used in this algorithm and the global convergence of it are given in Spellucci (1995).

Table 12.1 Performances of DONLP2

	n	*me*	*mc*	*iter*	*nf*	*nfg*	*nr*	*er*	*cpu*	*Vfo*
ELCH	10	3	0	40	89	39	335	0.22e-15	0.02	−47.170828
ALKI	10	3	8	62	127	62	1720	0.68e-9	0.04	−1768.8069
PREC	8	0	6	18	26	18	216	0.55e-15	0.01	3.9511635
PPSE	9	6	0	12	29	11	216	0.19e-11	0.01	5055.0118
MSP3	13	0	15	262	843	260	16,241	0.82e-5	0.22	97.591032
MSP5	16	0	21	73	156	70	4335	0.89e-5	0.12	174.79630
POOL	34	20	0	12	11	9	319	0.15e-6	0.04	2785.800
TRAFO	6	0	2	11	14	11	47	0.44e-15	0.01	135.07595
LATHE	10	1	14	7	11	5	195	0.58e-8	0.01	−4428.4506

In Table 12.1 we have: $n =$ the number of variables, $me =$ the number of equality constraints, $mc =$ the number of inequality constraints, $iter =$ the number of iterations to get a solution, $nf =$ the number of evaluations of the objective function, $nfg =$ the number of evaluations of the gradient of the objective function, $nr =$ the number of evaluations of the constraints, $er =$ the satisfaction of constraints at solution, $cpu =$ the cpu computing time to get a solution (seconds), $vfo =$ the value of the objective function at solution.

Numerical Study

In Chapter 4 we have presented a number of 18 real nonlinear optimization applications. In Table 12.1 we present the numerical performances of DONLP2 in the implementation given by Spellucci (1995) for solving 9 nonlinear optimization applications.

Notes and References

DONLP is an active-set sequential quadratic programming algorithm elaborated by Spellucci (1985, 1993, 1994, 1995, 1998) for solving general continuous nonlinear optimization problems. The algorithm solves a sequence of equality quadratic programming sub-problems for search direction determination. The stepsize is computed by minimizing an exact penalty function as a merit function, namely, the l_1 penalty function. The multiplier estimates are computed by solving an equality-constrained quadratic programming sub-problem which reduces the solving of the corresponding KKT system. It uses a slightly modified version of the Pantoja-Mayne (1991) update for the Hessian of the Lagrangian, variable dual scaling, and an improved Armijo-type stepsize algorithm. Bounds on the variables are treated in a gradient projection-like manner. If we compare DONLP (Table 12.1) versus NLPQLP (Table 13.1), we see that for solving 8 nonlinear optimization applications from Chapter 4, DONLP needs 424 iterations, while NLPQLP only 308.

DONLP, like NLPQLP (presented in Chapter 13), illustrates the diversity of the active-set sequential quadratic programming methods. Both these algorithms (like SNOPT in Chapter 15) implement a *pure* sequential quadratic programming method with a lot of very sophisticated ingredients, without trying to integrate different procedures based on sequential linear-quadratic programming, interior point, or filter line search. These are the limitations of these algorithms.

Chapter 13
A SQP Algorithm with Successive Error Restoration: NLPQLP

Let us consider the general nonlinear optimization problem with equality and inequality constraints:

$$\min f(x)$$

subject to: (13.1)

$$\begin{aligned} c_i(x) &= 0, \quad i = 1, \ldots, m_e, \\ c_i(x) &\geq 0, \quad i = m_e + 1, \ldots, m, \end{aligned}$$

where it is assumed that the functions $f: \mathbb{R}^n \to \mathbb{R}$, and $c_i: \mathbb{R}^n \to \mathbb{R}$, $i = 1, \ldots, m$, are twice continuously differentiable. Also, assume that

$$X = \{x \in \mathbb{R}^n : c_i(x) = 0, \ i = 1, \ldots, m_e, \ c_i(x) \geq 0, i = m_e + 1, \ldots, m\} \neq \emptyset.$$

For solving this problem, we shall present the sequential quadratic programming with the successive error restoration algorithm NLPQLP elaborated by Klaus Schittkowski (1981, 1986, 2002, 2005, 2009, 2010). Since sequential quadratic programming is sensitive to the accuracy by which partial derivatives are provided, NLPQLP is stabilized by a nonmonotone line search and by internal and external restarts in case of errors when computing the search direction due to inaccurate derivatives.

Search Direction

The basic idea is to formulate and solve a sequence of quadratic programming sub-problems obtained by linearization of constraints and approximating the Lagrangian function:

N. Andrei, *Continuous Nonlinear Optimization for Engineering Applications in GAMS Technology*, Springer Optimization and Its Applications 121,
DOI 10.1007/978-3-319-58356-3_13

$$L(x, u) = f(x) - \sum_{i=1}^{m} u_i c_i(x) \tag{13.2}$$

quadratically, where $x \in \mathbb{R}^n$ is the vector of the primal variables and $u = [u_1, \ldots, u_m]^T \in \mathbb{R}^m$ is the vector of the dual variables or the vector of the Lagrange multipliers.

Let $v_k \in \mathbb{R}^m$ be an approximation of the Lagrange multipliers. To formulate the quadratic programming sub-problem at the current iterate x_k, the gradients $\nabla f(x_k)$ and $\nabla c_i(x_k)$, $i = 1, \ldots, m$, as well as an approximation $B_k \in \mathbb{R}^{n \times n}$ of the Hessian to the Lagrangian, are computed. Then we solve the quadratic programming sub-problem:

$$\begin{aligned} &\min \nabla f(x_k)^T p + \frac{1}{2} p^T B_k p \\ &\text{subject to:} \\ &\nabla c_i(x_k)^T p + c_i(x_k) = 0, \quad i = 1, \ldots, m_e, \\ &\nabla c_i(x_k)^T p + c_i(x_k) \geq 0, \quad i = m_e + 1, \ldots, m. \end{aligned} \tag{13.3}$$

Let p_k be the optimal solution and u_k the corresponding Lagrange multipliers of (13.3). A new iteration is computed as

$$\begin{bmatrix} x_{k+1} \\ v_{k+1} \end{bmatrix} = \begin{bmatrix} x_k \\ v_k \end{bmatrix} + \alpha_k \begin{bmatrix} p_k \\ u_k - v_k \end{bmatrix}, \tag{13.4}$$

where $\alpha_k \in (0, 1]$ is the stepsize.

Although we can guarantee that the matrix B_k is positive definite, it is quite possible for (13.3) not to be solvable because its constraints are inconsistent. To remedy this situation, Schittkowski (1986) suggested introducing an additional variable $\delta \in \mathbb{R}$, thus transforming (13.3) into a modified quadratic sub-problem with $(n+1)$ variables, with consistent constraints.

Another numerical difficulty with (13.3) is that the gradients of all constraints must be reevaluated at each iteration. But if x_k is close to solution, then the computation of the gradients of the inactive constraints is redundant. To avoid this redundant computational effort, define the set $I \triangleq \{i : m_e < i \leq m\}$, and given a constant $\varepsilon > 0$ small enough, define the sets

$$\bar{I}_1^{(k)} = \left\{ i \in I : c_i(x_k) \leq \varepsilon \quad \text{or} \quad v_k^{(i)} > 0 \right\}, \bar{I}_2^{(k)} = I \backslash \bar{I}_1^{(k)}, \tag{13.5}$$

where $v_k = \left[v_k^{(1)}, \ldots, v_k^{(m)} \right]^T$. With these developments, at each iteration we solve the following modified quadratic programming sub-problem:

$$\min_{p\in\mathbb{R}^n,\ \delta\in[0,1]} \nabla f(x_k)^{\mathrm{T}} p + \frac{1}{2} p^{\mathrm{T}} B_k p + \frac{1}{2}\rho_k \delta^2$$

subject to: (13.6)

$$\begin{aligned}
&\nabla c_i(x_k)^{\mathrm{T}} p + (1-\delta)c_i(x_k) = 0, \quad i = 1, \ldots, m_e, \\
&\nabla c_i(x_k)^{\mathrm{T}} p + (1-\delta)c_i(x_k) \geq 0, \quad i \in \bar{I}_1^{(k)}, \\
&\nabla c_i\left(x_{\kappa(k,i)}\right)^{\mathrm{T}} p + c_i(x_k) \geq 0, \quad i \in \bar{I}_2^{(k)}.
\end{aligned}$$

In (13.6), $\kappa(k,i) \leq k$ represents the previous iterations where the corresponding gradient has been evaluated the last time. We start with $\bar{I}_1^{(0)} = I$ and $\bar{I}_2^{(0)} = \varnothing$. At the subsequent iterations, only the constraint gradients belonging to the active-set $\bar{I}_1^{(k)}$ are reevaluated. The remaining rows of the Jacobian matrix remain filled with previous computed gradients. In (13.6) $\rho_k > 0$ is a parameter sufficiently small.

Let (p_k, δ_k, u_k) be the solution of (13.6), where u_k is the vector of multipliers and δ_k is the additional variable introduced to prevent the inconsistency of linear constraints. If the linear independency of constraints qualification (LICQ) is satisfied (see Remark 5.2), then $\delta_k < 1$. The matrix B_k is a positive definite approximation of the Hessian to the Lagrange function. For the global convergence analysis, any choice of B_k is appropriate as long as its eigenvalues are bounded away from zero. However, to guarantee a superlinear convergence rate, B_k is updated by the BFGS quasi-Newton method modified by a stabilization procedure to ensure the positive definiteness. The penalty parameter ρ_k is required to reduce the perturbation of the search direction by the additional variable δ as much as possible. A suitable choice is given in Schittkowski (1983).

Stepsize Computation

The stepsize parameter α_k is required in (13.4) in order to enforce the global convergence of the SQP method when the initial point x_0 is arbitrarily selected and $v_0 = 0$, $B_0 = I$. The stepsize α_k should satisfy at least a sufficient decrease condition of a merit function $\Phi_\sigma(\alpha)$ given by

$$\Phi_\sigma(\alpha) \triangleq P_\sigma\left(\begin{bmatrix} x \\ v \end{bmatrix} + \alpha \begin{bmatrix} p \\ u - v \end{bmatrix}\right), \tag{13.7}$$

where $P_\sigma(x, v)$ is a suitable penalty function, for example, the augmented Lagrange function:

$$P_\sigma(x, v) = f(x) - \sum_{i\in J}\left(v_i c_i(x) - \frac{1}{2}\sigma_i c_i(x)^2\right) - \frac{1}{2}\sum_{i\in K}\frac{v_i^2}{\sigma_i}, \tag{13.8}$$

where the sets J and K are defined by Schittkowski (1983) as:

$$J = \{1, \ldots, m_e\} \cup \{i : m_e < i \le m, \ \ c_i(x) \le v_i/\sigma_i\} \quad \text{and} \quad K = \{1, \ldots, m\} \backslash J.$$

The objective function is penalized as soon as an iterate leaves the feasible region. The penalty parameters σ_i, $i = 1, \ldots, m$, control the degree of constraint violation. To guarantee a descent direction of the merit function, σ_i are chosen such that:

$$\Phi'_{\sigma_k}(0) = \nabla P_{\sigma_k}(x_k, v_k)^{\mathrm{T}} \begin{bmatrix} p_k \\ u_k - v_k \end{bmatrix} < 0. \tag{13.9}$$

Usually, the stepsize α_k is computed by means of the Armijo's rule (1966), i.e., a sufficient descent condition of the merit function (13.8) which guarantees convergence to a stationary point. However, to take into consideration the curvature of the merit function, we need some kind of compromise between a polynomial interpolation, typically a quadratic one, and a reduction of the stepsize by a given factor until a stopping criterion has been reached (Schittkowski, 2010). Determination of the stepsize α_k is very important in the economy of NLPQLP algorithm. The idea is not to vast too many function calls in the procedure for α_k computation. Moreover, the behavior of the merit function becomes irregular in case of constrained optimization because of the very steep slopes at the border caused by large penalty terms. The implementation of a procedure for α_k computation is more complex if linear constraints and simple bounds on variables are to be satisfied during the line search. Since $\Phi_\sigma(0)$, $\Phi'_\sigma(0)$, and $\Phi_\sigma(\alpha_i)$ are known (α_i is the current value of the stepsize), then a procedure based on the quadratic interpolation for α_{k+1} computation is as in the following fragment of the algorithm:

Algorithm 13.1 Linear search in NLPQLP

1.	Choose the parameters $0 < \beta < 1$ and $0 < \mu < 1/2$. Consider $\sigma = \sigma_k$. Set $\alpha_0 = 1$ and $i = 0$
2.	If $\Phi_\sigma(\alpha_i) < \Phi_\sigma(0) + \mu\alpha_i\Phi'_\sigma(0)$, then stop; otherwise, go to step 3
3.	Compute: $\overline{\alpha}_i = \frac{1}{2} \frac{\alpha_i^2 \Phi'_\sigma(0)}{\alpha_i\Phi'_\sigma(0) - \Phi_\sigma(\alpha_i) + \Phi_\sigma(0)}$
4.	Let $\alpha_i = \max\{\beta\alpha_i, \overline{\alpha}_i\}$. Set $i = i + 1$ and go to step 2 ♦

The algorithm is well known; see, for example, (Powell, 1978b). $\overline{\alpha}_i$ is the minimizer of the quadratic interpolation polynomial. For checking the termination, the Armijo descent property is used. Computation of α_i in step 4 of the algorithm as $\alpha_i = \max\{\beta\alpha_i, \overline{\alpha}_i\}$ is required to avoid irregular values, since the minimizer of the quadratic interpolation polynomial can be outside the feasible region $(0, 1]$. Schittkowski mentions that the Algorithm 13.1 is equipped with some safeguards to prevent, for example, the violation of the bounds. In Algorithm 13.1, it is supposed that $\Phi_\sigma(1)$ is known before calling it. Moreover, the algorithm stops if sufficient descent is not observed after a certain number of iterations. If the tested stepsize falls below machine precision or the accuracy by which model function values are computed, then the merit function cannot decrease further on. It is possible that Algorithm 13.1 breaks down because of too many iterations. In this case, Schittkowski recommends to proceed from a descent direction of the merit function when $\Phi'_\sigma(0)$ is extremely small. In order to avoid the interruption of the whole iteration process of NLPQLP, another criterion is used instead of the criterion in step 2 of the Algorithm 13.1. The idea is to repeat the line search with another stopping criterion. Instead of testing (13.8), we accept a stepsize α_k as soon as the inequality

$$\Phi_{\sigma_k}(\alpha_k) \leq \max_{k-t(k)\leq j\leq k}\{\Phi_{\sigma_j}(0)\} + \mu\alpha_k\Phi'_{\sigma_k}(0), \tag{13.10}$$

has been satisfied, where $t(k) = \min\{k, \bar{t}\}$, and $\bar{t}$ is a given parameter. Thus, an increase of the reference value Φ_{σ_j} is allowed, i.e., an increase of the merit function value. To implement (13.10) it is necessary to store the previous values of the merit function from the first $t(k)$ iterations. The nonmonotone search goes back to Grippo, Lampariello and Lucidi (1986) and was extended to the trust-region methods for constrained optimization by Bonnans, Panier, Tits, and Zhou (1992), Deng, Xiao, and Zhou (1993), Grippo, Lampariello, and Lucidi (1989, 1991), Raydan (1993, 1997), Panier and Tits (1991), and Toint (1996, 1997).

Updating the Matrix B_k

To avoid calculation of the second derivatives and to obtain a final superlinear convergence rate, the standard approach used in NLPQLP is to update B_k by the quasi-Newton BFGS formula:

$$B_{k+1} = B_k + \frac{q_k q_k^T}{s_k^T q_k} - \frac{B_k s_k s_k^T B_k}{s_k^T B_k s_k}, \tag{13.11}$$

where $q_k = \nabla_x L(x_{k+1}, u_k) - \nabla_x L(x_k, u_k)$ and $s_k = x_{k+1} - x_k$. The algorithm is equipped with special safeguards guarantee that ensures the curvature condition $s_k^T q_k > 0$. In this case, if B_0 is positive definite, then all the matrices B_k generated by (13.11) remain positive definite along the iterations. The algorithm contains a scaling and a restart procedure to replace an actual B_k by $\gamma_k I$ before performing the update (13.11), where $\gamma_k = s_k^T q_k / s_k^T s_k$, (Liu & Nocedal, 1989).

The main steps of the NLPQLP algorithm are as follows.

Algorithm 13.2 NLPQLP (Schittkowski)

1.	Choose an initial point x_0, an initial approximation to the Hessian of the Lagrange function B_0 and tolerances ε_1 and ε_2. Set $k=0$
2.	Compute $f(x_k)$, $c(x_k)$, $\nabla f(x_k)$, and $\nabla c(x_k)$
3.	Compute p_k as solution of the quadratic programming sub-problem (13.6)
4.	Compute the stepsize by means of the Algorithm 13.1
5.	If the KKT optimality conditions are satisfied with tolerance ε_1 and the norm of the gradient of the Lagrange function is smaller than the tolerance ε_2, then stop; otherwise, go to step 6
6.	Using (13.11) update B_k
7.	Compute a new Cholesky factorization of B_k
8.	Set $k=k+1$ and continue with step 2 ♦

Numerical Study
In Table 13.1 we present the performances of NLPQLP for solving eight nonlinear optimization applications described in Chapter 4.

Table 13.1 Performances of NLPQLP

	n	*me*	*mc*	*iter*	*nf*	*ng*	*nq*	*KKT*	*cpu*	*vfo*
ELCH	10	3	0	21	30	21	21	0.11e-8	0.03	−47.76109
ALKI	10	3	8	53	69	53	53	0.13e-11	0.03	−1768.807
PREC	8	0	6	19	20	19	19	0.36e-8	0.02	3.9511635
PPSE	9	6	0	9	10	9	9	0.21e-8	0.03	5055.0118
MSP3	13	0	15	146	264	146	146	0.62e-8	0.04	79.631357
POOL	34	20	0	22	22	22	22	0.64e-13	0.07	2785.800
TRAFO	6	0	2	19	19	19	19	0.19e-10	0.03	135.0759
LATHE	10	1	14	19	19	19	19	0.27e-9	0.03	−4430.087

In Table 13.1 we have: $n=$ the number of variables, $me=$ the number of equality constraints, $mc=$ the number of inequality constraints, $iter=$ the number of iterations to get a local optimal solution, $nf=$ the number of evaluations of the functions of the problem, $ng=$ the number of the evaluations of the gradients of the functions, $nq=$ the number of the quadratic programming sub-problems solved to get a local optimal solution of the problem, $KKT=$ the norm of the KKT conditions, $cpu=$ the cpu computing time for solving the problem (seconds), $vfo=$ the value of the objective function at the solution.

Application ETA-MACRO
ETA-MACRO simulates a market economy through a dynamic nonlinear optimization process. To describe the production relationships within this economy, two dynamic sub-models are incorporated: (a) ETA, a process analysis for energy

technology assessment, and (b) a macroeconomic growth model providing for substitution between capital, labor, and energy inputs. The application is described in Manne (1977) (see also Bergman, 1988; Murtagh & Saunders, 1995, pp.98). Table 13.2 contains the performances of NLPQLP for solving the application ETA-MACRO for a different number of variables.

Table 13.2 Performances of NLPQLP for solving the application ETA-MACRO

n	*me*	*mc*	*iter*	*nf*	*ng*	*nq*	*KKT*	*cpu*	*vfo*
60	1	41	31	31	31	31	0.74e-8	0.17	5.3700593
120	1	81	103	103	103	13	0.83e-8	3.69	7.9532086
300	1	201	823	823	823	823	0.86e-8	454.90	9.3301684

Notes and References

NLPQLP is an extension of the sequential quadratic programming solver NLPQL (Schittkowski, 1985) that implements a nonmonotone line search to ensure global convergence. Our description of NLPQLP is based on the papers of Schittkowski (1986, 2002, 2005, 2009, 2010). It uses a quasi-Newton approximation to the Hessian of the Lagrangian, which is updated with the BFGS formula. To calculate the stepsize that minimizes an augmented Lagrangian merit function, a nonmonotone line search is used. NLPQLP illustrates the sequential quadratic programming paradigm also referred to as active-set methods, because it provides an estimate of the active set at every iteration. In NLPQLP the quadratic programming sub-problems are modified to avoid the redundant computational effort of evaluating the gradients corresponding to inactive constraints. If we compare NLPQLP (Table 13.1) versus KNITRO/ACTIVE (Table 14.1), we see that for solving 8 nonlinear optimization applications from Chapter 4, NLPQLP needs 308 iterations, while KNITRO/ACTIVE only needs 226.

Chapter 14
Active-set Sequential Linear-Quadratic Programming: KNITRO/ACTIVE

KNITRO represents one of the most elaborated algorithms (and Fortran package) for solving general large-scale nonlinear optimization problems (Byrd, Gould, Nocedal, & Waltz, 2004). This is characterized by great flexibility and robustness integrating two very powerful and complementary algorithmic approaches for nonlinear optimization: the *active-set sequential linear-quadratic approach* and the *interior point approach*. KNITRO includes a number of much studied algorithms for linear algebra, very carefully implemented in computing programs, able to solve a large variety of nonlinear optimization problems like special cases of unconstrained optimization, systems of nonlinear equations, least square problems, and linear and nonlinear programming problems.

KNITRO has two algorithmic options. The first one, known as KNITRO/ACTIVE, uses a new active-set method based on sequential linear-quadratic programming (SLQP) and the projected conjugate gradient iteration. The second, known as KNITRO/INTERIOR, uses the interior point methods in two implementations: KNITRO/INTERIOR-CG in which the algorithmic step is computed by means of an iterative conjugate gradient method and KNITRO/INTERIOR-DIRECT in which the step is (usually) computed via a direct factorization of the corresponding linear systems. These two approaches KNITRO/ACTIVE and KNITRO/INTERIOR communicate by the so-called crossover technique. This crossover procedure, used for the first time by Megiddo (1989) in linear programming, is implemented internally switching to ACTIVE algorithm after INTERIOR-DIRECT or INTERIOR-CG algorithm has solved the problem to a requested tolerance.

The difference between these two approaches is as follows. The active-set sequential linear-quadratic programming algorithm is similar in nature to a sequential quadratic programming method (see Chapter 11), but it uses linear programming sub-problems to estimate the active set at each iteration. This active-set approach may be useful when a good initial point can be provided. On the other hand, in the interior point methods (see Chapter 17), also known as barrier methods,

N. Andrei, *Continuous Nonlinear Optimization for Engineering Applications in GAMS Technology*, Springer Optimization and Its Applications 121,
DOI 10.1007/978-3-319-58356-3_14

the nonlinear programming problem is replaced by a sequence of barrier sub-problems controlled by a barrier parameter. The algorithm uses trust-region and a merit function to promote convergence. The algorithm performs one or more minimization steps on each barrier problem and then decreases the barrier parameter and repeats the process until the problem has been solved to the desired accuracy.

In this chapter we discuss only KNITRO/ACTIVE. This algorithm is described in Byrd, Gould, Nocedal, and Waltz (2004) and its global convergence in Byrd, Gould, Nocedal, and Waltz (2002).

KNITRO/INTERIOR will be described in Chapter 19 where the crossover technique is also presented.

Let us consider the general nonlinear optimization problem

$$\min f(x)$$

subject to (14.1)

$$c_E(x) = 0,$$
$$c_I(x) \geq 0,$$

where $f: \mathbb{R}^n \to \mathbb{R}$, $c_E: \mathbb{R}^n \to \mathbb{R}^l$, and $c_I: \mathbb{R}^n \to \mathbb{R}^m$ are twice continuously differentiable functions. Define $E \triangleq \{1, \ldots, l\}$and $I \triangleq \{1, \ldots, m\}$.

The active-set method implemented in KNITRO/ACTIVE does not follow a SQP approach because the cost of solving generally linear-constrained quadratic programming sub-problems imposes limitations on the size of problems that can be solved in practice. In addition, the incorporation of the second derivative information in SQP methods has proved to be difficult in practice. Instead, in KNITRO/ACTIVE a sequential linear-quadratic programming (SLQP) method is used (see Section 11.6). This SLQP computes a step in two phases. First, a linear programming problem (LP) is solved in order to identify a working set W. This is followed by an equality-constrained quadratic programming phase in which the constraints in the working set W are imposed as equalities. The total step of the algorithm is a combination of the previous steps obtained in linear programming and equality-constrained quadratic programming phases.

14.1 KNITRO/ACTIVE Algorithm

To achieve progress on both feasibility and optimality, the algorithm is designed to reduce the l_1 penalty function:

$$P(x, \sigma) \triangleq f(x) + \sigma \sum_{i \in E} |c_i(x)| + \sigma \sum_{i \in I} \max\{0, -c_i(x)\}, \tag{14.2}$$

where c_i, $i \in E$, represents the components of vector c_E, and c_i, $i \in I$, represents the components of c_I. The penalty parameter σ is chosen by an adaptive procedure described later.

In the LP phase, given an estimate x_k of the solution of the nonlinear optimization problem (14.1), the following linear programming problem is formed and solved:

$$\min_p \nabla f(x_k)^{\mathrm{T}} p \tag{14.3a}$$

subject to

$$\nabla c_i(x_k)^{\mathrm{T}} p + c_i(x_k) = 0, \quad i \in E, \tag{14.3b}$$

$$\nabla c_i(x_k)^{\mathrm{T}} p + c_i(x_k) \geq 0, \quad i \in I, \tag{14.3c}$$

$$\|p\|_\infty \leq \Delta_k^{LP}, \tag{14.3d}$$

where $\Delta_k^{LP} > 0$ is the trust-region radius. Observe that (14.3) differs from the classical sub-problem used in SQP methods only in that (14.3) does not contain the quadratic term $\frac{1}{2} p^{\mathrm{T}} B_k p$, where B_k is an approximation to the Hessian of the Lagrangian of (14.1). Now, since the constraints of (14.3) may be inconsistent, instead of it we solve the l_1 penalty reformulation of (14.3) as

$$\begin{aligned} \min_p l_\sigma(p) &\triangleq \nabla f(x_k)^T p + \sigma_k \sum_{i \in E} \left| \nabla c_i(x_k)^{\mathrm{T}} p + c_i(x_k) \right| \\ &+ \sigma_k \sum_{i \in I} \max\left\{0, \ -\nabla c_i(x_k)^{\mathrm{T}} p - c_i(x_k)\right\} \end{aligned}$$

subject to: (14.4)

$$\|p\|_\infty \leq \Delta_k^{LP}.$$

The solution of this linear program, denoted by p^{LP}, is computed by the simplex algorithm. Based on this solution, we define the working set W as a linearly independent subset of the active set at the LP solution p^{LP}:

$$\begin{aligned} A(p^{LP}) = &\left\{ i \in E : \nabla c_i(x_k)^{\mathrm{T}} p^{LP} + c_i(x_k) = 0 \right\} \\ &\cup \left\{ i \in I : \nabla c_i(x_k)^{\mathrm{T}} p^{LP} + c_i(x_k) = 0 \right\}. \end{aligned} \tag{14.5}$$

Similarly, define the set V of violated constraints as:

$$\begin{aligned} V(p^{LP}) = &\left\{ i \in E : \nabla c_i(x_k)^{\mathrm{T}} p^{LP} + c_i(x_k) \neq 0 \right\} \\ &\cup \left\{ i \in I : \nabla c_i(x_k)^{\mathrm{T}} p^{LP} + c_i(x_k) < 0 \right\}. \end{aligned} \tag{14.6}$$

To ensure the progress of the algorithm subject to the penalty function $P(x, \sigma)$ defined in (14.2), define the Cauchy step:

$$p^C = \alpha^{LP} p^{LP}, \tag{14.7}$$

where $\alpha^{LP} \in (0\,,1]$ is the stepsize that provides sufficient decrease in the following piecewise quadratic model of the penalty function $P(x, \sigma)$:

$$q_k(p) = l_\sigma(p) + \frac{1}{2} p^{\mathrm{T}} B(x_k, \lambda_k) p. \tag{14.8}$$

In (14.8), $l_\sigma(p)$ is defined by (14.4), and $B(x_k, \lambda_k)$ is the Hessian of the Lagrange function or an approximation of it.

Given the working set W_k the following equality-constrained quadratic program in variable p is solved, treating the constraints in W_k as equalities and ignoring all other constraints:

$$\min \frac{1}{2} p^{\mathrm{T}} B(x_k, \lambda_k) p + \left(\nabla f(x_k) + \sigma_k \sum_{i \in V} \gamma_i \nabla c_i(x_k) \right)^T p \tag{14.9a}$$

subject to

$$\nabla c_i(x_k)^{\mathrm{T}} p + c_i(x_k) = 0, \quad i \in E \cap W_k, \tag{14.9b}$$

$$\nabla c_i(x_k)^{\mathrm{T}} p + c_i(x_k) = 0, \quad i \in I \cap W_k, \tag{14.9c}$$

$$\|p\|_2 \leq \Delta_k, \tag{14.9d}$$

where γ_i is the algebraic sign of the ith constraint violated in x_k. Observe that (14.9d) is spherical and is distinct from the trust-region radius Δ_k^{LP} used in (14.3d). The solution of (14.9), denoted by p^Q, is obtained by the projected conjugate gradient algorithm described later.

The total step p of the SLQP method is computed as

$$p = p^C + \alpha^Q \left(p^Q - p^C \right), \tag{14.10}$$

where $\alpha^Q \in [0, 1]$ is the stepsize that approximately minimizes the model function (14.8). Now, we can present the KNITRO/ACTIVE algorithm as follows:

Algorithm 14.1 KNITRO/ACTIVE (Byrd, Gould, Nocedal and Waltz)

1.	Choose an initial point $x_0 \in \mathbb{R}^n$, as well as the parameters $\Delta_0 > 0$, $\Delta_0^{LP} > 0$, and $0 < \eta < 1$. Set $k = 0$
2.	If a stopping test for the nonlinear problem (14.1) is satisfied, stop; otherwise continue with step 3
3.	Update the penalty parameter σ_k and solve the linear program (14.4) to obtain the step p_k^{LP} and the working set W_k

(continued)

4.	Compute $\alpha_k^{LP} \in (0, 1]$ as an approximate minimizer of the quadratic model $q(\alpha p_k^{LP})$ given by (14.8) such that $\alpha_k^{LP} \lVert p_k^{LP} \rVert \le \Delta_k$. Set $p_k^C = \alpha_k^{LP} p_k^{LP}$
5.	Solve the equality quadratic programming sub-problem (14.9) to obtain the solution p_k^Q
6.	Compute $p_k^{CE} = p_k^Q - p_k^C$ as the segment line from the Cauchy point to the solution of the equality quadratic programming sub-problem
7.	Compute $\alpha_k^Q \in [0, 1]$ as an approximate minimizer of $q_k(p_k^C + \alpha p_k^{CE})$ with respect to α
8.	Compute $p_k = p_k^C + \alpha_k^Q p_k^{CE}$, and set $x_T = x_k + p_k$
9.	Compute the standard ratio: $$\rho_k = \frac{P(x_k, \sigma_k) - P(x_T, \sigma_k)}{q_k(0) - q_k(p_k)}.$$ If $\rho_k \ge \eta$, then set $x_{k+1} = x_T$; otherwise set $x_{k+1} = x_k$ and go to step 10
10.	Update Δ_{k+1}^{LP} and Δ_{k+1}. Set $k = k + 1$ and go to step 2 ♦

The trust-region radius Δ_k is updated based on the standard ratio ρ_k. The choice of Δ_k^{LP} is important because this determines the working set. The estimates of the multipliers λ_k used in the Hessian are the least squares estimates using the working set W_k and modified so that $\lambda_i \ge 0$ for $i \in I$.

14.2 Strategy for Penalty Parameter Update

The algorithm KNITRO/ACTIVE in step 3 requires an updating of the penalty parameter σ_k. This is done in the first phase where the linear program (14.3) is solved. For this, a piecewise linear model of constraint violation at current point x_k is defined as

$$m_k(p) = \sum_{i \in E} \left| \nabla c_i(x_k)^T p + c_i(x_k) \right| + \sum_{i \in I} \max\left\{0, -\nabla c_i(x_k)^T p - c_i(x_k)\right\}, \tag{14.11}$$

so that the objective (14.4) of the LP sub-problem can be written as:

$$l_\sigma(p) = \nabla f(x_k)^T p + \sigma_k m_k(p). \tag{14.12}$$

Given a value σ_k, then the solution of the LP sub-problem (14.4) is denoted as $p^{LP}(\sigma_k)$ to emphasize its dependence on the penalty parameter. Let $p^{LP}(\sigma_\infty)$ be the minimizer of $m_k(p)$ subject to the trust-region constraint $\|p\|_\infty \leq \Delta_k^{LP}$ from (14.4). Then, the following algorithm describes the computation of the LP step p_k^{LP} and the updating of the penalty parameter σ_k.

Algorithm 14.2 Penalty Update Algorithm

1.	Initialize the data: x_k, $\sigma_{k-1} > 0$, and $\Delta_k^{LP} > 0$. Choose the parameters $\varepsilon_1, \ \varepsilon_2 \in (0, 1]$
2.	Solve the sub-problem (14.4) with $\sigma = \sigma_{k-1}$ to obtain $p^{LP}(\sigma_{k-1})$
3.	If $m_k(p^{LP}(\sigma_{k-1})) = 0$, then set $\sigma^+ = \sigma_{k-1}$; otherwise, compute $p^{LP}(\sigma_\infty)$
4.	If $m_k(p^{LP}(\sigma_\infty)) = 0$, then find $\sigma^+ > \sigma_{k-1}$ such that $m_k(p^{LP}(\sigma^+)) = 0$. Otherwise, find $\sigma^+ \geq \sigma_{k-1}$ such that $$m_k(0) - m_k(p^{LP}(\sigma^+)) \geq \varepsilon_1[m_k(0) - m_k(p^{LP}(\sigma_\infty))]$$
5.	If necessary, increase the value of σ^+ to satisfy $$l_{\sigma^+}(0) - l_{\sigma^+}(p^{LP}(\sigma^+)) \geq \varepsilon_2\sigma^+[m_k(0) - m_k(p^{LP}(\sigma^+))].$$ Set $\sigma_k = \sigma^+$ and $p_k^{LP} = p^{LP}(\sigma^+)$ ♦

The selection of $\sigma^+ > \sigma_{k-1}$ is achieved by successive increasing of the current trial value by 10, for example, and by resolving the LP problem. The algorithm is simple and easy to be implemented when the *worm start* strategy is used. The penalty update algorithm above guarantees that σ is chosen large enough to ensure convergence to a stationary point (Byrd, Gould, Nocedal, & Waltz, 2002).

Since the Algorithm 14.1 is a penalty method, it can naturally deal with infeasibility. If a problem is infeasible, then the penalty parameter will be driven to infinity. Moreover, if the algorithm is converging to a stationary point, for our infeasibility measure, we have

$$m_k(0) - m_k\left(p^{LP}(\sigma_\infty)\right) \to 0$$

during the penalty update procedure, providing a clear indication of local infeasibility.

14.3 Iteration of Projected Conjugate Gradient Algorithm

One of the main modules of the Algorithm 14.1 is solving the equality quadratic programming sub-problem (14.9) in step 5. This module is shared by all the algorithms implemented in KNITRO (with active set or interior point). In KNITRO the equality quadratic programming sub-problems (14.9) are solved using the projected conjugate gradient method (Gould, Hribar, & Nocedal, 2001; Keller, Gould, & Wathen, 2000), which is well suited for large-scale problems. This is also suitable in case of large-scale sub-problems and can treat the situations in which the negative curvature appears without modifications of the Hessian of the

Lagrange function. A variant of this iteration in the context of sequential linear-quadratic programming is as follows.

Let us consider the following quadratic program:

$$\min \frac{1}{2} x^T G x + h^T x \tag{14.13a}$$

subject to

$$Ax = b, \tag{14.13b}$$

where G is supposed to be symmetric and positive definite on the null space of $A \in \mathbb{R}^{n \times l}$. One way to solve (14.13) is to apply a special form of the conjugate gradient (CG) iteration to the KKT system associated to (14.13), which is given by:

$$\begin{bmatrix} G & A^T \\ A & 0 \end{bmatrix} \begin{bmatrix} x \\ v \end{bmatrix} = \begin{bmatrix} -h \\ b \end{bmatrix}. \tag{14.14}$$

Although the coefficient matrix is not positive definite, we can apply the CG method to (14.14), provided that we precondition and project the CG method so that it effectively solves the positive definite reduced problem within the feasible manifold (14.13b). Denote the preconditioning matrix by P. The following algorithm gives the iterations of the projected conjugate gradient.

Algorithm 14.3 Preconditioned Projected Conjugate Gradient Algorithm

1.	Choose an initial point x_0 satisfying $Ax_0 = b$, as well as the tolerance $\varepsilon > 0$. Set $x = x_0$ and compute $r = Gx + h$. Set $z = Pr$ and $p = -z$
2.	If $\lVert z \rVert \le \varepsilon$, then stop; otherwise go to step 3
3.	Compute (in this order): $\alpha = r^T z / p^T G p, \quad x = x + \alpha p,$ $r^+ = r + \alpha G p, \quad z^+ = P r^+,$ $\beta = (r^+)^T z^+ / r^T z, \quad p = -z^+ + \beta p,$ $z = z^+, \quad r = r^+$ and continue with step 2 ♦

The Algorithm 14.3 has exactly the same form as the standard preconditioned CG algorithm for solving symmetric and positive definite systems of linear equations (Golub & Van Loan, 1996). The difference is that P is normally a symmetric and positive definite matrix, whereas in our case it represents a *projection and preconditioning matrix* defined as follows.

Given a vector r, compute $z = Pr$ as solution of the system:

$$\begin{bmatrix} D & A^T \\ A & 0 \end{bmatrix} \begin{bmatrix} z \\ w \end{bmatrix} = \begin{bmatrix} r \\ 0 \end{bmatrix}, \tag{14.15}$$

where D is a symmetric and positive definite matrix on the null space of A and w is an auxiliary vector. To be a practical preconditioning matrix, D must be a sparse matrix so that solving (14.15) is significantly less costly than solving (14.14). By construction, $z = Pr$ is in the null space of A and so are all the search directions generated by the Algorithm 14.3. Since $Ax_0 = b$, all subsequent iterates x also satisfy these linear constraints. Let $Z \in \mathbb{R}^{n \times (n-l)}$ be a basis for the null space of A from (14.13b). Then, in the null space of A, the solution of (14.13) may be expressed as $x = x_0 + Zu$, where $u \in \mathbb{R}^{n-l}$ is the solution of the linear system:

$$\left(Z^{\mathrm{T}} G Z\right) u = Z^{\mathrm{T}} (G x_0 + h). \tag{14.16}$$

The iterations generated by the Algorithm 14.3 are given by $x = x_0 + Zu$, where u are the iterations of the preconditioned CG method on the system (14.16) using the matrix $Z^{\mathrm{T}} D Z$ as a preconditioner. Therefore, the Algorithm 14.3 is a standard preconditioned CG algorithm as soon as G and D are positive definite on the null space of A (Gould, Hribar, & Nocedal, 2001).

The Algorithm 14.3 has two advantages over the reduced conjugate gradient approach. The first is that there is no need to compute a null space basis and consequently no risk for the ill-conditioning in Z to deteriorate the rate of convergence of the conjugate gradient iteration. The second benefit is that the projection matrix in (14.15) can also be used to compute the normal step and the Lagrange multipliers, and thus the extra cost of each of these computations is only one back solve involving the factors of this projection matrix (Byrd, Nocedal, & Waltz, 2006).

The Algorithm 14.3 assumes that an initial feasible point x_0 is provided (see step 1). The factorization of the system (14.15) allows us to compute such a point by solving

$$\begin{bmatrix} D & A^{\mathrm{T}} \\ A & 0 \end{bmatrix} \begin{bmatrix} w \\ x_0 \end{bmatrix} = \begin{bmatrix} 0 \\ b \end{bmatrix},$$

which is in fact the minimum norm solution in the norm weighted by D.

14.4 Hessian Options

KNITRO contains some options for the first and second derivatives. The user can supply these elements, which generally results in the greatest level of efficiency and robustness of the algorithms implemented in KNITRO. However, if these elements are not possible to be supplied, then KNITRO has other options on the second derivative: dense quasi-Newton BFGS, dense quasi-Newton SR1, finite difference Hessian-vector product, and limited-memory BFGS Hessian. The dense quasi-Newton BFGS option uses gradient information to compute a symmetric, positive definite approximation to the Hessian matrix. Typically, this method requires more iterations to converge than the exact Hessian version. The dense quasi-Newton SR1 approach builds an approximate Hessian by using gradient information, but in this case the SR1 Hessian approximation is not restricted to be positive definite. Therefore, the SR1 approximation may be a better approach than BFGS if there is a lot of negative curvature in the problem. In case of large-scale applications, the Hessian of the Lagrangian cannot be computed, or it is too large to store. In KNITRO the Hessian-vector product can be obtained by finite differences of gradients of the Lagrangian. Each Hessian-vector product requires one additional gradient evaluation.

Practical Hints in GAMS

KNITRO has plenty of facilities and options: initial point strategy, feasibility of iterates to some or all inequality constraints, control of barrier parameter update strategy, how to compute the gradients or the Hessian, second derivative options, special problem classes, etc. All these can be studied in the documentation elaborated by Waltz (2004) or in *GAMS – The Solver Manuals* (2005, pp.177–190). The KNITRO options file `knitro.opt` allows the user to easily set certain algorithmic options. Options are set by specifying a keyword and a corresponding value on a line in the `knitro.opt` file. Most user options can be specified with either a numeric value or a string value. For running the KNITRO/ACTIVE in `knitro.opt` file, the line *algorithm 3* is written. The individual user options and their possible numeric values are described in more details in GAMS documentation of KNITRO. For example, the option `hessopt` specifies how to compute the Hessian of the Lagrangian (BFGS, SR1, Hessian-vector product using finite difference, exact Hessian-vector product, limited-memory BFGS Hessian). The option `honorbnds` indicates whether or not to enforce satisfaction of the simple bounds throughout the optimization. `Honorbnds` $=$ `1` enforces that the initial point and all subsequent iterates satisfy the bounds on the variables. The option `initpt` indicates whether an initial point strategy is used. `Maxit` is another option specifying the maximum number of iterations before termination. The option `scale` indicates whether a scaling of the objective and of the constraints functions is performed based on their values at the initial point. If scaling is active, all internal computations including the stopping tests are based on the scaled values. If `scale` $=$ `1`, the objective function and the constraints may be scaled. Another important option is `soc` which indicates whether or not to use the second-order correction strategy. If `soc` $=$ `2`, then the second-order correction steps are always attempted if the original step is rejected and there are nonlinear constraints. Some other options need the double-precision values. All these are explained in *GAMS – The Solver Manuals* (2005, pp.177–190). Their values may change the behavior of KNITRO/ACTIVE for solving nonlinear optimization problems.

In the following let us present the performances of KNITRO/ACTIVE for solving some applications described in Chapter 4.

Numerical Study

Table 14.1 contains the performances of KNITRO/ACTIVE for solving 12 nonlinear optimization applications presented in Chapter 4 in GAMS language.

In Table 14.2 we see the performances of KNITRO/ACTIVE for solving six large-scale nonlinear optimization applications described in Chapter 4 in GAMS language.

Table 14.1 Performances of KNITRO/ACTIVE. Option 3. Small-scale nonlinear optimization applications

	n	*me*	*mc*	*it*	*itcg*	*nf*	*ng*	*nh*	*cpu*	*vfo*
ELCH	10	3	0	18	80	19	19	18	0.019	−47.761090
ALKI	10	3	8	22	33	58	23	31	0.022	−1768.8069
PREC	8	0	6	45	116	117	46	68	0.028	3.95116344
PPSE	9	6	0	6	5	11	7	7	0.016	5055.01174
MSP3	13	0	15	7	6	9	8	7	0.015	97.5910347
MSP5	16	0	21	21	34	54	22	32	0.032	174.786994
POOL	34	20	0	15	20	33	16	22	0.025	2569.800
TRAFO	6	0	2	30	88	100	31	60	0.026	135.07596
LATHE	10	1	14	83	127	242	84	126	0.056	−4429.3999
DES	150	50	0	24	378	79	25	41	0.369	1055.18231
CSTC	303	200	0	6	16	7	7	6	0.084	3.4800747
DIFF	396	324	0	1	0	2	1	0	0.090	0.00

Table 14.2 Performances of KNITRO/ACTIVE. Option 3. Large-scale nonlinear optimization applications

	n	*me*	*mc*	*it*	*itcg*	*nf*	*ng*	*nh*	*cpu*	*vfo*
HANG	1002	501	0	10	226	24	11	13	0.343	5.0685777
	2002	1001	0	10	226	24	11	13	0.982	5.0685101
	4002	2001	0	10	226	24	11	13	2.964	5.0684889
	8002	4001	0	10	226	24	11	13	8.288	5.068482
FLOW	1182	754	0	10	9	11	11	10	3.456	0.15e-11
FLOWO	1556	1005	0	17	16	18	18	17	7.017	0.79e-31
POL	4004	3000	0	193	255	363	194	257	46.793	9.8214896
	6004	4500	0	219	310	464	220	308	79.314	10.545559
	8004	6000	0	74	105	164	75	105	81.193	10.621358
	10,004	7500	0	150	216	337	151	214	134.66	10.291229
CAT	3003	2000	0	117	261	429	118	233	11.138	−0.048055
	6003	4000	0	155	333	551	156	299	33.039	−0.048055
	9003	6000	0	112	190	266	113	183	37.491	−0.048045
CONT	2505	2000	0	14	10	15	15	14	1.893	1.0132389
	5005	4000	0	14	11	15	15	14	4.947	1.0059224
	7505	6000	0	12	12	13	13	12	8.573	1.0045614
	10,005	8000	0	16	17	17	17	16	15.245	1.0040718

In Tables 14.1 and 14.2 we have: n = the number of variables, me = the number of equality constraints, mc = the number of inequality constraints, it = the number of iterations for solving the application, $itcg$ = the number of conjugate gradient iterations, nf = the number of evaluations of the functions, ng = the number of evaluations of the gradients, nh = the number of evaluations of Hessians, cpu = cpu computing time to get a solution (seconds), vfo = the value of the objective function at optimal solution.

Notes and References

The content of this chapter is based on the works described in Byrd, Gould, Nocedal, and Waltz (2002, 2004), Byrd, Nocedal, and Waltz (2003, 2006), Nocedal and Wright (2006), Waltz (2004), and Andrei (2015). KNITRO is working under the GAMS technology (see (*GAMS – The Solver Manuals*, 2005, pp.177–190)). KNITRO includes both sequential linear-quadratic programming and interior point methods. The active-set sequential linear-quadratic programming method described in this chapter to determine an estimate of the active set at each iteration solves a linear programming sub-problem that approximates the l_1 exact penalty sub-problem. The linear programming sub-problem has an additional infinity-norm trust-region constraint. The constraints in this linear programming sub-problem that are satisfied as equalities are marked as active, and they are used to set up an equality-constrained quadratic programming sub-problem whose objective is a quadratic approximation of the Lagrangian of the problem at the current iterate. This quadratic programming sub-problem includes an l_2-norm trust-region constraint. The equality quadratic programming sub-problem is solved by using a projected conjugate gradient algorithm. The penalty parameter is updated to ensure sufficient decrease toward feasibility.

Chapter 15
A SQP Algorithm for Large-Scale Constrained Optimization: SNOPT

The algorithm described in this chapter, elaborated by Gill, Murray, and Saunders (2002, 2005), is dedicated to solve nonlinear optimization problems of the following form:

$$\min_{x\in\mathbb{R}^n} f(x)$$

subject to (15.1)

$$l \leq \begin{bmatrix} x \\ c(x) \\ Ax \end{bmatrix} \leq u,$$

where f is the objective function (linear or nonlinear), $c(x)$ is the vector of nonlinear constraints, A is the matrix corresponding to the linear constraints, and l and u are bounds on variables. Suppose that the nonlinear functions are smooth and their first derivatives are available and the Jacobian of the constraints is a sparse matrix.

SNOPT is the implementation of a particular SQP algorithm that exploits sparsity in the constraint Jacobian and maintains a limited-memory quasi-Newton approximation B_k to the Hessian of the Lagrange function. To update B_k in the presence of the negative curvature, a new method is used. The quadratic programming sub-problems are solved using an inertia-controlling reduced Hessian active-set method (SQOPT). Other features include the treatment of infeasible nonlinear constraints using elastic programming, the use of a well-conditioned nonorthogonal basis for the null space of the quadratic programming working set, early termination of the quadratic programming sub-problems, the finite difference estimates of the missing gradients, etc. The method used by the quadratic programming solver SQOPT is based on solving a sequence of linear systems involving the reduced Hessian $Z^T B_k Z$, where Z is implicitly defined by using the sparse LU factorization.

N. Andrei, *Continuous Nonlinear Optimization for Engineering Applications in GAMS Technology*, Springer Optimization and Its Applications 121,
DOI 10.1007/978-3-319-58356-3_15

In the following we present the technique for treating the infeasible constraints, and then we see how SNOPT is working for solving the general nonlinear optimization problem with inequality constraints $\min\{f(x) : c(x) \geq 0\}$. Finally, some particularizations of these developments will be applied to problems with linear and nonlinear constraints (including simple bounds) given in (15.1).

15.1 Infeasible Constraints

In SNOPT the infeasible constraints are treated by means of l_1 penalty functions. At the very beginning, the infeasible linear constraints are identified by solving a problem of the following form:

$$\min_{x,v,w} e^{\mathrm{T}}(v + w)$$

subject to (15.2)

$$l \leq \begin{bmatrix} x \\ Ax - v + w \end{bmatrix} \leq u, \quad v \geq 0, \quad w \geq 0,$$

where e is a vector with all components equal to one. The problem (15.2) is to minimize the one-norm of the general linear constraint violations subject to the simple bounds. This problem is often called *elastic programming* (Conn, 1976; Bartels, 1980). If the linear constraints are infeasible ($v \neq 0$ and $w \neq 0$), then SNOPT stops without computing the nonlinear functions. Otherwise, all subsequent iterates satisfy the linear constraints. SNOPT then solves (15.1) by using the quadratic programming sub-problem based on linearization of nonlinear constraints. If a quadratic programming sub-problem is infeasible or unbounded (or if the Lagrange multiplier estimates for the nonlinear constraints become large), SNOPT enters the *nonlinear elastic mode* and solves the problem:

$$\min_{x,v,w} f(x) + \gamma e^{\mathrm{T}}(v + w)$$

subject to (15.3)

$$l \leq \begin{bmatrix} x \\ c(x) - v + w \\ Ax \end{bmatrix} \leq u, \quad v \geq 0, \quad w \geq 0,$$

where $f(x) + \gamma e^{\mathrm{T}}(v + w)$ is known as the composite objective function and the penalty parameter $\gamma\ (\gamma \geq 0)$ may take a finite sequence of increasing values. If (15.1) has a feasible solution and γ is large enough, then the solution of (15.1) and (15.3) are identical. On the other hand, if (15.1) has no feasible solution, then (15.3) will tend to determine an infeasible point if γ is sufficiently large.

15.2 The SQP Iteration for General Inequality Nonlinear Optimization

The SQP approach has been popularized mainly by Biggs (1972), Hann (1976), and Powell (1977, 1978b) and further developed by Schittkowski (1986) (NLPQLP), Gill, Murray, Saunders, and Wright (1986) (NPSOL), and Spellucci (1998) (DONLP). Under mild conditions these solvers typically find a local optimum for continuous general nonlinear optimization problems from arbitrary starting points, and they require relatively few evaluations of the functions defining the problem and their gradients compared, for example, to MINOS (Murtagh & Saunders, 1978, 1995). For large-scale problems with *equality* constraints the method of Lalee, Nocedal, and Plantenga (1998) uses either the exact Hessian of the Lagrangian or a limited-memory quasi-Newton approximation. The methods by Biegler, Nocedal, and Schmid (1995) use the reduced Hessian method. They maintain a dense approximation to the reduced Hessian based on quasi-Newton updates. For large problems with general *inequality* constraints as in (15.1), SQP methods have been proposed by Eldersveld (1991) (LSSQP) and Fletcher and Leyffer (1998, 2002) (filterSQP). In the following we shall present the algorithm SNOPT based on the works of Gill, Murray, and Saunders (2005). Consider the problem:

$$\min_x f(x)$$

subject to (15.4)

$$c(x) \geq 0,$$

where $x \in \mathbb{R}^n, f(x): \mathbb{R}^n \to \mathbb{R}, c(x): \mathbb{R}^n \to \mathbb{R}^m$ are twice continuously differentiable. Suppose that a KKT point (x^*, λ^*) exists where the first-order optimality conditions are satisfied:

$$c(x^*) \geq 0, \quad \lambda^* \geq 0, \quad c(x^*)^T\lambda^* = 0, \quad \nabla c(x^*)^T\lambda^* = \nabla f(x^*). \tag{15.5}$$

The sequential quadratic programming method involves *major* and *minor iterations*. The major iterations generate a sequence of iterates (x_k, λ_k) that converge to (x^*, λ^*). At each iteration a quadratic programming sub-problem is solved to generate a search direction toward the next iteration (x_{k+1}, λ_{k+1}). Solving the quadratic sub-problems is based on an iterative procedure involving the so-called minor iterations.

Let x_k and λ_k be estimates of x^* and λ^*. Gill, Murray, and Saunders (2005) use the following modified Lagrange function associated to (15.4):

$$L(x, x_k, \lambda_k) = f(x) - \lambda_k^T d_L(x, x_k), \tag{15.6}$$

which is defined in terms of *constraint linearization* and the *departure from linearity*:

$$c_L(x, x_k) = c(x_k) + \nabla c(x_k)(x - x_k), \tag{15.7a}$$

$$d_L(x, x_k) = c(x) - c_L(x, x_k). \tag{15.7b}$$

This idea is taken from the MINOS algorithm. In fact SNOPT is an improved extension of MINOS (Saunders, 2015b, 2015c). The first and second derivatives of the modified Lagrangian are:

$$\nabla L(x, x_k, \lambda_k) = \nabla f(x) - (\nabla c(x) - \nabla c(x_k))^{\mathrm{T}} \lambda_k, \tag{15.8a}$$

$$\nabla^2 L(x, x_k, \lambda_k) = \nabla^2 f(x) - \sum_{i=1}^{m} (\lambda_k)_i \nabla^2 c_i(x). \tag{15.8b}$$

Observe that for $x = x_k$, we have $L(x_k, x_k, \lambda_k) = f(x_k)$ and $\nabla L(x_k, x_k, \lambda_k) = \nabla f(x_k)$. Now, let

$$L_q(x, x_k, \lambda_k) = f(x_k) + \nabla f(x_k)^{\mathrm{T}}(x - x_k) + \frac{1}{2}(x - x_k)^{\mathrm{T}} \nabla^2 L(x_k, x_k, \lambda_k)(x - x_k)$$

be the quadratic approximation of L at x_k. If $(x_k, \lambda_k) = (x^*, \lambda^*)$, then the optimality conditions for the quadratic program:

$$\min_x L_q(x, x_k, \lambda_k)$$

subject to (15.9)

$$c_L(x, x_k) \geq 0$$

are identical to those of the original problem (15.4). This suggests that if B_k is an approximation to $\nabla^2 L$ at point (x_k, λ_k), then an improved estimate of the solution may be obtained from $\left(\widehat{x}_k, \widehat{\lambda}_k\right)$ as the solution of the following quadratic programming sub-problem:

$$\min_x f(x_k) + \nabla f(x_k)^{\mathrm{T}}(x - x_k) + \frac{1}{2}(x - x_k)^{\mathrm{T}} B_k (x - x_k)$$

subject to: (15.10)

$$c(x_k) + \nabla c(x_k)(x - x_k) \geq 0.$$

The optimality conditions of (15.10) are

$$c(x_k) + \nabla c(x_k)(\widehat{x}_k - x_k) = \widehat{s}_k, \qquad \widehat{\lambda}_k \geq 0, \quad \widehat{s}_k \geq 0,$$

$$\nabla f(x_k) + B_k(\widehat{x}_k - x_k) = \nabla c(x_k)^{\mathrm{T}} \widehat{\lambda}_k, \quad \widehat{\lambda}_k^{\mathrm{T}} \widehat{s}_k = 0,$$

where $\widehat{s}_k$ is the vector of slack variables for the linearized constraints. The triplet $\left(\widehat{x}_k, \widehat{\lambda}_k, \widehat{s}_k\right)$ is an estimate of (x^*, λ^*, s^*), where the slack variables s^* satisfy $c(x^*) - s^* = 0, s^* \geq 0$. The vector $\widehat{s}_k$ is needed in the line-search procedure.

The working set matrix The working set is important for both major and minor iterations and represents the set of constrains that are binding at a solution (see Definition 5.12). If the sub-problem (15.10) has been solved, then the algorithm which solves this problem returns an independent set of constraints that are active at the solution. This is the *optimal working set* for the sub-problem (15.10). The same constraint indices define the working set of (15.4) and of the next sub-problem (15.10) at iteration $k+1$. The corresponding gradients of these constraints form the rows of the *working set matrix* W_k, an $n_Y \times n$ full-rank sub-matrix of the Jacobian $\nabla c(x_k)$. Let Z_k be a $n \times n_Z$ full-rank matrix which is a basis of the null space of W_k. Then $W_k Z_k = 0$ and $n_Z = n - n_Y$. In SNOPT Z_k is defined by the LU factorization of W_k. The matrix Z_k is useful both in theoretical developments and in the construction of the *reduced Hessian* $Z_k^T B_k Z_k$ and of the *reduced gradient* $Z_k^T \nabla f(x_k)$.

The merit function and line search Let (x_k, λ_k, s_k) be the current solution of (15.10). As soon as we have a solution $\left(\widehat{x}_k, \widehat{\lambda}_k, \widehat{s}_k\right)$ for (15.10), a new estimation for the solution of (15.4) can be computed by means of a linear search from (x_k, λ_k, s_k) toward the optimal solution of (15.10). The purpose of linear search is a sufficient decrease of a merit function based on the augmented Lagrangian:

$$M_\sigma(x, \lambda, s) = f(x) - \lambda^T (c(x) - s) + \frac{1}{2}\sigma_i (c_i(x) - s_i)^2, \tag{15.11}$$

where $\sigma \in \mathbb{R}^m$ is the vector of the penalty parameters. For the stepsize $\alpha \in (0, 1]$, let $v(\alpha)$ be the line:

$$v(\alpha) = \begin{bmatrix} x_k \\ \lambda_k \\ s_k \end{bmatrix} + \alpha \begin{bmatrix} \widehat{x}_k - x_k \\ \widehat{\lambda}_k - \lambda_k \\ \widehat{s}_k - s_k \end{bmatrix}$$

and $\varphi_\sigma(\alpha) = M_\sigma(v(\alpha))$ which represents M_σ as a function by α. Observe that $\varphi'_\sigma(0)$ is the directional derivative of the merit function at the base point $\alpha = 0$ for a given σ. To determine the values of the penalty parameters, Gill, Murray, and Saunders (2005) recommend the following procedure. Let σ^* be the solution of the following least-squares problem:

$$\min \|\sigma\|^2$$

subject to

$$\varphi'_\sigma(0) = -\frac{1}{2}(\widehat{x}_k - x_k) B_k (\widehat{x}_k - x_k), \quad \sigma \geq 0. \tag{15.12}$$

The solution of this problem can be obtained analytically, and it can be shown that for any $\sigma \geq \sigma^*, \varphi'_\sigma(0) \leq -\frac{1}{2}(\widehat{x}_k - x_k) B_k (\widehat{x}_k - x_k)$, (Eldersveld, 1991; Gill, Murray, Saunders, & Wright, 1992).

It is important to allow the penalty parameters to decrease during the early major iterations. The reduction scheme involves a parameter $\Delta_\sigma \geq 1$. Let $\widehat{\sigma}$ be the value of the penalty parameter at the start of the iterate k. Then define the new parameter $\overline{\sigma}_i$

as the geometric mean of σ_i and σ_i^* as long as this mean is sufficiently positive and not too close to σ_i:

$$\overline{\sigma}_i = \max\{\sigma_i^*, \widehat{\sigma}_i\}, \tag{15.13}$$

where

$$\widehat{\sigma}_i = \begin{cases} \sigma_i, & \text{if } \sigma_i < 4(\sigma_i^* + \Delta_\sigma), \\ \sqrt{\sigma_i(\sigma_i^* + \Delta_\sigma)}, & \text{otherwise.} \end{cases}$$

Initially, for $k=0$, $\Delta_\sigma=1$. This value of Δ_σ can be increased by a factor of two in order to ensure a positive value for $\overline{\sigma}_i$ and not too close to σ_i.

Now, considering $\sigma = \overline{\sigma}$ in the merit function (15.11), a safeguarded line search is used to find α_{k+1} that reduces M_σ to give the next solution estimate $v(\alpha_{k+1}) = (x_{k+1}, \lambda_{k+1}, s_{k+1})$.With these, prior to the solution of (15.10) at iteration $k+1$, s_{k+1} is redefined to minimize the merit function as a function of s, (Gill, Murray, Saunders, & Wright, 1986; Eldersveld, 1991).

Bounding the constraint violation In the line search, the following condition is enforced for some vector $b > 0$:

$$c(x_k + \alpha_k(\widehat{x}_k - x_k)) \geq -b.$$

SNOPT uses $b_i = \tau_v \max\{1, -c_i(x_0)\}$, where τ_v is a constant, for example, $\tau_v = 10$. This defines a region in which the objective function is expected to be defined and bounded below (Spellucci, 1981). If $\alpha_k = 0$, then the objective is considered to be "unbounded" in the expanded region. In this case the elastic program is introduced.

The approximate Hessian As soon as the line search has been finished, we define the change in x and the gradient of the modified Lagrangian as

$$s_k = x_{k+1} - x_k, \quad y_k = \nabla L(x_{k+1}, x_k, \lambda) - \nabla L(x_k, x_k, \lambda),$$

for some vector λ. Then, an estimate of the curvature of the modified Lagrangian along s_k is incorporated in BFGS quasi-Newton update:

$$B_{k+1} = B_k + \frac{y_k y_k^T}{y_k^T s_k} - \frac{B_k s_k s_k^T B_k}{s_k^T B_k s_k}. \tag{15.14}$$

When B_k is positive definite, then B_{k+1} is positive definite if and only if the curvature $y_k^T s_k$ is positive. From (15.6) observe:

$$\begin{aligned} y_k &= \nabla L(x_{k+1}, x_k, \lambda_{k+1}) - \nabla L(x_k, x_k, \lambda_{k+1}) \\ &= \nabla f(x_{k+1}) - \nabla f(x_k) - (\nabla c(x_{k+1}) - \nabla c(x_k))^T \lambda_{k+1}. \end{aligned}$$

Maintaining positive definiteness of Hessian Since the Hessian of the modified Lagrangian need not be positive definite at a local minimizer, the approximate curvature $y_k^T s_k$ can be negative or very small at points arbitrarily close to (x^*, λ^*). The curvature is considered not sufficiently positive if

$$y_k^T s_k < \rho_k, \quad \rho_k = \alpha_k(1 - \eta)p_k^T B_k p_k, \tag{15.15}$$

where $\eta \in (0, 1)$ is a given constant $(0 < \eta < 1)$ and $p_k = \widehat{x}_k - x_k$ is the search direction defined by the quadratic programming sub-problem (15.10). In such cases, if the problem has nonlinear constraints, two attempts are made to modify the update: the first is modifying s_k and y_k, and the second is modifying only y_k. If neither modification provides sufficiently positive approximate curvature, then no update is made.

The first modification Define a new point z_k and evaluate the nonlinear functions in order to obtain the new values for $s_k = x_{k+1} - z_k$ and $y_k = \nabla L(x_{k+1}, x_k, \lambda_{k+1}) - \nabla L(z_k, x_k, \lambda_{k+1})$. The point z_k is chosen as the first feasible point $\bar{x}_k$ of (15.10). The search direction may be written as:

$$p_k = (\bar{x}_k - x_k) + (\widehat{x}_k - \bar{x}_k) \equiv p_R + p_N.$$

Consider $z_k = x_k + \alpha_k p_R$, thus obtaining $s_k = \alpha_k p_N$ and

$$y_k^T s_k = \alpha_k y_k^T p_N \cong \alpha_k^2 p_N^T \nabla^2 L(x_k, x_k, \lambda_k) p_N,$$

so that $y_k^T s_k$ approximates the curvature along p_N. If W_k is the final working set of (15.10) at x_k, it is also the working set at $\bar{x}_k$, then $W_k p_N = 0$. Therefore, $y_k^T s_k$ approximates the curvature for the reduced Hessian, which must be positive semidefinite at the minimizer of (15.4).

The second modification If (x_k, λ_k) is not close to (x^*, λ^*), then the modified approximate curvature $y_k^T s_k$ may not be sufficiently positive. Hence, the second modification may be necessary as follows. Choose Δy_k so that $(y_k + \Delta y_k)^T s_k = \rho_k$ and redefine y_k as $y_k + \Delta y_k$. This approach was suggested by Powell (1978b), who proposed redefining y_k as a linear combination of y_k and $B_k s_k$.

To get Δy_k consider the augmented modified Lagrangian:

$$L_A(x, x_k, \lambda_k) = f(x) - \lambda_k^T d_L(x, x_k) + \frac{1}{2} d_L(x, x_k)^T \Omega d_L(x, x_k), \tag{15.16}$$

where $\Omega = \mathrm{diag}(\omega_i), \omega_i \geq 0, i = 1, \ldots, m$, is a matrix of parameters to be determined. As we can see, the perturbation

$$\Delta y_k = (\nabla c(x_{k+1}) - \nabla c(x_k))^T \Omega d_L(x_{k+1}, x_k)$$

is equivalent to redefining the gradient difference as:

$$y_k = \nabla L_A(x_{k+1}, x_k, \lambda_{k+1}) - \nabla L_A(x_k, x_k, \lambda_{k+1}). \tag{15.17}$$

The elements $\omega_i, i = 1, \ldots, m$, are determined by the linearly constrained least-squares problem:

$$\min_\omega \|\omega\|_2^2$$

subject to (15.18)

$$a^T\omega = \beta, \quad \omega \geq 0,$$

where $\beta = \rho_k - y_k^T s_k$ and $a_i = v_i w_i, \quad i = 1, \ldots, m$, with $v = (\nabla c(x_{k+1}) - \nabla c(x_k))s_k$ and $w = d_L(x_{k+1}, x_k)$.The idea is to choose the smallest ω_i's that increase $y_k^T s_k$ to ρ_k (see (15.15)). If there is no solution for problem (15.18) or if $\|\omega\|$ is very large, no update of Hessian is made. It is worth mentioning that the second modification is not required in the neighborhood of a solution because $\nabla^2 L_A$ converges to $\nabla^2 L$ when $x \to x^*$, and the first modification will already have been successful. The second modification is related to updating an approximation of the Hessian of the Lagrangian suggested by Han (1976) and Tapia (1974).

Convergence tests A point (x, λ) is a satisfactory solution for (15.4) if it satisfies the first-order optimality conditions (15.5) within certain tolerance. In SNOPT two sufficiently small constants τ_P and τ_D are used, which define the quantities $\tau_x = \tau_P(1 + \|x\|_\infty)$ and $\tau_\lambda = \tau_D(1 + \|\lambda\|_\infty)$. The SQP algorithm terminates if

$$c_i(x) \geq -\tau_x, \quad \lambda_i \geq -\tau_\lambda, \quad c_i(x)\lambda_i \leq \tau_\lambda, \quad |d_j| \leq \tau_\lambda, \tag{15.19}$$

where $d = \nabla f(x) - \nabla c(x)^T\lambda$.

These conditions cannot be satisfied if (15.4) is infeasible. However, in these situations the problem (15.4) can be transformed in its elastic variant:

$$\min_{x,v} f(x) + \gamma e^T v$$

subject to (15.20)

$$c(x) + v \geq 0, \quad v \geq 0,$$

where γ takes increasing values $\{\gamma_l\}$ up to some maximum value. The first-order optimality conditions for (15.20) include:

$$0 \leq \lambda_i \leq \gamma, \quad (c_i(x) + v_i)\lambda_i = 0, \quad v_i(\gamma - \lambda_i) = 0. \tag{15.21}$$

The fact that $\|\lambda^*\|_\infty \leq \gamma$ at the solution of (15.20) leads to initiate the elastic mode if $\|\lambda_k\|_\infty$ exceeds some value, say γ_1, or if (15.10) is infeasible. SNOPT uses

$$\gamma_1 \triangleq \gamma_0 \|\nabla f(x_{k_1})\|_\infty, \quad \gamma_l = \gamma_1 10^{l(l-1)/2}, \quad l = 2, 3, \ldots, \tag{15.22}$$

where $\gamma_0 = 10^4$ and x_{k_1} is the iterate at which γ is first needed.

15.3 The Quadratic Programming Solver SQOPT

We now return to our problem given by (15.1) considered by SNOPT. The form of this problem requires solving (at every major iteration) of a quadratic programming sub-problem with $f(x)$ replaced by a convex quadratic function and with $c(x)$ replaced by its current linearization in the current point:

$$\min_x f(x_k) + \nabla f(x_k)^{\mathrm{T}}(x - x_k) + \frac{1}{2}(x - x_k)^{\mathrm{T}} B_k (x - x_k)$$

subject to: (15.23)

$$l \leq \begin{bmatrix} x \\ c(x_k) + \nabla c(x_k)(x - x_k) \\ Ax \end{bmatrix} \leq u.$$

In SNOPT this sub-problem is solved by the package SQOPT (Gill, Murray, & Saunders, 1997), which employs a two-phase active-set algorithm and implicitly implements elastic programming when necessary. The Hessian B_k may be positive semidefinite and is defined by a routine for forming products $B_k v$ for a given vector v.

At each minor iteration, when (15.23) is solved, the *active-set method* determines a search direction d satisfying the KKT system:

$$\begin{bmatrix} B_k & W^{\mathrm{T}} \\ W & 0 \end{bmatrix} \begin{bmatrix} d \\ y \end{bmatrix} = -\begin{bmatrix} g_q \\ 0 \end{bmatrix}, \tag{15.24}$$

where W is the current working set matrix and g_q is the gradient of the objective function from (15.23). SQOPT implements several null space methods as described in Gill, Murray, and Saunders (2005).

Null space approach One way to obtain a solution for the linear system (15.24) is to solve the reduced Hessian system:

$$Z^{\mathrm{T}} B_k Z d_Z = -Z^{\mathrm{T}} g_q, \quad d = Z d_Z, \tag{15.25}$$

where Z is a basis for the null space of W. SQOPT maintains Z in reduced gradient form as in MINOS. The idea is to use sparse LU factors of a square nonsingular matrix B, called basis, whose columns change as the working set Wchange:

$$W = \begin{bmatrix} B & S & N \\ 0 & 0 & I \end{bmatrix} P, Z = P^{\mathrm{T}} \begin{bmatrix} -B^{-1}S \\ I \\ 0 \end{bmatrix}, \tag{15.26}$$

where P is a permutation matrix such that the selected basis B is nonsingular. The variables associated with B and S are called basic and superbasic, respectively.

The other variables are called nonbasic. The number of superbasic variables n_Z is exactly the column dimension of S and Z. Products Zv and $Z^T g$ are obtained by solving the corresponding systems with B or B^T, respectively. If n_Z is small, then SQOPT uses dense Cholesky factorization $Z^T B_k Z = R^T R$. As the major iterations converge, the quadratic programming sub-problems require fewer changes to their working set, and with warm start they are eventually solved in one or two minor iterations. Hence, the work required by SQOPT becomes dominant by the computation of the reduced Hessian $Z^T B_k Z$ and its factor R. For this reason SQOPT can optionally maintain a quasi-Newton approximation $Z^T B_k Z \cong R^T R$, as in MINOS.

Conjugate gradient approach By construction, the Hessians B_k of (15.23) are positive definite or positive semidefinite. Therefore, the conjugate gradient method is a natural tool for solving very large systems (15.24). SQOPT includes a conjugate gradient option for finding approximate solutions to

$$\left(Z^T B_k Z + \delta^2 I\right) d_Z = -Z^T g_q, \tag{15.27}$$

where $\delta \cong 10^{-3}$ is a small regularization parameter to allow for singular $Z^T B_k Z$. When Z has many columns, then it is possible for the conjugate gradient methods to require many iterations to get an useful approximation to d_Z. The conjugate gradient methods require some sort of preconditioning which is much dependent on the system to be solved. However, in SQOPT no preconditioner is used. The explanation is that when looking at (15.27) we see that both B_k and $Z^T Z$ have similar structures: "a diagonal matrix plus a matrix of small rank."

The initial point To use a good starting point x_0, the algorithm SQOPT is applied to one of the proximal-point problems for initialization:

$$\min_x \left\{ \|\bar{x} - \bar{x}_0\|_1 : \ \bar{l} \le \begin{bmatrix} x \\ Ax \end{bmatrix} \le \bar{u} \right\} \tag{15.28}$$

or

$$\min_x \left\{ \|\bar{x} - \bar{x}_0\|_2^2 : \ \bar{l} \le \begin{bmatrix} x \\ Ax \end{bmatrix} \le \bar{u} \right\} \tag{15.29}$$

where $\bar{l}$ and $\bar{u}$ are the corresponding bounds for linear constraints and simple bounds from (15.23). $\bar{x}$ and $\bar{x}_0$ correspond to the nonlinear variables in x and x_0. In practice (15.28) is preferred because it is linear and can use SQOPT's implicit elastic bounds. The solution of (15.28) (or (15.29)) defines a new starting point x_0 for the SQP iteration.

SNOPT, as described by Gill, Murray, and Saunders (2005), is a complex algorithm with plenty of details which are not presented here. These refer to null space computation, choices for multipliers, large-scale Hessians in case of problems with few nonlinear variables, dense Hessians, limited-memory procedure to update Hessians, elastic bounds, inertia control, unbounded quadratic programming sub-problems, basis handling in SQOPT, threshold pivoting in LU factorization, basis repair in singular case, basis repair in rectangular case, undefined functions, early termination of quadratic programming sub-problems, linearly constrained problems, etc. Suppose that a starting point (x_0, λ_0) is available and that the reduced Hessian quadratic programming solver SQOPT is being used. The main steps of SNOPT are as follows.

Algorithm 15.1 SNOPT (Gill, Murray, and Saunders)

1.	Apply quadratic programming solver to problem (15.28) (or (15.29)) to find a point close to x_0 satisfying the linear constraints and the simple bounds on variables. If (15.28) is infeasible, then the problem (15.1) is infeasible. Otherwise, a working set matrix W_0 is obtained. Set $k=0$ and evaluate the functions and the gradients at x_0. Initialize penalty parameters $\sigma_i=0, i=1, \ldots, m$
2.	Factorize W_k
3.	Define s_k to minimize the merit function (15.11) as a function of the slacks s
4.	Determine $\bar{x}_k$ as a feasible solution for (15.23). This is an intermediate point for SQOPT solver, which also provides a working set matrix $\bar{W}_k$ as well as its null space $\bar{Z}_k$. If no feasible point exists, initiate elastic mode and restart SQOPT solver
5.	Compute the reduced Hessian $\bar{Z}_k^T B_k \bar{Z}_k$ and compute its Cholesky factor
6.	Using SQOPT, continue solving the quadratic programming sub-problem (15.23) to find $\left(\widehat{x}_k, \widehat{\lambda}_k\right)$. Observe that in step 4 only a feasible solution was obtained for (15.23). SQOPT provides a working set $\widehat{W}_k$ as well as its null space matrix $\widehat{Z}_k$ If the elastic mode has not been initiated, but $\left\Vert\widehat{\lambda}_k\right\Vert_\infty$ is "large," then enter the elastic mode and restart SQOPT solver If sub-problem (15.23) is unbounded and x_k satisfies the nonlinear constraints, the problem (15.1) is unbounded (f is unbounded below in the feasible region). Otherwise, if (15.23) is unbounded, go to step 8
7.	If the solution (x_k, λ_k) satisfies the convergence tests for (15.1) analogous to (15.19), then this is the optimal solution of the problem; stop If convergence tests similar to (15.19) are satisfied for sub-problem (15.20), then go to step 8. Otherwise, go to step 9
8.	If the elastic mode of (15.23) has not been initiated, then consider the elastic variant of the problem and repeat step 6. Otherwise, if γ has not reached its maximum value, increase γ and repeat step 6. Otherwise, declare the problem (15.1) infeasible

(continued)

9.	Update the penalty parameters as in (15.13)
10.	Find a stepsize α_k that gives a sufficient reduction in the merit function (15.11). Set $x_{k+1} = x_k + \alpha_k(\widehat{x}_k - x_k)$ and $\lambda_{k+1} = \lambda_k + \alpha_k\left(\widehat{\lambda}_k - \lambda_k\right)$. Evaluate the functions and the gradients at the new point x_{k+1}
11.	Compute $s_k = x_{k+1} - x_k$, $y_k = \nabla L(x_{k+1}, x_k, \lambda_{k+1}) - \nabla L(x_k, x_k, \lambda_{k+1})$. If $y_k^T s_k < \rho_k$, where ρ_k is computed as in (15.15), then recompute s_k and y_k with x_k redefined as $x_k + \alpha_k(\bar{x}_k - x_k)$.(This requires an extra evaluation of the derivatives.) If necessary, increase $y_k^T s_k$ (if possible) by increasing y_k with Δy_k and by redefining the difference of gradients of the augmented Lagrangian (see (15.16))
12.	If $y_k^T s_k \geq \rho_k$, then apply the BFGS formula to update B_k by using the pair $(B_k s_k, y_k)$
13.	Using $\widehat{W}_k$ redefine W_k, set $k = k+1$ and go to step 2 ♦

For both linearly and nonlinearly constrained problems, SNOPT applies a sparse sequential quadratic programming method (Gill, Murray, & Saunders, 2005), using limited-memory quasi-Newton approximations to the Hessian of the Lagrangian. The merit function for stepsize computation is an augmented Lagrangian, as in the dense SQP solver NPSOL (Gill , Murray, Saunders, & Wright, 1986). The computational effort is in steps 2 and 5 where the matrix W_k is factorized and the reduced Hessian $\bar{Z}_k^T B_k \bar{Z}_k$ is formed and Cholesky factorized.

Practical Hints in GAMS

SNOPT is able to solve a large diversity of optimization problems: linear programming (LP), nonlinear optimization (NLP), optimization problems with discontinuous or non-differentiable functions (DNLP), and nonlinear mixed integer optimization problems (RMINLP). These nonlinear optimization problems may be solved by SNOPT through options specified by the option statement. The GAMS system can be instructed to use SNOPT solver by incorporating the following option in the GAMS source file: `option NLP = SNOPT;` if we have to solve a nonlinear optimization problem, etc.

SNOPT options are specified in a file called `snopt.opt`. To tell SNOPT to read this file, use the statement `mymodel.optfile = 1;` in the GAMS source file, where `mymodel` is the name of the model introduced by the line `model mymodel/all/;`. SNOPT has plenty of options that are useful in a GAMS environment, for example, referring to check frequency, crash option, crash tolerance, factorization frequency, Hessian full memory, Hessian limited memory, minor iteration limit, and superbasic limit. All these options are described in *GAMS – The Solver Manuals* (2005, pp.393–424).

A model identifier in GAMS can have several suffices to which one can assign values. They are using the name of the model and suffices: `iterlim` for maximum number of iterations, `reslim` for resources or time limit, `scaleopt` to scale or not the model, and `workspace` for the amount of memory used by SNOPT.

Numerical Study

Table 15.1 contains performances of SNOPT for solving 13 applications of nonlinear optimization presented in Chapter 4 in GAMS language.

In Table 15.2 we present the performances of SNOPT for solving six large-scale nonlinear optimization applications of different dimensions, described in Chapter 4 in GAMS language.

SNOPT uses very many concepts defined in MINOS. A comparison between these two algorithms shows the importance of the sequential quadratic programming implemented in SNOPT versus minimizing the augmented Lagrangian subject to the linearization of the constraints used in MINOS. For all these 17 large-scale numerical experiments shown in Tables 9.5 (MINOS) and 15.2 (SNOPT), respectively, in Table 15.3 we present the total number of major iterations (*itMt*), the total number of minor iterations (*itmt*) and the total CPU computing time (*cput*), in seconds, to get a solution for all these applications.

Notes and References

SNOPT is a *pure* sequential quadratic programming algorithm. The description of SNOPT follows the paper of Gill, Murray, and Saunders (2005). SNOPT solves the quadratic programming sub-problems using SQOPT (Gill, Murray, and Saunders, 2006) which is a reduced Hessian active-set method. SNOPT uses a positive semidefinite quasi-Newton Hessian approximation B_k. If the number of nonlinear variables is moderate, then B_k is stored as a dense matrix. Otherwise, the Hessian of the Lagrangian is updated by using limited-memory quasi-Newton BFGS updates, with resets to the current diagonal at a specified frequency.

Table 15.1 Performances of SNOPT. Small-scale nonlinear optimization applications

	n	*me*	*mc*	*itM*	*itm*	*nf*	*nr*	*s*	*cpu*	*vfo*
ELCH	10	3	0	19	49	54	0	7	0.01	−47.761090
ALKI	10	3	8	16	47	28	28	1	0.01	−1768.8070
PREC	8	0	6	53	119	202	202	4	0.01	3.951163
PPSE	9	6	0	9	17	20	20	1	0.01	5055.01167
MSP3	13	0	15	71	141	367	367	1	0.02	97.5875581
MSP5	16	0	21	54	152	106	106	2	0.02	174.910868
POOL	34	20	0	5	36	8	8	4	0.01	2569.8000
TRAFO	6	0	2	30	38	36	36	4	0.01	135.075962
LATHE	10	1	14	33	86	55	55	0	0.01	−4430.0882
DES	150	50	0	104	204	130	130	100	0.16	1055.18231
CSTC	303	200	0	10	111	15	15	100	0.03	3.48007467
DIFF	396	324	0	0	60	0	0	0	0.01	0.0
PENICI	707	602	0	89	3291	192	192	33	1.03	113.9898

Table 15.2 Performances of SNOPT. Large-scale nonlinear optimization applications

	n	*me*	*mc*	*itM*	*itm*	*nf*	*s*	*cpu*	*vfo*
HANG	1002	501	0	59	1032	93	499	3.37	5.0685783
	2002	1001	0	60	2028	125	999	18.80	5.0685126
	4002	2001	0	45	2332	75	1998	133.96	5.0685036
	8002	4001	0	57	4482	114	3881	53.80	5.0684912
FLOW	1163	735	0	0	457	2	0	0.01	0.0
FLOWO	1556	1005	0	0	329	2	0	0.01	0.0
POL	4004	3000	0	57	1562	114	0	0.92	13.480213
	6004	4500	0	63	2272	128	0	1.86	13.469429
	8004	6000	0	63	2992	125	0	3.06	13.638381
	10,004	7500	0	73	3392	145	0	4.51	13.210731
CAT	3003	2000	0	11	272	1183	0	2.28	−0.048055
	6003	4000	0	18	536	2430	0	8.75	−0.048052
	9003	6000	0	23	796	3703	0	20.47	-0.048047
CONT	2505	2000	0	3	696	5	0	0.05	1.01323
	5005	4000	0	3	1394	5	2	0.11	1.00592
	7505	6000	0	3	2094	5	0	0.27	1.00456
	10,005	8000	0	3	1424	5	0	0.44	1.00407

In Tables 15.1 and 15.2 we have: *n* = the number of variables, *me* = the number of equality constraints, *mc* = the number of inequality constraints, *itM* = the number of major iterations, *itm* = the number of minor iterations, *nf* = the number of evaluations of the objective function, *nr* = the number of evaluations of the constraints, *s* = the number of superbasic variables, *cpu* = the CPU computing time to solve the problem (seconds), *vfo* = the value of the objective function at optimal point.

Table 15.3 Comparison: MINOS versus SNOPT

	itMt	*itmt*	*cput*
MINOS	1605	69,534	1369.61
SNOPT	541	28,091	252.67

The quadratic programming solver SQOPT works with a sequence of reduced Hessian systems of the form $Z^TB_kZd = -Z^Tg$, where Z is a basis for the null space of W, a rectangular matrix with n_z columns. SQOPT can deal with the reduced Hessian systems in various ways, depending on the size of n_z. If the constraints are currently active in the quadratic sub-problem and n_z is not excessively large, then SQOPT uses the dense Cholesky factorization $Z^TB_kZ = R^TR$. Alternatively, SQOPT can maintain a dense quasi-Newton approximation to avoid the cost of forming and of factorizing the reduced Hessian. SNOPT includes an option for using a projected conjugate gradient method rather than factoring the reduced Hessian. If a quadratic sub-problem is found to be infeasible or unbounded, then SNOPT tries to solve an elastic sub-problem that corresponds to a smooth reformulation of the l_1 exact penalty function. The solution of the quadratic sub-problem solved at the major iteration is used to obtain a search direction along which an augmented Lagrangian merit function is minimized. SNOPT uses a lot of procedures from MINOS. Both algorithms are working in GAMS technology.

Chapter 16
Generalized Reduced Gradient with Sequential Linearization: CONOPT

This algorithm is a very interesting and profitable combination of the generalized reduced gradient with the sequential linear programming and with the sequential quadratic programming. All these algorithms are imbedded in the generalized reduced gradient (GRG) scheme as described in Drud (1976, 1983, 1985, 1994, 1995, 1996, 2011).

CONOPT considers general nonlinear optimization problems of the following form:

$$\min f(x)$$

subject to $$\begin{aligned} c(x) &= b, \\ l \leq x &\leq u, \end{aligned} \tag{16.1}$$

where $x \in \mathbb{R}^n$, $f: \mathbb{R}^n \to \mathbb{R}$, and $c: \mathbb{R}^n \to \mathbb{R}^m$ are twice continuously differentiable functions and $b \in \mathbb{R}^m$ and $l, u \in \mathbb{R}^n$ are simple bounds on variables. Suppose that $n \geq m$. Assume that $l_i \leq u_i$, $i = 1, \ldots, n$. Some of the bound values may be minus or plus infinity. If the problem has inequality constraints, then these are transformed into equalities by *slack variables*. Moreover, in CONOPT, it is assumed that x includes slack and *artificial variables* for all constraints. This implies that the Jacobian of the constraints includes a unit matrix and, therefore, has full row rank.

The GRG algorithm was suggested by Abadie and Carpentier (1969), and it was developed inter alia by Abadie (1978, 1979), Abadie and Guigou (1970), Abadie and Haggag (1979), Lasdon, Fox, and Ratner (1974), Lasdon and Waren (1978, 1980, 1982), Lasdon, Waren, Jain, and Ratner (1978), Drud (1976, 1983, 1985, 1994, 1995, 1996), and Andrei (1985, 1987).

CONOPT works as follows. The set of m equations $c(x) = b$ defining the (nonlinear) constraints of the problem can be used to eliminate m variables. The vector x is partitioned in the sub-vectors $x_b \in \mathbb{R}^m$ and $x_n \in \mathbb{R}^{n-m}$. Therefore, the

N. Andrei, *Continuous Nonlinear Optimization for Engineering Applications in GAMS Technology*, Springer Optimization and Its Applications 121,
DOI 10.1007/978-3-319-58356-3_16

system of nonlinear constraints $c(x) = b$ can be rewritten as $x_b = c_1(x_n)$. It is quite clear that it is not possible to find an analytic expression for function c_1, except for some very particular cases like linear constraints. However, from the theory of implicit functions (see Theorem A2.4), we have the following result: If (x_b^0, x_n^0) satisfies the system $c(x_b^0, x_n^0) = 0$ and the Jacobian $\partial c / \partial x_b$ has rank m, i.e., it is nonsingular, then in a neighborhood of (x_b^0, x_n^0), it is possible to transform $c(x_b, x_n) = 0$ into $x_b = c_1(x_n)$. The function c_1 is differentiable, and its Jacobian is found by implicit differentiation as:

$$\frac{\partial c}{\partial x} dx = \frac{\partial c}{\partial x_b} dx_b + \frac{\partial c}{\partial x_n} dx_n = 0. \tag{16.2}$$

Hence:

$$\frac{\partial c_1}{\partial x_n} = \frac{\partial x_b}{\partial x_n} = -\left(\frac{\partial c}{\partial x_b}\right)^{-1} \frac{\partial c}{\partial x_n}. \tag{16.3}$$

In practice, the function $x_b = c_1(x_n)$ is computed by solving the nonlinear system $c(x_b, x_n) = 0$ with respect to x_b, x_n being known.

The equation $x_b = c_1(x_n)$ is introduced in the objective function to get

$$f(x_b, x_n) = f(c_1(x_n), x_n) \triangleq F(x_n),$$

where $F(x_n)$ is differentiable with derivative:

$$\frac{\partial F}{\partial x_n} = \frac{\partial f}{\partial x_b} \frac{\partial x_b}{\partial x_n} + \frac{\partial f}{\partial x_n} = \frac{\partial f}{\partial x_n} - \frac{\partial f}{\partial x_b} \left(\frac{\partial c}{\partial x_b}\right)^{-1} \frac{\partial c}{\partial x_n}. \tag{16.4}$$

With these developments, (16.1) can be reformulated as:

$$\begin{gathered} \min F(x_n) \\ \text{subject to:} \\ x_b = c_1(x_n), \\ l_b < x_b < u_b, \; l_n \leq x_n \leq u_n. \end{gathered} \tag{16.5}$$

Observe that the transformations used here are very similar to those from linear programming. x_b contains the basic variables. These variables are used to compensate for changes in the nonbasic variables x_n so that the constraints $c(x) = b$ can be satisfied. The derivative $\partial F / \partial x_n$ from (16.4) is similar to the vector of *reduced costs* from linear programming. It shows the influence on the objective function of changes in the nonbasic variables, taking into consideration the corresponding changes in the basic variables. Usually, $\partial F / \partial x_n$ is called the *reduced gradient*. The matrix $\partial c / \partial x_b$ is similar to the *basis matrix* from linear programming.

However, *the major difference between the nonlinear problem and the linear problem is that in the optimal solution, the nonlinear problem can have more than m variables strictly between the lower and the upper bounds.* Thus, an optimization method like the simplex algorithm that works only with basic solutions cannot be used in the nonlinear case. The main steps in GRG are as follows:

Algorithm 16.1 CONOPT (Drud)

1.	Find an initial feasible solution x^0
2.	Compute the Jacobian matrix $\partial c/\partial x$. Partition x in $x_b \in \mathbb{R}^m$ and $x_n \in \mathbb{R}^{n-m}$, so that $l_b < x_b < u_b$ and the submatrix $\partial c/\partial x_b$ are nonsingular
3.	Compute the inverse $(\partial c/\partial x_b)^{-1}$
4.	Compute the Lagrange multipliers as solution of the system: $$\left(\frac{\partial c}{\partial x_b}\right)u = -\frac{\partial f}{\partial x_b}$$
5.	Compute the reduced gradient $r = \frac{\partial f}{\partial x_n} + u\left(\frac{\partial c}{\partial x_n}\right)$
6.	Compute the projection h of the reduced gradient on space defined by the simple bounds of nonbasic variables: $$h_i = \begin{cases} 0, & \text{if } \mathrm{r}_i < 0 \text{ and } x_{ni}^0 = l_{ni}, \\ 0, & \text{if } \mathrm{r}_i > 0 \text{ and } x_{ni}^0 = u_{ni}, \\ r_i, & \text{otherwise} \end{cases}$$
7.	If $h = 0$, then stop
8.	One-dimensional search. Choose $\alpha > 0$ to minimize $F\left(x_n^0 + \alpha h\right)$ according to the following steps for different values of α: (i) $x_{ni} = \begin{cases} l_{ni}, & \text{if } x_{ni}^0 + \alpha h_i < l_{ni}, \\ u_{ni}, & \text{if } x_{ni}^0 + \alpha h_i > u_{ni}, \\ x_{ni}^0 + \alpha h_i, & \text{otherwise} \end{cases}$ (ii) Compute $x_b = c_1(x_n)$, i.e., solve the system $c(x_b, x_n) = b$ subject to x_b (iii) Compute $F\left(x_n^0 + \alpha h\right) = f(x_b, x_n)$
9.	Store the best solution in x^0 and go to step 2 ♦

The GRG algorithm is based on the same principle as the simplex algorithm for linear programming: elimination of the basic variables. Locally, the problem is thus reduced to an optimization problem in the nonbasic variables which are lower and upper bounded. The main aspect of CONOPT is that it uses sequential linearization. Sequential linearization methods are defined as methods based on solving a sequence of sub-problems with linearized constraints and an augmented Lagrangian as objective function.

Every step of the GRG algorithm contains more details from linear programming and nonlinear optimization. *The optimizing steps are specialized in several versions according to whether the model appears to be almost linear or not.* For almost linear models, some of the linear algebra work involving Jacobian and basis matrices can be avoided or done using cheap linear programming updating techniques. The second-order information is not needed. The line search can be improved by observing that, like in linear programming, the optimal step will almost always be determined by the first variable that reaches a bound. In this case, the algorithm uses the *sequential linear-quadratic programming*. If the model is fairly nonlinear, other aspects can be optimized: the set of basic variables will often remain constant over several iterations, and other parts of the sparse matrix algebra will take advantage of this. In this case, an improved search direction can be computed by using specialized *sequential quadratic programming*.

Some details on the algorithm

In step 1, firstly, as described by Drud (1994), CONOPT identifies equations that only involve one variable that is not fixed: solve for this variable, fix it, and remove the constraint. This procedure is applied repeatedly by removing the constraints. Secondly, we try to guess a set of basic variables away from bounds for the remaining constraints and use this basis in some cheap Newton iterations. If the process does not converge quickly, then constraints with large residuals or constraints that deviate from the linear model are taken out of the process. In this way, we usually get a point where most constraints are satisfied. Finally, a procedure in which the sum of the slacks in the remaining violated constraints using the GRG approach to the model with relaxed bounds on these slacks is minimized. This means that once a constraint has become satisfied, it remains satisfied for the rest of the optimization. Thus, by using this approach, we find a good initial basis that will lead to very few infeasible slacks and, therefore, to few iterations. In step 2 of the algorithm, the Jacobian matrix is computed at points where the nonlinear constraints are satisfied. Moreover, this evaluation of the Jacobian is made only at the beginning of each line search. This reduces the computing effort. Selection of variables involves many tactical decisions and has a crucial influence on the Jacobian matrix factorization in step 3. Step 4, where the Lagrange multipliers are determined, involves solving of a linear system which uses the factorization of the basic matrix from step 3. For each step in the line search, in step 8, CONOPT adjusts the basic variables to get a feasible solution. This involves repeated evaluations of the constraint residuals and solves with the basis factors. In addition, if one or more of the basic variables approach a bound, extra work is needed to find the exact stepsize at which the critical variable reaches the bound. To achieve feasibility, CONOPT uses a Newton process. If it makes a large step and the Newton procedure fails to converge, then it reduces the step and tries again. Since the derivatives are evaluated at a feasible point from which the search starts, the Newton process will always converge for small steps, ignoring degeneracy problems and taking only a few steps. In the following, let us describe some details on some steps of CONOPT.

Selection and factorization of the basis

This is a very important step in the economy of the algorithm. During the optimization, the values of the estimated basic variables are calculated and checked against their bounds. This process works when the values of the basic variables are within bounds for small changes in the nonbasic variables. Therefore, the basic variables should preferably be away from their bounds, and the basis matrix should be so that small changes in the numerical values corresponding to small movements do not change the solution of the system $c(x_b, x_n) = 0$ too much. In other words, the basis matrix should be well conditioned. The basis must be factored, and the factors must be updated when the matrix changes.

Intensive numerical tests referring to the selection and to the factorization of the basis have given the following results:

- The conditioning of the basis is more important than the sparsity. A well-conditioned basis will lead in general to fewer iterations because the Hessian of the reduced objective function also becomes better conditioned.
- The set of basic variables should not vary too much. In other words, from one iteration to another, the second-order information in the reduced Hessian approximation can be preserved if the basis before and after a basis change differs by one variable only.
- At degenerate points, the basis must be very carefully controlled to avoid cycling. Nonlinear models seem to behave like linear models: degeneracy becomes more accentuated as models grow in size.
- For poorly scaled problems, the factorization procedure by Hellerman and Rarick (1971, 1972) is slow because many proposed pivots appear to be too small relative to other elements in the column, resulting in many expensive column swaps.

Initially, in CONOPT, the basis is LU factorized using a Markowitz (1957) pivot selection procedure, similar to the one suggested by Suhl and Suhl (1990, 1991), which is an improvement of the Hellerman and Rarick, where only updated spike columns are separately stored. The remaining columns of the basis factors are represented by the original columns in the Jacobian as described in Kalan (1971). The main difference between the implementation given by Suhl and Suhl (1990) and the one given by Drud (1994) is that in CONOPT, a block decomposition of the bump is performed and to each individual block, the Markowitz criterion is applied.

The Jacobian is not changed in degenerate iterations. Therefore, a cheap simplex-like update of the basis factors similar to the one suggested by Reid (1982) is applied. This will reduce the length of the spikes during factorization.

The Jacobian will change during the nondegenerate iterations. To prevent unnecessary calculations of the Jacobian and expensive full factorizations after small steps, an estimation of the quality of the Jacobian is performed in CONOPT. This estimation is based on the speed of convergence of the Newton iterations in the line search and on the change in objective, relative to the change predicted by a linear model (see Drud (1994)).

CONOPT keeps the old set of basic variables after the iterations in which no basic variable reached a bound. If the step was limited by a superbasic variable reaching a bound, CONOPT estimates the quality of the Jacobian as mentioned above. If the Jacobian is good, then CONOPT keeps the factorization of the basis. If it is poor or if the step was determined by nonlinearities, then the Jacobian is recomputed. The subsequent factorization uses the same pivot sequence and sparsity pattern determined during the last full factorization.

When the line search is poorly behaved because the Newton iterations do not converge even for small steps, CONOPT tries to find a better conditioned basis. The procedure is to replace a basic variable with a better superbasic. All superbasic columns are updated with the active eta vectors, and the largest possible pivot is selected. To avoid cycling, the basis is changed only if the pivot is greater than one. The heuristic is based on the assumption that a large determinant indicates a well-conditioned basis, a reasonable assumption for well-scaled problems.

Line Search and Newton Iteration

The selection of the search direction in CONOPT involves two parts:

- The selection of the set of superbasic variables to be changed in the coming iteration
- The determination of a direction for superbasic variables, given by the reduced gradient

Concerning the first part, the variables between bounds are always superbasics. A variable at a bound is only made superbasic if its reduced gradient has the right sign, is greater than an absolute tolerance, and is large relative to other reduced gradients. In the second part, two modes are identified to determine a direction for superbasics: the *linear mode* and the *nonlinear mode*.

In the linear mode, it is expected that a linear approximation to the problem is a good one and that the stepsize in the coming line search will be determined by bounds. The superbasic variables are moved in the direction of the negative reduced gradient, i.e., in the steepest descent direction. We immediately try to go as far as possible until a basic or superbasic variable reaches a bound. This point is used as the initial point in the Newton iterations. The linear mode has two advantages. It helps us save the overhead of the quasi-Newton procedure in areas where it is not needed, and it can accelerate the line search significantly by going directly to the bound.

In the nonlinear mode, the line search is expected to be limited by the nonlinearity of the problem, i.e., it is expected that no variable will reach a bound. The search direction for the superbasic variables is determined from a quasi-Newton method that also determines the initial stepsize suggested to the line search. For generating search directions in the nonlinear mode, a quasi-Newton method based on BFGS is used. With the growing size of problems, some kind of conjugate gradient method is used to handle problems with a very large number of superbasics.

Firstly, the linear mode is used and switched from the linear mode to the nonlinear mode after a number of consecutive line searches in which no variable reached a bound. After that, switch back from the nonlinear to the linear mode after a number of consecutive line searches in which a variable reached a bound.

The Newton Iterations

In CONOPT the line search has the following three components: (i) the Newton algorithm that restores feasibility for given values of the superbasic variables and given estimate for the basic variables, (ii) computation of the initial estimates for the basic variables and management of the bounds on superbasic and basic variables, and (iii) computation of the stepsize and implementation of the termination criteria.

For each step in one-dimensional search, the nonlinear system $c(x_b, x_n) = b$ must be solved subject to x_b for x_n fixed. Since this system must be solved many times, this is the most expensive component of CONOPT. It is important not to solve this system more accurately than necessary and not to spend too much time trying to solve the system. Finally, the output of a line-search procedure consists of the value of the objective function and of the values of the basic variables. At all intermediate points, only the objective function value is used to determine a new stepsize. The values of the basic variables are used to determine new initial values for the next Newton iteration. Therefore, the main criterion for stopping the iteration is based on the error *df*, where *df* is the change that would occur in the objective function if the residuals of the system $c(x_b, x_n) = b$ were all reduced to zero. This can be approximated by $df = \Delta x_j = e_j^T \Delta x_b = e_j^T J_b^{-1} z = u^T z$, where u is the vector of Lagrange multipliers (computed in step 4 of CONOPT) and $z = b - c(x_b, x_n)$ is the current vector of residuals. The convergence rate of the Newton algorithm is monitored as follows. If $\|z\|_1 \leq 0.001n$ or $\|z\|_1 (\|z\|_1 / \|z_{old}\|_1)^p < 0.001n$, then the Newton iterates are convergent, and the line search continues ahead. Otherwise, the Newton algorithm terminates. (z_{old} is the residual from the previous iteration, and p is an integer parameter which is modified along the iterations.) If the residuals are small, then it is supposed that the approximation of *df* is good enough, and therefore the test of rapid convergence on *df* can be applied: if $|df| \leq rtobj$, then the Newton iterations are stopped with success. Otherwise, if $|df|(|df|/|df_{old}|)^p > rtobj$, then the Newton iterations are stopped with failure and the initial stepsize must be decreased (*rtobj* is a parameter computed by the algorithm). This will reduce the number of Newton iterations as well as the number of evaluations of the functions. The basic variables are computed as $x_b = x_b + \Delta x_b$, where Δx_b is the solution of the linear system $J_b \Delta x_b = z$ which uses the same inverse as in computation of the reduced gradient (see step 5). If this new x_b overflows a bound by more than a predefined threshold, then the Newton algorithm is stopped. No solution is expected to exist within the bounds, and therefore the algorithm returns to the stepsize procedure in which a new and smaller step is chosen. Otherwise the corresponding basic variable is fixed at its bound. Some details including the Newton algorithm used in CONOPT are given in Drud (1985).

The Linear Mode: Sequential Linear Programming

When the problem appears linear, CONOPT will often take many small steps, each determined by a new variable reaching a bound. Although the line search is fast in the linear mode, each step requires one or more evaluations of the nonlinear constraint, and the overall cost may become high relative to the progress. In order to avoid the many nonlinear constraint evaluations, CONOPT may replace the steepest descent direction with a *sequential linear programming* (SLP) technique to find a search direction that anticipates the bounds on all variables and therefore gives a larger expected change in the objective function in each line search. The search direction and the last basis from the SLP procedure are used in an ordinary GRG-type line search in which the solution is made feasible at each step. The SLP procedure is only used to generate good directions. CONOPT uses a test which determines whether SLP is used or not. This test is based on the progress of the algorithm to solution. In other words, in CONOPT, the feasibility is conserved at each iteration.

When the optimization is in the linear mode, then a steepest descent algorithm is used to determine the search direction. As an alternative, CONOPT implements the steepest edge algorithm (Reid, 1975; Goldfarb, 1976; Goldfarb & Reid, 1977; Andrei, 1999, pp. 154). The idea, taken from linear programming, is to scale the nonbasic variables according to the Euclidean norm of the updated column, the so-called *edge length*. A unit step for a nonbasic variable will give rise to changes in the basic variables proportional to the edge length. A unit step for a nonbasic variable with a large edge length will therefore give large changes in the basic variables. This step has two adverse effects relative to a unit step for a nonbasic variable with a small edge length: a basic variable is more likely to reach a bound after a very short step length and the large change in basic variables is more likely to give rise to larger nonlinear terms. The steepest edge algorithm has been very successful in linear programming (Goldfarb & Reid, 1977). Also, this technique is profitable in nonlinear optimization, leading to fewer iterations for most nonlinear problems (Drud, 2005). However, the cost for maintaining the edge lengths can be more expensive in the nonlinear case. Therefore, the steepest edge procedure is mainly useful during the linear mode iterations.

The Nonlinear Mode: Sequential Quadratic Programming

As we have already seen in GRG algorithm, when the progress toward solution is given by nonlinearities (and not by bounds), then it executes many small steps which determine small variations of the superbasic variables and also small variations of the reduced gradient. CONOPT can use the second-order information to make good progress and to determine if bounds should be active or not. The second-order information is used in the *sequential quadratic programming* (SQP) procedure that, like in the SLP procedure, finds a good search direction and a good basis while maintaining the feasibility of steps. Therefore, CONOPT is an algorithm which generates feasible iterations. Like in the SLP case, CONOPT uses a test which determines if it should use the SQP procedure or not. The SQP procedure can be inhibited by assigning a value to a special parameter. If the matrix of the second-order derivatives becomes too dense, then CONOPT skips over the SQP procedure. A dense matrix of second derivatives will need more memory and will make the SQP iterations so slow that the potential saving in the number of iterations is wasted in computing and manipulating of the Hessian. As in the linear mode where the sequential linear programming is used, in the nonlinear mode, the sequential quadratic programming generates a scaled search direction, and the expected step length in the following line search is therefore 1. However, the step length may be less than 1 for several reasons: the line search is ill behaved, a basic variable reaches a bound before being predicted by the linear constraints, the objective is much more nonlinear along the search direction than expected, and the optimal step is not 1. For superbasic variables, the search direction is determined using the quasi-Newton BFGS updating.

In Drud (2005, 2011) plenty of aspects on CONOPT are presented and discussed concerning hints on good model formulation, algebraic information on initial point determination, preprocessing, scaling, finding a feasible solution, linear and nonlinear modes, SLP procedure, SQP procedure, etc. A very detailed comparison between sequential linearization methods (e.g., MINOS) and CONOPT, with respect to finding the first feasible solution, maintaining the feasibility of iterations, computing the Jacobian matrix, choice and factorization of basis, Lagrange multipliers computation, line search and Newton iterations, etc., is presented in Drud (1994). CONOPT represents a very advanced combination of three algorithms which integrate sequential linear programming and sequential quadratic programming, both of them being imbedded in the generalized reduced gradient technology. The algorithm is improved with many computational procedures based on numerical observations obtained in the process of solving large classes of different nonlinear optimization problems of different dimensions and complexities.

Practical Hints in GAMS

The options in CONOPT may be introduced through `conopt.opt` file. If we want to solve a nonlinear optimization problem with the name `mymodel`, which is defined by declaration `model mymodel/all/;`, then we must tell the solver that we want to use an option file with the statement `mymodel.optfile = 1;` placed in the GAMS source file before the SOLVE command.

The assignment or set statement is used to assign a new value to internal CONOPT variables, so-called CR-Cells. The optional set verb, the name of the CR-Cell, and the value must be separated by blanks, tabs, commas, colons, and/or "equal" sign. The value must be written using Fortran format with a maximum of ten characters and must have the same type as the CR-Cell, i.e., real CR-Cells must be assigned real values, integer CR-Cells must be assigned integer values, logical CR-Cells must be assigned logical values, etc. CR-Cells starting on R assume real values, CR-Cells starting on LS assume logical values, and all other CR-Cells starting on L assume integer values. For example, `rtmaxv: = 1.e8;` describe the internal value of infinity, `lfscal: = 20;` is frequency of scaling, `lsismp: = false;` is the option for logical switch for ignoring small pivots, etc. The CR-Cells that GAMS users can access in CONOPT are described in *GAMS – The Solver Manual* (2005, pp.39–82). The documentation of CONOPT also contains some practical comments on how to formulate nonlinear optimization models so as to become easier to solve with CONOPT.

Numerical Study

In Chapter 4, we presented a list of 18 nonlinear optimization applications. Table 16.1 shows the performances of CONOPT in case of small-scale nonlinear optimization applications. Table 16.2 shows the performances of CONOPT for large-scale nonlinear optimization applications.

Table 16.1 Performances of CONOPT. Small-scale nonlinear optimization applications

	n	*me*	*mc*	*nj*	*njn*	*nhd*	*nhs*	*it*	*cpu*	*vfo*
ELCH	10	3	0	25	10	10	45	27	0.043	−47.7610909
ALKI	10	3	8	38	14	5	7	14	0.062	−1768.80696
PREC	8	0	6	18	14	8	8	27	0.037	3.9511634
PPSE	9	6	0	37	32	7	10	13	0.040	5055.011803
MSP3	13	0	15	50	31	3	12	36	0.038	97.5875096
MSP5	16	0	21	89	71	5	19	21	0.040	174.7869944
POOL	34	20	0	115	78	0	63	12	0.037	2569.80
TRAFO	6	0	2	19	18	4	15	16	0.038	135.0759628
LATHE	10	1	14	53	37	8	10	36	0.049	−4430.08793
DES	150	50	0	301	300	150	11175	37	0.169	1055.182314
CSTC	303	200	0	1304	1303	303	202	28	0.211	3.4800747
DIFF	396	324	0	1621	0	0	0	5	0.036	0.0
PENICI	707	602	0	3908	3305	303	1010	2243	20.07	113.990

In the tables 16.1 and 16.2, we have: n = the number of variables, me = the number of equality constraints, mc = the number of inequality constraints, nj = the number of nonzeros in Jacobian matrix, njn = the number of nonlinear nonzeros in Jacobian matrix, nhd = the number of nonzero diagonal elements of the Hessian to the Lagrangian, nhs = the number of nonzero sub-diagonal elements of the Hessian to the Lagrangian, it = the number of iterations, cpu = the cpu computing time for solving the problem (seconds), vfo = the value of the objective function at solution.

Table 16.2 Performances of CONOPT. Large-scale nonlinear optimization applications

	n	*me*	*mc*	*nj*	*njn*	*nhs*	*it*	*cpu*	*vfo*
HANG	1002	501	0	3504	1503	501	22	0.27	5.0685777
	2002	1001	0	7004	3003	1001	19	0.379	5.0685101
	4002	2001	0	14004	6003	2001	28	1.366	5.0684889
	8002	4001	0	28004	12003	4001	31	4.766	5.0684827
FLOW	1163	735	0	4979	312	0	5	0.058	0.0
FLOWO	1556	1005	0	6714	400	0	7	0.078	0.0
POL	4004	3000	0	18002	8000	0	33	0.232	14.139480
	6004	4500	0	27002	12000	0	38	0.525	14.142097
	8004	6000	0	36002	16000	0	41	0.760	14.143172
	10004	7500	0	45002	20000	0	45	1.055	14.141714
CAT	3003	2000	0	12003	12000	2002	82	0.763	−0.048055
	6003	4000	0	24003	24000	4002	81	2.136	−0.048055
	9003	6000	0	36003	36000	6002	32	1.242	−0.048053
CONT	2505	2000	0	10005	4	0	21	0.109	1.013238
	5005	4000	0	20005	4	0	30	0.270	1.005922
	7505	6000	0	30005	4	0	31	0.480	1.004561
	10005	8000	0	40005	4	0	33	0.767	1.004071

Table 16.3 Comparison between KNITRO/ACTIVE and CONOPT

	itt	*cput*
KNITRO/ACTIVE	1143	477.375
CONOPT	579	15.256

Table 16.4 Performances of CONOPT with SQP inhibited. Large-scale nonlinear optimization applications

	n	*me*	*mc*	*it*	*cpu*	*vfo*
HANG	1002	501	0	81	0.269	5.6857779
	2002	1001	0	151	0.838	5.0685101
	4002	2001	0	769	6.311	5.0684889
	8002	4001	0	1658	24.901	5.0684828
FLOW	1163	735	0	5	0.058	0.0
FLOWO	1556	1005	0	7	0.066	0.0
POL	4004	3000	0	33	0.250	14.139480
	6004	4500	0	38	0.507	14.142096
	8004	6000	0	41	0.759	14.143172
	10004	7500	0	45	1.060	14.141714
CAT	3003	2000	0	283	4.965	−0.048055
	6003	4000	0	157	3.108	−0.048055
	9003	6000	0	75	3.441	−0.048053
CONT	2505	2000	0	21	0.113	1.013238
	5005	4000	0	30	0.275	1.005922
	7505	6000	0	31	0.465	1.004561
	10005	8000	0	33	0.787	1.004071

It is important to make comparisons. The numerical experiments presented in Tables 14.2 (KNITRO/ACTIVE) and 16.2 (CONOPT) for solving all those 17 large-scale nonlinear optimization applications described in Chapter 4 are used in Table 16.3, where we present the total number of iterations (*itt*) and the total CPU computing time (*cput*) (seconds) necessary to get a solution of these applications.

It is worth seeing the behavior of CONOPT when the procedure of the search direction computed by the sequential quadratic programming is inhibited. (The sequential quadratic programming procedure is inhibited by writing in `conopt.opt` file the line "`lsesqp:= f.`") In other words, the search direction is determined by means of the reduced gradient. Table 16.4 contains the performances of CONOPT with sequential quadratic programming inhibited.

A comparison between CONOPT with sequential quadratic programming (Table 16.2) and CONOPT without sequential quadratic programming (Table 16.4) is given in Table 16.5, where *itt* is the total number of iterations and *cput* is the total CPU computing time (seconds) for solving all these 17 nonlinear optimization applications.

Observe that CONOPT equipped with sequential quadratic programming for line-search determination is much better than its variant without sequential

Table 16.5 CONOPT with SQP versus CONOPT without SQP

	itt	*cput*
CONOPT with SQP	579	15.256
CONOPT without SQP	3458	48.173

Table 16.6 Comparison: MINOS, KNITRO/ACTIVE, SNOPT, and CONOPT. CPU computing time (seconds)

MINOS	KNITRO/ACTIVE	SNOPT	CONOPT
1369.61	477.375	252.67	15.256

quadratic programming. A simple analysis of Tables 16.2 and 16.4 shows that the nonlinear optimization applications in the linear mode (POL and CONT) are indifferent subject to the second-order information used by the SQP procedure.

As a conclusion to these numerical studies, let us compare MINOS (Table 9.5), KNITRO/ACTIVE (Table 14.2), SNOPT (Table 15.2), and CONOPT with SQP (Table 16.2) subject to the total CPU computing time. Table 16.6 presents the performances of these algorithms for solving all these 17 large-scale nonlinear optimization applications with respect to the total CPU computing time.

Notes and References

CONOPT algorithm is a line-search algorithm, and it is described in a number of papers by Arne Drud (1976, 1983, 1985, 1994, 1995, 1996, 2005, 2011). Each of them gives different computational details on numerical linear algebra used in CONOPT. The algorithm implements three active-set methods. The first one is a gradient projection method in the frame of the generalized reduced gradient method that projects the gradient of the objective function onto a linearization of the constraints. The second one is a sequential linear programming algorithm, and the third is a sequential quadratic programming algorithm. CONOPT includes algorithmic switches that automatically detect which method is the best. The search direction is computed in two modes. In the linear mode, the progress toward solution is given by bounds. In the nonlinear mode, this progress is given by nonlinearities. CONOPT is imbedded in GAMS technology.

Chapter 17
Interior Point Methods

One of the most powerful methods for solving nonlinear optimization problems known as interior point methods is to be presented in this chapter. They are related to barrier functions. The terms "interior point methods" and "barrier methods" have the same significance and may be used interchangeably. The idea is to keep the current points in the interior of the feasible region. A method for remaining in the interior of the feasible region is to add a component to the objective function, which penalizes close approaches to the boundary. This method was first suggested by Frisch (1955) and developed both in theoretical and computational details by Fiacco and McCormick (1964, 1966, 1968).

Interior point methods know two implementations. The first one uses the line-search technique and the second uses filters. The interior point methods based on line search are also classified in two methods. The first one is a direct extension of interior point methods for linear programming. They use line searches to enforce convergence. The second one uses a quadratic model to define the step and incorporates a trust-region constraint to provide stability. For these methods, we follow the main developments from Nocedal and Wright (2006). The interior point method with filter uses a filter for selecting the stepsize in the direction of the calculated step (Fletcher & Leyffer, 2002).

In this chapter, we consider the interior point method for general nonlinear optimization problems in line search and trust-region paradigms, respectively. At the same time, a variant of the interior point algorithm is presented, which illustrates a *methodology for the theoretical development of interior point methods* and of their convergence analysis.

Let us consider the problem:

N. Andrei, *Continuous Nonlinear Optimization for Engineering Applications in GAMS Technology*, Springer Optimization and Its Applications 121,
DOI 10.1007/978-3-319-58356-3_17

$$\min f(x) \tag{17.1a}$$

subject to

$$c_E(x) = 0, \tag{17.1b}$$

$$c_I(x) - s = 0, \tag{17.1c}$$

$$s \geq 0, \tag{17.1d}$$

where $x \in \mathbb{R}^n$, $c_E : \mathbb{R}^n \to \mathbb{R}^l$, and $c_I : \mathbb{R}^n \to \mathbb{R}^m$ are twice continuously differentiable functions and $s \in \mathbb{R}^m$ is a slack variable introduced to transform the inequality constraints $c_I(x) \geq 0$ in equalities. Here $E \triangleq \{1, \ldots, l\}$ and $I \triangleq \{1, \ldots, m\}$.

As described in Nocedal and Wright (2006), in the following, we present two derivations of interior point methods, which emphasize the main idea and terminology for solving general nonlinear optimization problems (17.1). The first derivation is based on *continuation* or *homotopy methods*.[1] The second one is based on *barrier methods*.

Homotopy methods The KKT optimality conditions for problem (17.1) are as follows:

$$\nabla f(x) - \nabla c_E(x)^{\mathrm{T}} y - \nabla c_I(x)^{\mathrm{T}} z = 0, \tag{17.2a}$$

$$Sz - \mu e = 0, \tag{17.2b}$$

$$c_E(x) = 0, \tag{17.2c}$$

$$c_I(x) - s = 0, \tag{17.2d}$$

with $\mu = 0$, together with

$$s \geq 0, z \geq 0. \tag{17.2e}$$

In (17.2) $\nabla c_E(x)$ and $\nabla c_I(x)$ are the Jacobian matrices of equality and inequality constraints, respectively. The vectors y and z are the Lagrange multipliers associated to these constraints. Let us define S and Z as diagonal matrices whose diagonal elements are the components of the vectors s and z, respectively. Observe that (17.2) with μ strictly positive are the *perturbed KKT conditions*. Equation (17.2b), with $\mu = 0$, and the conditions (17.2e) introduce a combinatorial aspect of determining the optimal active set of (17.1). This is the main difficulty of this approach based on KKT conditions. The homotopy method consists in approximately solving the KKT conditions (17.2) for a sequence of positive parameters $\{\mu_k\}$ that converges to zero while maintaining $s, z > 0$. In the limit, we hope to obtain a point that

[1] By *homotopy method* we understand a continuously tranformation of a mathematical object into another one, in the sense that a simple (easy) problem can be continuously deformed into the given (hard) problem (Henri Poincaré). The solutions to the deformed problems are related and can be tracked as the deformation proceeds. The function describing the deformation is called a homotopy map.

satisfies the KKT conditions (17.2). Moreover, asking for the iterates to decrease a merit function or to be acceptable to a filter, we hope that the iterations converge to a minimizer of (17.1) and not simply to a KKT point. The justification of this method is as follows. In a neighborhood of (x^*, s^*, y^*, z^*) which satisfies the linear independence of the constraint qualification (Remark 5.1), the strict complementarity condition (see Definition 5.22) and the second-order sufficient conditions (see Theorem 5.16), then for all sufficiently small positive values of μ, the system (17.2) has a locally unique solution denoted by $(x(\mu), s(\mu), y(\mu), z(\mu))$. All these points describe the so-called *primal-dual central path*. When $\mu \to 0$, this converges to (x^*, s^*, y^*, z^*) solution of the KKT conditions given by (17.2).

Barrier methods In this approach of the derivation of the interior point methods, a barrier problem is associated to (17.1) as

$$\min_{x,s} f(x) - \mu \sum_{i=1}^{m} \log(s_i) \tag{17.3a}$$

subject to

$$c_E(x) = 0, \tag{17.3b}$$

$$c_I(x) - s = 0, \tag{17.3c}$$

where μ is a positive parameter. We see that the inequality $s \geq 0$ is not included in (17.3) because the minimization of the barrier term $-\mu \sum_{i=1}^{m} \log(s_i)$ from (17.3) determines the components of s not to become too close to zero. Observe that the reformulation by using (17.3) avoids the combinatorial aspect of nonlinear programs, but as we can see, its solution does not coincide with the solution of (17.1) for $\mu > 0$. In this context, the barrier approach consists in finding an approximate solution of the barrier problem (17.3) for a positive sequence $\{\mu_k\}$ that converges to zero.

These two approaches are worth comparing based on KKT conditions and on the barrier problem. Let us write the KKT conditions for (17.3) as:

$$\nabla f(x) - \nabla c_E(x)^{\mathrm{T}} y - \nabla c_I(x)^{\mathrm{T}} z = 0, \tag{17.4a}$$

$$-\mu S^{-1} e + z = 0, \tag{17.4b}$$

$$c_E(x) = 0, \tag{17.4c}$$

$$c_I(x) - s = 0. \tag{17.4d}$$

Looking at (17.2) and (17.4), observe that they differ only in the second equations (17.2b) and (17.4b). Notice that (17.4b) becomes very nonlinear near the solution as $s \to 0$. To eliminate this difficulty, it is advantageous for the Newton method to transform the rational equation (17.4b) into a quadratic one. This is done by multiplying (17.4b) by S, thus obtaining the quadratic term Sz. This is always possible because all the diagonal elements of S are positive. With this simple transformation, the KKT conditions for the barrier problem coincide with the perturbed KKT conditions given by (17.2). We emphasize that the presence of the quadratic term Sz in (17.4), the simplest possible nonlinearity introduced in KKT system, makes the barrier method with all its defects nontrivial for solving the general nonlinear optimization problems (17.1).

We emphasize that both the homotopy and the barrier interpretation of nonlinear optimization problems are useful. The homotopy gives rise to the definition of the primal-dual directions, whereas the barrier view is crucial in the design of globally convergent algorithms (Nocedal & Wright, 2006).

We mention that the interior point is derived from the fact that the early barrier methods introduced by Fiacco and McCormick (1968) did not use slack variables and assumed that the initial point is feasible with the inequality constraints. These methods used the barrier function $f(x) - \sum_{i \in I} \log(c_i(x))$ to prevent iterates from leaving the feasible region defined by inequalities (see the logarithmic penalty-barrier function (8.4) used in the penalty-barrier algorithm SPENBAR).

17.1 Prototype of Interior Point Algorithm

Let us consider the Lagrange function associated to problem (17.1) as:

$$L(x, s, y, z) = f(x) - y^{\mathrm{T}} c_E(x) - z^{\mathrm{T}}(c_I(x) - s). \tag{17.5}$$

Now, as in Nocedal and Wright (2006), applying the Newton method to the nonlinear system (17.2), in variables x, s, y, z, we obtain the following linear algebraic system:

$$\begin{bmatrix} \nabla_{xx}^2 L & 0 & -J_E^{\mathrm{T}}(x) & -J_I^{\mathrm{T}}(x) \\ 0 & Z & 0 & S \\ J_E(x) & 0 & 0 & 0 \\ J_I(x) & -I & 0 & 0 \end{bmatrix} \begin{bmatrix} p_x \\ p_s \\ p_y \\ p_z \end{bmatrix} = - \begin{bmatrix} \nabla f(x) - J_E^{\mathrm{T}}(x) y - J_I^{\mathrm{T}}(x) z \\ Sz - \mu e \\ c_E(x) \\ c_I(x) - s \end{bmatrix}, \tag{17.6}$$

where $J_E(x) = \nabla c_E(x)$ and $J_I(x) = \nabla c_I(x)$, are the Jacobian matrices of the equality and inequality constraints, respectively.

The system (17.6) is called the *primal-dual system*. If it is solved, we get a solution $p = (p_x, p_s, p_y, p_z)$ with which a new iteration (x^+, s^+, y^+, z^+) can be computed as

$$x^+ = x + \alpha_s^{\max} p_x, \quad s^+ = s + \alpha_s^{\max} p_s, \tag{17.7a}$$

$$y^+ = y + \alpha_z^{\max} p_y, \quad z^+ = z + \alpha_z^{\max} p_z, \tag{17.7b}$$

where

$$\alpha_s^{\max} = \max\{\alpha \in (0, 1] : s + \alpha p_s \geq (1 - \tau)s\}, \tag{17.8a}$$

$$\alpha_z^{\max} = \max\{\alpha \in (0, 1] : z + \alpha p_z \geq (1 - \tau)z\}, \tag{17.8b}$$

with $\tau \in (0, 1)$. The relations (17.8) for determining the stepsize, called the *fraction to the boundary rule*, are designed to prevent the variables s and z from approaching their bounds of zero too quickly. A typical value of τ is 0.995.

To establish the prototype of the interior point algorithm, an *error function* is introduced, based on the perturbed KKT system (17.2):

$$E(x, s, y, z, \mu) = \max\{\|\nabla f(x) - J_E(x)^{\mathrm{T}}y - J_I(x)^{\mathrm{T}}z\|, \\ \|Sz - \mu e\|, \|c_E(x)\|, \|c_I(x) - s\|\}, \tag{17.9}$$

where $\|.\|$ is an arbitrary vector norm.

This simple procedure described above is the basis of all modern interior point methods. However, to get efficient algorithms, various modifications are needed. For example, a major ingredient is the procedure for solving the primal-dual linear system (17.6). Another important ingredient is the procedure for choosing the sequence of barrier parameters $\{\mu_k\}$. Of course, the determination of the initial point for starting the algorithm or the procedure for the approximation of the Hessian of the Lagrangian $\nabla_{xx}^2 L$ is also crucial.

Algorithm 17.1 Prototype of the interior point algorithm

1.	Choose the initial elements x_0 and $s_0 > 0$. Compute the initial values of multipliers y_0 and $z_0 > 0$. Choose an initial value for the barrier parameter $\mu_0 > 0$, as well as the parameters $\sigma, \tau \in (0, 1)$. Set $k = 0$
2.	If a stopping test for the nonlinear optimization problem (17.1) is satisfied, stop; otherwise go to step 3
3.	If $E(x_k, s_k, y_k, z_k, \mu_k) > \mu_k$, then go to step 4; otherwise go to step 8
4.	Solve the primal-dual system (17.6) with respect to the searching direction $p = (p_x, p_s, p_y, p_z)$
5.	Using (17.8), compute $\alpha_s^{\max}$ and $\alpha_z^{\max}$
6.	Using (17.7), compute a new estimation of the solution $(x_{k+1}, s_{k+1}, y_{k+1}, z_{k+1})$
7.	Set $\mu_{k+1} = \mu_k$, $k = k + 1$ and go to step 3
8.	Choose $\mu_k \in (0, \sigma\mu_k)$ and go to step 2 ♦

Theorem 17.1 *Let $f(x)$, $c_E(x)$, and $c_I(x)$ be continuously differentiable functions. Suppose that* Algorithm 17.1 *generates a sequence of iterates $\{x_k\}$ and that $\{\mu_k\} \to 0$. Then all limit points $\widehat{x}$ of the sequence $\{x_k\}$ are feasible. Moreover, if any limit point $\widehat{x}$ of the sequence $\{x_k\}$ satisfies the linear independence constraint qualification (LICQ), then the first-order optimality conditions of problem (17.1) hold at $\widehat{x}$.*

Proof. To simplify the proof, as in Nocedal and Wright (2006), suppose that the problem contains only the inequality constraints $c_I(x)$, and denote them as $c(x)$. Let $\widehat{x}$ be a limit point of the sequence $\{x_k\}$ and consider a subsequence $\{x_{k_l}\}$ convergent to $\widehat{x}$, i.e., $\{x_{k_l}\} \to \widehat{x}$. If $\mu_k \to 0$, then the error function E given by (17.9) converges to zero, that is, $(c_{k_l} - s_{k_l}) \to 0$. By continuity of c, this implies that $\widehat{c} \triangleq c(\widehat{x}) \geq 0$, i.e., $\widehat{x}$ is feasible and $s_{k_l} \to \widehat{s} = \widehat{c}$. Now, suppose that the linear independence constraint qualification holds at $\widehat{x}$, let $A = \{i : \widehat{c}_i = 0\}$ be the set of active constraints at $\widehat{x}$. For any $i \notin A$, it follows that $\widehat{c}_i > 0$ and $\widehat{s}_i > 0$. Therefore, by the complementarity condition (17.2b), it follows that $[z_{k_l}]_i \to 0$. Since $\nabla f(x_{k_l}) - J(x_{k_l})^T z_{k_l} \to 0$, we obtain:

$$\nabla f(x_{k_l}) - \sum_{i \in A} [z_{k_l}]_i \nabla c_i(x_{k_l}) \to 0. \tag{17.10}$$

But by the constraint qualification hypothesis LICQ, the vectors $\{\nabla \widehat{c}_i : i \in A\}$ are linear independent. Hence, from (17.10) and the continuity of $\nabla f(.)$ and $\nabla c_i(.)$, $i \in A$, it follows that the positive sequence $\{z_{k_l}\}$ converges to a value $\widehat{z} \geq 0$. Taking now the limit in (17.10) we get:

$$\nabla f(\widehat{x}) = \sum_{i \in A} \widehat{z}_i \nabla c_i(\widehat{x}).$$

Moreover, we have $\widehat{c}^T \widehat{z} = 0$, thus completing the proof. ♦

Observe that the theoretical basis of interior point methods is very accessible. All that is required is the continuity of the functions of the problem and of their gradients, as well as the LICQ hypothesis. This is the reason why these methods have very good behavior for solving large-scale nonlinear optimization problems.

As we have already said, the interior point methods are classified in two classes. *The first one is based on* Algorithm 17.1 *which is completed by adding a linear search, a strict control of reduction of the slack variables s and of the multipliers z, as well as the modification of the primal-dual system (17.6) when negative curvature is encountered. The second class computes the step by minimizing a quadratic model of the barrier problem (17.3) with respect to a trust-region constraint.* Both these approaches have many ingredients in common. In the following lines, we describe some aspects of the algorithmic development. Many other details may be found in Nocedal and Wright (2006), Sun and Yuan (2006), Forsgren, Gill, and Wright (2002), or Wright (1991).

17.2 Aspects of Algorithmic Development

In this section, we present some modifications and extensions of Algorithm 17.1 that enable it to solve large-scale nonconvex problems starting from any initial point.

Solving the primal-dual system The system (17.6) can be rewritten in symmetric form as

$$\begin{bmatrix} \nabla_{xx}^2 L & 0 & J_E^T(x) & J_I^T(x) \\ 0 & \Sigma & 0 & -I \\ J_E(x) & 0 & 0 & 0 \\ J_I(x) & -I & 0 & 0 \end{bmatrix} \begin{bmatrix} p_x \\ p_s \\ -p_y \\ -p_z \end{bmatrix} = -\begin{bmatrix} \nabla f(x) - J_E^T(x)y - J_I^T(x)z \\ z - \mu S^{-1}e \\ c_E(x) \\ c_I(x) - s \end{bmatrix}, \tag{17.11}$$

where $\Sigma = S^{-1}Z$. Clearly, this symmetric form is suitable for using symmetric linear systems solvers which reduce the computational effort at each iteration. For solving the system (17.11), we can use the *direct methods* (factorization of the coefficient matrix, Schur complement method, or null space method), as well as the *iterative methods* (conjugate gradient method).

Observe that the system (17.11) can be further reduced by the elimination of the variable p_s,obtaining:

$$\begin{bmatrix} \nabla_{xx}^2 L & J_E^T(x) & J_I^T(x) \\ J_E(x) & 0 & 0 \\ J_I(x) & 0 & -\Sigma^{-1} \end{bmatrix} \begin{bmatrix} p_x \\ -p_y \\ -p_z \end{bmatrix} = -\begin{bmatrix} \nabla f(x) - J_E^T(x)y - J_I^T(x)z \\ c_E(x) \\ c_I(x) - \mu Z^{-1}e \end{bmatrix}. \tag{17.12}$$

This system can be factorized by the *symmetric indefinite factorization* method (Bunch & Parlett 1971; Bunch & Kaufman, 1977; Cheng, 1998; Cheng & Higham, 1998). Let us denote the matrix from (17.12) by M, and then this factorization computes $P^T MP = LBL^T$, where L is a lower triangular matrix, B is a block diagonal matrix with 1×1 or 2×2 blocks, and P is a permutation matrix for conserving the sparsity and ensuring the numerical stability.

The system (17.12) can further be reduced by the elimination of the variable p_z, thus obtaining:

$$\begin{bmatrix} \nabla_{xx}^2 L + J_I^T(x)\Sigma J_I(x) & J_E^T(x) \\ J_E(x) & 0 \end{bmatrix}. \tag{17.13}$$

It is worth observing that if the number of inequality constraints is large, then the dimension of the system (17.13) is smaller than the dimension of the system (17.12). Moreover, the presence of the matrix $J_I^T(x)\Sigma J_I(x)$ in (17.13) dramatically modifies the structure of the sparse matrix of Hessian $\nabla^2_{xx}L$, which can be very unwanted for large-scale problems. A particular, favorable case is when the matrix $J_I^T(x)\Sigma J_I(x)$ is diagonal, a case which appears when all the constraints are simple bounds.

We emphasize that the symmetric primal-dual system (17.11) or its reduced forms (17.12) or (17.13) are ill-conditioned. The ill-conditioning is given by the presence of the matrix $\Sigma = S^{-1}Z$. When $\mu \to 0$, some elements of this matrix tend to $+\infty$; others tend to zero. Nevertheless, because of the special form in which this ill-conditioning arises, the direction computed by a stable factorization method is usually accurate enough. Errors can appear only when the slacks s or the multipliers z become very close to zero or when the Hessian $\nabla^2_{xx}L$ or the Jacobian matrix $J_E(x)$ is almost rank deficient. This is the reason why the direct factorization methods are considered the most efficient and reliable for computing steps in interior point methods (Nocedal & Wright, 2006).

Inertia and singularity As we know, the inertia of a matrix is a triplet of integers which identify the number of positive, negative, and zero eigenvalues of the matrix. For the primal-dual system (17.11), the step p is a descent direction if the matrix

$$\begin{bmatrix} \nabla^2_{xx}L & 0 \\ 0 & \Sigma \end{bmatrix} \tag{17.14}$$

is positive definite on the null space of the matrix

$$\begin{bmatrix} J_E(x) & 0 \\ J_I(x) & -I \end{bmatrix}. \tag{17.15}$$

The positive definiteness condition holds if the inertia of the primal-dual matrix (17.11) is given by $(n+m, l+m, 0)$, where n is the number of variables, m is the number of inequality constraints, and l is the number of equality constraints of problem (17.1). *In other words, if this matrix has exactly $n+m$ positive eigenvalues, $l+m$ negative eigenvalues, and no zero eigenvalues, then the positive definiteness condition holds.*

If the primal-dual matrix (17.11) does not satisfy this inertia condition, then this can be modified as follows. Observe that the diagonal matrix Σ is positive definite by construction, but $\nabla^2_{xx}L$ can be indefinite. Therefore, $\nabla^2_{xx}L$ can be replaced by the matrix $\nabla^2_{xx}L + \delta I$, where $\delta > 0$ is a sufficiently large parameter to ensure that the inertia of the modified matrix is exactly $(n+m, l+m, 0)$.

Another problem we face is when the matrix of the primal-dual system (17.11) is singular, caused by the rank deficiency of the matrix J_E. Observe that the matrix $[J_I \;\; -I]$ is always of full rank. To skip over this singularity of the Jacobian J_E, in the matrix of the primal-dual system (17.11), a regularization parameter $\gamma > 0$ is introduced. Hence, taking into consideration the correction of inertia and the protection to singularity of the Jacobian of the equality constraints, the matrix of the primal-dual system (17.11) can be modified as:

$$\begin{bmatrix} \nabla_{xx}^2 L + \delta I & 0 & J_E^T(x) & J_I^T(x) \\ 0 & \Sigma & 0 & -I \\ J_E(x) & 0 & -\gamma I & 0 \\ J_I(x) & -I & 0 & 0 \end{bmatrix}. \tag{17.16}$$

Suppose that we have the current value of the barrier parameter μ, the constants $\eta > 0$ and $\beta < 1$, and the value of the perturbation parameter δ_{old} from the previous iteration, then a procedure for the selection of parameters γ and δ is as follows.

Algorithm 17.2 Inertia correction and regularization

1.	In matrix (17.16), set $\delta = \gamma = 0$. Factorize the matrix (17.16)
2.	If the matrix (17.16) is nonsingular and its inertia is $(n+m, l+m, 0)$, then compute the primal-dual step; stop
3.	If the matrix (17.16) has zero eigenvalues, then set $\gamma = 10^{-8}\eta\mu^{\beta}$
4.	If $\delta_{old} = 0$, then set $\delta = 10^{-4}$; otherwise set $\delta = \delta_{old}/2$
5.	Factorize the modified matrix (17.16)
6.	If the inertia of this matrix is $(n+m, l+m, 0)$,then set $\delta_{old} = \delta$. Compute the primal-dual step by solving the system (17.11) with modified matrix; stop. Otherwise, set $\delta = 10\delta$ and go to step 5 ♦

Algorithm 17.2 is a variant established by Nocedal and Wright (2006) of a more elaborated algorithm given by Wächter and Biegler (2006). The constants used in it are arbitrary. The purpose of the algorithm is to drastically avoid modifications of the Hessian $\nabla_{xx}^2 L$ while minimizing the number of factorizations. Severe modifications of the matrix of the primal-dual system (17.11) deteriorate the performances of the algorithm because the second-order information given by $\nabla_{xx}^2 L$ is modified, and in this case, the algorithm behaves like a steepest descent. The algorithm for inertia correction and regularization is used at each iteration of the interior point algorithm.

Updating the barrier parameter It is important to update the barrier parameter in a proper way. The sequence of $\{\mu_k\}$ must converge to zero, so in the limit, we can recover the solution of the problem (17.1). But, if μ_k is decreased too slowly, then a large number of iterations is required by the algorithm. On the other hand, if μ_k is decreased too quickly, then some slack variables s or multipliers z may approach zero prematurely, thus slowing the progress of the algorithm to the solution of the problem. In the following, let us present some strategies for barrier parameter selection which prove to be efficient in practice.

The *Fiacco-McCormick strategy* fixes the values of the barrier parameter until the perturbed KKT conditions (17.2) are satisfied to some accuracy. Then the barrier parameter is decreased as:

$$\mu_{k+1} = \sigma_k \mu_k, \sigma_k \in (0, 1). \tag{17.17}$$

The parameter σ_k can be selected as a constant value, for example, $\sigma_k = 0.2$. However, it is preferable to have a strategy for the selection of σ_k. One idea is to choose smaller and smaller values for σ_k when the solution is approached. If close to the solution, we consider $\sigma_k \to 0$, and the parameter τ in (17.8) converges to 1, and then the interior point algorithm has a superlinear convergence rate. The Fiacco-McCormick strategy works well on many problems, but it is sensitive to the initial point selection and to scale the problem.

The *adaptive strategies* for updating the barrier parameter are more robust in case of nonlinear problems. These strategies are based on complementarity exactly as in linear programming and modifying the barrier parameter at each iteration k as:

$$\mu_{k+1} = \sigma_k \frac{s_k^{\mathrm{T}} z_k}{m}. \tag{17.18}$$

Another strategy for σ_k selection in (17.18) is based on the interior point methods in linear programming. Firstly, with $\mu = 0$ in (17.11), we solve this system to get the *predictor direction* $(\Delta x^{af}, \Delta s^{af}, \Delta y^{af}, \Delta z^{af})$. With this, as in (17.8), determine α_s^{af} and α_z^{af} as the longest step that can be taken along the affine scaling direction before violating the nonnegativity conditions $(s, z) \geq 0$. These values are used to compute μ_{af} as the value of the complementarity along the affine scaling step, i.e.:

$$\mu_{af} = \frac{\left(s_k + \alpha_s^{af} \Delta s^{af}\right)^{\mathrm{T}} \left(z_k + \alpha_z^{af} \Delta z^{af}\right)}{m}. \tag{17.19}$$

With this, compute σ_k as:

$$\sigma_k = \left(\frac{\mu_{af}}{s_k^{\mathrm{T}} z_k / m}\right)^3. \tag{17.20}$$

Merit functions and filters for step acceptance Let us assume that we have the primal-dual direction. The next step is to determine a stepsize along this direction. For this the merit functions or filters can be used. Since the interior point methods can be considered as methods for solving the barrier problem (17.3), then it is quite natural to define the merit function Φ or the filter in terms of barrier functions. For example, we may use an exact merit function of the form

$$\Phi_\sigma(x, s) = f(x) - \mu \sum_{i=1}^{m} \log(s_i) + \sigma(\|c_E(x)\| + \|c_I(x) - s\|), \tag{17.21}$$

where the norm in (17.21) can be l_1 or l_2 and $\sigma > 0$ is a penalty parameter updated as it was described in Section 11.2.

An interior point algorithm with line search is as follows. After the primal-dual step p has been computed and the maximum step lengths $\alpha_s^{\max}$ and $\alpha_z^{\max}$ given by (17.8) have been determined, execute a line search by backtracking. This line search determines the stepsizes $\alpha_s \in (0, \alpha_s^{\max}]$ and $\alpha_z \in (0, \alpha_z^{\max}]$, which provide a sufficient decrease of the merit function or ensure acceptability by the filter. The new estimate of the solution is computed as:

$$x^+ = x + \alpha_s p_x, \quad s^+ = s + \alpha_s p_s, \tag{17.22a}$$

$$y^+ = y + \alpha_z p_y, \quad z^+ = z + \alpha_z p_z. \tag{17.22b}$$

In case we use the filter, the pairs of the filter are formed by using the values of the barrier function $f(x) - \mu \sum_{i=1}^{m} \log(s_i)$ on one side and the constraint violations $\|(c_E(x), c_I(x) - s)\|$ on the other side. A step is accepted only if it is not dominated by any element in the filter (see Chapter 18).

Quasi-Newton approximations of Hessian In the interior point algorithm, the Hessian matrix $\nabla_{xx}^2 L$ which appears in (17.11) (or in its reduced variants (17.12) and (17.13)) can be approximated by means of the quasi-Newton methods BFGS or SR1. For large-scale problems, the *limited-memory* BFGS method is recommended (Nocedal, 1980; Liu & Nocedal, 1989). We emphasize that it is the approximation of the Hessian to the Lagrange function (17.5) which is computed, not the Hessian of the barrier function (17.3), which is highly ill-conditioned and changes rapidly.

If we apply the BFGS update formula, then we must have the pairs of vectors $(\Delta x, \Delta l)$, where Δx and Δl are the variations of the solution and of the gradient of the Lagrange function. After computing a step from (x_k, s_k, y_k, z_k) to $(x_{k+1}, s_{k+1}, y_{k+1}, z_{k+1})$, these variations are defined as:

$\Delta x_k = x_{k+1} - x_k$, $\Delta l_k = \nabla_x L(x_{k+1}, s_{k+1}, y_{k+1}, z_{k+1}) - \nabla_x L(x_k, s_k, y_k, z_k)$.

The BFGS update of B_{k+1} is defined as:

$$B_{k+1} = B_k - \frac{B_k \Delta x_k (\Delta x_k)^T B_k}{(\Delta x_k)^T B_k \Delta x_k} + \frac{\Delta l_k (\Delta l_k)^T}{(\Delta l_k)^T \Delta x_k}. \tag{17.23}$$

Observe that the BFGS update (17.23) generates dense $n \times n$ matrices. Therefore, for large-scale problems, it is preferable to use the *compact representation* of the *limited-memory BFGS update*. This representation is based on the following theorem (Nocedal & Wright, 2006).

Theorem 17.2 *Let B_0 be a symmetric and positive definite matrix and assume that a number of k pairs of vectors* $(\Delta x_i, \Delta l_i)$, and $i = 0, \ldots, k-1$, *satisfy the condition* $(\Delta x_i)^T \Delta l_i > 0$, $i = 0, \ldots, k-1$. *Let B_k be the matrix obtained by applying k BFGS updates with these vector pairs to B_0, using (17.23). Then*

$$B_k = B_0 - \begin{bmatrix} B_0 M_k & N_k \end{bmatrix} \begin{bmatrix} M_k^T B_0 M_k & L_k \\ L_k^T & -D_k \end{bmatrix}^{-1} \begin{bmatrix} M_k^T B_0 \\ N_k^T \end{bmatrix}, \tag{17.24}$$

where M_k and N_k are $n \times k$ matrices defined as:

$$M_k = [\Delta x_0, \ldots, \Delta x_{k-1}] \in \mathbb{R}^{n \times k}, \tag{17.25a}$$

$$N_k = [\Delta l_0, \ldots, \Delta l_{k-1}] \in \mathbb{R}^{n \times k}, \tag{17.25b}$$

and matrices L_k and D_k are $k \times k$ dimensional of the following form:

$$(L_k)_{i,j} = \begin{cases} (\Delta x_{i-1})^T \Delta l_{j-1}, & \text{if} \quad i > j, \\ 0, & \text{otherwise}, \end{cases} \tag{17.26a}$$

$$D_k = diag\left[(\Delta x_0)^T \Delta l_0, \ \ldots, \ (\Delta x_{k-1})^T \Delta l_{k-1} \right]. \tag{17.26b}$$

This result is proved by induction. The conditions $(\Delta x_i)^T \Delta l_i > 0$, and $i = 0, \ldots, k-1$, ensure that the inverse of the matrix from (17.24) exists, thus this expression being well defined.

Like in L-BFGS, in this case, we keep the r most recent pairs $(\Delta x_i, \Delta l_i)$ and at each iteration; this set of vectors are updated by eliminating the oldest one and by adding a newly generated pair. During the first r iterations, the updating procedure described in Theorem 17.2 can be used without any modification. For iterations $k > r$, the updating procedure must be modified to reflect the changing of the set of vector pairs $(\Delta x_i, \Delta l_i)$, and $i = k - r, k - r + 1, \ldots, k - 1$. Now, let us define the $n \times r$matrices M_k and N_k, by:

$$M_k = [\Delta x_{k-r}, \ldots, \Delta x_{k-1}] \in \mathbb{R}^{n \times r}, \tag{17.27a}$$

$$N_k = [\Delta l_{k-r}, \ldots, \Delta l_{k-1}] \in \mathbb{R}^{n \times r}. \tag{17.27b}$$

Then, the matrix B_k is obtained by applying r updates to the basic matrix $B_0^{(k)} = \delta_k I$, where $\delta_k = 1/\gamma_k$ and $\gamma_k = (\Delta x_{k-1})^T \Delta l_{k-1} / (\Delta l_{k-1})^T \Delta l_{k-1}$, to get

$$B_k = \delta_k I - \begin{bmatrix} \delta_k M_k & N_k \end{bmatrix} \begin{bmatrix} \delta_k M_k^T M_k & L_k \\ L_k^T & -D_k \end{bmatrix}^{-1} \begin{bmatrix} \delta_k M_k^T \\ N_k^T \end{bmatrix}, \tag{17.28}$$

where L_k and D_k are $r \times r$ matrices defined by:

$$(L_k)_{i,j} = \begin{cases} (\Delta x_{k-r-1+i})^T \Delta l_{k-r-1+j}, & \text{if} \quad i > j, \\ 0, & \text{otherwise}, \end{cases} \tag{17.29a}$$

$$D_k = diag\left[(\Delta x_{k-r})^T \Delta l_{k-r}, \ \ldots, \ (\Delta x_{k-1})^T \Delta l_{k-1} \right]. \tag{17.29b}$$

After the new iteration x_{k+1} is computed, the matrix M_{k+1} is obtained by deleting Δx_{k-r} and by adding the new Δx_k. The matrix N_{k+1} is updated in a similar way. The advantage of this compact representation is that the $(2r \times 2r)$ matrix in the middle of (17.28) and which must be inverted is of small dimensions. Usually, r, the number of stored pairs, $(\Delta x_i, \Delta l_i)$, is very small: three, five, or seven (rarely nine). Therefore, the numerical effort for inverse computing this matrix is negligible.

Interior point algorithms are implemented using the compact representation of BFGS updating to the Hessian of the Lagrange function (17.5). Moreover, since B is positive definite, assuming that J_E is of full rank, then the matrix of the primal-dual system (17.11) is nonsingular. Hence, the solution of (17.11) can be simply obtained by the Sherman-Morrison-Woodbury formula.

Feasible interior point methods In many nonlinear optimization applications, we need all the iterates generated by the interior point algorithm to be feasible with respect to some or to all the inequality constraints. Interior point methods provide a natural framework for designing feasible algorithms. If the current iterate x satisfies the inequality constraint $c_I(x) > 0$, then it is easy to adapt the primal-dual iteration given by (17.11) so that the feasibility is conserved. After computing the step p, we let $x^+ = x + p_x$, redefine the slacks as $s^+ = c_I(x^+)$, and test whether the new point (x^+, s^+) is acceptable for the merit function. If so, we define this point to be the new iterate. Otherwise, we reject step p and compute a new one, a shorter trial step. Clearly, in a line-search algorithm, backtracking is used, and in a trust-region method, a new step is computed by reducing the trust-region radius.

17.3 Line-Search Interior Point Algorithm

To describe the line-search interior point algorithm as in Nocedal and Wright (2006), we denote by $D\Phi_\sigma(x, s, p)$ the directional derivative of the merit function Φ_σ in the direction p. The algorithm of this method is very close to the prototype Algorithm 17.1 which contains two imbricate loops, the first being responsible for testing the KKT optimality conditions and the second one for solving the primal-dual system (17.11). If in line search a quasi-Newton method is used, then it is necessary to choose an initial symmetric and positive definite matrix B_0 which is updated along the iteration by quasi-Newton techniques.

Algorithm 17.3 Line-search interior point algorithm

1.	Choose x_0 and $s_0 > 0$. Compute the initial values of the multipliers y_0 and $z_0 > 0$. Choose an initial value of the barrier parameter $\mu_0 > 0$ as well as the parameters $\sigma, \tau \in (0, 1)$ and tolerances ε_μ and ε_{TOL}. Select an $n \times n$ matrix, symmetric positive definite B_0. Set $k = 0$
2.	If $E(x_k, s_k, y_k, z_k, 0) \geq \varepsilon_{TOL}$, then continue with step 3; otherwise stop. The current point is the optimal solution of the problem

(continued)

3.	If $E(x_k, s_k, y_k, z_k, \mu) \geq \varepsilon_\mu$, then go to step 4; otherwise continue with step 10
4.	Solve the primal-dual system (17.11) for the primal-dual direction $p = (p_x, p_s, p_y, p_z)$
5.	Using (17.8), compute $\alpha_s^{\max}$ and $\alpha_z^{\max}$. Set $p_w = [p_x, p_s]$
6.	Compute the stepsizes α_s and α_z satisfying $\alpha_s \in \left(0, \alpha_s^{\max}\right]$ and $\alpha_z \in \left(0, \alpha_z^{\max}\right]$, as well as the condition $\Phi_\sigma(x_k + \alpha_s p_x, s_k + \alpha_s p_s) \leq \Phi_\sigma(x_k, s_k) + \eta\alpha_s D\Phi_\sigma(x_k, s_k, p_w)$
7.	Using (17.22), compute a new estimation of solution $(x_{k+1}, s_{k+1}, y_{k+1}, z_{k+1})$
8.	If a quasi-Newton approach is used, update B_k
9.	Set $k = k + 1$. Continue with step 3
10.	Set $\mu = \sigma\mu$ and update ε_μ. Continue with step 2 ♦

In step 4 of this algorithm, it is quite possible for the matrix of the system (17.11) to be modified as in (17.16). If the merit function can cause the Maratos effect, then a second-order correction or a nonmonotone strategy can be implemented. The barrier tolerance ε_μ can be selected as $\varepsilon_\mu = \mu$, as in the prototype Algorithm 17.1.

To be reliable and to ensure the convergence, the algorithm must be equipped with some protection mechanisms which substantially modify the line search in step 6. These mechanisms refer to a careful control of the decrease in the barrier parameter μ and in the (inner) convergence tolerance ε_μ, and they let the parameter τ in (17.8) converge to 1 sufficiently rapidly.

Another specific difficulty in interior point methods is the lack of coordination between the step computation and the satisfaction of the bounds on variables. An approach is to monitor the stepsizes α_s and α_z in (17.7). If they are smaller than a given threshold, then the primal-dual step is replaced by a step that guarantees progress in feasibility and improvement in optimality.

An alternative in using the merit function is to use a filter to perform the line search. In a filter method, when the stepsizes are very small, the feasibility restoration phase can be started (Fletcher & Leyffer, 2002).

17.4 A Variant of Line-Search Interior Point Algorithm

In the following, we present a *variant* of a line-search interior point algorithm for solving general nonlinear optimization problems which include simple bounds on variables. We also point out some very important details which emphasize the mathematical technology based on the interior point theory (Andrei, 1998b, 1998c).

Let us consider the problem:

$$\min f(x) \tag{17.30a}$$

subject to:

$$c_E(x) = 0, \tag{17.30b}$$

$$c_I(x) \geq 0, \tag{17.30c}$$

$$l \leq x \leq u, \tag{17.30d}$$

where $f: \mathbb{R}^n \to \mathbb{R}$, $c_E: \mathbb{R}^n \to \mathbb{R}^{me}$, and $c_I: \mathbb{R}^n \to \mathbb{R}^m$ are supposed to be twice continuously differentiable functions. The bounds $l \in \mathbb{R}^n$ and $u \in \mathbb{R}^n$ may have any values with $l \leq u$. The idea of the algorithm for solving (17.30) is to use the first-order KKT optimality conditions in an iterative computational scheme of Newton's type in which the inequality constraints are penalized in a barrier term. The updating of the barrier parameter is given by a merit function. The stepsize is based on the Wolfe line search combined with the interior point centrality and with the condition of limiting the convergence to zero of the complementarity against the pure primal-dual optimality condition. We mention that this mathematical technology is also found in Kortanek, Potra, and Ye (1991), Goldfarb, Liu, and Wang (1991), El-Bakry, Tapia, Tsuchiya, and Zhang (1996), Vanderbei and Shanno (1997), and Gay, Overton, and Wright (1997).

KKT optimality conditions Considering the slack variables $s \in \mathbb{R}^m$ associated to the functional inequality constraints (17.30c) and $w, v \in \mathbb{R}^n$ associated to simple bounds (17.30d), the problem (17.30) is reformulated as

$$\min f(x) \tag{17.31a}$$

subject to:

$$c_E(x) = 0, \tag{17.31b}$$

$$c_I(x) - s = 0, \tag{17.31c}$$

$$x - w = l, \tag{17.31d}$$

$$x + v = u, \tag{17.31e}$$

$$s \geq 0, w \geq 0, v \geq 0. \tag{17.31f}$$

Now, the inequality constraints (17.31f) are transferred into the objective function by a barrier term, thus obtaining the problem:

$$\min f(x) - \mu \sum_{i=1}^{m} \log(s_i) - \mu \sum_{j=1}^{n} \log(w_j) - \mu \sum_{j=1}^{n} \log(v_j) \tag{17.32a}$$

subject to:

$$c_E(x) = 0, \tag{17.32b}$$

$$c_I(x) - s = 0, \tag{17.32c}$$

$$x - w - l = 0, \tag{17.32d}$$

$$x + v - u = 0. \tag{17.32e}$$

The Lagrange function of (17.32) is

$$\begin{aligned} &L(x, s, w, v, y, z, p, q, \mu) \\ &= f(x) - \mu \sum_{i=1}^{m} \log(s_i) - \mu \sum_{j=1}^{n} \log(w_j) - \mu \sum_{j=1}^{n} \log(v_j) \\ &- y^{\mathrm{T}} c_E(x) - z^{\mathrm{T}}(c_I(x) - s) - p^{\mathrm{T}}(x - w - l) - q^{\mathrm{T}}(x + v - u), \end{aligned} \tag{17.33}$$

where y, z, p and q are the Lagrange multipliers. With these, the first-order optimality conditions are

$$\nabla f(x) - J_E(x)^{\mathrm{T}} y - J_I(x)^{\mathrm{T}} z - p + q = 0, \tag{17.34a}$$

$$c_E(x) = 0, \tag{17.34b}$$

$$c_I(x) - s = 0, \tag{17.34c}$$

$$x - w - l = 0, \tag{17.34d}$$

$$x + v - u = 0, \tag{17.34e}$$

$$SZe - \mu e = 0, \tag{17.34f}$$

$$WPe - \mu e = 0, \tag{17.34g}$$

$$VQe - \mu e = 0, \tag{17.34h}$$

where the matrices S, Z, W, P, V, and Q are diagonal with elements s_i, z_i, w_i, p_i, v_i, and q_i, respectively. e is a vector of all ones. $J_E(x)$ and $J_I(x)$ are the Jacobian matrices of the equality $c_E(x) = 0$ and of the inequality constraints $c_I(x) \geq 0$, respectively.

If in (17.34) the barrier parameter μ is set to zero, then we see that the last three equations are exactly the complementarity slackness conditions. Usually, these last three equations are called μ *complementarity* conditions. Observe that the KKT system (17.34) is a nonlinear algebraic one with $5n + 2m + me$ equations and a similar number of unknowns, parameterized by the barrier parameter μ. For solving this system, the Newton algorithm is fundamental, which is known to be very efficient near the solution. Supposing the system (17.34) has a solution, then for each $\mu > 0$ we get a solution $(x_\mu, s_\mu, w_\mu, v_\mu, y_\mu, z_\mu, p_\mu, q_\mu)$. The path $\{(x_\mu, s_\mu, w_\mu, v_\mu, y_\mu, z_\mu, p_\mu, q_\mu) : \mu > 0\}$ is called the *primal-dual central path*. The interior point algorithm which will be presented in this section is an iterative procedure. At each iteration, it attempts to move toward a point on the central path closer to the optimal point than where the current point is. If the barrier parameter μ is set to zero, then the KKT conditions (17.34) can be partitioned in two classes:

$$CO(t) = \begin{bmatrix} \nabla f(x) - J_E(x)^T y - J_I(x)^T z - p + q \\ c_E(x) \\ c_I(x) - s \\ x - w - l \\ x + v - u \end{bmatrix}, \tag{17.35}$$

$$CT(s, z, w, p, v, q) = \begin{bmatrix} SZe \\ WPe \\ VQe \end{bmatrix}. \tag{17.36}$$

It is quite clear that if we could have a point $t = [x, y, z, p, q, s, w, v]^T$ to satisfy the system

$$F(t) \triangleq \begin{bmatrix} CO(t) \\ CT(s, z, w, p, v, q) \end{bmatrix} = 0, \tag{17.37}$$

then the component x of this point would be the solution of the problem (17.30). The separation of the optimality conditions into these two classes is crucial for the elaboration of an efficient algorithm for solving (17.30) and also for proving its convergence.

For linear programming, the system (17.34) is much simpler. The only nonlinear expressions in (17.34) are simple multiplications of slack and dual variables, and the presence of these simple nonlinearities makes the subject of linear programming nontrivial.

The Newton system As above, denote $t = [x, y, z, p, q, s, w, v]^T$. Then, the Newton method applied to the system (17.37) consists in the determination of the direction $\Delta t = [\Delta x, \Delta y, \Delta z, \Delta p, \Delta q, \Delta s, \Delta w, \Delta v]^T$ as a solution of the following perturbed system:

$$F'(t)\Delta t = -F(t) + \mu \widehat{e}, \tag{17.38}$$

where $\widehat{e} \in \mathbb{R}^{5n+2m+me}$ is a vector with zero components except for the last $2n+m$ ones which are all equal to one. $F'(t)$ is the Jacobian of function $F(t)$ computed at the current point t. The new point is computed as

$$t^+ = t + \alpha \Delta t, \tag{17.39}$$

where α is the stepsize.

Two problems are critical with this algorithm: the choice of α at each iteration and possibly the modification of the system (17.38) in order to find a local solution to it. For linear or convex quadratic programming, the modification of (17.38) is not necessary, and the stepsize at each iteration is determined by standard ratio tests (Andrei, 2011a; Lustig Marsten, & Shanno, 1990, 1991, 1992, 1994; Vanderbei,

1990, 1994, 2001; Ye, 1997; Wright, 1997). In case of general nonlinear optimization problems, it is well known that for a poor initial estimate of the solution, the Newton method may diverge, and so the linear system (17.38) needs to be modified once again. Besides, the technique for choosing the stepsize α is more complex. In order to achieve the convergence to a solution of (17.34), El-Bakry, Tapia, Tsuchiya, and Zhang (1996) introduced a merit function and showed that for a proper choice of the barrier parameter μ, there is a stepsize α such that the algorithm (17.39) is convergent to a solution of (17.30), provided that the Jacobian $F'(t)$ from (17.38) remains nonsingular at each iteration (Shanno, Breitfeld, & Simantiraki, 1996; Shanno & Simantiraki, 1997). A modification of the merit function introduced by El-Bakry, Tapia, Tsuchiya, and Zhang (1996) was considered by Vanderbei and Shanno (1997). They used the l_2 norm of the constraints multiplied by a parameter $\beta > 0$ and proved that there is a value of β such that the direction given by the Newton system is a descent direction for their merit function. The corresponding algorithm was implemented as a nonlinear version of LOQO by Vanderbei (1995).

The Newton direction determination The Newton system for direction determination involves the computation of the Jacobian matrix of the function $F(t)$ defined in (17.37). From (17.34) we get:

$$\begin{bmatrix} K(x,y,z) & -J_E(x)^T & -J_I(x)^T & -I & I & 0 & 0 & 0 \\ J_E(x) & 0 & 0 & 0 & 0 & 0 & 0 & 0 \\ J_I(x) & 0 & 0 & 0 & 0 & -I & 0 & 0 \\ I & 0 & 0 & 0 & 0 & 0 & -I & 0 \\ I & 0 & 0 & 0 & 0 & 0 & 0 & I \\ 0 & 0 & S & 0 & 0 & Z & 0 & 0 \\ 0 & 0 & 0 & W & 0 & 0 & P & 0 \\ 0 & 0 & 0 & 0 & V & 0 & 0 & Q \end{bmatrix} \begin{bmatrix} \Delta x \\ \Delta y \\ \Delta z \\ \Delta p \\ \Delta q \\ \Delta s \\ \Delta w \\ \Delta v \end{bmatrix} = \begin{bmatrix} \chi \\ -c_E(x) \\ -c_I(x)+s \\ l+w-x \\ u-v-x \\ -SZe+\mu e \\ -WP+\mu e \\ -VQ+\mu e \end{bmatrix}. \tag{17.40}$$

This system is not symmetric, but it is easy to be written in a symmetric form by multiplying the second and the third equations by -1 and the last three equations by S^{-1}, W^{-1}, and V^{-1}, respectively. After completing this very simple algebraic operation and after eliminating the variables Δs, Δw, and Δv as

$$\Delta s = \mu Z^{-1}e - Se - Z^{-1}S\Delta z, \tag{17.41a}$$

$$\Delta w = \mu P^{-1}e - We - P^{-1}W\Delta p, \tag{17.41b}$$

$$\Delta v = \mu Q^{-1}e - Ve - Q^{-1}V\Delta q, \tag{17.41c}$$

we get the following *reduced Newton system*:

$$\begin{bmatrix} K(x,y,z) & -J_E(x)^{\mathrm{T}} & -J_I(x)^{\mathrm{T}} & -I & I \\ -J_E(x) & 0 & 0 & 0 & 0 \\ -J_I(x) & 0 & -Z^{-1}S & 0 & 0 \\ -I & 0 & 0 & -P^{-1}W & 0 \\ I & 0 & 0 & 0 & -Q^{-1}V \end{bmatrix} \begin{bmatrix} \Delta x \\ \Delta y \\ \Delta z \\ \Delta p \\ \Delta q \end{bmatrix} = \begin{bmatrix} \chi \\ c_E(x) \\ c_I(x) - \mu Z^{-1}e \\ x - l - \mu P^{-1}e \\ u - x - \mu Q^{-1}e \end{bmatrix}, \tag{17.42}$$

where:

$$K(x,y,z) = \nabla^2 f(x) - \sum_{i=1}^{me} y_i \nabla^2 c_{Ei}(x) - \sum_{i=1}^{m} z_i \nabla^2 c_{Ii}(x), \tag{17.43}$$

$$\chi = -\nabla f(x) + J_E(x)^{\mathrm{T}} y + J_I(x)^{\mathrm{T}} z + p - q. \tag{17.44}$$

As above, $J_E(x) = \nabla c_E(x)$ and $J_I(x) = \nabla c_I(x)$ are the Jacobian matrices of the equality constraints $c_E(x) = 0$ and of the inequality constraints $c_I(x) \geq 0$, respectively.

Observe that the matrix of the reduced Newton system (17.42) is symmetric of order $(3n + m + me)$. Solving this system and using (17.41), we get the search direction $\Delta t = [\Delta x, \Delta y, \Delta z, \Delta p, \Delta q, \Delta s, \Delta w, \Delta v]^{\mathrm{T}}$. The existence of a solution for (17.42) implies the existence of the inverse of the diagonal matrices Z, P, and Q, which assumes a certain condition of initialization of the algorithm.

The standard Newton method assumptions are as follows (Dennis & Schnabel, 1983; Andrei, 1998c, 2009):

(*i*) There exists a solution for problem (17.30) and the associated dual variables satisfy the KKT conditions (17.34).

(*ii*) The Hessian matrices, $\nabla^2 f(x)$, $\nabla^2 c_{Ei}(x), i = 1, \ldots, me$, $\nabla^2 c_{Ii}(x)$, $i = 1, \ldots, m$, exist and are locally Lipschitz continuous at x^*.

(*iii*) The set of vectors $\{\nabla c_{E1}(x^*), \ldots, \nabla c_{Eme}(x^*)\} \cup \{\nabla c_{Ii}(x^*), i \in A(x^*)\}$ is linearly independent, where $A(x^*)$ is the set of active inequality constraints at x^*.

(*iv*) For every vector $d \neq 0$ satisfying $\nabla c_{Ei}(x^*)^{\mathrm{T}} d = 0$, $i = 1, \ldots, me$, $\nabla c_{Ii}(x^*)^{\mathrm{T}} d = 0$, and $i \in A(x^*)$, we have $d^{\mathrm{T}} K(x^*, y^*, z^*) d > 0$.

(*v*) For $i=1,\ldots,m$, $z_i c_{Ii}(x^*)>0$ and for $j=1,\ldots,n$, $p_j^*\left(x_j^*-l_j\right)>0$ and $q_j^*\left(u_j-x_j^*\right)>0$.

Proposition 17.1 *Suppose that the conditions (i)–(v) hold and* $s^*=c_I(x^*)$, $w^*=x^*-l$, *and* $v^*=u-x^*$. *Then the Jacobian matrix* $F'(t^*)$ *of function* $F(t)$ *given by (17.40) is nonsingular.*

Proof. Consider the reduced problem in which only the inequality constraints active in the minimizer point are used. Then, from the theory of equality constraints optimization, it follows that the matriceal block given by the first five rows and columns of the matrix (17.40) is nonsingular. Hence, the nonsingularity of the matrix from (17.40) follows from the strict complementarity condition (*v*) and from the nonsingularity of the matriceal block from (17.40). ♦

Generally, in current implementations, the reduced Newton system (17.42) is not used in the form in which it appears. Usually, the reduction is continued by the elimination from (17.42) of the variables Δz, Δp, and Δq as:

$$\Delta z=\mu S^{-1}e-S^{-1}Zc_I(x)-S^{-1}ZJ_I(x)\Delta x, \tag{17.45a}$$

$$\Delta p=\mu W^{-1}e-W^{-1}P(x-l)-W^{-1}P\Delta x, \tag{17.45b}$$

$$\Delta q=\mu V^{-1}e-V^{-1}Q(u-x)+V^{-1}Q\Delta x, \tag{17.45c}$$

Hence, the system (17.42) is further reduced as

$$\begin{bmatrix} \bar{K} & -J_E(x)^{\mathrm{T}} \\ J_E(x) & 0 \end{bmatrix}\begin{bmatrix} \Delta x \\ \Delta y \end{bmatrix}=\begin{bmatrix} r+\mu\bar{r} \\ c_E(x) \end{bmatrix}, \tag{17.46}$$

where:

$$\bar{K}=K(x,y,z)+J_I(x)^{\mathrm{T}}S^{-1}ZJ_I(x)+W^{-1}P+V^{-1}Q, \tag{17.47a}$$

$$r=\chi-J_I(x)^{\mathrm{T}}S^{-1}Zc_I(x)-W^{-1}P(x-l)+V^{-1}Q(u-x), \tag{17.47b}$$

$$\bar{r}=J_I(x)^{\mathrm{T}}S^{-1}e+W^{-1}e-V^{-1}e. \tag{17.47c}$$

Now, by solving (17.46), we get the variables Δx and Δy with which the rest of variables are immediately obtained from (17.45) and (17.41). Notice that, as in linear programming, we must solve a symmetric undefined algebraic system, but in this case, the Jacobian and the Hessian matrices of the functions defining the problem are involved, which is much more complicated.

The merit function By solving the system (17.46) and by using (17.45) and (17.41), we get a direction Δt pointing to the central path. Now we must determine the stepsize along this direction. This is done by a merit function associated to the optimality conditions (17.34). The idea of the merit function is to enable a progress toward a local minimizer of the problem by conserving the feasibility of the solution. In our algorithm, the merit function used for the line search is the squared l_2 norm of the KKT conditions (17.34), that is:

$$\Phi(t) = \|F(t)\|_2^2 = F(t)^T F(t). \tag{17.48}$$

Denote $\Phi_k = \Phi(t_k)$ as the value of the merit function at the iterate t_k and $\Phi_k(\alpha) = \Phi(t_k + \alpha\Delta t_k)$, which illustrates the dependences of the merit function on the stepsize α. Clearly, $\Phi_k(0) = \Phi(t_k) = \Phi_k$. Therefore:

$$\Phi_k(\alpha) = F(t_k + \alpha\Delta t_k)^{\mathrm{T}} F(t_k + \alpha\Delta t_k). \tag{17.49}$$

The following proposition shows that the search direction given by the perturbed Newton system (17.38):

$$\Delta t_k = [F'(t_k)]^{-1}[-F(t_k) + \mu\hat{e}] \tag{17.50}$$

is a descent one for the merit function $\Phi(t)$.

Proposition 17.2 *The direction Δt_k given by (17.50), solution of the perturbed Newton system (17.38), is a descent direction for the merit function (17.48).*

Proof. Consider the derivative of $\Phi_k(\alpha)$ at $\alpha = 0$, we get

$$\begin{aligned}\Phi'(0) &= 2F(t_k)^{\mathrm{T}} F'(t_k)\Delta t_k = 2F(t_k)^{\mathrm{T}} F'(t_k)[F'(t_k)]^{-1}[-F(t_k) + \mu\hat{e}] \\ &= 2F(t_k)^{\mathrm{T}}[-F(t_k) + \mu\hat{e}] = -2\|F(t_k)\|_2^2 + 2\mu F(t_k)^{\mathrm{T}}\hat{e},\end{aligned}$$

that is, $\Phi'(0) < 0$ if and only if $2\mu F(t_k)^{\mathrm{T}}\hat{e} \leq 2\|F(t_k)\|_2^2$. This determines the following estimation of the barrier parameter μ to be used in the logarithmic barrier function (17.32):

$$\mu \leq \frac{\|F(t_k)\|_2^2}{F(t_k)^{\mathrm{T}}\hat{e}} = \frac{\|F(t_k)\|_2^2}{s_k^{\mathrm{T}} z_k + w_k^{\mathrm{T}} p_k + v_k^{\mathrm{T}} q_k}. \tag{17.51}$$

Therefore, choosing at every iteration the value of the barrier parameter μ as in (17.51), the direction Δt_k given by (17.50) is indeed a descent direction for the merit function $\Phi(t_k)$. ♦

The stepsize determination As we have learned from linear programming (see Wright, 1997), for the nonlinear optimization problem (17.30), the distance from centrality is also given by:

$$\xi_k = \frac{\min_i\{s_i z_i, w_i p_i, v_i q_i\}}{\frac{s_k^{\mathrm{T}} z_k + w_k^{\mathrm{T}} p_k + v_k^{\mathrm{T}} q_k}{2n+m}}. \tag{17.52}$$

Clearly, $0 < \xi_k \leq 1$ and $\xi_k = 1$ if and only if, for all i, $s_i z_i$, $w_i p_i$, and $v_i q_i$ are equal to a constant. To specify a value for α, the following function is firstly introduced:

$$\Theta^I(\alpha) = \frac{\min_i\{s_i(\alpha)z_i(\alpha), w_i(\alpha)p_i(\alpha), v_i(\alpha)q_i(\alpha)\}}{\frac{s(\alpha)^T z(\alpha) + w(\alpha)^T p(\alpha) + v(\alpha)^T q(\alpha)}{2n+m}} - \gamma\tau_1, \tag{17.53}$$

where τ_1 is the initial distance from centrality:

$$\tau_1 = \frac{\min_i\{s_i^0 z_i^0, w_i^0 p_i^0, v_i^0 q_i^0\}}{\frac{s^{0T}z^0 + w^{0T}p^0 + v^{0T}q^0}{2n+m}}, \tag{17.54}$$

and $\gamma \in (0, 1)$ is a constant by which we can modify the distance from centrality. Observe that for $t = t_0$ and $\gamma = 1$, it follows that $\Theta^I(0) = 0$. Besides, $\Theta^I(\alpha)$ is a piecewise quadratic function. To choose at each iteration the stepsize α_k, it is necessary that α_k satisfies $\Theta^I(\alpha) \geq 0$ for all $\alpha \in [0, \alpha_k]$, i.e.:

$$\frac{\min_i\{s_i(\alpha)z_i(\alpha), w_i(\alpha)p_i(\alpha), v_i(\alpha)q_i(\alpha)\}}{\frac{s(\alpha)^T z(\alpha) + w(\alpha)^T p(\alpha) + v(\alpha)^T q(\alpha)}{2n+m}} \geq \gamma_k\tau_1, \tag{17.55}$$

where the variables of the problem are considered at the iteration k and $\gamma_k \in (0, 1)$. Since $\Theta^I(\alpha)$ is a piecewise quadratic function, from (17.55), it follows that α_k can be easily computed.

Considering the merit function (17.48), the Wolfe conditions for its minimization are

$$\Phi(t_{k+1}) \leq \Phi(t_k) + \beta\alpha_k \nabla\Phi(t_k)^T \Delta t_k, \tag{17.56}$$

$$\nabla\Phi(t_{k+1})^T \Delta t_k \geq \delta\nabla\Phi(t_k)^T \Delta t_k, \tag{17.57}$$

where $0 < \beta < \delta < 1$ are parameters responsible for the reduction of the merit function, as well as the rate of decreasing this function along the direction Δt_k. Having in view that $\Phi(t_{k+1}) = \Phi_k(\alpha)$, the first Wolfe condition (17.56) is equivalent to:

$$\Phi_k(\alpha) \leq \Phi_k(0) + \beta\alpha_k\Phi_k'(0). \tag{17.58}$$

Proposition 17.3 *For the merit function (17.48), taking*

$$\mu_k = \sigma_k \frac{s_k^T z_k + w_k^T p_k + v_k^T q_k}{2n+m} \tag{17.59}$$

where $\sigma_k \in (0, 1)$, *we have:*

$$\Phi_k'(0) = -2\left[\Phi_k(0) - \frac{\sigma_k}{2n+m}\left(s_k^T z_k + w_k^T p_k + v_k^T q_k\right)^2\right]. \tag{17.60}$$

Proof. By direct computation we get:

$$\begin{aligned}\Phi_k'(0) &= -2\Phi_k(0) + 2\mu F(t_k)^{\mathrm{T}}\widehat{e} = -2\Phi_k(0) + 2\mu\left[s_k^{\mathrm{T}}z_k + w_k^{\mathrm{T}}p_k + v_k^{\mathrm{T}}q_k\right] \\ &= -2\Phi_k(0) + 2\sigma_k\frac{s_k^{\mathrm{T}}z_k + w_k^{\mathrm{T}}p_k + v_k^{\mathrm{T}}q_k}{2n+m}\left[s_k^{\mathrm{T}}z_k + w_k^{\mathrm{T}}p_k + v_k^{\mathrm{T}}q_k\right] \\ &= -2\left[\Phi_k(0) - \frac{\sigma_k}{2n+m}\left(s_k^{\mathrm{T}}z_k + w_k^{\mathrm{T}}p_k + v_k^{\mathrm{T}}q_k\right)^2\right].\end{aligned}$$

♦

Proposition 17.4 *At every iteration:*

$$\frac{\left(s_k^{\mathrm{T}}z_k + w_k^{\mathrm{T}}p_k + v_k^{\mathrm{T}}q_k\right)^2}{2n+m} \le \Phi_k(0). \tag{17.61}$$

Proof. By simple algebraic manipulation we have:

$$\begin{aligned}&\frac{\left(s_k^{\mathrm{T}}z_k + w_k^{\mathrm{T}}p_k + v_k^{\mathrm{T}}q_k\right)^2}{2n+m} \le \|SZe\|_2^2 + \|WPe\|_2^2 + \|VQe\|_2^2 \\ &\le \|SZe\|_2^2 + \|WPe\|_2^2 + \|VQe\|_2^2 + \|CO(t_k)\|_2^2 = \Phi_k(0).\end{aligned}$$

♦

It is easy to show that the estimation of the barrier parameter μ_k given by (17.59) with $\sigma_k \in (0,1)$ is smaller than the estimation recommended by (17.51). Therefore, μ_k given by (17.59) ensures the descent character of the search direction Δt_k. Moreover, the following proposition gives an estimation of the reduction of the values of the merit function.

Proposition 17.5 *The direction Δt_k, solution of the perturbed Newton system (17.38) with μ given by (17.59), is a descent direction for the merit function $\Phi(t)$ at every t_k. Moreover, if the first Wolfe condition (17.56) is satisfied, then*

$$\Phi_k(\alpha_k) \le [1 - 2\alpha_k\beta(1-\sigma_k)]\Phi_k(0). \tag{17.62}$$

Proof. As we know:

$$\Phi_k'(0) = -2\|F(t_k)\|_2^2 + 2\mu_k\left(s_k^{\mathrm{T}}z_k + w_k^{\mathrm{T}}p_k + v_k^{\mathrm{T}}q_k\right).$$

Taking μ_k as in (17.59), it follows that

$$\begin{aligned}\Phi_k'(0) &= -2\Phi_k(0) + 2\sigma_k\frac{\left(s_k^{\mathrm{T}}z_k + w_k^{\mathrm{T}}p_k + v_k^{\mathrm{T}}q_k\right)^2}{2n+m} \\ &\le -2\Phi_k(0) + 2\sigma_k\Phi_k(0) = -2\Phi_k(0)(1-\sigma_k) < 0,\end{aligned}$$

proving the descent character of the direction given by the perturbed Newton system. Moreover, taking into consideration the above propositions (17.3 and 17.4), from (17.58), we have

$$\begin{aligned}\Phi_k(\alpha) &\le \Phi_k(0) + \beta\alpha_k\Phi_k'(0)\\ &= \Phi_k(0) + \beta\alpha_k\left[-2\left(\Phi_k(0) - \sigma_k\frac{\left(s_k^T z_k + w_k^T p_k + v_k^T q_k\right)^2}{2n+m}\right)\right]\\ &\le \Phi_k(0) - 2\beta\alpha_k\Phi_k(0) + 2\beta\alpha_k\sigma_k\Phi_k(0)\\ &= [1 - 2\alpha_k\beta(1-\sigma_k)]\Phi_k(0),\end{aligned}$$

which proves the proposition. ♦

This proposition shows that the sequence $\{\Phi_k\}$ is monotonous and nonincreasing; therefore for all k: $\Phi_k \le \Phi_0$. Moreover, if the sequence of the stepsize $\{\alpha_k\}$ is bounded away from zero and the parameter σ_k is bounded away from one at every iteration, then the merit function is linearly convergent to zero. Hence, the above inequality (17.62) is equivalent to:

$$\frac{\|F(t_{k+1})\|_2}{\|F(t_k)\|_2} \le [1 - 2\alpha_k\beta(1-\sigma_k)]^{1/2}. \tag{17.63}$$

Some numerical examples illustrate that a problem that may lead to the non-convergence of the algorithm is that the sequence $\{\|CT(s_k, z_k, w_k, p_k, v_k, q_k)\|\}$ converges to zero faster than the sequence $\{\Phi(t_k)\}$. In such a case, the sequence of the stepsizes $\{\alpha_k\}$ is decreasing to zero, thus determining the non-convergence of the algorithm. To avoid this situation, let us introduce the following function:

$$\Theta^{II}(\alpha) = s(\alpha)^T z(\alpha) + w(\alpha)^T p(\alpha) + v(\alpha)^T q(\alpha) - \gamma\tau_2\|CO(t(\alpha))\|_2, \tag{17.64}$$

where

$$\tau_2 = \frac{s_0^T z_0 + w_0^T p_0 + v_0^T q_0}{\|CO(t_0)\|_2}, \tag{17.65}$$

and $\gamma \in (0, 1)$ is a constant, the same as in (17.53). Observe that for $t = t_0$ and $\gamma = 1$, $\Theta^{II}(0) = 0$. In general, $\Theta^{II}(\alpha)$ is a nonlinear function. For choosing the stepsize α_k at every iteration, it is necessary that α_k satisfies:

$$\Theta^{II}(\alpha_k) \ge 0. \tag{17.66}$$

Proposition 17.6 *Let $\{t_k\}$ be a sequence generated as solution of (17.38). Then:*

$$\min\{1, 0.5\tau_2\}\,\Phi(t_k) \le \left(s_k^{\mathrm{T}} z_k + w_k^{\mathrm{T}} p_k + v_k^{\mathrm{T}} q_k\right)^2 \le (2n+m)\Phi(t_k). \qquad (17.67)$$

Proof. The second inequality follows from Proposition 17.4. Hence, we prove only the first one. Since $\Theta^i(\alpha_k) \ge 0$,for $i = 1, 2$, from (17.64) with $\gamma_k \ge 1/2$, we have:

$$\left(s_k^{\mathrm{T}} z_k + w_k^{\mathrm{T}} p_k + v_k^{\mathrm{T}} q_k\right) \ge (1/2)\tau_2 \|CO(t_k)\|_2.$$

Therefore:

$$\begin{aligned}
&\left(s_k^{\mathrm{T}} z_k + w_k^{\mathrm{T}} p_k + v_k^{\mathrm{T}} q_k\right) \\
&\ge \frac{1}{2}\left[\|SZe\|_2 + \|WPe\|_2 + \|VQe\|_2 + 0.5\tau_2\|CO(t_k)\|_2\right] \\
&\ge \frac{1}{2}\min\{1, 0.5\tau_2\}\|F(t_k)\|_2,
\end{aligned}$$

which completes the proof. ♦

This proposition shows that the complementarity conditions are bounded, and γ_k must be selected as a decreasing sequence with $1/2 \le \gamma_k \le \gamma_{k-1}$.

Having in view all these developments, at each iteration the stepsize α_k is computed as a solution of the following system of algebraic inequalities:

$$l \le x_k + \alpha_k \Delta x_k \le u, \qquad (17.68a)$$
$$s(\alpha_k), z(\alpha_k), w(\alpha_k), p(\alpha_k),\ v(\alpha_k), q(\alpha_k) > 0, \qquad (17.68b)$$
$$\Theta^I(\alpha) \ge 0, \alpha \in [0, \alpha_k], \qquad (17.68c)$$
$$\Theta^{II}(\alpha_k) \ge 0, \qquad (17.68d)$$
$$\Phi_k(\alpha_k) \le \Phi_k(0) + \alpha_k \beta \Phi_k'(0), \qquad (17.68e)$$
$$\nabla\Phi_k(\alpha_k)^{\mathrm{T}} \Delta t_k \ge \delta \nabla\Phi_k(0)^{\mathrm{T}} \Delta t_k, \qquad (17.68f)$$

where $0 < \beta < \delta < 1$.

To determine α_k satisfying (17.68), a strategy of *reducing the interval* can be used. The first two relations (17.68a) and (17.68b) are easy to implement. As in linear programming, the corresponding ratio test is performed, thus obtaining a value α_m which maintains the positivity of the variables as well as the simple bounds on variables. Then, a value $\alpha_k \in (0, \alpha_m]$ which satisfies the conditions (17.68c)-(17.68f) is selected. However, selection of α_k from the interval $(0, \alpha_m]$ is not simple.

Primal-dual interior point algorithm The interior point algorithm for solving (17.30) based on the above developments has three main parts which refer to the computations of the search direction, the barrier parameter, and the stepsize.

Algorithm 17.4 Primal-dual interior point algorithm (PDIP)

1	*Initialization.* Choose an initial point $t_0 = [x_0, y_0, z_0, p_0, q_0, s_0, w_0, v_0]$, such that $l \le x_0 \le u$, $(s_0, z_0) > 0$, $(w_0, p_0) > 0$, and $(v_0, q_0) > 0$. Choose the parameters $\varepsilon > 0$, $\beta \in (0, 1/2]$, $\gamma_{k-1} = 1$, and $\rho \in (0, 1)$. Set $k = 0$
2.	*Convergence test.* Compute the value of the merit function $\Phi(t_k) = F(t_k)^T F(t_k)$. If $\Phi(t_k) \le \varepsilon$, stop; otherwise go to step 3
3.	*Evaluation of the barrier parameter.* Choose $\sigma_k \in (0, 1)$. Using (17.59), compute the barrier parameter μ_k
4.	*Computing the search direction.* Determine Δt_k by solving the reduced Newton system (17.46) and by using (17.45) and (17.41)
5.	*Stepsize determination.* Use the Armijo technique: (a) Choose $1/2 \le \gamma_k \le \gamma_{k-1}$ (b) Compute α_m as the longest value of α_k which satisfies (17.68a) and (17.68b) (c) Determine $\alpha_k^I \in (0, \alpha_m]$ as the smallest positive root, such that $\Theta^I(\alpha) \ge 0$ for all $\alpha \in (0, \alpha_k^I]$ (d) Determine $\alpha_k^{II} \in (0, \alpha_k^I]$, such that $\Theta^{II}(\alpha_k^{II}) \ge 0$ (e) Set $\overline{\alpha}_k = \min\{\alpha_k^I, \alpha_k^{II}\}$ (f) Set $\alpha_k = \rho^j \overline{\alpha}_k$, such that α_k satisfies (17.68e), where $j = 0, 1, \ldots$ is the smallest integer with this property
6.	*Updating the variables.* Set $t_{k+1} = t_k + \alpha_k \Delta t_k$ and $k = k + 1$ and go to step 2 ♦

Some comments are as follows (details can be found in Andrei (1998c)).

1. (*Search direction*) In Proposition 17.2, we showed that the step direction Δt given by (17.50) is descent for the merit function, provided that the value of the barrier parameter is selected as in (17.51).

Proposition 17.7 *Suppose that f, c_E, and c_I are twice continuously differentiable, the derivative of $CO(t)$ defined by (17.35) is Lipschitz continuous, and the set of gradients $\{\nabla c_{E1}(x_k), \ldots, \nabla c_{Eme}(x_k)\} \cup \{\nabla c_{Ii}(x_k), i \in A(x^*)\}$ is linearly independent for k that is sufficiently large, where $A(x^*)$ is the set of active inequality constraints at x^*. Then the sequence $\{\Delta t_k\}$ generated by the algorithm PDIP is bounded.*

Proof. By permuting the rows and columns, $F'(t_k)$ can be rearranged as

$$F'(t_k) = \begin{bmatrix} A & B \\ B^{\mathrm{T}} & C \end{bmatrix},$$

where

$$A = \begin{bmatrix} K(x_k, y_k, z_k) & -J_E(x_k)^{\mathrm{T}} & -J_I(x_k)^{\mathrm{T}} \\ -J_E(x_k) & 0 & 0 \\ -J_I(x_k) & 0 & 0 \end{bmatrix},$$

the matrices B and C being very easily identified from $F'(t)$. By assumptions of the proposition, the matrix A is invertible and $\|A^{-1}\|$ is uniformly bounded. But

$$\begin{bmatrix} A & B \\ B^{\mathrm{T}} & C \end{bmatrix}^{-1}$$
$$= \begin{bmatrix} A^{-1} + A^{-1}B\left(C - B^{\mathrm{T}}A^{-1}B\right)^{-1}B^{\mathrm{T}}A^{-1} & -A^{-1}B\left(C - B^{\mathrm{T}}A^{-1}B\right)^{-1} \\ -\left(C - B^{\mathrm{T}}A^{-1}B\right)^{-1}B^{\mathrm{T}}A^{-1} & \left(C - B^{\mathrm{T}}A^{-1}B\right)^{-1} \end{bmatrix},$$

which is bounded, since every matrix involved is bounded. Hence, $[F'(t_k)]^{-1}$ is uniformly bounded, proving the proposition. ♦

2. (*Stepsize*) A crucial point of the algorithm is the stepsize computation. The following proposition shows that the sequence $\{\overline{\alpha}_k\}$ generated in step 5e of the algorithm PDIP is bounded away from zero.

Proposition 17.8 *Suppose that the functions $f(x)$, $c_E(x)$, and $c_I(x)$ are twice continuously differentiable and the derivative of $CO(t)$ is Lipschitz continuous with constant L. If the sequence $\{\sigma_k\}$ is bounded away from zero, then the sequence $\{\overline{\alpha}_k\}$ generated by the algorithm is also bounded away from zero.*

Proof. Since $\overline{\alpha}_k = \min\{\alpha_k^I, \alpha_k^{II}\}$, it suffices to show that the sequences $\{\alpha_k^I\}$ and $\{\alpha_k^{II}\}$ are bounded away from zero. Let us suppress the subscript k. As in El-Bakry, Tapia, Tsuchiya, and Zhang (1996), let us define the vectors $a(\alpha) = [s(\alpha), w(\alpha), v(\alpha)]^{\mathrm{T}} \in \mathbb{R}^{m+2n}$ and $b(\alpha) = [z(\alpha), p(\alpha), q(\alpha)]^{\mathrm{T}} \in \mathbb{R}^{m+2n}$. Then, the function $\Theta^I(\alpha)$ from (17.53) can be written as:

$$\Theta^I(\alpha) = \frac{\min_i\{a_i(\alpha)b_i(\alpha)\}}{\frac{a(\alpha)^{\mathrm{T}}b(\alpha)}{2n+m}} - \gamma\tau_1.$$

From the definition of α^I (step 5c of the algorithm PDIP), we see that α is the largest number in $[0, \alpha_m]$ such that

$$a_i(\alpha)b_i(\alpha) - \gamma\tau_1 a(\alpha)^{\mathrm{T}}b(\alpha)/(2n+m) \geq 0,$$

for every $i = 1, \ldots, 2n+m$ and $\alpha \in [0, \alpha^I]$. Now, define:

$$\eta_i = \left| \Delta a_i \Delta b_i - \gamma \tau_1 \frac{\Delta a^{\mathrm{T}} \Delta b}{2n+m} \right|.$$

From Proposition 17.7, it follows that Δt is bounded, and then there is a positive constant M such that $\eta_i \leq M$. Straightforward computation shows that for $\alpha \in [0, \alpha^I]$, we have:

$$a_i(\alpha) b_i(\alpha) - \gamma \tau_1 \frac{a(\alpha)^{\mathrm{T}} b(\alpha)}{2n+m}$$

$$= \left[a_i b_i - \gamma \tau_1 \frac{a^{\mathrm{T}} b}{2n+m} \right] + \alpha \left[a_i \Delta b_i + b_i \Delta a_i - \gamma \tau_1 \frac{a^{\mathrm{T}} \Delta b + b^{\mathrm{T}} \Delta a}{2n+m} \right]$$

$$+\alpha^2 \left[\Delta a_i \Delta b_i - \gamma \tau_1 \tfrac{\Delta a^{\mathrm{T}} \Delta b}{2n+m} \right] \geq \alpha(1 - \gamma \tau_1)\mu - \alpha^2 \left| \Delta a_i \Delta b_i - \gamma \tau_1 \tfrac{\Delta a^{\mathrm{T}} \Delta b}{2n+m} \right|$$

$$= \alpha(1 - \gamma \tau_1)\mu - \eta_i \alpha^2 \geq \alpha(1 - \gamma \tau_1)\mu - M\alpha^2 \geq 0.$$

Hence:

$$\alpha^I \geq (1 - \gamma \tau_1)\mu / M.$$

But as we know, $\mu = \sigma \frac{a^{\mathrm{T}} b}{2n+m}$, and for σ bounded away from zero, it follows that μ is bounded below. Hence, α^I is bounded away from zero.

Now let us show that the sequence $\{\alpha^{II}\}$ generated by step 5 of the algorithm PDIP is bounded away from zero. According to the mean-value theorem for vector-valued functions (Dennis & Schnabel, 1983, Chapter 4), we have

$$CO(t + \alpha \Delta t) = CO(t) + \alpha \left[\int_0^1 \nabla CO(t + \xi \alpha \Delta t) d\xi \right] \Delta t$$

$$= CO(t) + \alpha \nabla CO(t) \Delta t + \alpha \left[\int_0^1 (\nabla CO(t + \xi \alpha \Delta t) - \nabla CO(t)) d\xi \right] \Delta t$$

$$= (1 - \alpha) CO(t) + \alpha \left[\int_0^1 (\nabla CO(t + \xi \alpha \Delta t) - \nabla CO(t)) d\xi \right] \Delta t,$$

where the last equality is from (17.38). Now, having in view that the derivative of $CO(t)$ is Lipschitz continuous with constant L, we obtain:

$$\| CO(t + \alpha \Delta t) \| \leq \| CO(t) \| |1 - \alpha| + L \alpha^2 \| \Delta t \|^2.$$

Using this inequality, we have:

$$\begin{aligned}\Theta^{II}(\alpha) &= a(\alpha)^{\mathrm{T}} b(\alpha) - \gamma\tau_2 \|CO(t+\alpha\Delta t)\| \\ &= a^{\mathrm{T}} b + \alpha\left[a^{\mathrm{T}}\Delta b + b^{\mathrm{T}}\Delta a\right] + \alpha^2 \Delta a^{\mathrm{T}}\Delta b - \gamma\tau_2\|CO(t+\alpha\Delta t)\| \\ &\geq |1-\alpha| a^{\mathrm{T}} b + \alpha\sigma a^{\mathrm{T}} b + \alpha^2\Delta a^{\mathrm{T}}\Delta b - \gamma\tau_2\left(\|CO(t)\||1-\alpha| + L\alpha^2\|\Delta t\|^2\right) \\ &= |1-\alpha|\left(a^{\mathrm{T}} b - \gamma\tau_2\|CO(t)\|\right) + \alpha\sigma a^{\mathrm{T}} b + \alpha^2\left(\Delta a^{\mathrm{T}}\Delta b - \gamma\tau_2 L\|\Delta t\|^2\right) \\ &\geq \alpha\left[\sigma a^{\mathrm{T}} b - \alpha\left|\Delta a^{\mathrm{T}}\Delta b - \gamma\tau_2 L\|\Delta t\|^2\right|\right].\end{aligned}$$

From Proposition 17.7, there is a constant $N > 0$ such that:

$$\left|\Delta a^{\mathrm{T}}\Delta b - \gamma\tau_2 L\|\Delta t\|^2\right| \leq N.$$

Hence $\Theta^{II}(\alpha) \geq \alpha[\sigma a^{\mathrm{T}} b - \alpha N]$. From condition (17.66), it follows that $\alpha^{II} \geq \sigma a^{\mathrm{T}} b/N$. Since the sequence $\{\sigma_k\}$ is bounded away from zero, then the sequence $\{\alpha_k^{II}\}$ is also bounded away from zero. ♦

3. (*Convergence*) The convergence of the algorithm PDIP is proved as follows.

Theorem 17.3 *Suppose that the functions $f(x)$, $c_E(x)$, and $c_I(x)$ are twice continuously differentiable and the derivative of $CO(t)$ is Lipschitz continuous. Let $\{t_k\}$ be the sequence generated by the algorithm PDIP, where $\{\sigma_k\} \subset (0,1)$ is bounded away from zero and one. Then the sequence $\{F(t_k)\}$ converges to zero, and for any limit point $t^* = [x^*, y^*, z^*, p^*, q^*, s^*, w^*, v^*]^{\mathrm{T}}$, x^* is a KKT point of the problem (17.30).*

Proof. Firstly, note that the sequence $\{\|F(t_k)\|\}$ is monotonously decreasing; hence it is convergent. By contradiction, suppose that the sequence $\{\|F(t_k)\|\}$ is not convergent to zero. Then, from Proposition 17.5, we have:

$$\Phi_k(\alpha_k)/\Phi_k(0) \leq 1 - 2\alpha_k\beta(1-\sigma_k).$$

Therefore, from Proposition 17.8, it follows that the sequence $\{\Phi_k\}$ converges linearly to zero. This gives a contradiction. On the other hand, from Proposition 17.2, we have $\nabla\Phi(t_k)\Delta t_k = -2F(t_k)^{\mathrm{T}} F(t_k) + 2\mu_k F(t_k)^{\mathrm{T}}\widehat{e}$. Since the sequence $\{\alpha_k\}$ is bounded away from zero, it follows that the backtracking line search used in step 5 of the algorithm PDIP produces:

$$\frac{\nabla\Phi(t_k)\Delta t_k}{\|\Delta t_k\|} = \frac{-2\left[F(t_k)^{\mathrm{T}} F(t_k) - \mu_k F(t_k)^{\mathrm{T}}\widehat{e}\right]}{\|\Delta t_k\|} \to 0.$$

Therefore, from Proposition 17.7, Δt_k is bounded away from zero, i.e.:

$$\Phi(t_k) - \mu_k\left[s_k^T z_k + w_k^T p_k + v_k^T q_k\right] \to 0.$$

However:

$$\Phi(t_k) - \mu_k\left[s_k^T z_k + w_k^T p_k + v_k^T q_k\right] \geq (1 - \sigma_k)\Phi(t_k).$$

Therefore, it must hold that $\Phi(t_k) \to 0$, because the sequence $\{\sigma_k\}$ is bounded away from one. Again this leads to a contradiction. Hence, the sequence $\{\|F(t_k)\|\}$ must be convergent to zero. Since the KKT conditions for the problem (17.30) are satisfied by t^*, it follows that x^* is a KKT point for (17.30). ♦

Example 17.1 Let us consider the following nonlinear optimization problem (Hock & Schittkowski, 1981; Andrei, 1998c):

$$\min x_1 x_4(x_1 + x_2 + x_3) + x_3$$

subject to:

$$x_1 x_2 x_3 x_4 - 25 \geq 0,$$

$$x_1^2 + x_2^2 + x_3^2 + x_4^2 - 40 = 0,$$

$$1 \leq x_i \leq 5, \quad i = 1, 2, 3, 4.$$

The evolution of some elements corresponding to a simple variant of the algorithm PDIP is presented in Table 17.1.

In this variant of the PDIP algorithm, the condition (17.68d) was not implemented. Although this condition was not implemented, however, in the last part of the optimization process, we have $\|CO(t_k)\|_2 < \|CT(.)\|_2$. This behavior ensures the convergence of the algorithm. Table 17.2 shows the evolution of the algorithm parameters.

The parameter γ_k is updated as $\gamma_{k+1} = 0.5 + (\gamma_k - 0.5)/2$. Some other updating formula may be imagined, so that $\gamma_k \in [0.5, \gamma_{k-1}]$. The parameter σ_k is computed as

Table 17.1 Evolution of some elements of PDIP

k	$f(x_k)$	$\Phi(t_k)$	$\|CO(t_k)\|_2$	$\|CT(.)\|_2$
0	16.0000	534.076	465.076	69.0000
1	16.9494	11.2666	8.97539	2.291261
2	17.1584	0.253076	8.97539	0.135262
3	16.9951	0.020296	1.41485	0.61477e-2
4	17.0168	0.45552e-3	0.18875e-3	0.26677e-3
5	17.0158	0.11076e-4	0.52654e-6	0.10549e-4
6	17.0144	0.42452e-6	0.56172e-9	0.42387e-6
7	17.0141	0.16245e-7	0.65172e-9	0.16244e-7
8	17.0140	0.24373-10	0.1999e-14	0.2437e-10

Table 17.2 Evolution of parameters of PDIP

k	γ_k	σ_k	$\alpha_{\max}$	α_k	μ_k
0	0.750000	0.1	1.090858	1.039720	0.233333
1	0.625000	0.2	0.992283	0.926926	0.90664e-1
2	0.562500	0.2	1.170237	1.0	0.22643e-1
3	0.531250	0.2	1.210422	1.0	0.48802e-2
4	0.515625	0.2	1.165723	1.0	0.10580e-2
5	0.507812	0.2	1.220644	1.0	0.21612e-3
6	0.503906	0.19530	1.238018	1.0	0.42383e-4
7	0.501953	0.03823	1.039720	1.0	0.16244e-5

Table 17.3 Evolution of some elements of PDIP

k	$f(x_k)$	$\Phi(t_k)$	$\|CO(t_k)\|_2$	$\|CT(.)\|_2$
0	130.8000	7422.758	7009.758	413.0000
1	38.03691	26.02877	17.60484	8.423930
2	30.49207	0.278046	0.40382e-3	0.2776424
3	29.96742	0.85973e-2	0.37688e-5	0.85935e-2
4	29.91532	0.51103e-3	0.86534e-4	0.42450e-3
5	29.89686	0.26626e-4	0.90809e-5	0.17545e-4
6	29.89512	0.59983e-6	0.49498e-8	0.59488e-6
7	29.89452	0.26198e-7	0.23005e-8	0.23898e-7
8	29.89439	0.18476e-9	0.12652e-9	0.58238e-10
9	29.89438	0.34199e-12	0.34150e-12	0.49357e-15

$$\sigma_k = \begin{cases} \eta_1, & \text{if} \quad \sigma_k \leq \eta_2\left(s_k^T z_k + w_k^T p_k + v_k^T q_k\right), \\ \eta_2\left(s_k^T z_k + w_k^T p_k + v_k^T q_k\right), & \text{if} \quad \sigma_k > \eta_2\left(s_k^T z_k + w_k^T p_k + v_k^T q_k\right), \end{cases}$$

where $\eta_1 = 0.1$ and $\eta_2 = 100$. Observe that the stepsize $\alpha_k \to 1$, exactly as in the "pure" Newton method.

Solution of the problem is $x^* = [1, 5, 5, 1]$. The objective value is 17.0140173. The SPENBAR (Andrei, 1996a, 1996b, 1996c, 1996d, 1996e) gives the same solution involving 8 major iterations, 143 minor iterations, and 591 evaluations of the functions defining the problem.

Example 17.2 Consider the problem (Andrei, 2015, pp.777):

$$\min 24.55x_1 + 26.75x_2 + 39x_3 + 40.5x_4$$

subject to:

$$2.3x_1 + 5.6x_2 + 11.1x_3 + 1.3x_4 - 5 \geq 0,$$

$$12x_1 + 11.9x_2 + 41.8x_3 + 52.1x_4 - 21$$

$$-1.645\left(0.28x_1^2 + 0.19x_2^2 + 20.5x_3^2 + 0.62x_4^2\right)^{1/2} \geq 0,$$

$$x_1 + x_2 + x_3 + x_4 - 1 = 0,$$

$$0 \leq x_i \leq 10, \quad i = 1, 2, 3, 4.$$

Table 17.4 Evolution of parameters of PDIP

k	γ_k	σ_k	$\alpha_{\max}$	α_k	μ_k
0	0.750000	0.1	0.9581337	0.9498877	0.51
1	0.625000	0.2	1.098776	1.0	0.1158525
2	0.562500	0.2	1.032449	1.0	0.0282029
3	0.531250	0.2	1.058755	1.0	0.56835e-2
4	0.515625	0.2	1.006443	0.9938785	0.11798e-2
5	0.507812	0.2	1.106218	1.0	0.24329e-3
6	0.503963	0.2	1.244491	1.0	0.48777e-4
7	0.501953	0.048885	1.051240	1.0	0.23898e-5
8	0.500976	0.0024132	1.002405	1.0	0.58237e-8

The primal-dual interior point algorithm PDIP gives the results from Tables 17.3 and 17.4. Again observe that the algorithm has the same behavior along the iterations. Even if the condition $\Theta^{II}(\alpha_k) \geq 0$ was not implemented in PDIP, we see that after very few iterations, $\|CO(t_k)\|_2 < \|CT(.)\|_2$. This ensures the convergence of the algorithm. Again note that the stepsize converges to 1, like in the pure Newton method.

Solution to this problem is $x^* = [0.6355, 0.286E-7, 0.3127, 0.05177]$. MINOS gives the same solution in 5 major iterations, 18 minor iterations, and 23 evaluations of the function defining the problem. On the other hand, SPENBAR needs 10 major iterations, 455 minor iterations, and 2083 evaluations of the functions of the problem, giving the same solution. NLPQLP gives the same solution, and it needs 14 iterations and 17 evaluations of the functions.

Application ELCH In Chapter 4 we presented this application. The algorithm PDIP gives a solution with the value of the objective function equal to −47.7610908 in 10 iterations and 12 evaluations of the functions defining the problem. The norm of the KKT conditions is equal to 0.315467E-8.

Application ALKI For this application, also described in Chapter 4, the algorithm PDIP needs 18 iterations and 20 evaluations of the functions defining the problem, giving a solution for which the objective function value is equal to −1763.524394. The norm of the KKT conditions is 0.112423E-8.

17.5 Trust-Region Interior Point Algorithm

There are two main differences between the trust-region interior point and the line-search interior point algorithms. The first one is that the trust-region interior point is not a real primal-dual method. Unlike the line-search interior point method in which both the primal and the dual variables are simultaneously computed, the trust-region interior point algorithm firstly computes a step in variables (x, s), and after that the estimates of the multipliers (dual variables) are computed. The second difference is that the trust-region interior point algorithm uses a scaling of the variables that discourage movement toward the boundary of the feasible region. This makes the trust-region interior point algorithms have better convergence properties than those of line-search interior point.

In the following, like in Nocedal and Wright (2006), the trust-region algorithm for finding approximate solutions of a fixed barrier problem is presented. Then a trust-region interior point algorithm is described, where the barrier parameter is driven to zero.

An Algorithm for the Barrier Problem Let us consider the barrier problem (17.3). This is an equality-constrained optimization problem that can be solved by using a sequential quadratic programming method with trust region. However, a direct application of this method is not possible because it generates steps that tend to violate the positivity of the slack variables and therefore cut frequently short the trust-region constraint. To skip over this difficulty, Nocedal and Write (2006) suggest designing another sequential quadratic programming method.

At iterate (x, s), for a given value of the barrier parameter μ, the estimates of the Lagrange multipliers (dual variables) (y, z) are firstly computed, and then the step $p=(p_x, p_s)$ is determined, which approximately solves the following sub-problem:

$$\min_{p_x,p_s} \nabla f(x)^T p_x + \frac{1}{2} p_x^T \nabla_{xx}^2 L p_x - \mu e^T S^{-1} p_s + \frac{1}{2} p_s^T \Sigma p_s \tag{17.69a}$$

subject to:

$$J_E(x)p_x + c_E(x) = r_E, \tag{17.69b}$$

$$J_I(x)p_x - p_s + (c_I(x) - s) = r_I, \tag{17.69c}$$

$$\left\| \left(p_x, S^{-1}p_s\right) \right\|_2 \le \Delta, \tag{17.69d}$$

$$p_s \ge -\tau s. \tag{17.69e}$$

In this problem, $\Sigma = S^{-1}Z$ and the scalar parameter $\tau \in (0, 1)$ is chosen close to 1, (e.g., $\tau = 0.995$). The inequality (17.69e) has the same role as the fraction to the boundary rule given by (17.8). The constraints (17.69b) and (17.69c) are linear. Clearly, we would like to set $r = (r_E, r_I) = 0$, but this can cause the constraints (17.69b)–(17.69d) to be incompatible or to give a step p that makes little progress toward feasibility. Therefore, the parameter r is selected by an auxiliary computation like in sequential quadratic programming with trust region.

The choice of the objective function (17.69a) is motivated by the fact that the first-order optimality conditions of (17.69a)–(17.69c) are exactly those given by (17.2) in which (17.2b) is scaled by S^{-1}.

The trust-region constraint (17.69d) guarantees that the problem (17.69) has a finite solution even when $\nabla_{xx}^2 L(x, s, y, z)$ is not positive definite, and therefore this Hessian needs never to be modified. Besides, the trust-region formulation ensures that adequate progress is made at every iteration.

The scaling S^{-1} used in (17.69d) is motivated by the crucial fact that the shape of the trust region must take into account the requirement that the slacks do not approach zero prematurely. The scaling S^{-1} restricts those components i of p_s for which s_i are close to their lower bound of zero.

In the following we present the sequential quadratic programming algorithm with trust region for the barrier problem. This means that at every iteration the sub-problem (17.69) is solved. The condition of stopping the iterations is defined by the error function (17.9) which uses the perturbed KKT system (17.2), as well as by the sufficient reduction of the merit function Φ_σ defined by (17.21). The algorithm is as follows.

Algorithm 17.5 Trust-region algorithm for barrier problems

1.	Choose the initial point x_0 and $s_0 > 0$. Compute the initial values of multipliers y_0 and $z_0 > 0$. Select an initial value of the barrier parameter $\mu > 0$. Consider a tolerance ε_μ and set $k = 0$
2.	If $E(x_k, s_k, y_k, z_k, \mu) \geq \varepsilon_\mu$, then continue with step 3; otherwise, stop; the current point is the solution of the problem
3.	Compute $p = (p_x, p_s)$ as an approximate solution of the sub-problem (17.69)
4.	If p provides sufficient decrease in the merit function Φ_σ, then set $x_{k+1} = x_k + p_x$, and $s_{k+1} = s_k + p_s$, compute the new multiplier estimates y_{k+1}, and $z_{k+1} > 0$, and set $\Delta_{k+1} \geq \Delta_k$. Continue with step 5. Otherwise, define $x_{k+1} = x_k$, and $s_{k+1} = s_k$ and set $\Delta_{k+1} < \Delta_k$. Continue with step 3
5.	Set $k = k + 1$ and go to step 2 ♦

Now, let us discuss some aspects regarding the determination of an approximate solution of the sub-problem (17.69), as well as the estimates (y_{k+1}, z_{k+1}) of the Lagrange multipliers.

Solving the sub-problem (17.69) This problem is approximately solved. At the very beginning, a change of variable is made, which transforms the trust-region constraint (17.69d) into a ball. By defining

$$\tilde{p} = \begin{bmatrix} p_x \\ \tilde{p}_s \end{bmatrix} = \begin{bmatrix} p_x \\ S^{-1} p_s \end{bmatrix}, \tag{17.70}$$

then the sub-problem (17.69) becomes

$$\min_{p_x, \tilde{p}_s} \nabla f(x)^{\mathrm{T}} p_x + \frac{1}{2} p_x^{\mathrm{T}} \nabla_{xx}^2 L p_x - \mu e^{\mathrm{T}} \tilde{p}_s + \frac{1}{2} \tilde{p}_s^{\mathrm{T}} S \Sigma S \tilde{p}_s \tag{17.71a}$$

subject to:

$$J_E(x)p_x + c_E(x) = r_E, \tag{17.71b}$$

$$J_I(x)p_x - S\tilde{p}_s + (c_I(x) - s) = r_I, \tag{17.71c}$$

$$\|(p_x, \tilde{p}_s)\|_2 \leq \Delta, \tag{17.71d}$$

$$\tilde{p}_s \geq -\tau e. \tag{17.71e}$$

To compute the vectors r_E and r_I, consider the following sub-problem in variables $v = (v_x, v_s)$:

$$\min_v \|J_E(x)v_x + c_E(x)\|_2^2 + \|J_I(x)v_x - Sv_s + (c_I(x) - s)\|_2^2 \tag{17.72a}$$

subject to:

$$\|(v_x, v_s)\|_2 \leq 0.8\Delta, \tag{17.72b}$$

$$v_s \geq -\frac{\tau}{2}e. \tag{17.72c}$$

Now, if the constraint (17.72c) is ignored, then the problem (17.72) has a standard form of a trust-region problem, and therefore an approximate solution of it can be obtained by the known techniques. If the solution of this problem violates the bounds (17.72c), then a backtracking can be used to satisfy them.

Having a solution (v_x, v_s) for (17.72), the vectors r_E and r_I are computed as:

$$r_E = J_E(x)v_x + c_E(x), \tag{17.73a}$$

$$r_I = J_I(x)v_x - Sv_s + (c_I(x) - s). \tag{17.73b}$$

With these developments, we are ready to compute an approximate solution $\tilde{d}$ for (17.71) as follows. From (17.73) we see that v is a particular solution of the linear constraints (17.71b)–(17.71c). Therefore, the sub-problem (17.71a)–(17.71c), which is a quadratic programming problem with equality constraints, can be solved using, for example, the projected conjugate gradient algorithm. During the solving process, the satisfaction of the trust-region constraint (17.71d) is monitored. If the boundary of this region is reached or if a negative curvature is detected or if an approximate solution is obtained, then the algorithm is stopped. If the solution obtained by the projected conjugate gradient algorithm does not satisfy the bounds (17.71e), then a backtracking is used until all these are satisfied. After the step $(p_x, \tilde{p}_s)$ has been computed, the vector p is recovered from (17.70).

It should be mentioned that at each iteration, the projected conjugate gradient algorithm requires solving a linear system in order to perform the projection operation. For the quadratic programming (17.71a, 17.71b and 17.71c), this projection matrix is:

$$\begin{bmatrix} I & \widehat{A}^{\mathrm{T}} \\ \widehat{A} & 0 \end{bmatrix}, \text{where } \widehat{A} = \begin{bmatrix} J_E(x) & 0 \\ J_I(x) & -S \end{bmatrix}. \tag{17.74}$$

We see that although this trust-region approach requires the solution of an augmented linear system, the matrix (17.74) is simpler than the primal-dual matrix from (17.11). Moreover, the advantage is that the Hessian $\nabla_{xx}^2 L$ needs never to be factored because the conjugate gradient approach requires only products of this matrix with vectors. Besides, the matrix $S\Sigma S$ from (17.71a) has a much tighter distribution of eigenvalues than Σ. Therefore, the conjugate gradient method will run much better in the presence of ill-conditioning, being viable for solving the quadratic problem (17.71a)–(17.71c) (Nocedal and Wright 2006).

Lagrange multiplier estimates and step acceptance As we have already seen in step 4 of Algorithm 17.5, estimates of the Lagrange multipliers y_{k+1} and $z_{k+1} > 0$ are computed. This is done as follows: At the iterate (x, s), the (y, z) are chosen as the least-square multipliers corresponding to the sub-problem (17.71a, 17.71b and 17.71c), i.e.:

$$\begin{bmatrix} y \\ z \end{bmatrix} = \left(\widehat{A}\widehat{A}^{\mathrm{T}}\right)^{-1} \widehat{A} \begin{bmatrix} \nabla f(x) \\ -\mu e \end{bmatrix}, \tag{17.75}$$

where $\widehat{A}$ is the matrix from (17.74). It is quite possible that the estimates z obtained from (17.75) may not always be positive. Therefore, to enforce positivity of this multiplier, the following computational scheme can be used:

$$z_i = \min\{10^{-3}, \mu/s_i\}, \quad i = 1, \ldots, m. \tag{17.76}$$

Observe that if all the components of z are defined as in (17.76), then the matrix Σ reduces to μS^{-2}. The quantities μ/s_i are called the ith *primal multiplier estimate* of multipliers.

As in the standard trust-region method, step p is accepted if

$$\text{ared}(p) \geq \eta \text{pred}(p), \tag{17.77}$$

where:

$$\text{ared}(p) = \Phi_\sigma(x, s) - \Phi_\sigma(x + p_x, s + p_s), \tag{17.78}$$

$$\text{pred}(p) = q_\sigma(0) - q_\sigma(p), \tag{17.79}$$

$$q_\sigma(p) = \nabla f(x)^{\mathrm{T}} p_x + \frac{1}{2} p_x^{\mathrm{T}} \nabla_{xx}^2 L p_x - \mu e^{\mathrm{T}} S^{-1} p_s + \frac{1}{2} p_s^{\mathrm{T}} \Sigma p_s + \sigma m(p), \tag{17.80}$$

$$m(p) = \left\| \begin{bmatrix} J_E(x)p_x + c_E(x) \\ J_I(x)p_x - p_s + c_I(x) - s \end{bmatrix} \right\|_2. \tag{17.81}$$

In (17.77) η is a constant in (0, 1), for example, $\eta = 10^{-8}$.

To determine a corresponding value for the penalty parameter σ, we impose that this is large enough that

$$\text{pred}(p) \geq \rho\sigma(m(0) - m(p)), \tag{17.82}$$

where ρ is a parameter in (0, 1). Observe that this condition is exactly as (11.25). Therefore, the value of σ can be computed by the procedure described in (11.26) from the sequential quadratic programming.

Description of the trust-region interior point algorithm For updating the barrier parameter, the Fiacco-McCormick strategy is used. The condition of stopping the algorithm uses the error function E defined by (17.9). If the quasi-Newton approach is used, then the Hessian $\nabla^2_{xx}L$ is replaced by a symmetric approximation (BFGS or limited memory BFGS).

Algorithm 17.6 Trust-region interior point algorithm

1.	Choose the value for the parameters $\eta > 0$, $\tau \in (0, 1)$, $\sigma \in (0, 1)$, and $\xi \in (0, 1)$. Choose the tolerances ε_μ and ε_{TOL} small enough. If the quasi-Newton approach is used, choose a symmetric $n \times n$ matrix B_0. Choose the initial values of the parameters $\mu > 0$ and Δ_0 as well as the initial points $x_0, s_0 > 0$. Set $k = 0$
2.	If $E(x_k, s_k, y_k, z_k, 0) > \varepsilon_{TOL}$, then go to step 3; otherwise stop
3.	If $E(x_k, s_k, y_k, z_k, \mu) > \varepsilon_\mu$, then go to step 4; otherwise go to step 13
4.	Using (17.75 and 17.76), compute the Lagrange multipliers
5.	Compute $\nabla^2_{xx}L(x_k, s_k, y_k, z_k)$ or update a quasi-Newton approximation B_k and define $\Sigma_k = S_k^{-1}Z_k$
6.	Compute the normal step $v_k = (v_x, v_s)$.
7.	Compute $\tilde{p}_k$ by applying the projected conjugate gradient algorithm to problem (17.71).
8.	Using (17.70) compute the total step p_k
9.	Update v_k to satisfy (17.82)
10.	Compute pred(p_k) and ared(p_k) by (17.79) and (17.78), respectively
11.	If ared$(p_k) \geq \eta$pred(p_k), then set $x_{k+1} = x_k + p_x$ and $s_{k+1} = s_k + p_s$, and choose $\Delta_{k+1} \geq \Delta_k$. Otherwise, set $x_{k+1} = x_k$, and $s_{k+1} = s_k$ and choose $\Delta_{k+1} < \Delta_k$
12.	Set $k = k + 1$ and go to step 3
13.	Set $\mu = \sigma\mu$, update ε_μ, and go to step 2 ♦

The tolerance on the barrier parameter can be defined as $\varepsilon_\mu = \mu$, or an adaptive strategy can be used for its updating. It is quite clear that the merit function can lead to the Maratos effect. In this case, the second-order corrections or a nonmonotone strategy can be used. Algorithm 17.6 is implemented in KNITRO/INTERIOR (Byrd, Hribar and Nocedal, 1999), which uses the exact Hessian or a quasi-Newton approximation of this matrix. It will be presented in Chapter 19.

Notes and References

This chapter is based on the papers by Nocedal and Wright (2006) and Andrei (1998c). The interior point methods were first developed for linear programming by Ilya Dikin (1936–2008) (1967, 1974) and then by Karmarkar (1984). The extension to the quadratic and nonlinear programming was quite natural. There is a vast literature on nonlinear interior point methods. For a comprehensive list of references, we recommend the papers by Forsgren, Gill, and Wright (2002) and Gould, Orban, and Toint (2005). The book by Conn, Gould, and Toint (2000) also gives a thorough presentation of several interior point algorithms. The papers by Ulbrich, Ulbrich, and Vicente (2004) and Wächter and Biegler (2005a, 2005b) present an analysis of interior point algorithms that use filter globalization. The primal barrier methods for nonlinear optimization were originally proposed by Frisch (1955) and were further analyzed by Fiacco and McCormick (1968). In this chapter, we first present the prototype of interior point algorithms. After that a variant of the line-search interior point algorithm is deeply discussed, where a methodology of the interior point algorithm development and analysis is given. The conclusion is that for having an efficient and robust interior point algorithm, some ingredients based on sequential linearization combined with line-search or trust-region globalization strategies must be introduced.

A history of barrier function methods was given by Nash (1998). Shanno (2012) presents interesting comments on "who invented the interior point algorithm." His conclusion is that the history of the interior point method goes back to Fiacco and McCormick. It seems that the terms *interior point* and *primal central path* appear in the book by Fiacco and McCormick.

This chapter represents the fundamentals for some very much elaborated algorithms for nonlinear optimization, algorithms which combine different strategies based on the sequential linear or quadratic programming or on the filter line search in the frame of the interior point method: KNITRO/INTERIOR (see Chapter 19) and IPOPT (see Chapter 20).

Chapter 18
Filter Methods

The filter methods developed by Fletcher and Leyffer (2002) as a new technique for globalization the nonlinear optimization algorithms are described in this chapter. The idea is motivated by the aim of avoiding the need to choose penalty parameters in penalty functions or augmented Lagrangan functions and their variants. Let us consider the nonlinear optimization problems with inequality constraints:

$$\min f(x)$$

subject to

$$c_i(x) \leq 0, \; i = 1, \ldots, m, \tag{18.1}$$

where the objective function $f: \mathbb{R}^n \to \mathbb{R}$ and the functions $c_i: \mathbb{R}^n \to \mathbb{R}$ $i = 1, \ldots, m$ defining the constraints are supposed to be twice continuously differentiable. The methods for solving this problem are based on the Newton method. Given an estimate x_k of the solution x^* of (18.1), a linear or quadratic approximation of (18.1) is solved, thus obtaining a new estimate x_{k+1} which we hope to be better as the previous one. Near a solution, this approach is guaranteed to be convergent. However, far away from the solution, the sequence $\{x_k\}$ generated by the above procedure may not converge. In this situation, away from the solution, the idea is to use the Newton method again but considering the penalty or merit functions. The penalty functions or the merit functions are a combination of the objective function and a measure of constraints violation such as $h(x) = \|c(x)^+\|_\infty$, where $c(x) = [c_1(x), \ldots, c_m(x)]^T$ and $c_i^+ = \max\{0, c_i\}$. A very well-known example is the l_1 exact penalty function $p(x, \sigma) = f(x) + \sigma h(x)$, where $\sigma > 0$ is the penalty parameter. If σ is sufficiently large, then this penalty function can be minimized in an iterative procedure to ensure progress to the solution.

This approach based on penalty or merit functions is interesting but not without problems. Unfortunately, a suitable value of the penalty parameter depends on the solution of the problem (18.1), that is $\sigma > \|\lambda^*\|_D$, where λ^* is the vector of the Lagrange multipliers and $\|.\|_D$ is the dual norm. Hence, determination of a suitable

N. Andrei, *Continuous Nonlinear Optimization for Engineering Applications in GAMS Technology*, Springer Optimization and Its Applications 121,
DOI 10.1007/978-3-319-58356-3_18

penalty parameter is difficult. More than this, if the penalty parameter is too large, then any monotonic method would be forced to follow the nonlinear constraint manifold very closely, resulting in much shortened Newton steps and therefore slow convergence. All these deficiencies are very well known, and for their avoidance some computational schemes have been imagined, like methods based on the augmented Lagrange function (modified or not) or interior point methods. In this section we present the methods for solving nonlinear optimization problems known as filters methods (Fletcher & Leyffer, 1998, 1999, 2002); (Fletcher, Leyffer, & Toint, 2002, 2006).

The idea of a filter method is not of combining the objective and the constraints violation into a single function. Instead, (18.1) is viewed as a *biobjective optimization problem* that minimizes $f(x)$ and $h(x)$. However, as we can see immediately, the second objective is more important because we must ensure that $h(x^*)=0$. For this, Fletcher and Leyffer (2002) introduced the concept of *domination* taken from the multi-objective optimization. *A point x_k dominates a point x_l if and only if $f(x_k) \le f(x_l)$ and $h(x_k) \le h(x_l)$*. Therefore, *a filter is defined as a list of pairs $(h(x_l), f(x_l))$ such that no pair dominates another pair.*

This concept can be implemented in different ways. The filter methods use *sequential linear programming* or *sequential quadratic programming* in the context of trust-region methods. At iteration $k=0$, the filter is initialized as $F_k=\{(u,-\infty)\}$ where u is an upper bound on the acceptable constraint violation. Solving the linear programming problem or the quadratic programming problem, a solution s is obtained. If this point is acceptable by the filter, then we set $x_{k+1}=x_k+s$ and possibly increase the trust-region radius and update the filter (adding the previous point and removing any dominated entries). On the other hand, if the step is dominated by the current filter, then we reject it, set $x_{k+1}=x_k$, reduce the trust-region radius, and solve the linear or quadratic programming problem in a new context. To be convergent, this general description of the filter method needs some refinements:

(1) *Filter envelope*. To avoid convergence to an infeasible limit point where $h(x^*)>0$, an envelope around the current filter is added (Chin, 2002; Chin & Fletcher, 2003). A new iterate is acceptable if for all $l \in F_k, h_{k+1} \le \beta h_l$, or $f_{k+1} \le f_l - \gamma h_{k+1}$ where $f_k = f(x_k)$, $h_k = h(x_k)$, and $0<\beta, \gamma<1$ are constants.

(2) *Sufficient reduction*. The filter alone cannot ensure convergence to stationary points. For example, if the sequence $\{x_k\}$ satisfies the condition $h_{k+1} \le \beta h_k$, then the iterates could converge to an arbitrary feasible point. Therefore, if the constraint violation becomes small, a sufficient reduction condition can be imposed as follows: denote the predicted reduction by Δq_k as the value of the objective function of the linear or quadratic problem and introduce the following *switching condition*: if $\Delta q_k>0$, then check $f_k - f_{k+1} \ge \sigma \Delta q_k$, where $\sigma \in (0,1)$ is a constant.

(3) *Feasibility restoration*. By reducing the trust-region radius of the linear or quadratic problems, these may become inconsistent. That is, the current point is too far from the feasibility region to get a significant progress to the solution. In this case a sequential linear or quadratic programming problem is invoked, which minimizes the constraint violation $h(x)$. This restoration phase is left, once an acceptable point has been obtained.

The filter algorithm contains an *inner* and an *outer* iteration. During the inner iteration, the trust-region radius is reduced until either an acceptable point is found or the restoration phase is initiated. The aim of the restoration phase is to find an acceptable iterate x_{k+1} such that the linear (or quadratic) programming problem is compatible for some smaller values of the trust-region radius. The iterate and the filter are updated in the outer iteration, which also initializes the trust-region radius to a given lower bound. The filter is updated by adding (h_k, f_k) to F_k which corresponds to an *h-type iteration* after we move to x_{k+1}. Fletcher and Leyffer (2002) motivate the switching condition as follows. When close to a feasible point, it is expected that the linear or the quadratic model of the problem predicts a decrease in the objective function, that is, $\Delta q_k > 0$. However, far away from a feasible point, the predicted reduction is usually negative, that is, $\Delta q_k < 0$. Iterations that satisfy the switching condition are called *f-type iteration*. All the other steps are called *h-type iteration*. Observe that if $h_k = 0$ at a nonstationary point, then $\Delta q_k > 0$ thereby implying that only f-typed steps can be accepted. In other words, no points are added to the filter for which $h_k = 0$. This ensures that the restoration phase generates only filter-acceptable points.

18.1 Sequential Linear Programming Filter Algorithm

Suppose that the constraints from (18.1) include some linear constraints, for example, simple bounds on variables that define a nonempty bounded region X. In the current point x for problem (18.1), we can associate the following linear programming sub-problem modified with the trust-region radius $\rho > 0$, as

$$\min_{d \in \mathbb{R}^n} g^T d$$

subject to: (18.2)

$$c_i + a_i^T d \leq 0, \; i = 1, \ldots, m,$$
$$\|d\|_\infty \leq \rho,$$

where $g = \nabla f(x)$, $c_i = c_i(x)$, and $a_i = \nabla c_i(x)$. Observe that the l_∞ norm is used to define the trust-region because it is very simple to be implemented by adding simple bounds to the linear programming sub-problem. If it exists, let d be the solution of (18.2). Denote

$$\Delta l = -g^T d \tag{18.3}$$

as the *predicted reduction* in $f(x)$ and with

$$\Delta f = f(x) - f(x + d), \tag{18.4}$$

as the *actual reduction* in $f(x)$. The measure of constraint infeasibility which is used in this text is given by

$$h(c) = \|c^+\|_\infty, \tag{18.5}$$

where $c_i^+ = \max\{0, c_i\}$, $i = 1, \ldots, m$.

Now, more exactly, the filter method can be introduced as follows (Fletcher & Leyffer, 2002). A filter is defined by pairs of values (h,f) obtained by evaluating $h(c(x))$ and $f(x)$ for different values of x. *A pair* (h_i,f_i) *dominates another pair* (h_j,f_j) *if and only if both* $h_i \leq h_j$ *and* $f_i \leq f_j$ *are satisfied, thus indicating that the former point is at least as good as the latter in respect to both measures.* The filter is defined to be a list of pairs (h_i,f_i) such that no pair dominates any other. A point x is acceptable for inclusion in the filter if its (h,f) pair is not dominated by any entry in the filter. This is the condition that

$$\text{either } h < h_i \text{ or } f < f_i \tag{18.6}$$

for all $i \in F$ where F denotes the current set of filter entries. We may include a point in the filter by which we mean that its pair (h,f) is added to the list of pairs in the filter, and any pairs in the filter that are dominated by the new pair are deleted from the filter.

Observe that this definition of the filter is not suitable for proving the convergence because it allows points to accumulate in the neighborhood of a filter entry that has $h_i > 0$. However, this may be corrected by defining a small envelope around the current filter in which points are accepted. Therefore, the condition for a point to be *acceptable* to the filter is that its pair (h,f) satisfies

$$\text{either } h \leq \beta h_i \text{ or } f \leq f_i - \gamma h_i, \tag{18.7}$$

for all $i \in F$, where β and γ are parameters such that $1 > \beta > \gamma > 0$, with β close to one and γ close to zero. Observe that the first inequality from (18.7) is exactly the reduction in h. The dependence on h_i of the second inequality is a small artifact to enable convergence of the algorithm. Since γ is small, the extra term has little practical impact.

It is also convenient to introduce an upper bound

$$h(c(x)) \leq \beta u \tag{18.8}$$

$(u > 0)$ on the constraint infeasibility, which is readily implemented by initializing the filter with entry $(u, -\infty)$.

As we know, a common feature in a trust-region algorithm for unconstrained minimization is the use of a sufficient reduction criterion

$$\Delta f \geq \sigma \Delta l, \tag{18.9}$$

where Δl is positive and $\sigma \in [0, 1)$ is a given parameter. However, in the trust-region algorithm, Δl may be negative or even zero, in which case this test is no longer appropriate. Fletcher and Lyeffer (2002) suggest using (18.9) with $\sigma \geq \gamma$ only when Δl is sufficiently positive, which can be achieved by testing the inequality

$$\Delta l \geq \delta h^2, \tag{18.10}$$

where h refers to $h(c(x))$ evaluated at the current point and $\delta > 0$ is a given parameter close to zero. Again, the dependence of the right-hand side of (18.10) on h is an artifact to enable convergence to be proved.

Let us denote the linear programming sub-problem (18.2) by $LP(x, \rho)$, then the sequential linear programming filter algorithm is as follows.

Algorithm 18.1 Sequential linear programming filter algorithm – filterSD

1.	Choose an initial point $x \in X$ and initialize the filter with $(u, -\infty)$. Set $k = 1$
2.	Consider the restoration phase to find a point $x_k \in X$ acceptable for inclusion in the filter such that $LP(x_k, \widetilde{\rho})$ is compatible for a $\widetilde{\rho} \geq \rho^0$ and initialize $\rho = \widetilde{\rho}$
3.	Solve the sub-problem $LP(x_k, \rho)$
4.	If $LP(x_k, \rho)$ is infeasible, then place (h_k, f_k) in the filter (*h-type* iteration). Set $k = k+1$ and go to step 2
5.	If $LP(x_k, \rho)$ has the solution d then if $d = 0$, stop; the current point is a KKT point for the problem
6.	If $d \neq 0$, then evaluate $f(x_k + d)$ and $c(x_k + d)$
7.	If $x_k + d$ is not acceptable for inclusion in the filter, then set $\rho = \rho/2$ and go to step 3
8.	If $x_k + d$ is acceptable for inclusion in the filter and $\Delta f < \sigma \Delta l$ and $\Delta l \geq \delta (h_k)^2$, then set $\rho = \rho/2$ and go to step 3
9.	If $x_k + d$ is acceptable for inclusion in the filter and $\Delta f < \sigma \Delta l$ and $\Delta l \geq \delta (h_k)^2$ are not satisfied, then set $\rho_k = \rho$, $d_k = d$, $\Delta l_k = \Delta l$, and $\Delta f_k = \Delta f$
10.	If $\Delta l_k < \delta (h_k)^2$, then the pair (h_k, f_k) is included in the filter (*h-type* iteration)
11.	Set $x_{k+1} = x_k + d_k$, $k = k+1$, initialize $\rho \geq \rho^0$, and go to step 3 ♦

Some important details and the convergence of the filter algorithm with sequential linear programming are given in Fletcher, Leyffer, and Toint (1999). Mainly, the convergence theorem is based on the necessary optimality condition by Fritz John and on the following standard hypothesis: *the set X defined by the linear constraints of the problem (18.1) is nonempty and bounded, and the functions of the problem f(x) and c(x) are twice continuously differentiable on X*. When the algorithm is applied, one of four different possible outcomes can occur:

(*i*) The restoration phase iterates infinitely and fails to find a point x which is acceptable to the filter and for which $LP(x, \rho)$ is compatible for some $\rho \geq \rho^0$.
(*ii*) A KKT point is found ($d=0$ solves $LP(x_k, \rho)$ for some k).
(*iii*) All iterations are *f-type* iterations for k sufficiently large.
(*iv*) There exists an infinite subsequence of *h-type* iterations.

The following theorem proves the convergence of the sequential linear programming filter algorithm as specified in Fletcher, Leyffer, and Toint (1999).

Theorem 18.1 *If the standard hypothesis holds, then for the sequential linear programming filter algorithm, either (i) or (ii) occurs, or the sequence of iterates generated by the algorithm has an accumulation point that satisfies the Fritz John necessary optimality conditions.* ♦

More discussions about the sequential linear programming filter algorithm (filterSD) can be found in Fletcher and Leyffer (1998, 2002). In the following let us present some numerical results with this algorithm.

Example 18.1 To illustrate the running of the filterSD algorithm, consider the nonlinear optimization problem (Duran & Grossmann, 1986):

$$\min(10x_1 - 18\ln(x_2+1) - 7x_3 + 5x_4 + 6x_5 + 8x_6 \\ -19.2\ln(x_1 - x_2 + 1) + 10)$$

subject to:

$$0.8\ln(x_2+1) + 0.96\ln(x_1 - x_2 + 1) - 0.8x_3 \geq 0,$$
$$\ln(x_2+1) + 1.2\ln(x_1 - x_2 + 1) - x_3 - 2x_6 \geq -2,$$
$$x_2 - x_1 \leq 0,$$
$$x_2 - 2x_4 \leq 0,$$
$$x_1 - x_2 - 2x_5 \leq 0,$$
$$x_4 + x_5 \leq 1,$$

where

$$0 \leq x_1 \leq 2, \quad 0 \leq x_2 \leq 2, \quad 0 \leq x_3 \leq 1, \quad 0 \leq x_4 \leq 1, \quad 0 \leq x_5 \leq 1, \quad 0 \leq x_6 \leq 1.$$

Let $x_0 = [0,0,0,0,0,0]^T$ be the initial point. The filterSD algorithm gives the following results. The objective function value in initial point is $f(x_0) = 10$. The values of the constraints in initial point are $c_i(x_0) = 0$ and $i = 1, \ldots, 6$. The optimal solution is x^*, and the values of the constraints in the optimal point $c(x^*)$ are:

$$x^* = \begin{bmatrix} 1.146514 \\ 0.546592 \\ 1 \\ 0.273296 \\ 0.299960 \\ 0 \end{bmatrix} \quad c(x^*) = \begin{bmatrix} 0 \\ 0 \\ -0.5999216 \\ 0 \\ 0 \\ 0.57325729 \end{bmatrix}.$$

Table 18.1 Performances of filterSD

	n	*me*	*mc*	*it*	*nf*	*ng*	*vfo*
ALKI	10	3	8	10	89	66	−1768.8069
ELCH	10	3	0	4	293	238	−47.761090
PREC	8	0	6	12	383	301	3.95116351
PPSE	9	6	0	5	27	24	5055.01180
MSP3	13	0	15	9	45	42	97.5482075
MSP5	16	0	21	21	195	165	174.786994
POOL	34	20	0	3	7	7	2785.80000
LATHE	10	1	14	12	43	43	−4434.0019

In Table 18.1 we have: $n =$ the number of variables, $me =$ the number of equality constraints, $mc =$ the number of inequality constraints, $it =$ the number of iterations to get a solution, $nf =$ the number of evaluations of the functions defining the problem, $ng =$ the number of the evaluations of the gradients of the functions, $vfo =$ the value of the objective function in the optimal point.

The value of the objective function in the optimal point is equal to $f(x^*) = 0.7592843922$. This solution was obtained in 5 iterations, 29 evaluations of the functions of the problem, and 23 evaluations of the gradients of the functions. Along the iterations the value of the trust-region radius was $\rho = 1000$. ♦

Numerical Study

Table 18.1 presents the performances of an implementation of the filterSD algorithm given by Fletcher for solving eight nonlinear optimization applications described in Chapter 4.

Table 18.2 presents a comparison of filterSD (Table 18.1) versus DONLP (Table 12.1), NLPQLP (Table 13.1), and KNITRO/ACTIVE (Table 14.1) for solving some nonlinear optimization problems described in Chapter 4, subject to the number of iteration (*it*) and the number of evaluations of the functions defining the problem (*nf*).

In filterSD, at each iteration, for solving linear programming sub-problems with simple bounds on variables, a recursive form of an active-set method is used. To solve degeneracy, the Wolfe method is used. A limited memory-reduced gradient sweep method is used for minimization in the null space, so the KKT point is usually a local minimizer. Observe that filterSD has very good numerical performances for solving nonlinear optimization problems. The main advantage of this algorithm is that it avoids using second derivatives and also avoids storing an approximate reduced Hessian matrix by taking a limited memory spectral gradient approach based on Ritz values (Fletcher, 2011).

18.2 Sequential Quadratic Programming Filter Algorithm

In this section we give some details on the filter algorithm with sequential quadratic programming for solving the nonlinear optimization problem (18.1). The filter method views (18.1) as a biobjective nonlinear optimization problem: the first

Table 18.2 Comparison of filterSD versus DONLP, NLPQLP, and KNITRO

	filterSD		DONLP		NLPQLP		KNITRO (3)	
	it	*nf*	*it*	*nf*	*it*	*nf*	*it*	*nf*
ELCK	4	293	40	89	21	30	18	19
ALKI	10	89	62	127	53	69	22	58
PREC	12	383	18	26	19	20	45	117
LATHE	12	43	7	11	19	19	83	242
PPSE	5	27	12	29	9	10	6	11
MSP3	9	45	269	843	146	264	7	9
POOL	3	7	12	11	22	22	15	33
TOTAL	55	887	420	1136	289	434	196	489

one is minimization of the objective function f and the second is the satisfaction of the constraints. These two conflicting aims can be formalized as

$$\min f(x) \tag{18.11}$$

and

$$\min h(c(x)), \tag{18.12}$$

where

$$h(c(x)) \triangleq \left\| c(x)^+ \right\|_1 \triangleq \sum_{i=1}^{m} c_i(x)^+ \tag{18.13}$$

is the l_1 norm of the constraint violation and $c_i(x)^+ = \max\{0, c_i(x)\}, i = 1, \ldots, m$. Observe that the problem of satisfiability has been written as a minimization problem. Hence, (18.11) and (18.12) represent a biobjective nonlinear optimization problem. Here the l_1 norm is used because it has some convenient features that can be exploited in defining the algorithm.

The fundamental idea of the algorithm is to use the filter as a criterion for accepting or rejecting a step in the sequential quadratic programming method. In other words, given the current point x_k, solution of the quadratic programming sub-problem

$$\min_d \frac{1}{2} d^T W_k d + d^T g_k$$

subject to (18.14)

$$A_k^T d + c_k \le 0,$$
$$\|d\|_\infty \le \rho,$$

generates a step d_k. Here $g(x) = \nabla f(x)$, W_k is (an approximation to) the Hessian of the Lagrangian $L(x, \lambda) = f(x) + \lambda^T c(x)$, A_k is the Jacobian of the constraints, and c_k is the value of the constraints at point x_k. Set $x_{k+1} = x_k + d_k$. The new point x_{k+1} is accepted by the filter if the corresponding pair (h_{k+1}, f_{k+1}) is not dominated by any other point in the filter. Otherwise, the step is rejected and the trust-region radius ρ is reduced. The use of the l_∞ norm in (18.14) ensures that (18.14) remains tractable as a quadratic programming.

The above strategy based on solving a sequence of quadratic programming sub-problems must be completed in order to cope with the stuations in which the sub-problems are infeasible. In a sequential quadratic programming trust-region method, reducing the trust-region radius will give rise to an infeasible quadratic programming sub-problem if the current point is infeasible in the nonlinear optimization problem. Usually, a trust-region algorithm arrives at this situation after rejecting a number of consecutive steps. Thus, it is not sufficient to simply increase the trust-region radius to regain feasibility. An infeasible quadratic programming sub-problem also occurs when the linearized constraints are themselves inconsistent. To skip over this situation, Fletcher and Leyffer (2002) chose the strategy to minimze $h(c(x))$. This is referred to as the *restoration phase*, which tries to get close to the feasible region of the nonlinear optimization problem. A possible outcome is that the restoration phase finds a nonzero minimum of $h(c(x))$ which is taken as an indication that the nonlinear optimization problem is infeasible. Otherwise, we can assume that the algorithm finds a point at which the quadratic programming sub-problem is feasible, and therefore the above method can be continued from this point.

Having in view all these developments, the basic sequential quadratic programming filter algorithm starts the iterations with an initial guess x_0 and an initial estimate of the Lagrange multipliers λ_0. On the subsequent iterations the Lagrange multipliers are updated whenever the sequential quadratic programming sub-problem generates a new point that is accepted by the filter.

Algorithm 18.2 Basic sequential quadratic programming filter – filterSQP

1.	Choose an initial point x_0 a value for the parameter ρ and set $k=1$
2.	Test a criterion for stopping the iterations
3.	Solve the quadratic programming sub-problem (18.14)
4.	If (18.14) is infeasible, find a new point x_{k+1} using the restoration phase; otherwise determine the step d_k and provisionally set $x_{k+1}=x_k+d_k$ and continue with step 5
5.	If the pair (h_{k+1},f_{k+1}) is acceptable to the filter, then: Accept x_{k+1} and add the pair (h_{k+1},f_{k+1}) to the filter Remove the points dominated by the pair (h_{k+1},f_{k+1}) Possibly increase the trust-region radius Otherwise Set $x_{k+1}=x_k$ and reduce the trust-region radius
6.	Set $k=k+1$ and go to step 2 ♦

Algorithm 18.2 is the basic filter algorithm with sequential quadratic programming. It is very easy to be implemented, but some algorithmic extensions are required to exclude those situations in which it might fail or might converge slowly. Fletcher and Leyffer (1998, 2002) present some extesions and techniques for its protection against certain difficulties for solving nonlinear optimization problems. Some of these extensions are as follows.

Second-order correction step One of the most important property of any sequential quadratic programming algorithm is that it usually has second-order convergence near solution. However, the use of a non-differentiable penalty function can preclude the acceptance of the unit step arbitrarily close to the solution, thus preventing second-order convergence. This is known as the Maratos effect. This difficulty can be avoided by computing a correction to the step that eliminates second-order contributions of the nonlinear constraints. This is the second-order correction step. In the filter algorithm, it is quite possible for a sequential quadratic programming unit step to increase *both the objective and the constraint violation functions.*

Second-order correction step is computed whenever $x_{k+1} = x_{(k+1,0)}$ is rejected by the filter in Algorithm 18.2. In this case the quadratic programming sub-problem that is solved is defined as

$$\min_d \frac{1}{2} d^{\mathrm{T}} W_k d + d^{\mathrm{T}} g_k$$

subject to (18.15)

$$A_k^{\mathrm{T}} d + c_{(k+1,l)} - A_k^{\mathrm{T}} d_{(k,l)} \leq 0,$$
$$\|d\|_\infty \leq \rho,$$

for $l = 0, 1, \ldots,$ where $c_{(k+1,0)} = c_{k+1}$ and $d_{(k,0)} = d_k$. Let $\widehat{d}_k \triangleq d_{(k,l+1)}$ be the solution of (18.15). The new point $x_{(k+1,l)} = x_k + d_{(k,l+1)}$ is tested in the filter. If the pair $(h_{(k+1,l)}, f_{(k+1,l)})$ is acceptable to the filter, then the step $d_{(k,l+1)}$ is accepted and the trust-region radius may be increased. Otherwise, a sequence of second-order correction steps is performed, generating a sequence of trial points $x_{(k+1,l)}$, $l = 0, 1, \ldots,$ until one of the following holds:

(*i*) An acceptable trial point $x_{(k+1,j)}$ is found.
(*ii*) An infeasible quadratic programming problem is detected.
(*iii*) The rate of convergence of the second-order correction steps defined by:

$$r \triangleq \frac{h_{(k+1,l)}}{h_{(k+1,l-1)}}$$

is considered to be too slow.

(*iv*) An almost feasible point with $h_{(k+1,l)} < \varepsilon$ is generated, where ε is a tolerance used in solving the problem (18.1).

In case (*i*) the next iteration is $x_{k+1} = x_{(k+1,j)}$ and the filter with sequential quadratic programming continues from step 2. In all the other cases, the steps are rejected and the trust-region radius is reduced. The last case (*iv*) ensures the finiteness of this process. The best second-order correction step is stored by using a penalty function estimate to rank the steps.

Feasibility restoration phase This is the more complex part of the algorithm. One difficulty in using the trust-region approach is that the reduction of the trust-region radius may cause the quadratic programming sub-problem to become infeasible. Besides, possibly, the linearizations of the nonlinear constraints may themselves be inconsistent. The strategy of Fletcher and Leyffer (2002) for dealing with this situation is to enter a restoration phase in which the purpose is to get closer to the feasible region by minimizing $h(c(x))$.

If an infeasible quadratic programming problem is detected, then the solver exits with a solution of the linear programming problem:

$$\min_{d} \sum_{j\in J} \nabla c_j(x_k)^{\mathrm{T}} d$$

subject to: (18.16)

$$\nabla c_j(x_k)^{\mathrm{T}} d + c_j(x_k) \le 0, \quad j \in J^{\perp},$$
$$\|d\|_{\infty} \le \rho.$$

In other words, the quadratic programming solver partitions the constraints into two index sets: $J \subset \{1, 2, \ldots, m\}$ and its complement $J^{\perp} = \{1, 2, \ldots, m\} \backslash J$. The set J contains infeasible constraints at point x_k: $c_j(x_k) + \nabla c_j(x_k)^{\mathrm{T}} d > 0$ $j \in J$ whose l_1 sum is minimized at the solution to the linear programming problem (18.16), subject to the constraints in $J^{\perp}$ being satisfied.

The strategy in the restoration phase is to apply a sequential quadratic programming trust-region method to the nonlinear problem

$$\min_{x} \sum_{j\in J} c_j(x)^{+}$$

subject to (18.17)

$$c_j(x) \le 0, \quad j \in J^{\perp},$$

that is defined by this partitioning into the sets J and $J^{\perp}$. However, the difficulty is that it is quite possible for J and $J^{\perp}$ to change from iteration k to iteration $k+1$, thus making it difficult to enforce convergence. Therefore, the restoration phase consists of a sequence of sequential quadratic programming iterations that continue whilst the sub-problem (18.14) is infeasible. At each iteration, first check the feasibility of the system:

$$\nabla c_j(x_k)^{\mathrm{T}} d + c_j(x_k) \le 0, \quad j = 1, \ldots, m, \tag{18.18}$$
$$\|d\|_{\infty} \le \rho.$$

If the system (18.18) has a feasible solution, then the restoration phase ends. Otherwise, the sets J and $J^{\perp}$ which solve (18.16) are determined. Next a quadratic programming sub-problem is constructed by adding the second-order term $(1/2)d^{\mathrm{T}} W_k d$, into the objective function of (18.16), where

$$W(x, \lambda) = \sum_{j\in J} \nabla^2 c_j(x) + \sum_{j\in J^{\perp}} \lambda_j \nabla^2 c_j(x). \tag{18.19}$$

The multipliers $\lambda_j\, j \in J^{\perp}$ are those obtained from the solution of (18.16). Note that for solving (18.16), a warm start can be used, i.e., the solution of (18.16) can be used to initialize the solution of the new quadratic programming problem. Iterations in the restoration phase are continued until *either* a feasible quadratic programming problem is encountered (case in which the algorithm can return to the basic sequential quadratic programming algorithm) *or* an infeasible KKT point of (18.17) is found for some sets J and $J^{\perp}$ (case in which the algorithm terminates with indication that (18.1) is locally infeasible).

For solving (18.17) Fletcher and Leyffer (2002) suggest using the filter method by introducing the so-called *restoration filter* (or *phase I filter*) defined by

$$h_J \triangleq h(c_J(x)) \triangleq \sum_{j \in J} c_j(x)^+$$

and similarly $h_{J^{\perp}}$. A phase I filter is a list of pairs $(h_J, h_{J^{\perp}})$ such that no pair dominates any other.

Fletcher and Leyffer (2002) present a computational situation which leads to the introduction of the *blocking entry* in the filter. Consider the situation in which (18.1) has a global solution x^{**} and a worse local solution x^*, as well as that x_0 is a feasible point fairly close to x^{**}, but the subsequent iterates x_k are convergent to x^*. Then, if $f(x_0) \leq f(x^*)$, then the filter entry (h_0, f_0) prevents the sequential quadratic programming iteration from converging to x^*. In this case we would like to backtrack to x_0, but our decision not to store x_0 with the filter information precludes this. The pair (h_0, f_0) is called *blocking entry* in the filter. Some more details are found in Fletcher and Leyffer (2002). The algorithm for the restoration phase is as follows.

Algorithm 18.3 Feasibility restoration algorithm

1.	Consider: the current point x_k, the trust-region radius ρ, and a constraint upper bound u which measures the violation of constraints
2.	Test a criterion for stopping the iterations
3.	Solve (18.16) and determine the sets J and $J^{\perp}$, which remain fixed for this iteration
4.	If (18.16) is feasible, then return to a normal sequential quadratic programming and clear the restoration filter. Otherwise, add the quadratic term (18.19) to the objective function of (18.16) and solve phase I of the quadratic programming algorithm. Let d_k be the solution and set $x_{k+1} = x_k + d_k$
5.	If $\left(h_J^{k+1}, h_{J^{\perp}}^{k+1}\right)$ is acceptable to the phase I filter, then: Accept x_{k+1} and add the pair $\left(h_J^{k+1}, h_{J^{\perp}}^{k+1}\right)$ to the restoration filter Remove the points dominated by $\left(h_J^{k+1}, h_{J^{\perp}}^{k+1}\right)$ from the filter Possibly increase the trust-region radius ρ Otherwise, solve a sequence of quadratic programming second-order correction problems to get the step $\widehat{d}_k$. Set $\widehat{x}_{k+1} = x_k + \widehat{d}_k$ and go to step 6

(continued)

6.	If $\left(\widehat{h}_J^{k+1}, \widehat{h}_{J^\perp}^{k+1}\right)$ is acceptable to the phase I restoration filter, then: Accept $\widehat{x}_{k+1}$ and add $\left(\widehat{h}_J^{k+1}, \widehat{h}_{J^\perp}^{k+1}\right)$ to the restoration filter Remove the points dominated by $\left(\widehat{h}_J^{k+1}, \widehat{h}_{J^\perp}^{k+1}\right)$ from the restoration filter Possibly increase the trust-region radius ρ Otherwise, if J has changed from the iteration $k-1$ and $\widehat{h}_{(k,L)} < u$, then: Accept the best second-order correction step, i.e., set $x_{k+1} = \widehat{x}_{k+1}$ Remove all the blocking entries from the restoration filter Reduce the upper bound u to $u = \max\left\{\widehat{h}_{(k,L)}, u/10\right\}$ Otherwise: Reject the step, i.e., set $x_{k+1} = x_k$ Reduce the trust-region radius ρ
7.	Set $k = k+1$ and go to step 2 ♦

In the algorithm 18.3 $\widehat{h}_{(k,L)}$ corresponds to $\widehat{x}_{(k,L)}$ which is the best trial point of iteration k, i.e. the point with lowest l_1 norm.

Observe that the trust-region radius is changed in the feasibility restoration algorithm. At its return to the sequential quadratic programming, the current value of ρ reflects how well a first-order Taylor series approximates the nonlinear constraints in x_k. Therefore, this value of ρ is used in the next step of the sequential quadratic programming.

Sufficient reduction A new iterate x_{k+1} is acceptable to the filter if

$$h_{k+1} \le \beta h_l \tag{18.20}$$

or if

$$f_{k+1} \le f_l - \max\{\alpha_1 \Delta q_l, \alpha_2 h_l \mu_l\} \tag{18.21}$$

are satisfied for all filter entries l. Here β, α_1, and α_2 are positive constants which can be taken to be $\beta = 0.99$, $\alpha_1 = 0.25$, and $\alpha_2 = 0.0001$, respectively. An estimation of the penalty parameter μ_l is computed as the least power of ten but larger than $\|\lambda_l\|_\infty$ and cuts this value off so as to lie in the interval $[10^{-6}, 10^6]$. The value of $h_l\mu_l$ can be used as a predicted reduction in f. Both Δq_l and μ_l are stored along with f_l and h_l. Observe that μ_l measures the marginal effect of the changes in f due to changes in h. This is the reason for using the term $\alpha_2 h_l \mu_l$ in (18.21). A quadratic programming step predicts a reduction of h_l to zero, equivalent to a predicted reduction in f of $h_l\mu_l$.

Beyond the extreme points of the filter Mainly, the filter is a reliable oracle for deciding whether or not to accept a step as long as the new constraint violation lies within the range of constraint violations recorded in the filter. However, the current heuristics do not exclude the possibility of generating a sequence of filter entries in which $\{f_k\}$ is monotonically increasing and $\{h_k\}$ is monotonically decreasing, without converging to a KKT point. But, as long as the sufficient reduction criterion in h is satisfied by the sequence, the points will be accepted by the filter. To overcome this possibility, a new additional heuristic is introduced when h_{k+1} provides sufficient reduction from h_1, where h_1 now refers to the leftmost entry in the filter. An overestimate $\mu = 1000\mu_1$ of the penalty parameter is considered. It requires x_{k+1} providing a reduction in the resulting exact penalty function. The value of the exact penalty function corresponding to the leftmost entry f_1, h_1 is $f_1 + \mu h_1$. The new point is then accepted if, in addition to the sufficient reduction from h_1, it satisfies $f_{k+1} + \mu h_{k+1} \leq f_1 + \mu h_1$. Fletcher and Leyffer (2002) called this the *North-West corner rule*. A similar rule is applied in the *South-East corner of the filter*.

Now, the complete sequential quadratic programming algorithm based on filter concept can be presented as follows.

Algorithm 18.4 Filter sequential quadratic programming – filterSQP

1.	Choose the initial point x_0 and the initial trust-region radius ρ. Set $k = 0$
2.	Test a criterion for stopping the iterations
3.	Solve (18.14) for a step d_k. If (18.14) is infeasible, then continue with the restoration phase (see Algorithm 18.3) and return to normal sequential quadratic programming when x_{k+1} is found, whose corresponding quadratic programming is feasible. Otherwise, set $x_{k+1} = x_k + d_k$ and go to step 4
4.	If (h_{k+1}, f_{k+1}) is acceptable to the filter, then: Accept x_{k+1} and add (h_{k+1}, f_{k+1}) to the filter Remove the points dominated by $(h_{k+1}, f_{k+1})$from the filter Possibly increase the trust-region radius ρ Otherwise, go to step 5
5.	If $h_{k+1} > 0$, then solve a sequence of quadratic programming problems for the second-order correction step $\widehat{d}_k$. Set $\widehat{x}_{k+1} = x_k + \widehat{d}_k$ If $\left(\widehat{h}_{k+1}, \widehat{f}_{k+1}\right)$ is acceptable to the filter, then: Accept $\widehat{x}_{k+1}$ and $\left(\widehat{h}_{k+1}, \widehat{f}_{k+1}\right)$ to the filter Remove the points dominated by $\left(\widehat{h}_{k+1}, \widehat{f}_{k+1}\right)$ from the filter Possibly increase the trust-region radius ρ Otherwise, go to step 6

(continued)

6.	If no acceptable point in the filter, then: If the algorithm is in the first iteration after the restoration phase, then: Accept the best second-order correction step, i.e., set $x_{k+1} = \widehat{x}_{k+1}$ Remove all the blocking entries from the filter Reduce the upper bound to: $u = \max\left\{\widehat{h}_{k+1}, u/10\right\}$ Otherwise, reject step $\widehat{d}_k$, i.e., set $x_{k+1} = x_k$, and reduce the trust-region radius ρ
7.	Set $k = k+1$ and go to step 2 ♦

Some more details regarding a specific implementation of this algorithm can be found in Fletcher and Leyffer (2002).

Example 18.2 Let us consider the nonlinear optimization problem from example 18.1 solved by means of the filter with sequential linear programming filterSD. The filterSQP algorithm in the implementation given by Fletcher and Leyffer (1999) with initial point $x_0 = [0,0,0,0,0,0]^T$ gives the following results.

The value of the objective function in the initial point is $f(x_0) = 10$. FilterSQP determines the solution $x^* = [1.14651,\ 0.54659,\ 1,\ 0.27329,\ 0.29996,\ 0]^T$for which the value of the objective is $f(x^*) = 0.75928439$. This solution was obtained in four iterations, five evaluations of the functions defining the problem, five evaluations of the gradients of the functions, and five evaluations to the Hessian of the Lagrange function. The number of quadratic programming sub-problems is four. The value of the norm of constraints at the solution is $\|c(x^*)\| = 1.3222E-10$. The final trust-region radius is $\rho = 10$. The norm of the KKT optimality conditions is equal to $1.393E-8$. ♦

The global convergence of the trust-region sequential quadratic programming filter algorithm, filterSQP, under reasonable conditions (the iterates x_k are in a compact set, the functions $f(x)$ and $c(x)$ are twice continuously differentiable, the Hessian of the Lagrange function associated to the problem is bounded), has been proved by Fletcher, Gould, Leyffer, Toint, and Wächter (2002). The idea of the proof is to decompose the step into its normal and tangential components. This allows for an approximate solution of the quadratic sub-problem and incorporates the safeguarding test described above. The normal step is responsible for the feasibility of the linearized constraints from (18.14). The tangential step reduces the value of the objective function model while continuing to satisfy the constraints from (18.14).

Notes and References

The filter method was suggested by Fletcher at a SIAM Conference in 1999. The filter method is motivated by the aim of avoiding the need to choose penalty parameters, as it would occur with the use of l_1 penalty functions or augmented Lagrangan functions. The filter methods know two implementations: with sequential linear programming and with sequential quadratic programming. The global convergence of the filter method with sequential linear programming was proved by Fletcher, Leyffer, and Toint (1999). Later on, the global convergence of the filter method with sequential quadratic programming was presented by Fletcher, Leyffer, and Toint (2002), Ulbrich (2004), and Fletcher, Leyffer, and Shen (2009). These filter methods were independently suggested by Surry, Radcliffe, and Boyd (1995) and Lemaréchal, Nemirovskii, and Nesterov (1995). Ulbrich, Ulbrich, and Vicente (2004) considered a trust-region filter method based on the acceptance of the trial steps using the norm of the optimality conditions. On the other hand, Benson, Shanno, and Vanderbei (2002a, 2002b) proposed several heuristics based on the idea of the filter method, for which improved efficiency has been reported versus the merit function approach. A comprehensive description of an algorithm called IPOPT, including the feasibility restoration phase for the filter method, the second-order corrections, the inertia correction of the KKT matrix, and heuristics that allow faster convergence was given by Wächter and Biegler (2006) (see Chapter 20).

Chapter 19
Interior Point Sequential Linear-Quadratic Programming: KNITRO/INTERIOR

In Chapter 14 the KNITRO/ACTIVE algorithm based on the active-set sequential linear-quadratic programming method has been presented. In this chapter the KNITRO/INTERIOR algorithm is being described, together with its numerical performances for solving large-scale general continuously nonlinear optimization problems. KNITRO/INTERIOR provides two procedures for computing the steps within the interior point approach. In the version INTERIOR-CG, each step is computed using a projected conjugate gradient iteration. It factors a projection matrix and uses the conjugate gradient method to approximately minimize a quadratic model of the barrier problem. In the version INTERIOR-DIRECT, the algorithm attempts to compute a new iterate by solving the primal-dual KKT system using direct linear algebra. In case this step cannot be guaranteed to be of good quality or if a negative curvature is detected, then the new iterate is computed by the INTERIOR-CG algorithm. The description of the KNITRO/INTERIOR-CG algorithm is given by Byrd, Hribar, and Nocedal (1999), and its global convergence theory is presented by Byrd, Gilbert, and Nocedal (2000). The method implemented in the KNITRO/INTERIOR-DIRECT algorithm is described by Waltz, Morales, Nocedal, and Orban (2003).

Consider the problem

$$\min_{x \in \mathbb{R}^n} f(x) \tag{19.1a}$$

subject to

$$c_E(x) = 0, \tag{19.1b}$$

$$c_I(x) \geq 0, \tag{19.1c}$$

where $f: \mathbb{R}^n \to \mathbb{R}$, $c_E : \mathbb{R}^n \to \mathbb{R}^l$, and $c_I : \mathbb{R}^n \to \mathbb{R}^m$ are twice continuously differentiable functions.

N. Andrei, *Continuous Nonlinear Optimization for Engineering Applications in GAMS Technology*, Springer Optimization and Its Applications 121,
DOI 10.1007/978-3-319-58356-3_19

The interior point or the barrier methods implemented in KNITRO associate the following barrier problem to (19.1)

$$\min_{x \in \mathbb{R}^n, s \in \mathbb{R}^m} f(x) - \mu \sum_{i=1}^{m} \log(s_i) \tag{19.2a}$$

subject to:

$$c_E(x) = 0, \tag{19.2b}$$

$$c_I(x) - s = 0, \tag{19.2c}$$

where $s \in \mathbb{R}^m$ is a vector of slack variables and $\mu > 0$ is the barrier parameter. *The interior point methods consist in finding the approximate solutions of the barrier problem* (19.2) *for a sequence of positive barrier parameters* $\{\mu_k\}$ *that converges to zero.*

The KKT optimality conditions for (19.2) are as follows:

$$\nabla f(x) - J_E^{\mathrm{T}}(x)y - J_I^{\mathrm{T}}(x)z = 0, \tag{19.3a}$$

$$-\mu e + Sz = 0, \tag{19.3b}$$

$$c_E(x) = 0, \tag{19.3c}$$

$$c_I(x) - s = 0, \tag{19.3d}$$

where $e = [1, \ldots, 1]^{\mathrm{T}}$, $S = \mathrm{diag}(s_1, \ldots, s_m)$, $J_E(x)$, and $J_I(x)$ are the Jacobian matrices corresponding to the equality and inequality constraints vectors, respectively. The vectors y and z are the Lagrange multipliers (dual variables) associated to the equality and inequality constraints. We also have $s, z > 0$.

In the line-search approach, the Newton method is applied to (19.3), backtracking if necessary, so that the variables s and z remain positive, and a merit function is sufficiently reduced. In the trust-region approach, a quadratic programming problem is associated to (19.2), and the step of the algorithm is an approximate solution of this quadratic programming sub-problem. These two approaches are implemented in KNITRO-INTERIOR/DIRECT and KNITRO-INTERIOR/CG algorithms, respectively.

A very important ingredient in interior point methods is the procedure for choosing the sequence of barrier parameters $\{\mu_k\}$. KNITRO contains several options. In the *Fiacco-McCormic monotone strategy*, the barrier parameter μ is held fixed for a series of iterations until the KKT (19.3) are satisfied to some accuracy. In the *adaptive strategy*, the barrier parameter is updated at every iteration using different rules: the rule implemented in LOQO based on the deviation of the minimum complementarity pair from the average (Vanderbei & Shanno 1999); a probing strategy that uses Mehrotra's predictor step to select a target value for μ; a quality-function approach, and some other rules described by Nocedal, Wächter, and Waltz (2005).

To control the quality of the steps, both interior point algorithms implemented in KNITRO make use of a non-differentiable merit function

$$\Phi_\sigma(x,s) = f(x) - \mu \sum_{i=1}^{m} \log(s_i) + \sigma\|c_E(x)\|_2 + \sigma\|c_I(x) - s\|_2, \tag{19.4}$$

where $\sigma > 0$. A step is acceptable only if it provides a sufficient decrease of the merit function (19.4). Now, let us present the interior point algorithms implemented in KNITRO.

19.1 KNITRO/INTERIOR-DIRECT Algorithm

In this algorithm, the search direction is determined by direct solving of the Newton system associated to the nonlinear system given by the KKT optimality conditions (19.3). To obtain global convergence in the presence of nonconvexity or Hessian or Jacobian singularities, this step may be replaced, under certain circumstances, by a safeguarding trust-region step. KNITRO/INTERIOR-DIRECT is described by Byrd, Nocedal, and Waltz (2006). See also Waltz, Morales, Nocedal, and Orban (2003).

By applying the Newton method to system (19.3) in the variables x, s, y, z, we get:

$$\begin{bmatrix} \nabla^2_{xx}L & 0 & -J_E^T(x) & -J_I^T(x) \\ 0 & Z & 0 & S \\ J_E(x) & 0 & 0 & 0 \\ J_I(x) & -I & 0 & 0 \end{bmatrix} \begin{bmatrix} d_x \\ d_s \\ d_y \\ d_z \end{bmatrix} = - \begin{bmatrix} \nabla f(x) - J_E^T(x)y - J_I^T(x)z \\ Sz - \mu e \\ c_E(x) \\ c_I(x) - s \end{bmatrix}, \tag{19.5}$$

where, as usual, L represents the Lagrange function:

$$L(x,s,y,z) = f(x) - y^T c_E(x) - z^T(c_I(x) - s). \tag{19.6}$$

If the inertia of the matrix in (19.5) is

$$(m+n, l+m, 0), \tag{19.7}$$

then the step d determined as solution of (19.5) can be guaranteed to be a descent direction for the merit function (19.4). In this case compute the scalars:

$$\alpha_s^{\max} = \max\{\alpha \in (0,1] : s + \alpha d_s \geq (1-\tau)s\}, \tag{19.8a}$$

$$\alpha_z^{\max} = \max\{\alpha \in (0,1] : z + \alpha d_z \geq (1-\tau)z\}, \tag{19.8b}$$

where $\tau = 0.995$. If $\min\{\alpha_s^{\max}, \alpha_z^{\max}\}$ is not too small, then perform a backtracking line search that computes the stepsizes

$$\alpha_s \in \left(0, \alpha_s^{\max}\right], \quad \alpha_z \in \left(0, \alpha_z^{\max}\right], \tag{19.9}$$

which achieve a sufficient decrease of the merit function (19.4).

The new iterate is computed as:

$$x^+ = x + \alpha_s d_x, \quad s^+ = s + \alpha_s d_s, \tag{19.10a}$$

$$y^+ = y + \alpha_z d_y, \quad z^+ = z + \alpha_z d_z. \tag{19.10b}$$

On the other hand, if the inertia of the matrix in (19.5) is not as in (19.7) or if the stepsize α_s or α_z is less than a given threshold, then the step d solution of (19.5) is rejected. In this case, the search direction d is not a descent one for the merit function (19.4). The algorithm reverts to the trust-region method implemented in KNITRO/INTERIOR-CG, algorithm described in the next section which is guaranteed to provide a successful step even in the presence of a negative curvature or singularity.

It is worth mentioning that this strategy implemented in KNITRO-INTERIOR/DIRECT is different from the line-search interior point algorithms IPOPT (Wächter & Biegler, 2005a, 2005b, 2006) and LOQO (Vanderbei & Shanno, 1999). Whenever the inertia condition is not satisfied, then IPOPT and LOQO modify the Hessian $\nabla_{xx}^2 L$. Instead, KNITRO-INTERIOR/DIRECT implements the above strategy because this permits to compute a step using a null-space approach without modifying the Hessian $\nabla_{xx}^2 L$. Moreover, using the trust-region step guarantees progress in case the line-search approach can fail (Wächter & Biegler, 2000; Byrd, Marazzi, & Nocedal, 2004).

It is known that the step lengths α_s or α_z in (19.10) converge to zero when the line-search iterations converge to nonstationary points. In this case the algorithm monitors these step lengths. If one of them is smaller than a given threshold, then the iterate given by (19.5) and (19.10) is discarded and replaced with the trust-region step.

Therefore, we see that the algorithm monitors two criteria: the *inertia* of the matrix in (19.5) and the *step lengths* (19.9). If these two criteria are satisfied, then we continue the iterations by using the Newton method for solving (19.5) by the direct method.

For initialization of the algorithm, it is necessary to compute the initial values of the multipliers y_0 and z_0. The multipliers are computed as the least-squares solution of the system (19.3a) and (19.3b). If the line search is discarded, then one or more KNITRO/INTERIOR-CG iterations are executed until one of them provides sufficient reduction in the merit function. The KNITRO-INTERIOR/DIRECT algorithm presented below uses $D\Phi_\sigma(x, s, d)$, which is the directional derivative of the merit function Φ_σ along a direction d.

Algorithm 19.1 KNITRO-INTERIOR/DIRECT (Byrd, Hribar, and Waltz)

1.	Choose the initial point x_0, $s_0 > 0$, and the parameters $\eta > 0$ and $0 < \alpha_{\min} < 1$. Compute the initial values of multipliers y_0 and $z_0 > 0$. Choose a value of the trust-region radius $\Delta_0 > 0$ and a value for the barrier parameter $\mu > 0$. Set $k = 0$
2.	If a test for stopping the iterations is satisfied, then stop; otherwise go to step 3
3.	If the perturbed KKT optimality conditions (19.3) are approximately satisfied, then go to step 9; otherwise go to step 4
4.	Factorize the matrix of the primal-dual system (19.5), and compute *neig* as the number of the negative eigenvalues of this matrix
5.	Set *LineSearch* = *False*
6.	If $neig \leq l + m$, then: Solve the system (19.5) to obtain the direction $d = (d_x, d_s, d_y, d_z)$ Define $w = (x_k, s_k)$ and $d_w = (d_x, d_s)$ Compute $\alpha_s^{\max}$ and $\alpha_z^{\max}$ as in (19.8) If $\min\{\alpha_s^{\max}, \alpha_z^{\max}\} > \alpha_{\min}$, then: Update the penalty parameter σ_k(see the merit function below) Compute a stepsize $\alpha_s = \overline{\alpha}\alpha_s^{\max}$, $\overline{\alpha} \in (0, 1]$, such that $\Phi_\sigma(w + \alpha_s d_w) \leq \Phi_\sigma(w) + \eta\alpha_s D\Phi_\sigma(w, d_w)$ If $\alpha_s > \alpha_{\min}$, then: Set $\alpha_z = \overline{\alpha}\alpha_z^{\max}$ Compute $(x_{k+1}, s_{k+1}, y_{k+1}, z_{k+1})$ as in (19.10) Set *LineSearch* = *True*
7.	If *LineSearch* = *False*, then compute $(x_{k+1}, s_{k+1}, y_{k+1}, z_{k+1})$ using the algorithm KNITRO-INTERIOR/CG described in the next section
8.	Compute Δ_{k+1}. Set $k = k + 1$ and go to step 3
9.	Choose a new smaller value for the barrier parameter μ and go to step 2 ♦

Observe that the algorithm computes and maintains a trust-region radius Δ_k at every iteration in case it needs to revert to the trust-region KNITRO-INTERIOR/CG algorithm in step 7.

19.2 KNITRO/INTERIOR-CG Algorithm

The second algorithm implemented in KNITRO computes the search directions using a quadratic model and trust regions. This strategy permits great freedom in the choice of the Hessian and provides a mechanism for coping with Jacobian and Hessian singularities. However, the iterations of this algorithm are more complex. The KNITRO-INTERIOR/CG algorithm is described by Byrd, Hribar, and Nocedal (1999) and analyzed by Byrd, Gilbert, and Nocedal (2000).

To describe this algorithm, observe that the barrier problem (19.2) is an equality constrained optimization problem which can be solved by using a sequential quadratic programming (SQP) method with trust regions. However, a straightforward application of the SQP method to the barrier problem leads to inefficient steps that tend to violate the positivity of the slack variables and frequently cuts short the trust-region constraint. To overcome this difficulty, Byrd, Nocedal, and Waltz (2006) suggest the following SQP method associated to the barrier problem (19.2).

At the current iterate (x_k, s_k) and for a given value of the barrier parameter μ, first compute the Lagrange multiplier estimates (y_k, z_k), and then the step $d=(d_x, d_s)$ is computed as solution of the following sub-problem:

$$\begin{aligned} \min_{d_x,d_s} \nabla f(x_k)^{\mathrm{T}} d_x + \frac{1}{2} d_x^{\mathrm{T}} \nabla_{xx}^2 L(x_k, s_k, y_k, z_k) d_x \\ -\mu e^{\mathrm{T}} S_k^{-1} d_s + \frac{1}{2} d_s^{\mathrm{T}} \Sigma_k d_s \end{aligned} \tag{19.11a}$$

subject to:

$$c_E(x_k) + J_E(x_k) d_x = r_E, \tag{19.11b}$$

$$c_I(x_k) + J_I(x_k) d_x - d_s - s_k = r_I, \tag{19.11c}$$

$$\left\| d_x, S_k^{-1} d_s \right\|_2 \le \Delta_k, \tag{19.11d}$$

$$d_s \ge -\tau s, \tag{19.11e}$$

where $\Sigma_k = S_k^{-1} Z_k$ and $\tau = 0.995$. Ideally, the residues r_E and $r_I$should be zero, i.e., $r=(r_E, r_I)=0$, but since this value can determine the constraints to be incompatible or produce a poor step, it is better to choose r as the smallest vector such that the constraints (19.11b, 19.11c and 19.11d) should be consistent.

The choice of the objective function (19.11a) is motivated by the fact that the first-order optimality conditions of (19.11a)–(19.11c) are given by (19.3) with the second block of equations scaled by S^{-1}. Therefore, the steps computed by using (19.11) are related to those of the line-search algorithm described above, in the previous section. The trust-region constraint (19.11d) guarantees that (19.11) has a finite solution even if the Hessian of the Lagrangian $\nabla_{xx}^2 L(x_k, s_k, y_k, z_k)$ is not positive definite. Hence, the Hessian matrix need never be modified in this algorithm. The scaling S_k^{-1} used in the trust-region constraint is crucial in the economy of the algorithm.

Step Computation

The presence of the nonlinear constraints (19.11d) and of the bounds (19.11e) makes solving of (19.11) with high accuracy difficult. However, useful inexact solutions can be computed at a moderate cost. KNITRO uses the null-space approach in which the step d is computed as a sum of a *normal step* v that attempts to satisfy the linear constraints (19.11b)–(19.11c) with $r=0$ as well as possibly the trust-region and a *tangential step* that lies on the tangent space of the constraints and that tries to achieve optimality.

The normal step $v=(v_x, v_s)$ is the solution of the following sub-problem:

$$\min_v \|J_E v_x + c_E\|_2^2 + \|J_I v_x - v_s + c_I - s\|_2^2 \tag{19.12a}$$

subject to:

$$\left\|\left(v_x, S^{-1}v_s\right)\right\|_2 \le 0.8\Delta, \tag{19.12b}$$

in which the arguments of the functions appearing in (19.12) have been omitted. An inexact solution to the sub-problem (19.12) is computed by using a dogleg approach, which minimizes (19.12a) along a piecewise linear path composed of a steepest descent step in the norm used in (19.12b) and a minimum-norm Newton step with respect to the same norm. The scaling $S^{-1}v_s$ in the norm from (19.12b) tends to limit the extent to which the bounds on the slack variables are violated.

Once the normal step v has been computed, the vectors r_E and r_I from (19.11b) and (19.11c), respectively, are computed as the residuals, namely:

$$r_E = J_E v_x + c_E, \quad r_I = J_I v_x - v_s + (c_I - s).$$

With the normal step $v=(v_x, v_s)$ computed, the sub-problem (19.11) can be written as:

$$\min_{d_x, d_s} \nabla f^T d_x - \mu e^T S^{-1} d_s + \frac{1}{2}\left(d_x^T \nabla_{xx}^2 L d_x + d_s^T \Sigma d_s\right) \tag{19.13a}$$

subject to:

$$J_E d_x = J_E v_x, \tag{19.13b}$$

$$J_I d_x - d_s = J_I v_x - v_s, \tag{19.13c}$$

$$\left\|\left(d_x, S^{-1}d_s\right)\right\|_2 \le \Delta, \tag{19.13d}$$

which is called the *tangential sub-problem*. Now, to find an approximate solution d of sub-problem (19.13), firstly introduce the scaling

$$\tilde{d}_s = S^{-1} d_s, \tag{19.14}$$

which transforms (19.13d) into a sphere. Then the projected conjugate gradient (CG) method to the transformed quadratic program is applied, where all iterates are in the linear manifold defined by (19.13b)–(19.13c). While solving by CG, the strategy of Steihaug is used, by monitoring the satisfaction of the trust-region constraint (19.13d). The iterations are stopped if the boundary of this region is reached or if a negative curvature is detected. Finally, if necessary, the step d is truncated to satisfy (19.11e). The algorithm KNITRO-INTERIOR/CG is as follows.

Algorithm 19.2 KNITRO-INTERIOR/CG (Byrd, Hribar, and Waltz)

1.	Choose x_0, $s_0 > 0$, and $\Delta_0 > 0$. Choose a value for $\eta > 0$. Set $k = 0$
2.	If a test for stopping the algorithm is satisfied, stop; otherwise go to step 3
3.	If the perturbed KKK system (19.3) is approximately satisfied, go to step 11; otherwise go to step 4
4.	Compute the normal step $v_k = (v_x, v_s)$
5.	Compute the Lagrange multipliers y_k and $z_k > 0$
6.	Compute the total step d_k by applying the projected conjugate gradient algorithm to the sub-problem (19.13a, 19.13b and 19.13c)
7.	Update the penalty parameter σ_k (see the merit function below)
8.	Compute $ared_k(d_k)$ and $pred_k(d_k)$ using the relations below
9.	If $ared_k(d_k) \geq \eta\ pred_k(d_k)$, then set $x_{k+1} = x_k + d_x$, $s_{k+1} = s_k + d_s$, and update Δ_{k+1}; otherwise, set $x_{k+1} = x_k$ and $s_{k+1} = s_k$ and choose $\Delta_{k+1} < \Delta_k$
10.	Set $k = k + 1$ and go to step 3
11.	Choose a smaller value for the barrier parameter μ and go to step 2 ♦

In algorithm 19.2

$$ared(d) = \Phi_\sigma(x, s) - \Phi_\sigma(x + d_x, s + d_s) \tag{19.15}$$

is the actual reduction in the merit function. $pred(d)$ is immediately defined.

The Lagrange multiplier estimates (y_k, z_k) are computed by a least-squares approximation to the equations (19.3a)–(19.3b) evaluated at point x_k and truncated to ensure positivity of z_k.

Observe that the interior point Algorithm 19.2 is asymptotically equivalent to standard line-search interior point algorithms, but it is significantly different in two respects. First, it is not a fully primal-dual method because the multipliers are computed as a function of the primal variables (x, s) as opposed to the formulation (19.5) in which the primal and the dual variables are computed simultaneously from their previous values. Secondly, the trust-region method uses a scaling of the variables that avoids moves toward the boundary of the feasible region. Therefore, the algorithm generates steps very different from those produced by a line-search method.

Merit Function

The role of the merit function (19.4) is to determine whether a step can be accepted. Byrd, Hribar, and Nocedal (1999) show that the efficiency of the algorithm depends on the choice of the penalty parameter σ. In both interior point algorithms implemented in KNITRO/INTERIOR, at every iteration, σ is chosen so that the decrease in the quadratic model of the merit function produced by a step d is proportional with σ multiplied by the decrease in linearized constraints.

More exactly, suppose that either the KNITRO-INTERIOR/DIRECT or the KNITRO-INTERIOR/CG algorithms has generated a step d. Then define the following linear/quadratic model of the merit function Φ_σ:

$$Q_\sigma(d) = \nabla f^{\mathrm{T}} d_x - \mu e^{\mathrm{T}} S^{-1} d_s + \frac{\xi}{2}\left(d_x^{\mathrm{T}} \nabla_{xx}^2 L d_x + d_s^{\mathrm{T}} \Sigma d_s\right) + \sigma(m(0) - m(d)), \tag{19.16}$$

where

$$m(d) = \left\| \begin{bmatrix} J_E d_x + c_E \\ J_I d_x - d_s + c_I - s \end{bmatrix} \right\|_2, \tag{19.17}$$

represents the first-order violation of the constraints, and ξ is a parameter to be discussed below.

With these, define the predicted decrease in the merit function as:

$$pred(d) = Q_\sigma(0) - Q_\sigma(d). \tag{19.18}$$

In all cases the penalty parameter σ is chosen large enough such that

$$pred(d) \geq \rho\sigma(m(0) - m(d)), \tag{19.19}$$

where $0 < \rho < 1$ (e.g., $\rho = 0.1$).

If the value of σ from the previous iteration satisfies (19.19), then σ is left unchanged. Otherwise, σ is increased so that it satisfies the inequality (19.19) with some accuracy.

For the trust-region method as implemented in KNITRO-INTERIOR/CG, set $\xi = 1$ in (19.16). On the other hand, in the KNITRO-INTERIOR/DIRECT parameter, ξ is defined as:

$$\xi = \begin{cases} 1, & \text{if} \quad d_x^{\mathrm{T}} \nabla_{xx}^2 L d_x + d_s^{\mathrm{T}} \Sigma d_s > 0, \\ 0, & \text{otherwise.} \end{cases} \tag{19.20}$$

This choice of ξ guarantees the directional derivative of Φ_σ in the direction d is negative.

Computational Aspects

KNITRO-INTERIOR contains many algorithmic options and features that are listed and discussed in the documentation of the package (Waltz, 2004). In the following we shall present some of these options implemented in KNITRO.

Initial point strategy It is well known that any interior point algorithm is very sensitive to the choice of the initial point. In KNITRO several different strategies are implemented. One strategy is as follows. At an initial point x_0 specified by the user, an affine scaling step $d^A = \left(d_x^A, d_s^A, d_y^A, d_z^A\right)$ is computed by solving the system (19.5) with $\mu = 0$. Then, $s_1 = \max\{1, |s_0 + d_s^A|\}$ and $z_1 = \max\{1, |z_0 + d_z^A|\}$ are defined, where the operators max and absolute are componentwise applied. The primal variables x and the multipliers y associated to the equality constraints are not altered, i.e., $(x_1, y_1) = (x_0, y_0)$. Finally, the initial value of the barrier parameter is computed as $\mu_1 = s_1^{\mathrm{T}} z_1 / m$.

Hessian options There are some options in KNITRO for using the second derivatives. One possibility is that the user can supply the first and the second derivatives, which generally results in the greatest level of robustness and efficiency for all three algorithms implemented in KNITRO. In some particular applications, the Hessian $\nabla_{xx}^2 L$ cannot be computed or is too large to store. In these cases the KNITRO/INTERIOR-CG and ACTIVE algorithms allow the user to provide the products of Hessian to vectors at every iteration. Another variant implemented in all three algorithms in KNITRO is to approximate $\nabla_{xx}^2 L$ by quasi-Newton updates: BFGS, memory-less BFGS, or SR1. For example, BFGS is implemented as

$$B_{k+1} = B_k + \frac{q_k q_k^{\mathrm{T}}}{s_k^{\mathrm{T}} q_k} - \frac{B_k s_k s_k^{\mathrm{T}} B_k}{s_k^{\mathrm{T}} B_k s_k}, \tag{19.21}$$

where $q_k = \nabla_x L(x^+, s^+, y^+, z^+) - \nabla_x L(x, s^+, y^+, z^+)$ and $s_k = x_{k+1} - x$.

Feasible iterations In some applications it is desirable for all the iterates generated by the optimization algorithm to be feasible with respect to some or to all of the inequality constraints. KNITO uses the following procedure to generate feasible iterates. If the current iteration x satisfies the constraints $c_I(x) > 0$, then after computing the step d, set $x^+ = x + d_x$. The slack variables are redefined as $s^+ = c_I(x^+)$ and tested whether the point (x^+, s^+) is acceptable for the merit function Φ_σ. If the case, this point is defined as the new iterate; otherwise, this point is rejected and a new shorter one is computed (in a line-search method by backtracking and in a trust-region method by reducing the trust-region radius). The justification of this strategy is as follows. If at a trial point we have $c_i(x^+) \leq 0$ for some inequality constraint, then the value of the merit function is $+\infty$, and this trial point is rejected. This strategy also rejects the steps $x + d_x$ that are too close to the boundary of the feasible region because such steps increase the barrier term $-\mu \sum_{i=1}^{m} \log(s_i)$ in the merit function.

Crossover technique The situation is that the interior point methods provide only an approximate estimate of the solution and of the optimal active set. However, in many practical applications, it is useful to know with high accuracy which constraints are active. Moreover, quite often, it is important to know highly accurate estimates of the Lagrange multipliers. They are important in the sensitivity analysis of the solution to some variations of some parameters in the problem (see Chapter 5). Increasing the accuracy of the solution or of the active set can be done by switching from the interior point method to an active-set iteration, a process which is called *crossover*. This technique was introduced for the first time in linear programming by Megiddo (1989). In linear programming, the crossover technique involves two stages: identifying the active constraints and moving from a nonbasic optimal solution to a nearby basic one. However, in nonlinear programming, this form of crossover cannot be used because we cannot expect the set of active constraints to correspond to a basic solution. In nonlinear optimization, the crossover was introduced for the first time by Byrd, Nocedal, and Waltz (2006). It seeks to identify a set of active constraints (with linearly independent constraint gradients) and computes a solution at which those constraints are satisfied with near equality, solution which satisfies the Lagrangian stationarity using only these constraints.

The crossover procedure internally commutes to the KNITRO/ACTIVE algorithm *after* the KNITRO/INTERIOR-DIRECT or the KNITRO/INTERIOR-CG algorithms have solved the problem to the required tolerance. Firstly, solve the equality quadratic programming (14.9) to generate a new solution estimate. If this step does not solve the problem immediately, then the full KNITRO/ACTIVE algorithm is started with an initial LP trust-region radius Δ_k^{LP} computed on the basis of the active-set estimate. The idea is to choose Δ_k^{LP} small enough to exclude all the inactive constraints but large enough to include the active ones. The crossover algorithm as described in Byrd, Nocedal, and Waltz (2006) is as follows.

Algorithm 19.3 KNITRO crossover algorithm

1.	The interior point algorithms DIRECT or CG terminates with a solution (x_k, s_k, y_k, z_k) which satisfies the stopping tolerance ε_{TOL}
2.	Estimate the set of active constraints A
3.	Using this active-set estimate, generate a step d^Q by solving the equality quadratic programming problem (14.9). Perform a line search to determine the stepsize α^Q. If $x_k + \alpha^Q d^Q$ satisfies the stopping tolerances, then the algorithm stops with this point and with the corresponding multipliers; otherwise, go to step 4

(continued)

4.	Determine the initial LP trust-region radius Δ_0^{LP} and the penalty parameter σ_0 for the KNITRO/ACTIVE (Algorithm 14.1) as: $$\Delta_0^{LP} = \min\left\{\frac{c_i(x_k, s_k)}{\|\nabla c_i(x_k, s_k)\|} : i \notin A\right\},$$ $$\sigma_0 = 10\|(y_k, z_k)\|_\infty$$
5.	Start KNITRO/ACTIVE algorithm using the initial point (x_k, s_k, y_k, z_k), Δ_0^{LP}, and σ_0 ♦

In step 3 of algorithm 19.3, the active set is estimated by using a tolerance test rather than by solving the linear programming problem (14.4). This is because, in some difficult problems, the cost of solving the linear programming sub-problem can be non-trivial and, as motivated by Byrd, Nocedal, and Waltz (2006), the cost of the crossover procedure shall be a small part of the overall solution time. Therefore, it is not necessary to solve (14.4) to identify the optimal active set. If strict complementarity holds at the solution, the initial estimate of the active set based on the simple tolerance test will be correct, and the crossover will succeed in one iteration without solving (14.4).

The formula for computing the initial trust-region radius Δ_0^{LP} from step 4 of the algorithm 19.3 guarantees that if the active set estimate is correct, then this initial trust-region radius will be small enough to exclude all the inactive constraints. As motivated by the theory of l_1 exact penalty function, the penalty parameter is initialized to be a little larger than the Lagrange multiplier of the largest magnitude at the interior point solution.

Practical Hints in GAMS

The KNITRO options file `knitro.opt` allows the user to set the algorithmic options. The list of options in KNITRO is described in (*GAMS – The Solver Manuals* 2005, pp.177–190). KNITRO implements four algorithms. To select one of them in `knitro.opt` file, the option `algorithm` must be introduced with the following values:

`algorithm 0`	KNITRO will automatically choose the best algorithm based on the problem characteristics
`algorithm 1`	KNITRO will use the INTERIOR-DIRECT algorithm
`algorithm 2`	KNITRO will use the INTERIOR-CG algorithm
`algorithm 3`	KNITRO will use the ACTIVE algorithm (see Chapter 14)

Another option active in interior point algorithms implemented in KNITRO and which can be written in `knitro.opt` file is `barrule`, which controls the barrier parameter update strategy. This option has the following values:

`barrule 0`	KNITRO will automatically choose the rule for updating the barrier parameter
`barrule 1`	KNITRO will monotonically decrease the barrier parameter
`barrule 2`	KNITRO will use an adaptive rule based on the complementarity gap to determine the value of the barrier parameter at every iteration
`barrule 3`	KNITRO will take a probing (affine-scaling) step to dynamically determine the barrier parameter value at every iteration
`barrule 4`	KNITRO uses a Mehrotra predictor-corrector type rule to determine the value of the barrier parameter with safeguards on the corrector step
`barrule 5`	KNITRO uses a Mehrotra predictor-corrector type rule to determine the value of the barrier parameter without safeguards on the corrector step

Numerical Study

In the following, Tables 19.1, 19.2, and 19.3 present the performances of KNITRO with option 0 (KNITRO will automatically try to choose the best algorithm based on the problem characteristics), option 1 (KNITRO will use the INTERIOR-DIRECT algorithm), and KNITRO with option 2 (KNITRO will use the INTERIOR-CG algorithm) for solving 13 nonlinear optimization applications described in Chapter 4.

The performances of KNITRO/ACTIVE (option 3) for solving small-scale nonlinear optimization applications are presented in Table 14.1. Table 19.4 presents the total number of iterations (*itt*), the total number of evaluations of the functions defining the problems (*nft*), the total number of evaluations of the gradients of the functions (*ngt*), the total number of evaluations of the Hessian (*nht*), and the total cpu time (*cput*) in seconds for solving 13 nonlinear optimization applications

Table 19.1 Performances of KNITRO. Option 0. Small-scale nonlinear optimization applications

	n	*me*	*mc*	*it*	*itcg*	*nf*	*ng*	*nh*	*cpu*	*vfo*
ELCH	10	3	0	6	0	7	7	6	0.143	−47.761090
ALKI	10	3	8	10	0	13	11	10	0.011	−1768.8069
PREC	8	0	6	10	9	11	11	10	0.081	3.95116321
PPSE	9	6	0	8	0	9	9	8	0.009	5055.01179
MSP3	13	0	15	26	7	36	27	26	0.044	97.5875095
MSP5	16	0	21	23	10	31	24	23	0.045	174.786942
POOL	34	20	0	9	0	10	10	9	0.026	2569.800
TRAFO	6	0	2	16	18	19	17	16	0.009	135.07592
LATHE	10	1	14	23	5	50	24	23	0.020	−4430.0879
CAM	20	9	11	16	0	17	17	16	0.013	−43.859948
DES	150	50	0	39	350	99	40	39	0.479	1055.1823
CSTC	303	200	0	4	0	5	5	4	0.035	3.4800745
DIFF	396	324	0	2	0	3	1	0	0.016	0

Table 19.2 Performances of KNITRO-INTERIOR/DIRECT. Option 1. Small-scale nonlinear optimization applications

	n	*me*	*mc*	*it*	*itcg*	*nf*	*ng*	*nh*	*cpu*	*vfo*
ELCH	10	3	0	6	0	7	7	0	0.130	−47.761090
ALKI	10	3	8	10	0	13	11	10	0.009	−1768.8069
PREC	8	0	6	10	9	11	11	10	0.012	3.95116321
PPSE	9	6	0	8	0	9	9	8	0.008	5055.01179
MSP3	13	0	15	26	7	36	27	26	0.012	97.5875095
MSP5	16	0	21	23	10	31	24	23	0.014	174.786942
POOL	34	20	0	9	0	10	10	9	0.010	2569.800
TRAFO	6	0	2	16	18	19	17	16	0.009	135.07592
LATHE	10	1	14	23	5	50	24	23	0.017	−4430.0879
CAM	20	9	11	23	0	24	24	23	0.017	−43.859947
DES	150	50	0	39	350	99	40	39	0.470	1055.1823
CSTC	303	200	0	4	0	5	5	4	0.011	3.4800745
DIFF	396	324	0	2	0	3	1	0	0.020	0

Table 19.3 Performances of KNITRO-INTERIOR/CG. Option 2. Small-scale nonlinear optimization applications

	n	*me*	*mc*	*it*	*itcg*	*nf*	*ng*	*nh*	*cpu*	*vfo*
ELCH	10	3	0	6	27	7	7	6	0.008	−47.761090
ALKI	10	3	8	20	79	21	21	20	0.011	−1768.8065
PREC	8	0	6	22	124	24	23	22	0.057	4.2867607
PPSE	9	6	0	5	13	6	6	5	0.008	5055.01300
MSP3	13	0	15	36	237	37	37	36	0.019	97.5875197
MSP5	16	0	21	84	325	146	85	84	0.035	174.795082
POOL	34	20	0	17	59	18	18	17	0.011	2569.800
TRAFO	6	0	2	37	103	52	38	37	0.017	135.076250
LATHE	10	1	14	27	118	33	28	27	0.016	−4430.0806
CAM	20	9	11	20	122	21	21	20	0.014	−43.859868
DES	150	50	0	28	262	79	29	28	0.237	1055.1823
CSTC	303	200	0	4	10	5	5	4	0.019	3.4800745
DIFF	396	324	0	1	0	2	1	0	0.024	0

In the above tables, we have: n = the number of variables, me = the number of equality constraints, mc = the number of inequality constraints, it = the number of iterations to obtain a solution, $itcg$ = the number of conjugate gradient iterations, nf = the number of evaluations of the functions defining the problem, ng = the number of evaluations of the gradients of the functions defining the problem, nh = the number of evaluations of the Hessian, cpu = the cpu computing time for obtaining a solution of the problem (seconds), vfo = the value of the objective function at optimal point.

considered in this numerical study with option 0 (Table 19.1), option 1 (Table 19.2), option 2 (Table 19.3), and option 3 (Table 14.1), respectively.

Tables 19.5, 19.6, and 19.7 show the numerical performances of KNITRO with option 0, option 1, and option 2, respectively, for solving six large-scale nonlinear optimization applications of different dimensions, presented in Chapter 4.

Table 19.4 Performances of KNITRO algorithms. Small-scale nonlinear optimization applications

Option	*itt*	*nft*	*ngt*	*nht*	*cput*
0	192	310	203	190	0.931
1 (DIRECT)	199	317	210	191	0.739
2 (CG)	307	451	319	306	0.476
3 (ACTIVE)	283	737	295	423	0.796

Table 19.5 Performances of KNITRO. Option 0. Large-scale nonlinear optimization applications

	n	*me*	*mc*	*it*	*itcg*	*nf*	*ng*	*cpu*	*vfo*
HANG	2002	1001	0	21	73	44	22	0.349	5.0685101
	4002	2001	0	81	328	291	82	3.039	5.0684889
FLOW	1182	754	0	5	0	6	6	0.160	0.311e-11
FLOWO	1556	1005	0	10	19	11	11	1.031	0.107e-6
POL	4004	3000	0	27	35	29	28	1.203	14.009791
	6004	4500	0	26	35	27	27	1.836	13.990035
	8004	6000	0	28	36	29	29	2.568	14.009035
	10004	7500	0	30	39	31	31	3.454	14.007359
CAT	3003	2000	0	8	0	9	9	0.301	−0.048052
	6003	4000	0	8	0	9	9	0.674	−0.048048
	9003	6000	0	7	0	8	8	0.993	-0.048045
CONT	2505	2000	0	9	0	10	10	0.109	1.0132439
	5005	4000	0	10	0	11	11	0.403	1.0059324
	7505	6000	0	11	0	12	12	0.743	1.0045764
	10005	8000	0	11	0	12	12	0.947	1.0040918

Table 19.6 Performances of KNITRO/DIRECT. Option 1. Large-scale nonlinear optimization applications

	n	*me*	*mc*	*it*	*itcg*	*nf*	*ng*	*cpu*	*vfo*
HANG	2002	1001	0	21	73	44	22	0.425	5.068510
	4002	2001	0	81	328	291	82	3.068	5.0684889
FLOW	1182	754	0	5	0	6	6	0.160	3.85e-11
FLOWO	1556	1005	0	10	19	11	11	1.088	1.071e-5
POL	4004	3000	0	27	35	29	28	1.291	14.009791
	6004	4500	0	26	35	27	27	1.839	13.990035
	8004	6000	0	28	36	29	29	2.586	14.009035
	10004	7500	0	30	39	31	31	3.481	14.007359
CAT	3003	2000	0	8	0	9	9	0.353	−0.048052
	6003	4000	0	8	0	9	9	0.650	−0.048048
	9003	6000	0	7	0	8	8	1.055	-0.048045
CONT	2505	2000	0	9	0	10	10	0.190	1.0132439
	5005	4000	0	10	0	11	11	0.408	1.0059324
	7505	6000	0	11	0	12	12	0.734	1.0045764
	10005	8000	0	11	0	12	12	0.968	1.0040918

Performances of KNITRO/ACTIVE (option 3) for solving large-scale nonlinear optimization applications are presented in Table 14.2. Table 19.8 contains a comparison of these four variants of KNITRO subject to the total number of iterations (*itt*), the total number of conjugate gradient iterations (*itcgt*), the total number of evaluations of the function defining the problem (*nft*), the total number of evaluations of gradients of the functions defining the problem (*ngt*), and the total cpu computing time to obtain a solution (*cput*) for solving 15 nonlinear optimization applications considered in this numerical study, with option 0 (Table 19.5), option 1 (Table 19.6), option 2 (Table 19.7), and option 3 (Table 14.2), respectively.

Application Penici

This application is described in Chapter 4. Table 19.9 presents the performances of the algorithms implemented in KNITRO.

Table 19.7 Performances of KNITRO/CG. Option 2. Large-scale nonlinear optimization applications

	n	*me*	*mc*	*it*	*itcg*	*nf*	*ng*	*cpu*	*vfo*
HANG	2002	1001	0	6	141	7	7	0.211	5.0685100
	4002	2001	0	6	142	7	7	0.538	5.0684889
FLOW	1182	754	0	29	174	30	30	1.080	1.831e-11
FLOWO	1556	1005	0	48	731	49	49	4.104	1.8483e-5
POL	4004	3000	0	44	179	46	45	1.546	14.202608
	6004	4500	0	43	353	45	44	2.866	14.243107
	8004	6000	0	19	197	23	20	1.938	14.265885
	10004	7500	0	19	153	23	20	2.208	14.260745
CAT	3003	2000	0	10	46	11	11	0.379	−0.048047
	6003	4000	0	6	18	7	7	0.571	−0.047980
	9003	6000	0	7	25	8	8	1.069	−0.047954
CONT	2505	2000	0	13	92	14	14	0.357	1.0134407
	5005	4000	0	13	98	14	14	0.758	1.006334
	7505	6000	0	14	95	15	15	1.211	1.005174
	10005	8000	0	13	98	14	14	1.532	1.004447

Table 19.8 Performances of KNITRO algorithms. Large-scale nonlinear optimization applications

Option	*itt*	*itcgt*	*nft*	*ngt*	*cput*
0	292	565	539	307	17.81
1 (DIRECT)	292	565	539	307	18.296
2 (CG)	290	2542	313	305	20.367
3 (ACTIVE)	1123	2197	2711	1138	468.705

Table 19.9 Performances of KNITRO. Application PENICI ($n = 707, l = 602, m = 0$)

Option	*it*	*itcg*	*nf*	*ng*	*cpu*
0	397	2922	433	398	7.374
1 (DIRECT)	397	2922	433	398	7.357
2 (CG)	307	3994	559	308	4.786
3 (ACTIVE)	Too many iterations				

Notes and References

This chapter is based on the papers of Byrd, Hribar, and Nocedal (1999), Byrd, Gilbert, and Nocedal (2000), Waltz, Morales, Nocedal, and Orban (2003), Byrd, Nocedal, and Waltz (2006), and Andrei (2015). KNITRO is imbedded in the GAMS technology (see *GAMS – The Solver Manuals* 2005, pp. 177–190). KNITRO/INTERIOR implements two methods: trust-region and line-search interior point/barrier methods. The idea of the algorithm is to solve a sequence of barrier sub-problems for a decreasing sequence of barrier parameters. For accepting the step and for ensuring the global convergence, it uses a merit or a penalty function. The barrier sub-problems are solved by a sequence of linearized primal-dual equations. For solving the primal-dual system, KNITRO implements two options: the direct factorization of the system and the preconditioned conjugate gradient method. The preconditioned conjugate gradient method solves the indefinite primal-dual system by projecting onto the null space of the equality constraints. KNITRO uses the crossover technique to obtain an active set from the solution of the interior point method. In Chapter 21 we have the computational evidence that KNITRO is one of the best algorithms implemented in GAMS for solving nonlinear continuous optimization problems (see Figuers 21.1 and 21.2).

Chapter 20
Interior Point Filter Line Search: IPOPT

In this chapter, an implementation of an interior point filter line-search algorithm for large-scale nonlinear programming proposed by Wächter and Biegler (2005a, 2005b) is presented. As we know, to allow convergence from poor starting points and to enforce progress to the solution, interior point methods both in trust region and in line-search frameworks with exact penalty merit function have been developed. For example, KNITRO uses the l_1 exact penalty function (Byrd, Hribar, & Nocedal, 1999; Byrd, Gilbert, & Nocedal, 2000). On the other hand, Fletcher and Leyffer (2002) proposed filter methods as an alternative to merit functions, as a tool for global convergence guarantee in algorithms for nonlinear optimization. The idea of filter is that trial points are accepted if they improve the objective function value or improve the constraint violation instead of a combination of these two measures defined by the merit function. Even if the filter methods include different heuristics, they have been adapted to barrier methods in a number of ways. For example, Ulbrich, Ulbrich, and Vicente (2004) considered a trust-region filter method and accept the trial step on the basis of the norm of the optimality conditions. Benson, Shanno, and Vanderbei (2002a) proposed several heuristics using the idea of filter methods, for which improved efficiency is reported compared to their previous merit function approach. The global convergence of an interior point algorithm with a filter line search was analyzed by Wächter and Biegler (2001). Here, the assumptions made for the analysis of the global convergence are less restrictive than those made for line-search interior point methods for nonlinear programming developed, for example, by El-Bakry, Tapia, Tsuchiya, and Zhang (1996), Yamashita (1998), or Tits, Wächter, Bakhtiari, Urban, and Lawrence (2003).

In this context, IPOPT is an interior point filter line-search algorithm for solving large-scale nonlinear optimization problems developed by Wächter and Biegler (2001, 2006). The idea of this algorithm is to combine the primal-dual interior point algorithms with filter line search for solving problems of the form:

N. Andrei, *Continuous Nonlinear Optimization for Engineering Applications in GAMS Technology*, Springer Optimization and Its Applications 121,
DOI 10.1007/978-3-319-58356-3_20

$$\min f(x) \tag{20.1a}$$

subject to

$$c(x) - 0, \tag{20.1b}$$

$$x_L \leq x \leq x_U, \tag{20.1c}$$

where $x \in \mathbb{R}^n$, $x_L \in \mathbb{R}^n$, and $x_U \in \mathbb{R}^n$ are lower and upper bounds on variables, and f: $\mathbb{R}^n \to \mathbb{R}$ and $c : \mathbb{R}^n \to \mathbb{R}^m$, with $m \leq n$ are assumed to be twice continuously differentiable. Problems with general nonlinear inequality constraints can be reformulated in the above form by introducing slack variables.

In the following, we shall present the primal-dual interior point algorithm IPOPT which involves the step computation, the line search with filter, the second-order corrections, the restoration of feasibility, as well as some heuristics which improve the performances of the algorithm. Finally, numerical results for solving some nonlinear optimization applications described in Chapter 4 together with comparisons are presented.

20.1 Basic Algorithm IPOPT

The Primal-Dual Barrier Approach

To simplify the presentation, at the very beginning, the method will be presented for solving the problem:

$$\min f(x) \tag{20.2a}$$

subject to:

$$c(x) = 0, \tag{20.2b}$$

$$x \geq 0. \tag{20.2c}$$

Later on, we will see the method for solving the general nonlinear optimization problems with simple bound on variables. As a barrier method, the proposed algorithm computes an approximate solution of a sequence of barrier sub-problems:

$$\min_{x \in \mathbb{R}^n} \varphi_\mu(x) \triangleq f(x) - \mu \sum_{i=1}^{n} \log\left(x^i\right) \tag{20.3a}$$

subject to

$$c(x) = 0, \tag{20.3b}$$

for a decreasing sequence of barrier parameters $\{\mu_j\}$ converging to zero, where x^i is the ith component of vector x. This problem can be interpreted as applying a homotopy method to the primal-dual equations:

$$\nabla f(x) + \nabla c(x)\lambda - z = 0, \tag{20.4a}$$

$$c(x) = 0, \tag{20.4b}$$

$$XZe - \mu e = 0, \tag{20.4c}$$

where the homotopy parameter μ is driven to zero (Byrd, Liu, & Nocedal, 1997). In (20.4) $\lambda \in \mathbb{R}^m$ and $z \in \mathbb{R}^n$ are the Lagrange multipliers associated to (20.2b) and (20.2c), respectively. Observe that the equations (20.4) with $\mu = 0$ and the constraints $x, z \geq 0$ are exactly the KKT optimality conditions for (20.2). Clearly, if the constraint qualifications (see Remark 5.2) are satisfied, then these are the first-order optimality conditions.

The strategy for solving (20.2) is to compute an approximate solution to the barrier problem (20.3) for a fixed value of μ and then decrease the barrier parameter and continue the solution of the next barrier problem from the approximate solution of the previous one, etc.

From the primal-dual equations (20.4), we see that the optimality error for the barrier problem can be defined as

$$E_\mu(x, \lambda, z) = \max\left\{ \frac{\|\nabla f(x) + \nabla c(x)\lambda - z\|_\infty}{s_d}, \|c(x)\|_\infty, \frac{\|XZe - \mu e\|_\infty}{s_c} \right\}, \tag{20.5}$$

where the scaling parameters s_d, $s_c \geq 1$ are defined below. $E_0(x, \lambda, z)$ is the value of (20.5) for $\mu = 0$. This measures the optimality error for the original problem (20.2). With this, we can consider that the algorithm terminates with an approximate solution (x^*, λ^*, z^*) if

$$E_0(x^*, \lambda^*, z^*) \leq \varepsilon_{TOL}, \tag{20.6}$$

where $\varepsilon_{TOL} > 0$ is a tolerance for stopping the algorithm.

Even if the problem is well scaled, the multipliers λ and z might become very large, for example, when the gradients of the active constraints are nearly linear dependent at the solution of (20.2). In this case, the algorithm faces major numerical difficulties. In order to adapt the termination criteria to handle such situations, in (20.5), the scaling factors are introduced as:

$$s_d = \max\left\{1, \frac{\|\lambda\|_1 + \|z\|_1}{(m+n)s_{\max}}\right\}, \qquad s_c = \max\left\{1, \frac{\|z\|_1}{n s_{\max}}\right\}.$$

In this way, a component of the optimality error is scaled whenever the average value of the multipliers becomes larger than a fixed number $s_{\max} \geq 1$ (e.g., $s_{\max} = 100$).

The algorithm has two types of iterations: *major iterations*, denoted by j, in which the value of the barrier parameter μ_j is modified, and *minor iterations*, denoted by k, in which the barrier problem is solved for a fixed value μ_j of the barrier parameter. In order to achieve the superlinear convergence of the algorithm, i.e., to get a local solution of (20.2) satisfying the strong second-order sufficient optimality conditions (see Theorem 5.13), Wächter and Biegler (2006) follow the approach proposed by Byrd, Liu, and Nocedal (1997). Therefore, at the major iteration j, it is required that the approximate solution $\left(x_{j+1}^*, \lambda_{j+1}^*, z_{j+1}^*\right)$ of the barrier problem (20.3) for a given value of the barrier parameter μ_j satisfies the tolerance test:

$$E_{\mu_j}\left(x_{j+1}^*, \lambda_{j+1}^*, z_{j+1}^*\right) \le \kappa_\varepsilon \mu_j \tag{20.7}$$

before the algorithm continues with the solution of the next barrier problem, where $\kappa_\varepsilon > 0$ is a given constant. The barrier parameter is updated as

$$\mu_{j+1} = \max\left\{\frac{\varepsilon_{TOL}}{10}, \ \min\left\{\kappa_\mu \mu_j, \mu_j^{\theta_\mu}\right\}\right\}, \tag{20.8}$$

with constants $\kappa_\mu \in (0,1)$ and $\theta_\mu \in (1,2)$. In this way, the barrier parameter is decreased at a superlinear rate. On the other hand, given the desired tolerance ε_{TOL}, the updating given by (20.8) does not allow μ to become smaller than necessary, thus avoiding numerical difficulties at the end of the optimization procedure.

Solving the Barrier Problem

Now, for solving the barrier problem (20.3) in which the barrier parameter is fixed at the value μ_j, the algorithm uses the Newton method applied to the primal-dual equations (20.4). As we have already said, the minor iterations necessary for solving the barrier problem are denoted by k. Therefore, when given an iteration (x_k, λ_k, z_k) with $x_k, z_k > 0$, then the search direction $\left(d_k^x, d_k^\lambda, d_k^z\right)$ is obtained as solution of the linearization of (20.4) at (x_k, λ_k, z_k), namely:

$$\begin{bmatrix} W_k & J_k & -I \\ J_k^{\mathrm{T}} & 0 & 0 \\ Z_k & 0 & X_k \end{bmatrix} \begin{bmatrix} d_k^x \\ d_k^\lambda \\ d_k^z \end{bmatrix} = -\begin{bmatrix} \nabla f(x_k) + J_k \lambda_k - z_k \\ c(x_k) \\ X_k Z_k e - \mu_j e \end{bmatrix}. \tag{20.9}$$

In (20.9), $J_k = \nabla c(x_k)$ and W_k is the Hessian of the Lagrange function associated to the problem (20.2):

$$L(x, \lambda, z) = f(x) + c(x)^{\mathrm{T}} \lambda - z. \tag{20.10}$$

Instead of solving the nonsymmetric linear system (20.9), the proposed method computes the solution equivalently by first solving the smaller symmetric linear system:

$$\begin{bmatrix} W_k + \Sigma_k & J_k \\ J_k^{\mathrm{T}} & 0 \end{bmatrix} \begin{bmatrix} d_k^x \\ d_k^\lambda \end{bmatrix} = - \begin{bmatrix} \nabla \varphi_{\mu_j}(x_k) + J_k \lambda_k \\ c(x_k) \end{bmatrix} \tag{20.11}$$

derived from (20.9) by eliminating the last block row, where $\Sigma_k = X_k^{-1} Z_k$. The vector d_k^z is then computed as:

$$d_k^z = \mu_j X_k^{-1} e - z_k - \Sigma_k d_k^x. \tag{20.12}$$

As we know, in most line-search methods, we have to ensure that the matrix $W_k + \Sigma_k$ from (20.11), projected onto the null space of the constraint Jacobian J_k^{T}, must be positive definite. This is necessary to guarantee certain descent properties of the filter line search used in the algorithm described below. Also, if J_k does not have full rank, the iteration matrix in (20.11) is singular, i.e., the system (20.11) does not have solutions. Therefore, as a protection, it might be necessary to modify the matrix of the system (20.11). Wächter and Biegler (2006) suggest the following modification of the system (20.11):

$$\begin{bmatrix} W_k + \Sigma_k + \delta_w I & J_k \\ J_k^{\mathrm{T}} & -\delta_c I \end{bmatrix} \begin{bmatrix} d_k^x \\ d_k^\lambda \end{bmatrix} = - \begin{bmatrix} \nabla \varphi_{\mu_j}(x_k) + J_k \lambda_k \\ c(x_k) \end{bmatrix}, \tag{20.13}$$

where $\delta_w,\ \delta_c \geq 0$. The choice of these constants is to be discussed.

Having a solution to the systems (20.13) and (20.12), we must determine the stepsizes. Since x and z are both positive at an optimal solution of the barrier problem (20.3), then this property is maintained for all iterations. Therefore, in order to achieve this property, we first compute

$$\alpha_k^{\max} = \max\{\alpha \in (0,1] : x_k + \alpha d_k^x \geq (1 - \tau_j) x_k\}, \tag{20.14a}$$

$$\alpha_k^z = \max\{\alpha \in (0,1] : z_k + \alpha d_k^z \geq (1 - \tau_j) z_k\}, \tag{20.14b}$$

where the parameter $\tau_j \in (0,1)$ is computed as

$$\tau_j = \max\left\{\tau_{\min},\ 1 - \mu_j\right\}, \tag{20.15}$$

$\tau_{\min} \in (0,1)$ being its minimum value. In interior point methods, τ_j is known as the parameter which represents the *fraction to the boundary* of the feasible region. With these, the next iteration is computed as

$$x_{k+1} = x_k + \alpha_k d_k^x, \tag{20.16a}$$

$$\lambda_{k+1} = \lambda_k + \alpha_k d_k^\lambda, \tag{20.16b}$$

$$z_{k+1} = z_k + \alpha_k^z d_k^z, \tag{20.16c}$$

where α_k^z is computed as in (20.14b) and, in order to ensure the global convergence, the stepsize $\alpha_k \in \left(0, \alpha_k^{\max}\right]$ for the remaining variables is determined by a backtracking line search procedure exploring a decreasing sequence of trial steps: $\alpha_{k,l} = 2^{-l}\alpha_k^{\max}$, (with $l = 0, 1, \ldots$), of Armijo (1966) type.

Observe the difference between the computation of z_{k+1} and the rest of variables. We see that the parameter which represents the fraction to the boundary is used only in computing the variables x_{k+1} and z_{k+1}.

In Wächter and Biegler (2001), for achieving the superlinear convergence of the algorithm, it is proved that the *primal-dual barrier term Hessian* Σ_k does not much deviate arbitrarily from the *primal Hessian* $\mu_j X_k^{-2}$. This requirement is achieved by reseting the components z_{k+1}^i of z_{k+1} as

$$z_{k+1}^i = \max\left\{\min\left\{z_{k+1}^i, \frac{\kappa_\Sigma \mu_j}{x_{k+1}^i}\right\}, \frac{\mu_j}{\kappa_\Sigma x_{k+1}^i}\right\}, \quad i = 1, \ldots, n, \tag{20.17}$$

where $\kappa_\Sigma \geq 1$ is a fixed parameter. This strategy guarantees that each component σ_{k+1}^i of the diagonal matrix Σ_{k+1} is in the interval:

$$\sigma_{k+1}^i \in \left[\frac{1}{\kappa_\Sigma} \frac{\mu_j}{\left(x_k^i\right)^2}, \ \kappa_\Sigma \frac{\mu_j}{\left(x_k^i\right)^2}\right]. \tag{20.18}$$

In IPOPT $\kappa_\Sigma = 10^{10}$.

Line-Search Filter Method

The idea of this method is to interpret the solving of the barrier problem (20.3) for μ_j fixed, as a biobjective optimization problem with two goals: minimizing the objective function $\varphi_{\mu_j}(x)$ and minimizing the constraint violation defined by $\theta(x) \triangleq \|c(x)\|$. As seen in Chapter 18, for solving this biobjective minimization problem, the emphasis is on minimizing the constraint violation.

In this context, in line search with backtracking, a trial point $x_k(\alpha_{k,l}) \triangleq x_k + \alpha_{k,l} d_k^x$ is considered as acceptable if it leads to sufficient progress toward either goal compared to the current iterate, i.e., if

$$\theta(x_k(\alpha_{k,l})) \leq (1 - \gamma_\theta)\theta(x_k), \tag{20.19a}$$

or

$$\varphi_{\mu_j}(x_k(\alpha_{k,l})) \leq \varphi_{\mu_j}(x_k) - \gamma_\varphi \theta(x_k), \tag{20.19b}$$

holds for fixed constants $\gamma_\theta, \ \gamma_\varphi \in (0, 1)$.

However, as we can see, each of the above criteria is complicated enough and therefore is replaced by requiring sufficient progress in the barrier objective function whenever for the current iterate we have $\theta(x_k) \le \theta^{\min}$ for some constant $\theta^{\min} \in (0, \infty]$ and the following *switching condition*:

$$\nabla\varphi_{\mu_j}(x_k)^T d_k^x < 0, \text{and } \alpha_{k,l}\left[-\nabla\varphi_{\mu_j}(x_k)^T d_k^x\right]^{s_\varphi} > \delta[\theta(x_k)]^{s_\theta}, \tag{20.20}$$

with constants $\delta > 0, \; s_\theta > 1, \; s_\varphi \ge 1$, is satisfied. Observe that if $\theta(x_k) \le \theta^{\min}$ and (20.20) is true for the current stepsize $\alpha_{k,l}$, then the trial point has to satisfy the Armijo condition:

$$\varphi_{\mu_j}(x_k(\alpha_{k,l})) \le \varphi_{\mu_j}(x_k) + \eta_\varphi \alpha_{k,l} \nabla\varphi_{\mu_j}(x_k)^T d_k^x \tag{20.21}$$

instead of condition (20.19) for the trial point to be acceptable. In (20.21) $\eta_\varphi \in (0, 1/2)$ is a constant.

The algorithm maintains a filter, i.e., a set $F_k \subseteq \{(\theta, \varphi) \in \mathbb{R}^2 : \; \theta \ge 0\}$ computed for each iteration k. The filter F_k contains those combinations of constraint violation values θ and the objective function values φ, which are prohibited for a successful trial point at iteration k, i. e., during the line search, a trial point $x_k(\alpha_{k,l})$ is rejected if $\left(\theta(x_k(\alpha_{k,l})), \varphi_{\mu_j}(x_k(\alpha_{k,l}))\right) \in F_k$. We say that the trial point is not acceptable to the current filter. At the beginning of the optimization process, the filter is initialized to

$$F_0 = \left\{(\theta, \varphi) \in \mathbb{R}^2 : \; \theta \ge \theta^{\max}\right\}, \tag{20.22}$$

for some $\theta^{\max}$, so that the algorithm will never allow trial points that have a constraint violation larger than $\theta^{\max}$ to be accepted. Later, after every iteration, the filter is augmented using the update formula:

$$F_{k+1} = F_k \cup \left\{(\theta, \varphi) \in \mathbb{R}^2 : \theta \ge (1 - \gamma_\theta)\theta(x_k) \quad \text{and} \quad \varphi \ge \varphi_{\mu_j}(x_k) - \gamma_\varphi \theta(x_k)\right\}, \tag{20.23}$$

where the accepted trial stepsize does not satisfy the switching condition (20.20) or in which the Armijo condition (20.21) does not hold. This ensures that the iterates cannot return to the neighborhood of x_k. On the other hand, if both (20.20) and (20.21) hold for the accepted stepsize, the filter remains unchanged. This procedure ensures that the algorithm cannot cycle indefinitely, for example, between two points that alternatively decrease the constraint violation and the barrier objective function.

Finally, in some cases, it is not possible to find a trial stepsize $\alpha_{k,l}$ that satisfies the above criteria. In these cases, a minimum desired stepsize is approximated by using linear models of the involved functions. For this, Wächter and Biegler (2006) define the function:

$$\alpha_k^{\min} \triangleq \gamma_\alpha \begin{cases} \min\left\{\gamma_\theta, \dfrac{\gamma_\varphi \theta(x_k)}{-\nabla \varphi_{\mu_j}(x_k)^{\mathrm{T}} d_k^x}, \dfrac{\delta[\theta(x_k)]^{s_\theta}}{\left[-\nabla \varphi_{\mu_j}(x_k)^{\mathrm{T}} d_k^x\right]^{s_\varphi}}\right\}, \\ \text{if} \quad \nabla \varphi_{\mu_j}(x_k)^{\mathrm{T}} d_k^x < 0 \quad \text{and} \quad \theta(x_k) \le \theta^{\min}, \\ \min\left\{\gamma_\theta, \dfrac{\gamma_\varphi \theta(x_k)}{-\nabla \varphi_{\mu_j}(x_k)^{\mathrm{T}} d_k^x}\right\}, \\ \text{if} \quad \nabla \varphi_{\mu_j}(x_k)^{\mathrm{T}} d_k^x < 0 \quad \text{and} \quad \theta(x_k) > \theta^{\min}, \\ \gamma_\theta, \\ \text{otherwise}, \end{cases} \tag{20.24}$$

where $\gamma_\alpha \in (0, 1]$ is a "safety factor." If the backtracking line search encounters a trial stepsize with $\alpha_{k,l} \le \alpha_k^{\min}$, then the algorithm reverts to a feasibility restoration phase. That is, the algorithm tries to find a new iterate $x_{k+1} > 0$ which is acceptable to the current filter and for which the condition (20.19) holds by reducing the constraint violation using some iterative method. Observe that the restoration phase might not be able to produce a new iterate for the filter line-search method, for example, when the problem is infeasible.

Second-Order Corrections

Many methods for nonlinear optimization use second-order corrections to improve the proposed step in case a trial point was rejected. A second-order correction (SOC) for some step d_k^x aims at reducing the infeasibility by applying an additional Newton step for the constraints at the point $x_k + d_k^x$ using the Jacobian J_k^{T} evaluated at x_k. The details are given in Wächter and Biegler (2006) and are as follows:

If the first trial, the stepsize $\alpha_{k,0}$ has been rejected, and if $\theta(x_k(\alpha_{k,0})) \ge \theta(x_k)$, then a second-order correction $d_k^{x,soc}$ is computed, such that:

$$J_k^{\mathrm{T}} d_k^{x,soc} + c\left(x_k + \alpha_{k,0}\, d_k^x\right) = 0. \tag{20.25}$$

The new corrected search direction is obtained as:

$$d_k^{x,cor} = d_k^{x,soc} + \alpha_{k,0}\, d_k^x. \tag{20.26}$$

The condition (20.25) does not uniquely define the second-order correction, and different choices would be possible. In order to avoid additional matrix factorization, the proposed method uses the same matrix as in (20.13) to compute the overall corrected step (20.26) as solution of the system:

$$\begin{bmatrix} W_k + \Sigma_k + \delta_w I & J_k \\ J_k^{\mathrm{T}} & -\delta_c I \end{bmatrix} \begin{bmatrix} d_k^{x,cor} \\ d_k^{\lambda} \end{bmatrix} = -\begin{bmatrix} \nabla\varphi_{\mu_j}(x_k) + J_k\lambda_k \\ c_k^{soc} \end{bmatrix}, \tag{20.27}$$

where

$$c_k^{soc} = \alpha_{k,0}c(x_k) + c\left(x_k + \alpha_{k,0}d_k^x\right) \tag{20.28}$$

is obtained from (20.13), (20.25), and (20.26).

As soon as the corrected search direction $d_k^{x,cor}$ has been obtained, the fraction-to-the-boundary rule is applied to get

$$\alpha_k^{soc} \triangleq \max\left\{\alpha \in (0,1] : \ x_k + \alpha d_k^{x,cor} \geq (1-\tau_j)x_k\right\} \tag{20.29}$$

and tests if the resulting trial point $x_k^{soc} \triangleq x_k + \alpha_k^{soc} d_k^{x,cor}$ is accepted in the filter. Note that the original search direction d_k^x is still used in (20.20) and in the right-hand side of (20.21). Also, x_k^{soc} replaces $x(\alpha_k)$ in (20.21). If this trial point passes the tests, then it is acceped as the new iterate. Otherwise, additional second-order corrections are applied until the correction step has not decreased the constraint violation by a fraction $\kappa_{soc} \in (0,1)$ or until a maximum number $p^{\max}$ of second-order corrections have been performed. In this case, the original search direction d_k^x is restored, and the regular backtracking line search is restarted with a shorter stepsize $\alpha_{k,1} = \alpha_{k,0}/2$.

The Algorithm

Now, the following filter line-search algorithm for solving the barrier sub-problem (20.3) can be presented.

Algorithm 20.1 Line-search filter barrier algorithm (Wächter and Biegler)

1.	Choose an initial point (x_0, λ_0, z_0) with $x_0, z_0 > 0$, an initial value for the barrier parameter $\mu_0 > 0$, and the constants ε_{TOL}, $s_{\max} \geq 1$, $\kappa_\varepsilon > 0$, $\kappa_\mu \in (0,1)$, $\theta_\mu \in (1,2)$, $\tau_{\min} \in (0,1)$, $\kappa_\Sigma > 1$, $\theta_{\max} \in (\theta(x_0), \infty]$, $\theta_{\min} > 0$, γ_θ, $\gamma_\varphi \in (0,1)$, $\delta > 0$, $\gamma_\alpha \in (0,1]$, $s_\theta > 1$, $s_\varphi \geq 1$, $\eta_\varphi \in (0, 1/2)$, $\kappa_{soc} \in (0,1)$, and $p^{\max} \in \{0,1,2,\ldots\}$
2.	*Initialization*. Initialize $j=0$, $k=0$, as well as the filter F_0 as in (20.22). Compute τ_0 as in (20.15)
3.	*Test convergence for the overall problem*. If $E_0(x_k, \lambda_k, z_k) \leq \varepsilon_{TOL}$, then stop.
4.	*Test convergence for the barrier problem*. If $E_{\mu_j}(x_k, \lambda_k, z_k) \leq \kappa_\varepsilon \mu_j$, then go to step 5
5.	Compute μ_{j+1} and τ_{j+1} as in (20.8) and (20.15), respectively

(continued)

6.	Re-initialize the filter $F_k = \{(\theta, \varphi) \in \mathbb{R}^2 : \theta \geq \theta^{\max}\}$
7.	If $k = 0$, then continue with step 4; otherwise go to step 8
8.	*Compute the search direction.* Compute $\left(d_k^x, d_k^\lambda, d_k^z\right)$ as solution of the system (20.13), where the parameters δ_w and δ_c are computed from the algorithm IC for correction of inertia
9.	*Backtracking. Initialization of line search.* Set $\alpha_{k,0} = \alpha_k^{\max}$, where $\alpha_k^{\max}$ is computed as in (20.14a) and set $l = 0$
10.	*Backtracking. Compute the new trial point.* Set $x_k(\alpha_{k,l}) = x_k + \alpha_{k,l} d_k^x$
11.	*Backtracking. Test on acceptability to the filter.* If $\left(\theta(x_k(\alpha_{k,l})),\ \varphi_{\mu_j}(x_k(\alpha_{k,l}))\right) \in F_k$, then reject the trial step and go to step 13
12.	*Backtracking. Test on sufficient decrease with respect to the current iterate.* Case I: $\theta(x_k) \leq \theta^{\min}$ and (20.20) holds. If (20.21) holds, then accept the trial step $x_{k+1} = x_k(\alpha_{k,l})$ and continue with step 19; otherwise go to step 13 Case II: $\theta(x_k) > \theta^{\min}$ or (20.20) is not satisfied. If (20.19) holds, then accept the trial step $x_{k+1} = x_k(\alpha_{k,l})$ and go to step 19; otherwise continue with step 13
13.	*Backtracking. Initialize the second-order correction.* If $l > 0$ or $\theta(x_{k,0}) < \theta(x_k)$, then skip the second-order correction and go to step 18; otherwise, initialize the second-order correction by setting $p = 1$ and c_k^{soc} as in (20.28). Initialize $\theta_{old}^{soc} = \theta(x_k)$
14.	*Backtracking. Compute the second-order correction.* Compute $d_k^{x,cor}$ and d_k^λ as solutions of the system (20.27). Compute α_k^{soc} from (20.29) and then set $x_k^{soc} = x_k + \alpha_k^{soc} d_k^{x,cor}$
15.	*Backtracking. Test on acceptability to the filter (in SOC).* If $\left(\theta\left(x_k^{soc}\right),\ \varphi_{\mu_j}\left(x_k^{soc}\right)\right) \in F_k$, then reject the trial stepsize and go to step 18
16.	*Backtracking. Test on sufficient reduction with respect to the current iterate in SOC.* Case I: $\theta(x_k) \leq \theta^{\min}$ and (20.20) holds for $\alpha_{k,0}$. If (20.21) holds with $x_k(\alpha_{k,l})$, replaced by x_k^{soc}, then the trial step $x_{k+1} = x_k^{soc}$ is accepted and go to step 19; otherwise, continue with step 17 Case II: $\theta(x_k) > \theta^{\min}$ or (20.20) is not satisfied for $\alpha_{k,0}$. If (20.19) holds for $x_k(\alpha_{k,l})$, replaced by x_k^{soc}, then the trial step $x_{k+1} = x_k^{soc}$ is accepted, and go to step 19; otherwise, continue with step 17

(continued)

17.	*Backtracking. The next second-order correction.* If $p = p^{\max}$ or $\theta(x_k^{soc}) > \kappa_{soc}\theta_{old}^{soc}$, then abort second-order correction and go to step 18; otherwise, set $p = p+1$, $c_k^{soc} = \alpha_k^{soc} c_k^{soc} + c(x_k^{soc})$, and $\theta_{old}^{soc} = \theta(x_k^{soc})$. Go back to step 14
18.	*Backtracking. Choose a new trial stepsize.* Set $\alpha_{k,l+1} = \alpha_{k,l}/2$ and $l = l+1$. If the trial stepsize becomes too small, i.e., $\alpha_{k,l} < \alpha_k^{\min}$, where $\alpha_k^{\min}$ is defined as in (20.24), then continue with the feasibility restoration phase, i.e., step 22; otherwise go back to step 10
19.	*Accept the trial point.* Set $\alpha_k = \alpha_{k,l}$ (or $\alpha_k = \alpha_k^{soc}$ if the SOC point was accepted in step 16) and update the Lagrange multiplier estimates λ_{k+1} and z_{k+1} using (20.16b) and (20.16c), respectively, with α_k^z computed as in (20.14b). If necessary, apply (20.17) to correct z_{k+1}
20.	*Augment the filter.* If (20.20) or (20.21) does not hold for α_k, then augment the filter as in (20.23); otherwise, the filter is unchanged, i.e., $F_{k+1} = F_k$
21.	*Continue with the next iteration.* Set $k = k+1$ and go back to step 3
22.	*Feasibility restoration phase.* Augment the filter as in (20.23) and compute a new iterate $x_{k+1} > 0$ by decreasing the infeasibility measure $\theta(x)$, so that x_{k+1} is acceptable to the augmented filter, i.e., $\left(\theta(x_{k+1}), \varphi_{\mu_j}(x_{k+1})\right) \notin F_{k+1}$. Then continue with the regular iteration in step 21 ♦

At every iteration, at least one trial point will be tested before the algorithm may switch to the feasibility restoration phase in step 22. Also, the condition in step 7 ensures that eventually, at least one step is taken for each decreased value of the barrier parameter.

Observe that the algorithm is complex, with a multitude of parameters which specify the optimization conditions. Wächter and Biegler (2006) give the values of the constants used in their implementation as $\kappa_e = 10$, $\kappa_\mu = 0.2$, $\theta_\mu = 1.5$, $\tau_{\min} = 0.99$, $\gamma_\theta = 10^{-5}$, $\gamma_\varphi = 10^{-5}$, $\delta = 1$, $\gamma_\alpha = 0.05$, $s_\theta = 1.1$, $s_\varphi = 2.3$, $\eta_\varphi = 10^{-4}$, $\kappa_{soc} = 0.99$, and $p^{\max} = 4$, as well as $\mu_0 = 0.1$, $\theta^{\max} = 10^4 \max\{1, \theta(x_0)\}$, and $\theta^{\min} = 10^{-4} \max\{1, \theta(x_0)\}$, where x_0 is the starting point. At the same time, they present some implementation details which lead to the IPOPT algorithm. In the following, we will only present some of the most interesting details in order to complete Algorithm 20.1 for it to be as close as possible to IPOPT described in Wächter (2002) or Wächter and Biegler (2006).

20.2 Implementation Details

General Lower and Upper Bounds

For simplicity, Algorithm 20.1 is dedicated to solving nonlinear optimization problems (20.2). However, this algorithm can be immediately generalized for solving the general problems (20.1). In particular, for the problem (20.1) with simple bounds on variables $x_L \le x \le x_U$, the barrier problem (20.3) becomes

$$\min_{x\in\mathbb{R}^n} \varphi_{\mu_j}(x) = f(x) - \mu_j \sum_{i\in I_L} \log\left(x^i - x_L^i\right) - \mu_j \sum_{i\in I_U} \log\left(x_U^i - x^i\right) \qquad (20.30a)$$

subject to:

$$c(x) = 0, \qquad (20.30b)$$

where $I_L = \{i: \ x_L^i \ne -\infty\}$ and $I_U = \{i: \ x_U^i \ne +\infty\}$. At the same time, for the multipliers z, the simple bounds z_L^i and z_U^i are introduced for all finite lower and upper bounds, and the primal-dual Hessian of the barrier terms Σ_k is defined as the sum of the matrices $\Sigma_k^L = \operatorname{diag}\left(\sigma_{k,1}^L, \ldots, \sigma_{k,n}^L\right)$ and $\Sigma_k^U = \operatorname{diag}\left(\sigma_{k,1}^U, \ldots, \sigma_{k,n}^U\right)$, where:

$$\sigma_{k,i}^L = \begin{cases} z_{L,k}^i / \left(x_k^i - x_L^i\right), & \text{if } i \in I_L, \\ 0, & \text{otherwise}, \end{cases}$$

$$\sigma_{k,i}^U = \begin{cases} z_{U,k}^i / \left(x_U^i - x_k^i\right), & \text{if } i \in I_U, \\ 0, & \text{otherwise}. \end{cases}$$

Moreover, for $i \notin I_L$, define $z_{L,k}^i = 0$ and for $i \notin I_U$, define $z_{U,k}^i = 0$.

Initialization

Since Algorithm 20.1 requires the iterates to strictly satisfy the bound constraints (20.1c), it follows that it is often necessary to modify the user-provided initial point so that it is sufficiently away from the boundary. For this purposes, each component i of the initial point which has only one (say, a lower) bound is modified by

$$x_0^i = \max\left\{x_0^i, \ x_L^i + \kappa_1 \max\left\{1, \left|x_L^i\right|\right\}\right\},$$

where $\kappa_1 > 0$ is a constant. Similar modifications can be made for the variables only bounded above. The initial value of a variable x^i bounded on two sides is projected into the interval $\left[x_L^i + p_L^i, \ x_U^i - p_U^i\right]$, where the perturbations p_L^i and p_U^i are computed as

$$p_L^i \triangleq \min\{\kappa_1 \max\{1, |x_L^i|\},\ \kappa_2(x_U^i - x_L^i)\},$$
$$p_U^i \triangleq \min\{\kappa_1 \max\{1, |x_U^i|\},\ \kappa_2(x_U^i - x_L^i)\},$$

where $\kappa_2 \in (0, 1/2)$ (e.g., $\kappa_1 = \kappa_2 = 10^{-2}$).

The dual variables corresponding to the bound constraints are initialized to one componentwise. Finally, using the possibly modified initial point x_0 and the initial bound multipliers, the multipliers λ_0 for the equality constraints are obtained as least-square solutions for the dual infeasibility (20.4a), i.e., by solving the linear system:

$$\begin{bmatrix} I & J(x_0) \\ J(x_0)^{\mathrm{T}} & 0 \end{bmatrix} \begin{bmatrix} w \\ \lambda_0 \end{bmatrix} = -\begin{bmatrix} \nabla f(x_0) - z_{L,0} + z_{U,0} \\ 0 \end{bmatrix}, \tag{20.31}$$

where w are temporary variables discarded after this computation. However, if λ_0 obtained in this way is too large, i.e., $\|\lambda_0\|_\infty > \lambda_{\max}$ (e.g., with $\lambda_{\max} = 10^3$), then the least-square estimate λ_0 is discarded and set $\lambda_0 = 0$ ($J(x_0) = \nabla c(x_0)$).

Handling Unbounded Solution Sets

In some cases, the optimal point set for (20.1) does not consist of isolated points but contains an unbounded connected component. Then, the objective function of the corresponding barrier problem (20.30) for a fixed value of the barrier parameter $\mu_j$is unbounded below over the feasible set, since a log-barrier term converges to $-\infty$ as its argument goes to infinity. Therefore, the method for solving the barrier problem might fail to converge, even though the original problem is well posed. To prevent this behavior, linear damping terms for all variables with exactly one finite bound are added to the barrier objective function (20.30a), which then becomes

$$\varphi_{\mu_j}(x) = f(x) - \mu_j \sum_{i \in I_L} \log(x^i - x_L^i) - \mu_j \sum_{i \in I_U} \log(x_U^i - x^i)$$
$$+ \kappa_d \mu_j \sum_{i \in I_L \setminus I_U} (x^i - x_L^i) + \kappa_d \mu_j \sum_{i \in I_U \setminus I_L} (x_U^i - x^i),$$

where $\kappa_d > 0$ is a constant independent of μ_j (e.g., $\kappa_d = 10^{-4}$). In this way, divergence of variables that have only one bound is penalized. On the other hand, the effect of the damping term is reduced as μ_j decreases.

Inertia Correction

In order to be able to compute the search direction, we need to ensure that the matrix of the system (20.11) is nonsingular. In addition, the filter line-search method requires the matrix $W_k + \Sigma_k$ projected onto the null space of the constraint Jacobian J_k^{T} to be positive definite. But these conditions are satisfied if the matrix of the system (20.11) has the inertia equal to $(n, m, 0)$, i.e., if this matrix has exactly

n positive, m negative, and *no* zero eigenvalues. Therefore, if the inertia of this matrix is not $(n, m, 0)$, then the linear system (20.13) is resolved with different trial values for the scalars δ_w, $\delta_c \geq 0$ until the inertia is as desired. Observe that the desired inertia is obtained if δ_w is sufficiently large and the constraint Jacobian J_k^T has full rank. If J_k^T is rank deficient, the matrix of the system (20.13) is singular as long as δ_c is zero. Therefore, a positive value for δ_c and a sufficiently large value of δ_w ensure the correct eigenvalue signatures for the matrix from (20.13). However, in practice, the matrix of the system (20.13) can become so ill-conditioned that the factorization cannot be successfully performed, even with very large values of δ_w and some positive values for δ_c. In this case, Algorithm 20.1 switches directly to the feasibility restoration phase in step 22. These observations motivate the following heuristic for choosing the parameters δ_c and δ_w (Wächter & Biegler, 2005a).

Algorithm 20.2 Inertia correction algorithm

1.	Choose the constants $0 < \bar{\delta}_w^{\min} < \bar{\delta}_w^0 < \bar{\delta}_w^{\max}$, $\bar{\delta}_c > 0$, $\kappa_c \geq 0$, $0 < \kappa_w^- < 1 < \kappa_w^+ < \bar{\kappa}_w^+$. Initialize $\delta_w^{last} = 0$
2.	Attempt to factorize the unmodified matrix of the system (20.13) with $\delta_w = \delta_c = 0$. If the matrix is nonsingular and its inertia is $(n, m, 0)$, then use the resulting search direction in the line search obtained by solving the system (20.13). Otherwise, continue with step 3
3.	If the matrix from (20.13) has zero eigenvalues, then set $\delta_c = \bar{\delta}_c \mu^{\kappa_c}$; otherwise set $\delta_c = 0$
4.	If $\delta_w^{last} = 0$, then set $\delta_w = \bar{\delta}_w^0$; otherwise, set $\delta_w = \max\left\{\bar{\delta}_w^{\min}, \kappa_w^- \delta_w^{last}\right\}$
5.	Attempt to factorize the matrix from (20.13) modified with these values of δ_c and δ_w. If inertia is equal to $(n, m, 0)$, then set $\delta_w^{last} = \delta_w$ and use the resulting search direction in the line search. Otherwise, continue with step 6
6.	If $\delta_w^{last} = 0$, then set $\delta_w = \bar{\kappa}_w^+ \delta_w$; otherwise, set $\delta_w = \kappa_w^+ \delta_w$
7.	If $\delta_w > \bar{\delta}_w^{\max}$, then abort the search direction computation, skip the backtracking line search, and switch directly to the restoration phase in step 22 of Algorithm 20.1. Otherwise, continue with step 5 ♦

The above heuristic procedure was established by Wächter and Biegler (2006). In step 2, it first checks if the unmodified matrix has the desired inertia so that the pure Newton search direction is used whenever possible. If this is the case, then apply the Newton method for search direction computation; otherwise, increase δ_w. The first trial value is based on δ_w^{last} which stores the perturbation value from the last modification of the matrix from (20.13). In this way, the smallest perturbation is found necessary to avoid the factorization in step 5 of the algorithm for too small values of δ_w. The reason for using a much larger factor $\bar{\kappa}_w^+$ in step 6 of the algorithm for the first necessary correction rather than for the correction in later iterations is that we want to avoid a high number of trial factorizations when the scale of the problem and the order of magnitude for a successful correction are not known yet. Selection of a nonzero value for δ_c in step 3 is based on the assumption that the singularity is caused by the rank-deficient constraint Jacobian. Note that the non-zero value for δ_c in step 3 converges to zero as $\mu \to 0$ if $\kappa_c > 0$, so that the perturbation δ_c is smaller when a solution of the problem is approached.

Automatic Scaling of the Problem

The Newton steps for the primal-dual system (20.11) for solving the nonlinear system (20.4) are invariant to scaling the variables, the objective, and the constraints functions, i.e., to replace x, f, and c by $\tilde{x} = D_x x$, $\tilde{f}(x) = d_f f(x)$, and $\tilde{c}(x) = D_c c(x)$ for some $d_f > 0$ and for positive definite diagonal matrices $D_x = \text{diag}(d_x^1, \ldots, d_x^n)$ and $D_c = \text{diag}(d_c^1, \ldots, d_c^m)$. However, the optimization algorithm with its initialization procedures, globalization strategy, and stopping criteria usually behaves very differently for different scaling factors, particularly if the scaling factors are very large or very small (Wächter & Biegler, 2006).

The idea of scaling in IPOPT is that changing a variable by a given amount has a comparable effect on any function which depends on this variable, or in other words, the nonzero elements of the function gradients are of the same order of magnitude.

Another scaling procedure available in IPOPT as an option consists in applying an equilibration algorithm to the first derivative matrix:

$$M_0 = \begin{bmatrix} \nabla_x c(x_0)^T \\ \nabla_x f(x_0)^T \end{bmatrix}$$

to obtain scaling matrices D_x and $D_{cf} = \text{diag}(D_c, d_f)$ so that the nonzero elements in $D_{cf} M_0 D_x^{-1}$ are of order one. (The algorithm for equilibrating the matrices is the one implemented in subroutines MC19 and MC29 from the Harwell library (AEA Technology 2002)) Similarly, another procedure for scaling computes the scaling factors so that the matrix

$$\begin{bmatrix} D_x^{-1} & 0 \\ 0 & D_c \end{bmatrix} \begin{bmatrix} \nabla_{xx}^2 f(x_0) & \nabla_x c(x_0) \\ \nabla_x c(x_0)^T & 0 \end{bmatrix} \begin{bmatrix} D_x^{-1} & 0 \\ 0 & D_c \end{bmatrix}$$

has nonzero entries close to one. These procedures are available to users as options in IPOPT. Even if these strategies seem to work well in certain instances, the performances of the algorithm are not conclusive with scaling.

In IPOPT, Wächter and Biegler (2006) apply an automatic scaling procedure. Given a threshold value $g_{\max}$, for example, $g_{\max} = 100$, then the scaling factors are chosen as

$$d_f = \min\{1, g_{\max}/\|\nabla_x f(x_0)\|_\infty\},$$
$$d_c^j = \min\{1, g_{\max}/\|\nabla_x c^j(x_0)\|_\infty\}, \; j = 1, \ldots, m,$$

and $D_x = I$. Observe that this procedure will never multiply a function by a number larger than one and that all the gradient components in the scaled problem are at most of size $g_{\max}$ at the starting point.

The scaling factors are computed only at the beginning of the optimization at x_0.

Feasibility Restoration Phase

A very important ingredient of the filter line-search algorithm is the feasibility restoration phase (see step 22 of Algorithm 20.1). The task of the restoration phase is to compute a new iteration acceptable to the augmented filter F_{k+1} by decreasing the infeasibility whenever the regular backtracking line-search procedure cannot make sufficient progress and when the stepsize becomes too small (see step 18 of Algorithm 20.1). More than this, when the inertia is corrected as in Algorithm 20.2, the method switches to the restoration phase whenever the linear system (20.13) is very ill-conditioned and cannot be successfully factorized despite the modifications of the matrix from (20.13). The feasibility restoration phase has another significant purpose, namely, to detect the local infeasibility. If the problem is infeasible, then the algorithm is not able to generate sufficient progress in the regular backtracking line-search procedure and reverts to the restoration phase. In Wächter and Biegler (2001), the global convergence of the filter line-search method is proved under the hypothesis that in the neighborhood of feasible points, the gradients of the active constraints are linearly independent. Therefore, as a consequence, the algorithm does not switch to the feasibility restoration phase at feasible points. However, this assumption might be violated in practice, and the restoration phase might be called at a point with very small value of θ. In IPOPT, two procedures for feasibility restorations are used. (In order to avoid confusion, the overbars are used to denote quantities referring to the restoration phase, and the subscript t is used for the restoration phase iteration counter.)

Minimization of the constraint violation This is the first algorithm of the feasibility restoration phase. The purpose of this method is to return a new iterate $x_{k+1} > 0$ with $\left(\theta(x_{k+1}), \; \varphi_{\mu_j}(x_{k+1})\right) \notin F_{k+1}$ for step 22 of Algorithm 20.1 or to converge to a nonzero minimizer (or at least to a stationary point) of some norm of the constraint violation. The restoration phase algorithm applies the primal-dual interior point filter line-search algorithm to a smooth reformulation of the optimization problem:

$$\min_{\bar{x}\in\mathbb{R}^n} \|c(\bar{x})\|_1 + \frac{\xi}{2}\|D_R(\bar{x}-\bar{x}_R)\|_2^2 \tag{20.32a}$$

subject to:

$$\bar{x} \geq 0. \tag{20.32b}$$

In (20.32a), a term is included in the objective function that penalizes the deviation from a reference point $\bar{x}_R$, where $\xi > 0$ is a weighting parameter and the scaling matrix D_R is defined by:

$$D_R = diag\left(\min\{1, 1/|\bar{x}_R^1|\}, \ldots, \min\{1, 1/|\bar{x}_R^n|\}\right).$$

The reference point $\bar{x}_R$ is chosen to be the iterate x_k at which the restoration phase is called in step 22 of Algorithm 20.1. In this way, we seek to decrease the constraint violation and try to avoid a large deviation from $\bar{x}_R$, which determines undesired significant increase in the barrier objective function φ_{μ_j}.

A smooth reformulation of problem (20.32) is obtained by introducing nonnegative variables $\bar{p}, \bar{n} \in \mathbb{R}^m$ that capture the positive and negative parts of the constraints:

$$\min_{\bar{x}\in\mathbb{R}^n, \bar{p}, \bar{n}\in\mathbb{R}^m} \sum_{i=1}^{m}\left(\bar{p}^i + \bar{n}^i\right) + \frac{\xi}{2}\left\|D_R\left(\bar{x}-\bar{x}_R\right\|_2^2\right. \tag{20.33a}$$

subject to:

$$c(\bar{x}) - \bar{p} + \bar{n} = 0, \tag{20.33b}$$

$$\bar{x}, \bar{p}, \bar{n} \geq 0. \tag{20.33c}$$

We see that this optimization problem is of the form given in (20.2). Therefore, the filter line-search interior point algorithm can be applied to solve a sequence of barrier problems of the following form:

$$\begin{aligned} \min_{\bar{x}\in\mathbb{R}^n, \bar{p}, \bar{n}\in\mathbb{R}^m} \ & \rho\sum_{i=1}^{m}\left(\bar{p}^i + \bar{n}^i\right) + \frac{\xi}{2}\left\|D_R\left(\bar{x}-\bar{x}_R\right\|_2^2\right. \\ & -\bar{\mu}\sum_{i=1}^{n}\log\left(\bar{x}^i\right) - \bar{\mu}\sum_{i=1}^{m}\log\left(\bar{p}^i\right) - \bar{\mu}\sum_{i=1}^{m}\log\left(\bar{n}^i\right) \end{aligned} \tag{20.34a}$$

subject to:

$$c(\bar{x}) - \bar{p} + \bar{n} = 0. \tag{20.34b}$$

Observe that an additional scaling parameter $\rho > 0$ has been introduced. This is to allow a relative scaling of the overall objective function (20.34a) with respect to the constraints (20.34b). Moreover, if the regularization parameter $\xi > 0$ is chosen sufficiently small, the optimization problem (20.33) is the exact penalty formulation of the problem: *find the feasible point that is closest in a weighted norm to the reference point* $\bar{x}_R$:

$$\min_{\bar{x}\in\mathbb{R}^n} \|D_R(\bar{x}-\bar{x}_R)\|_2^2$$

subject to:

$$c(\bar{x}) = 0, \bar{x} \geq 0.$$

In addition to the original variables $\bar{x}$, the barier problem (20.34) contains the variables $\bar{p}$ and $\bar{n}$. The corresponding primal-dual equations (similar to (20.4)) include their dual variables, say $\bar{z}_p$ and $\bar{z}_n$. The search directions for the line search are obtained by linearization of these equations and can be written as

$$\begin{bmatrix} \bar{W} + \xi D_R^2 + \bar{\Sigma} & \nabla c(\bar{x}) \\ \nabla c(\bar{x})^{\mathrm{T}} & -\bar{\Sigma}_p^{-1} - \bar{\Sigma}_n^{-1} \end{bmatrix} \begin{bmatrix} \bar{d}^x \\ \bar{d}^\lambda \end{bmatrix} = -\begin{bmatrix} \xi D_R^2(\bar{x}-\bar{x}_R) + \nabla c(\bar{x})\bar{\lambda} - \bar{\mu}\bar{X}^{-1}e \\ c(\bar{x}) - \bar{p} + \bar{n} + \rho \bar{Z}_p^{-1}(\bar{\mu}e - \bar{p}) + \rho \bar{Z}_n^{-1}(\bar{\mu}e - \bar{n}) \end{bmatrix}, \tag{20.35}$$

where $\bar{W} = \sum_{i=1}^m \bar{\lambda}^i \nabla_{xx}^2 c(\bar{x})$, $\bar{\Sigma} = \bar{X}^{-1}\bar{Z}$, $\bar{\Sigma}_p = \bar{P}^{-1}\bar{Z}_p$, and $\bar{\Sigma}_n = \bar{N}^{-1}\bar{Z}_n$. Subsequently, $\bar{d}^p$, $\bar{d}^n$, $\bar{d}^{z_p}$, and $\bar{d}^{z_n}$ are obtained from:

$$\bar{d}^p = \bar{Z}_p^{-1}\left(\bar{\mu}e + \bar{P}\left(\bar{\lambda} + \bar{d}^\lambda\right) - \rho\bar{p}\right), \qquad \bar{d}^{z_p} = \bar{\mu}\bar{P}^{-1}e - \bar{z}^p - \bar{\Sigma}_p\bar{d}^p,$$

$$\bar{d}^n = \bar{Z}_n^{-1}\left(\bar{\mu}e + \bar{N}\left(\bar{\lambda} + \bar{d}^\lambda\right) - \rho\bar{n}\right), \qquad \bar{d}^{z_n} = \bar{\mu}\bar{N}^{-1}e - \bar{z}^n - \bar{\Sigma}_n\bar{d}^n,$$

$$\bar{d}^z = \bar{\mu}\bar{X}^{-1}e - \bar{z} - \bar{\Sigma}\bar{d}^x.$$

Observe that the structure of the nonzero elements of the linear system (20.35) is identical to the one in (20.13). Therefore, for solving (20.35), the same symbolic factorization may be used.

The filter line-search method applied to the problem (20.34) might itself revert to a restoration phase. If this occurs, then the optimal solution of (20.34) is computed *for a fixed value of* $\bar{x}$, namely, the current iterate $\bar{x}_t$, and this solution is used as the "result" of the restoration phase. In this case, since (20.34) becomes separable, this can be done by solving a quadratic equation for each $(\bar{p}^i, \bar{n}^i)$, i.e.:

$$\bar{n}^i = \frac{\bar{\mu} - \rho c^i(\bar{x})}{2\rho} + \sqrt{\left(\frac{\bar{\mu} - \rho c^i(\bar{x})}{2\rho}\right)^2 + \frac{\bar{\mu} c^i(\bar{x})}{2\rho}}.$$

$$\bar{p}^i = c^i(\bar{x}) + \bar{n}^i.$$

Reducing the KKT error This method tries to achieve reduction in the norm of the primal-dual equations by using the regular iteration steps. If in step 22 of Algorithm

20.1 the feasibility restoration phase is invoked, then the algorithm does not immediately revert to the feasibility restoration phase. Instead, the reduction in the norm of the primal-dual equations is tried. To describe this procedure, let $F_\mu(x, \lambda, z)$ denote the left-hand side of the nonlinear system (20.4). Then, given a constant $\kappa_F \in (0, 1)$, for example, $\kappa_F = 0.999$, the following algorithm reduces the error in the KKT system.

Algorithm 20.3 KKT error reduction algorithm

1.	Initialize the feasibility restoration phase with parameter $t = 0$ and choose the initial point $(\bar{x}_0, \bar{\lambda}_0, \bar{z}_0) = (x_k, \lambda_k, z_k)$
2.	Compute a search direction $\left(\bar{d}_t^x, \bar{d}_t^\lambda, \bar{d}_t^z\right)$ using the system (20.11) and (20.12)
3.	Apply the fraction-to-the-boundary rule: $\bar{\beta}_t \triangleq \max\{\beta \in (0, 1) : \bar{x}_t + \beta \bar{d}_t^x \geq (1 - \tau_j)\bar{x}_t, \quad \bar{z}_t + \beta \bar{d}_t^z \geq (1 - \tau_j)\bar{z}_t\}$
4.	Test whether $\left\|F_\mu(\bar{x}_{t+1}, \bar{\lambda}_{t+1}, \bar{z}_{t+1})\right\|_1 \leq \kappa_F \left\|F_\mu(\bar{x}_t, \bar{\lambda}_t, \bar{z}_t)\right\|_1$, where $(\bar{x}_{t+1}, \bar{\lambda}_{t+1}, \bar{z}_{t+1}) = (\bar{x}_t, \bar{\lambda}_t, \bar{z}_t) + \bar{\beta}_t\left(\bar{d}_t^x, \bar{d}_t^\lambda, \bar{d}_t^z\right)$. If this condition of reduction is not satisfied, then discard the trial point and switch to the procedure for minimizing the constraint violation (described above) by using $\bar{x}_t$ as a reference point. Otherwise, continue with step 5
5.	If $\left(\theta(\bar{x}_{t+1}), \ \varphi_{\mu_j}(\bar{x}_{t+1})\right) \notin F_{k+1}$, then continue the regular interior point Algorithm 20.1 from the point $(x_{t+1}, \lambda_{t+1}, z_{t+1}) = (\bar{x}_{t+1}, \bar{\lambda}_{t+1}, \bar{z}_{t+1})$. Otherwise, set $t = t + 1$ and go to step 2 ♦

In a neighborhood of a strict local solution satisfying the second-order sufficient optimality conditions (see Theorem 5.6) for the barrier problem, the projection of the Hessian $W_t + \Sigma_t$ onto the null space of the constraint Jacobian $\nabla c(\bar{x}_t)^T$ is positive definite. Therefore, no modification of the matrix of the system (20.13), as described in the inertia correction procedure, is applied. As a consequence, the search directions computed by solving the system (20.11) and the relations (20.12) are the Newton steps applied to the nonlinear system (20.4). Hence, the above algorithm will accept those steps and quickly converge toward this solution, provided it is started sufficiently close.

Practical Hints in GAMS

The following statement can be used inside the GAMS program to specify the use of IPOPT: `Option NLP = IPOPT;`. The above statement should appear before the solve statement. If IPOPT was specified as the default solver during GAMS installation, the above statement is not necessary.

IPOPT has many options that can be adjusted for the algorithm. The options are all identified by a string name, and their values can be of one of the three types: number (real), integer, or string. Number options are used for things like tolerances. Integer options are used for things like maximum number of iterations. String options are used for setting algorithm details, like the NLP scaling method. The options can be set by creating an `ipopt.opt` file placed in the directory where IPOPT is run. The `ipopt.opt` file is read line by line, and each line should contain the option name followed by whitespace and then the value. Comments can be included with the # symbol. GAMS/IPOPT currently understands the following GAMS parameters: `reslim` (time limit), `iterlim` (iteration limit), and `domlim` (domain violation limit). List of IPOPT options includes the barrier parameter update, the convergence, the Hessian approximation, the initialization, the line search, the linear_solver, NLP, NLP scaling, the output, the restoration phase, the step calculation, the warm start, etc.

The performance and robustness of IPOPT on larger models heavily rely on the used solver for sparse symmetric indefinite linear systems. The linear solver is chosen by the `linear_solver` option.

Numerical Study

The performances of IPOPT for solving the applications described in Chapter 4 are illustrated in Tables 20.1 and 20.2.

Table 20.1 Performances of IPOPT. Small-scale nonlinear optimization application

	n	*me*	*mc*	*it*	*nf*	*ng*	*nh*	*cpu*	*vfo*
ELCH	10	3	0	10	11	11	10	0.143	−47.761090
ALKI	10	3	8	13	14	14	13	0.011	−1768.8069
PREC	8	0	6	16	17	17	16	0.081	3.95116343
PPSE	9	6	0	11	12	12	11	0.009	5055.01180
MSP3	13	0	15	19	20	20	19	0.044	97.5875091
MSP5	16	0	21	28	29	29	28	0.045	174.786994
POOL	34	20	0	13	14	14	13	0.026	2569.7999
TRAFO	6	0	2	9	10	10	9	0.009	135.07596
LATHE	10	1	14	27	28	28	27	0.020	−4430.0875
DES	150	50	0	88	137	89	88	0.479	1055.18231
CSTC	303	200	0	5	6	6	5	0.035	3.4800741
DIFF	396	324	0	1	2	2	1	0.016	0
PENICI	707	602	0	405	673	395	405	4.626	113.98977

Table 20.2 Performances of IPOPT. Large-scale nonlinear optimization application

	n	*me*	*mc*	*it*	*nf*	*ng*	*nh*	*cpu*	*vfo*
HANG	2002	1001	0	6	7	7	6	0.180	5.0685100
	4002	2001	0	7	8	8	7	0.310	5.0684889
FLOW	1182	754	0	9	10	10	9	0.272	0.226e-10
FLOWO	1556	1005	0	19	20	20	19	0.597	0.689e-7
POL	4004	3000	0	66	67	67	66	3.845	14.216290
	6004	4500	0	54	55	55	54	3.741	14.198908
	8004	6000	0	67	68	68	67	6.337	14.190847
	10,004	7500	0	179	180	180	179	21.750	13.981795
CAT	3003	2000	0	15	16	16	15	0.252	−0.048055
	6003	4000	0	15	16	16	15	0.639	−0.048055
	9003	6000	0	15	16	16	15	0.909	−0.048055
CONT	2505	2000	0	14	15	15	14	0.412	1.0132389
	5005	4000	0	13	14	14	13	0.706	1.0059224
	7505	6000	0	15	16	16	15	0.962	1.0045614
	10,005	8000	0	15	16	16	15	1.264	1.0040718

In Tables 20.1 and 20.2 we have: n = the number of variables, me = the number of equality constraints, mc = the number of inequality constraints, it = the number of iterations to obtain a solution, nf = the number of evaluations of the functions defining the problem, ng = the number of evaluations of the gradients of the functions defining the problem, nh = the number of evaluations of the Hessian, cpu = the cpu computing time for obtaining a solution of the problem (seconds), vfo = the value of the objective function at optimal point.

Notes and References

The content of this chapter is based on the papers of Wächter and Biegler (2005a, 2005b, 2006), where the IPOPT algorithm is described. This is a line-search filter interior point algorithm with two loops. The outer loop approximately minimizes a sequence of nonlinearly equality-constrained barrier sub-problems for a decreasing sequence of barrier parameters. The inner loop uses a line-search filter sequential quadratic programming problem to approximately solve each barrier sub-problem. The global convergence of each barrier sub-problem is enforced through a line-search filter method (Fletcher & Leyffer, 1999, 2002). The filter is reset after each barrier parameter update. The steps are computed by solving a primal-dual system corresponding to the KKT conditions of the barrier problem using the subroutines from HSL mathematical software library collection of Fortran codes for large scientific computation (MA77, MA86, MA97). The algorithm controls the inertia of the primal-dual system by adding a positive parameter to the Hessian of the Lagrangian, thus ensuring the descent properties. The inner iterations include the second-order correction steps and some mechanisms for switching to a feasibility restoration if the stepsize becomes too small. To approximate the Hessian of the Lagrangian, IPOPT has an option for using the limited memory BFGS updates. IPOPT is imbedded in GAMS technology.

Chapter 21
Numerical Studies: Comparisons

As already seen, for solving nonlinear optimization applications we have a multitude of algorithms and codes integrated in GAMS which involve the theoretical concepts of the augmented Lagrangian, of sequential linear-quadratic or quadratic programming, of generalized reduced gradient, of interior point methods with line-search or interior point methods with trust region or filter, etc. All these algorithms work in conjunction with advanced linear algebra concepts, especially for solving positive definite or indefinite large-scale linear algebraic systems. Always, the performances of optimization algorithms crucially depend on the capabilities of linear algebra algorithms, on line search, on filter, etc. On the other hand, for solving nonlinear optimization applications from different domains of activity, the present technology involves the usage of algebraic-oriented languages. At present, plenty modeling and optimization technologies are known (Andrei, 2013b). The most used, both in academic tests and in real practical applications, are GAMS (Brooke, Kendrick, Meeraus, Raman, & Rosenthal, 1998; Rosenthal, 2011), AMPL (Fourer, Gay, & Kernighan, 2002), MPL (Kristjansson, 1993), LINDO (Schrage, 1997), TOMLAB (Holmström, 1997), etc.

The algorithms described in this monograph have some virtues which may only be identified in the process of solving nonlinear optimization applications with different structures and complexities. Therefore, in this chapter we present the performances of some nonlinear optimization algorithms, which have been detailed in the chapters of this monograph and integrated in the GAMS environment. In the first part of this numerical study, we consider the performances of algorithms implemented in the GAMS technology for solving the applications described in Chapter 4: MINOS (Tables 9.4 and 9.5), SNOPT (Tables 15.1 and 15.2), CONOPT (Tables 16.1 and 16.2), KNITRO/ACTIVE (option 3) (Tables 14.1 and 14.2), KNITRO/INTERIOR (options 0, 1, 2) (Tables 19.1, 19.2 and 19.3, 19.5, 19.6 and 19.7), and IPOPT (Tables 20.1 and 20.2).

Table 21.1 includes the name of the applications presented in Chapter 4 and solved in the GAMS environment, as well as a synthesis assembled from the

N. Andrei, *Continuous Nonlinear Optimization for Engineering Applications in GAMS Technology*, Springer Optimization and Its Applications 121,
DOI 10.1007/978-3-319-58356-3_21

Table 21.1 Performances of MINOS, SNOPT, CONOPT, KNITRO, and IPOPT

			MINOS		SNOPT		CONOPT		KNITRO		IPOPT	
	n	*m*	*itM*	*cpu*	*itM*	*cpu*	*it*	*cpu*	*it*	*cpu*	*it*	*cpu*
ELCH	10	3	1	0.02	19	0.01	27	0.043	6	0.143	10	0.143
ALKI	10	3	12	0.02	16	0.01	14	0.062	10	0.011	13	0.011
PREC	8	0	17	0.02	53	0.01	27	0.037	10	0.081	16	0.081
PPSE	9	6	6	0.01	9	0.01	13	0.040	8	0.009	11	0.009
MSP3	13	0	7	0.01	71	0.02	36	0.038	26	0.044	19	0.044
MSP5	16	0	10	0.01	54	0.02	21	0.040	23	0.045	28	0.045
POOL	34	20	5	0.01	5	0.01	12	0.037	9	0.026	13	0.026
TRAFO	6	0	8	0.00	30	0.01	16	0.038	16	0.009	9	0.009
LATHE	10	1	23	0.02	33	0.01	36	0.049	23	0.020	27	0.020
DES	150	50	66	2.75	104	0.16	37	0.169	39	0.479	88	0.479
CSTC	303	200	6	0.11	10	0.03	28	0.211	4	0.035	5	0.035
DIFF	396	324	1	0.02	0	0.01	5	0.036	2	0.016	1	0.016
HANG	2002	1001	209	27.49	60	18.80	19	0.379	21	0.349	6	0.180
	4002	2001	447	205.64	45	133.96	28	1.366	81	3.039	7	0.310
FLOW	1182	754	1	0.19	0	0.01	5	0.058	5	0.160	9	0.272
FLOWO	1556	1005	1	0.25	0	0.01	7	0.078	10	1.031	19	0.597
POL	4004	3000	28	5.29	57	0.92	33	0.232	27	1.203	66	3.845
	6004	4500	40	12.87	63	1.86	38	0.525	26	1.836	54	3.741
	8004	6000	53	22.45	63	3.06	41	0.760	28	2.568	67	6.337
	10,004	7500	66	36.52	73	4.51	45	1.055	30	3.454	179	21.750
CAT	3003	2000	11	2.18	11	2.28	82	0.763	8	0.301	15	0.252
	6003	4000	18	9.06	18	8.75	81	2.136	8	0.674	15	0.639
	9003	6000	23	20.73	23	20.47	32	1.242	7	0.993	15	0.909
CONT	2505	2000	1	0.09	3	0.05	21	0.109	9	0.109	14	0.412
	5005	4000	1	0.20	3	0.11	30	0.270	10	0.403	13	0.706
	7505	6000	1	0.69	3	0.27	31	0.480	11	0.743	15	0.962
	10,005	8000	1	1.33	3	0.44	33	0.767	11	0.947	15	1.264

Table 21.2 Global performances of MINOS, SNOPT, CONOPT, KNITRO, and IPOPT

	MINOS	SNOPT	CONOPT	KNITRO	IPOPT
itt	1063	829	798	468	749
cput	347.980	195.810	11.055	18.728	43.094

abovementioned tables, where n is the number of variables, m is the number of constraints of the applications considered in the above numerical studies, *it* is the number of iterations by CONOPT, KNITRO, and IPOPT to get a solution, *itM* is the number of major iterations required by MINOS and SNOPT to get a solution, and *cpu* is the CPU computing time to obtain a (local) solution (in seconds). Observe that the packages (MINOS, KNITRO, SNOPT, CONOPT, and IPOPT) included in the GAMS technology are able to solve a large variety of nonlinear optimization applications of different dimensions, structures, and complexities.

From Table 21.1 we can immediately obtain the global performances of these algorithms for solving the considered nonlinear optimization applications, performances given in Table 21.2. In Table 21.2 *itt* is the total number of iterations, and *cput* is the total CPU computing time for solving the applications.

From Table 21.2 we see that KNITRO and CONOPT are the algorithms with the best performances with respect to the number of iterations and CPU computing time metrics. The algorithms MINOS and SNOPT have modest performances versus KNITRO, CONOPT, or IPOPT. Observe that the interior point algorithms based on sequential quadratic programming or filter have better performances than MINOS, which is using sequential linear programming or than SNOPT, which is a "pure" sequential quadratic programming algorithm.

In (Andrei, 2013b) we presented a collection of 82 nonlinear optimization applications expressed in the GAMS language. These applications are taken from different domains of activity, such as certain mathematical algorithms and problems; nonlinear systems of equations; mechanical, electrical, or chemical engineering; heat transfer and fluid dynamics; economic development; water management in river systems; robust stability; and optimal control. All these applications represent a good collection of optimization problems to be used on testing the algorithms integrated in the GAMS technology. Therefore, in Table 21.3 of the second part of this numerical study, we show a comparison among the algorithms CONOPT, KNITRO (option 0), and IPOPT with respect to the number of iterations (*it*) and to the CPU computing time in seconds (*cpu*), subject to solving 100 nonlinear optimization applications of different dimensions taken from (Andrei, 2013b). Table 21.3 contains the performances of these algorithms, where n is the number of variables and m is the number of constraints without the simple bound on variables. Selection of CONOPT, KNITRO, and IPOPT is motivated by the fact that these algorithms use different optimization techniques based on the generalized reduced gradient combined with sequential linear or quadratic programming (CONOPT), the integration of the active-set interior point method with direct or conjugate gradient methods (KNITRO), and the interior point method combined with the filter method (IPOPT), respectively. Note that the efficient optimization

Table 21.3 Performances of CONOPT, KNITRO, and IPOPT

			CONOPT		KNITRO		IPOPT	
	n	*m*	*it*	*cpu*	*it*	*cpu*	*it*	*cpu*
0	1	2	3	4	5	6	7	8
Polygon	201	5050	375	7.927	35	2.839	60	5.348
	401	20,100	1242	64.711	38	9.641	71	30.177
	601	45,150	2115	246.188	150	156.227	155	171.249
	801	80,200	3337	628.270	177	331.113	163	339.489
Circle	3	100	6	0.104	11	0.018	14	0.121
	3	500	7	0.117	13	0.038	13	0.182
	3	1000	7	0.114	12	0.063	12	0.255
	3	2000	8	0.154	11	0.106	10	0.328
Hexagon	18	21	16	0.104	7	0.009	14	0.112
Surface	441	0	27	0.150	5	0.031	10	0.121
Minsurf	1682	1	15	1.790	10	0.207	19	0.209
	2602	1	24	7.823	19	0.705	18	0.273
	3722	1	26	0.451	18	0.922	27	0.536
	5042	1	31	0.647	14	0.976	16	0.412
	10,202	1	32	1.613	12	1.768	78	4.858
System2	2	2	21	0.107	3	0.005	4	0.103
cpar	5	5	30	0.100	8	0.009	10	0.108
cpa	10	10	[a]		11	0.017	[a]	
Speed	7	11	6	0.102	6	0.009	10	0.105
Vessel	4	4	5	0.099	5	0.008	9	0.108
Spring	3	4	16	0.083	14	0.017	35	0.129
gtc	3	0	15	0.106	11	0.017	13	0.110
gpf	2	0	4	0.116	5	0.017	7	0.108
Brake	4	6	5	0.097	16	0.017	14	0.114
Truss1	2	3	13	0.098	5	0.008	11	0.111
Truss2	4	1	4	0.100	3	0.007	7	0.106
tub	2	2	13	0.101	6	0.008	11	0.108
Fly	3	2	17	0.102	8	0.008	20	0.120
Lathe	10	15	36	0.114	23	0.017	27	0.127
htb	11	10	20	0.103	12	0.017	16	0.115
Robot2	8	8	19	0.101	3	0.008	16	0.116
Torsion	1686	5	30	1.145	11	0.351	17	0.209
	10,409	5	61	170.980	11	1.822	19	1.287
	40,406	5	110	134.517	[b]		19	6.458
Bearing	1682	1	30	2.412	10	0.175	15	0.206
	10,202	1	36	2.192	10	1.246	18	1.113
	40,402	1	54	15.660	10	5.550	19	6.163
	90,602	1	80	74.376	10	13.830	18	15.156
Chain	403	202	17	0.139	26	0.100	7	0.108
	1003	502	22	0.359	54	0.525	88	1.186

(continued)

Table 21.3 (continued)

			CONOPT		KNITRO		IPOPT	
	n	*m*	*it*	*cpu*	*it*	*cpu*	*it*	*cpu*
0	1	2	3	4	5	6	7	8
	2003	1002	19	0.956	21	0.342	6	0.193
	2403	1202	23	1.807	60	1.394	6	0.207
	4003	2002	28	1.490	81	3.063	7	0.313
Rocket	3008	2503	2218	5.822	112	8.400	135	73.540
	6008	5003	1407	10.696	110	21.574	b	
Camshape	1600	1601	45	0.376	66	0.894	58	733.451
	2000	2001	54	0.605	65	1.120	b	
Steering	2007	1601	178	1.494	10	0.210	10	0.315
	4007	3201	137	5.691	7	0.503	10	0.618
	6007	4801	104	7.310	7	0.673	11	0.875
	10,007	8001	82	10.941	7	1.303	12	1.622
Glider	2616	2410	709	8.595	298	8.471	40	2.311
	5216	4810	1078	16.453	224	11.804	1645	137.524
	10,416	9610	1066	36.083	34	3.361	212	56.255
	13,016	12,010	1331	59.946	519	78.750	c	
Trafo	6	2	17	0.105	5	0.008	8	0.084
Circuit	10	17	21	0.207	13	0.009	22	0.115
Electron	151	51	37	0.243	39	0.464	88	0.567
	301	101	60	0.999	47	2.576	135	3.082
	1501	501	354	175.360	62	126.873	144	188.198
Static	9	6	13	1.159	8	0.096	11	0.092
Dynamic	15	29	6	0.106	12	0.017	11	0.108
Reactor	8	4	27	0.020	10	0.009	16	0.094
Refrigeration	14	15	46	0.119	16	0.017	28	0.037
Netreactor	6	5	11	0.104	5	0.008	7	0.083
Alkilation	7	14	29	0.110	31	0.018	34	0.105
Pool1	7	6	3	0.106	14	0.017	22	0.095
Pool2	33	20	10	0.105	95	0.143	209	0.529
Separation1	24	18	4	0.098	11	0.017	18	0.089
Separation2	62	53	a		20	0.033	52	0.171
Catmix	1204	801	60	0.242	9	0.135	17	0.172
	2404	1601	83	0.523	8	0.200	15	0.251
	3004	2001	82	0.851	8	0.260	15	0.265
	6004	4001	81	2.308	8	0.673	15	0.617
	9004	6001	32	1.433	7	1.216	15	0.913
Control2	304	201	58	0.364	8	0.029	14	0.104
	1504	1001	78	1.611	8	0.124	15	0.292
	3004	2004	83	2.983	8	0.317	15	0.372
Hed	8	6	22	0.109	17	0.017	12	0.089
Diffusion1	397	325	5	0.112	2	0.024	1	0.072

(continued)

Table 21.3 (continued)

			CONOPT		KNITRO		IPOPT	
	n	*m*	*it*	*cpu*	*it*	*cpu*	*it*	*cpu*
0	1	2	3	4	5	6	7	8
Flow	1183	755	5	0.136	5	0.159	9	0.233
Flowobs	1557	1006	7	0.161	10	1.030	19	0.608
Ramsey	201	151	29	0.205	8	0.132	21	0.298
Robust	6	9	32	0.109	8	0.132	11	0.245
Benz	5	7	94	0.110	74	0.126	28	0.260
Fiat	18	25	17	0.110	30	0.129	90	0.380
Control1	60	39	11	0.104	1	0.008	1	0.063
Penicillin	707	602	2212	20.547	399	7.137	424	4.069
	1407	1202	4232	68.863	188	6.093	449	8.067
	2807	2402	6925	276.809	[b]		454	20.777
Ethanol	707	602	425	1.124	94	1.121	54	0.580
	1407	1202	1243	3.959	138	2.929	50	0.899
pol	4004	3000	33	0.376	27	1.199	66	3.787
	8004	6000	41	0.993	28	2.570	67	6.422
cont	2505	2000	21	0.199	9	0.188	14	0.380
	5005	4000	30	0.380	10	0.411	13	0.637
	10,005	8000	33	1.016	11	0.959	15	1.251
msp3	13	15	36	0.105	26	0.035	19	0.089
msp5	16	21	21	0.106	23	0.036	28	0.101
elch	10	3	27	0.105	6	0.031	10	0.087

[a]The problem is infeasible
[b]The solver hit the CPU computing time limit
[c]Failure in feasibility restoration phase in IPOPT

algorithms are based not only on advanced linear algebra techniques (e.g., LUSOL in MINOS and SNOPT, Harwell subroutines MA27-MA86 in IPOPT or a very sophisticated inverse representation of a matrix used in CONOPT), but they also integrate different strategies based on the augmented Lagrangian, on the sequential linear or quadratic programming, on filter or merit functions, etc. Besides, the nonlinear optimization algorithms considered in this numerical study always include a preprocessing phase in which the model of the problem is analyzed to get its characteristics, and it is eventually transfigured to an equivalent form.

All these algorithms are integrated in the GAMS technology. Even if MINOS and SNOPT are integrated in the GAMS technology, they are not included in these comparisons.

Table 21.3 shows that both CONOPT and KNITRO are the most robust algorithms used in this numerical study. Each of them can solve 98 applications from the 100 considered. On the other hand, IPOPT can solve only 96 applications from 100. For the application "glider," IPOPT has difficulties in the feasibility restoration phase. For the applications "torsion" and "penicillin," KNITRO hits the CPU computing time limit.

Table 21.4 Global performances of CONOPT, KNITRO, and IPOPT

	CONOPT	KNITRO	IPOPT
itt	23,276	3275	5518
cput	1614.058	726.544	1812.891

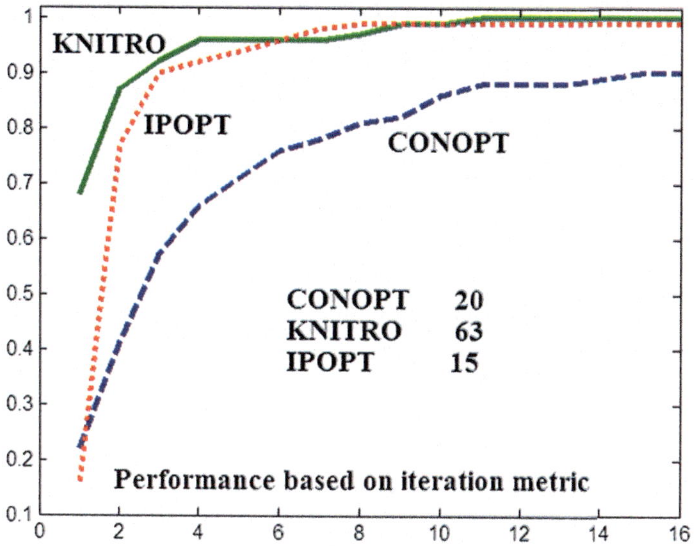

Figure 21.1 Performance profile based on the iteration metric

All the algorithms can solve large-scale problems. For example, the application "bearing" has the largest number of variables: 90,602. On the other hand, the application "polygon" has the largest number of constraints: 80,200. The largest application solved by all the algorithms is "steering," with 10,007 variables and 8001 constraints. Observe that for solving the application "polygon," all the algorithms need the largest resources of CPU computing time. Moreover, for this set of applications, both KNITRO and IPOPT need the same CPU computing time resources.

The global performances of these algorithms are presented in Table 21.4, where *itt* is the total number of iterations for solving all 93 nonlinear optimization applications, and *cput* is the total CPU computing time (seconds).

From Table 21.4 we see that in comparison with CONOPT and IPOPT, with respect to the number of iterations and to CPU computing time metrics, KNITRO is the most efficient algorithm, at least for this set of nonlinear optimization applications. The second in order is IPOPT. Even if CONOPT is on the last position, we immediately see that this is a very good algorithm, able to solve complex nonlinear optimization problems.

Figures 21.1 and 21.2 show the performance profiles of Dolan and Moré (2002) subject to the number of iterations and to CPU computing time metrics, respectively, for solving all the 93 nonlinear optimization applications. The performance

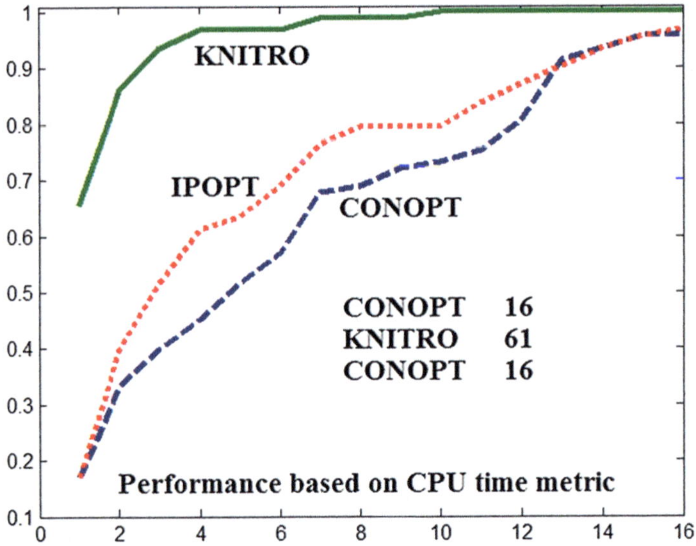

Figure 21.2 Performance profile based on the CPU time metric

profile plot subject to the number of iterations or to the CPU computing time of an algorithm shows the fraction of problems for which the given algorithm is within a factor τ of the best number of iteration or the best CPU time. The percentage of the problems for which an algorithm is the fastest is given on the left axis of the plot. The right side of the plot gives the percentage of the problems that were successfully solved by the algorithms. In other words, the left part of the plot represents the *efficiency* of an algorithm, while the right part represents the *robustness* of the algorithm. Usually, τ takes reasonable values in the interval $\tau \in [1, 16]$. Larger values do not give additional information of the performances of an algorithm.

From Figures 21.1 and 21.2, we see that KNITRO is the most efficient and the most robust algorithm subject to the number of iterations and CPU computing time metrics. With respect to both these metrics, CONOPT and IPOPT have the same efficiency, IPOPT being slightly more efficient than CONOPT subject to the number of iterations metric.

In Figure 21.1 we see that CONOPT solved 20 applications (out of 93) using the minimum number of iterations to get a local optimal solution. Similarly, KNITRO was more efficient; it solved 63 applications in a minimum number of iterations, etc. In Figure 21.2 we see the performances of these algorithms subject to the CPU computing time metric. The tables inserted in these figures show the performances of these algorithms for $\tau = 1$. We conclude that the algorithms working in the GAMS technology can solve large-scale nonlinear optimization applications.

Table 21.5 presents the list of nonlinear constrained optimization applications from Chapter 4, solved by the packages described in the book.

Table 21.5 List of nonlinear-constrained optimization applications from Chapter 4 solved by the packages described in the book

		SPENBAR	MINOS	DONLP	NLPQLP	KNITRO (3)	SNOPT	CONOPT	filter SD	KNITRO (0)	KNITRO (1)	KNITRO (2)	IPOPT
1	ELCH	*	*	*	*	*	*	*	*	*	*	*	*
2	ALKI	*	*	*	*	*	*	*	*	*	*	*	*
3	PREC	*	*	*	*	*	*	*	*	*	*	*	*
4	TRAFO	*	*	*	*	*	*	*		*	*	*	*
5	LATHE	*	*	*	*	*	*	*	*	*	*	*	*
6	PPSE	*	*	*	*	*	*	*	*	*	*	*	*
7	MSP3	*	*	*	*	*	*	*	*	*	*	*	*
8	MSP5	*	*	*		*	*	*	*	*	*	*	*
9	POOL	*	*	*	*	*	*	*	*	*	*	*	*
10	DES	*	*			*	*	*		*	*	*	*
11	HANG	*	*			*	*	*		*	*	*	*
12	CAT		*			*	*	*		*	*	*	*
13	CSTC		*			*	*	*		*	*	*	*
14	DIFF		*			*	*	*		*	*	*	*
15	FLOW/O		*			*	*	*		*	*	*	*
16	PENICI						*	*		*	*	*	*
17	CONT		*			*	*	*		*	*	*	*
18	POL		*			*	*	*		*	*	*	*

In this numerical study, we compared only CONOPT, IPOPT, and KNITRO with respect to the number of iterations and to the CPU computing time metrics. The other algorithms MINOS and SNOPT, which also work in the GAMS technology, have not been compared. This is because MINOS and SNOPT emphasize both the major and the minor iterations. However, both these packages are very good, being reliable to get a solution for large-scale and complex nonlinear optimization problems. For example, for the application "rocket" with $n = 6008$ and $m = 5003$, IPOPT hits the CPU time computing resources. However, MINOS solves this application in 34 major iterations (20,777 minor iterations) and 19.610 s. On the other hand, for this application SNOPT gives an optimal solution in 17 major iterations (17,287 minor iterations) and 18.971 s.

Notes, References, and Discussions

Obviously, plenty of methods are known for solving large-scale nonlinear optimization problems. Each of them uses different computational structures which have as a final point the satisfaction of the first-order KKT optimality conditions. Some of these computational structures are able to get the essence of the optimality conditions, thus generating efficient and robust optimization algorithms. These include the augmented (modified) Lagrangian, the sequential quadratic programming, and the interior point methods.

Mainly, the augmented (modified) Lagrangian and the sequential quadratic programming methods suggest generating and solving a sequence of linear-constrained optimization sub-problems with simple bounds on variables. On the other hand, the interior point methods solve a sequence of nonlinear algebraic systems of equations by the Newton method. This is the motivation why the interior point methods are more efficient and more robust versus *the augmented (modified) Lagrangian and the sequential quadratic programming methods.*

It is interesting to note that the *pure* computational schemes, for example, the augmented Lagrangian or the sequential quadratic programming, do not generate efficient and robust algorithms for solving complex nonlinear optimization problems. The most notable is SNOPT (Gill, Murray, and Saunders, 2002) as a pure sequential quadratic programming algorithm. Even if it is equipped with a lot of computational ingredients which facilitate the solving process of the quadratic programming sub-problems, it is not more efficient or more robust than some other nonlinear optimization algorithms, particularly KNITRO, CONOPT, or IPOPT.

The most powerful nonlinear optimization algorithms combine different optimization techniques and include advanced computational linear algebra techniques. For example, the augmented Lagrangian combined with the penalty functions lead to SPENBAR (Andrei, 1996a, 1996b, 1996c, 1996d, 1996e, 1998a) and to MINOS (Murtagh & Saunders, 1980). The combination of the sequential quadratic programming with the penalty functions generate the algorithms DONLP (Spellucci, 1995), NLPQLP (Schittkowski, 2002, 2009), KNITRO (Byrd Nocedal, & Waltz, 2006), CONOPT (Drud, 1994, 1995, 1996). On the other hand, the interior point methods combined with the line search and with the filters generate the very efficient and robust nonlinear optimization algorithms KNITRO-INTERIOR (Byrd, Nocedal, & Waltz, 2006) and IPOPT (Wächter & Biegler, 2001, 2006). Among these combined methods, KNITRO proves to be the most efficient and the most robust algorithm for solving a large variety of nonlinear optimization problems. KNITRO implements the crossover technique which comes from linear programming (Megiddo, 1989), another very important ingredient. The idea is to switch to the active-set algorithm as soon as the interior point algorithm has generated an approximate solution by solving the KKT nonlinear system using the direct or the conjugate gradient methods.

In this class of algorithms, a special position is held by CONOPT, elaborated by Drud (1994, 1995, 1996). *This algorithm, imbedded into the general structure of the generalized reduced gradient algorithm given by Abadie and Carpentier* (1969), *introduces the sequential linear or the quadratic programming techniques according to the type of the model of the problem to be solved.* Characteristic for CONOPT is that it includes very diversified and complex computational procedures obtained by solving large classes of nonlinear optimization problems with different structures and complexities and a very deep monitoring of the behavior of the algorithm with respect to the properties of the problem.

Besides the optimization techniques like the augmented (modified) Lagrangian, the sequential quadratic programming, or the interior point methods used in different combinations with line search or filters, efficient nonlinear optimization algorithms (and packages) include advanced linear algebra techniques (Meyer, 2000). These refer to *factorization of large-scale sparse matrices* (Hellerman & Rarick, 1971, 1972; Helgason & Kennington, 1980, 1982; Duff, 1977; Suhl & Suhl, 1990, 1991), *LU factorization* (O'Sullivan & Saunders, 2002; Saunders, 2015b), *updating the basis* (Bartels & Golub, 1969; Reid, 1976, 1982; Saunders, 2015a), *direct methods for solving indefinite linear algebraic systems of equations* (Bunch & Kaufman, 1977; Bunch & Parlett, 1971; Paige & Saunders, 1975; Cheng & Higham, 1998; Cheng, 1998), *advanced conjugate gradient methods* (Andrei, 1995; Reid, 1971; Saad, 1996), *projected conjugate gradient methods* (Golub & Van Loan, 1996; Gould, Hribar, & Nocedal, 2001), *inertia control* (Gill, Murray, and Saunders, 2005), *advanced limited memory techniques for updating the Hessian* (Liu & Nocedal, 1989; Byrd, Nocedal, & Schnabel, 1994; Gill & Leonard, 2003; Morales, 2002), *singularity of basis determination* (Higham, 1988), etc.

Appendix A: Mathematical Review

A1. Elements of Linear Algebra

The space of the real vectors of length n is denoted by $\mathbb{R}^n$. The space of the real $m \times n$ matrices is denoted by $\mathbb{R}^{m \times n}$. The *transpose* of x is denoted by x^T. Given the vectors $x, y \in \mathbb{R}^n$, the *scalar product* is defined by $x^T y = \sum_{i=1}^{n} x_i y_i$. The vectors x, $y \in \mathbb{R}^n$ are *orthogonal* (*perpendicular*) if $x^T y = 0$. This is denoted by writing $x \perp y$. The *magnitude* of a vector x is $\|x\| = (x^T x)^{1/2}$. The *angle* between nonzero vectors $x, y \in \mathbb{R}^n$ is defined to be the number $\theta \in [0, \pi]$ such that $\cos\theta = x^T y / \|x\| \, \|y\|$.

A matrix $A \in \mathbb{R}^{m \times n}$ is specified by its elements $a_{ij}, i = 1, \ldots, m$, and $j = 1, \ldots, n$. The *transpose* of $A \in \mathbb{R}^{m \times n}$ denoted by A^T is the $n \times m$ matrix with elements a_{ji}. The matrix A is *squared* if $m = n$. A squared matrix is *symmetric* if $A = A^T$. The *identity matrix* denoted by I is the square diagonal matrix whose diagonal elements are all 1. A square matrix $A = [a_{ij}]$ is said to be *lower triangular* if $a_{ij} = 0$ for $i < j$. Similarly, A is said to be *upper triangular* if $a_{ij} = 0$ for $i > j$.

A set of vectors $v_1, \ldots, v_k$ is said to be *linearly dependent* if there are scalars $\lambda_1, \ldots, \lambda_k$, not all zero, such that $\sum_{i=1}^{k} \lambda_i v_i = 0$. If no such set of scalars exists, then the vectors are said to be *linearly independent*. A *linear combination* of the vectors $v_1, \ldots, v_k$ is a vector of the form $\sum_{i=1}^{k} \lambda_i v_i$. The set of vectors that are linear combinations of $v_1, \ldots, v_k$ is the set *spanned* by the vectors. A linearly independent set of vectors that span $\mathbb{R}^n$ is said to be a *basis* for $\mathbb{R}^n$.

For a linear function $f: \mathbb{R}^n \to \mathbb{R}^m$, let $R(f)$ denote the *range* of f, that is, $R(f) = \{f(x) : x \in \mathbb{R}^n\} \subseteq \mathbb{R}^m$ is the set of all "images" as x varies over $\mathbb{R}^n$. The *range of a matrix* $A \in \mathbb{R}^{m \times n}$ is defined to be the subspace $R(A)$ of $\mathbb{R}^m$ that is generated by the range of $f(x) = Ax$, that is, $R(A) = \{Ax : x \in \mathbb{R}^n\} \subseteq \mathbb{R}^m$. The dimension of $R(A)$ is the rank of A, denoted rank(A). The rank of A can never be greater than the minimum of m and n.

The *null-space of a matrix* $A \in \mathbb{R}^{m \times n}$ is the set $N(A) = \{x : Ax = 0\} \subseteq \mathbb{R}^n$. In other words, $N(A)$ is the set of all solutions to the homogeneous system $Ax = 0$.

N. Andrei, *Continuous Nonlinear Optimization for Engineering Applications in GAMS Technology*, Springer Optimization and Its Applications 121,
DOI 10.1007/978-3-319-58356-3

The rank of a matrix A is equal to the maximum number of linearly independent columns in A. This number is also equal to the maximum number of linearly independent rows in A. The $m \times n$ matrix A is said to be of *full rank* if the rank of A is equal to the minimum of m and n.

A squared $n \times n$ matrix A is *nonsingular* if for any vector $b \in \mathbb{R}^n$ there exists $x \in \mathbb{R}^n$ such that $Ax = b$. For nonsingular matrix A, there exists a unique $n \times n$ matrix B such that $AB = BA = I$. The matrix B is denoted by A^{-1} and is called the *inverse* of A. For nonsingular matrices A and B, the following properties hold: $(A^{-1})^{-1} = A$, the product AB is nonsingular, $(AB)^{-1} = B^{-1}A^{-1}$, $(A^{\mathrm{T}})^{-1} = (A^{-1})^{\mathrm{T}}$.

A square matrix Q is *orthogonal* if it has the property $QQ^{\mathrm{T}} = Q^{\mathrm{T}}Q = I$. Therefore, the inverse of an orthogonal matrix is its transpose.

Suppose that $\|u\| = 1$, and let $u^{\perp}$ denote the space consisting of all vectors that are perpendicular to u. $u^{\perp}$ is called the *orthogonal complement* of u. The matrix $P = I - uu^T$ is the *orthogonal projector* onto $u^{\perp}$ in the sense that P maps each x to its orthogonal projection in $u^{\perp}$.

A scalar value λ is an *eigenvalue* of the $n \times n$ matrix A if there exists a nonzero vector u such that $Au = \lambda u$. The vector u is called an *eigenvector* of A. A matrix A is nonsingular if all its eigenvalues are different from zero. The eigenvalues of symmetric matrices are all real numbers. The nonsymmetric matrices may have imaginary eigenvalues. If a matrix is symmetric and positive definite, then its eigenvalues are all positive real numbers. We can test that a symmetric matrix is positive definite by computing its eigenvalues and by verifying that they are all positive or by performing a Cholesky factorization. If λ_i, $i = 1, \ldots, n$, are the eigenvalues of $A \in \mathbb{R}^{n \times n}$, then the *determinant* and the *trace* of A can be expressed as $\det(A) = \lambda_1 \lambda_2 \cdots \lambda_n$ and $tr(A) = \lambda_1 + \lambda_2 + \cdots + \lambda_n$.

A square matrix A is *positive definite* if and only if $x^{\mathrm{T}}Ax > 0$ for every nonzero $x \in \mathbb{R}^n$. For real symmetric matrix A, the following statements are equivalent.

- All eigenvalues of A are positive.
- $A = B^{\mathrm{T}}B$ for some nonsingular B. While B is not unique, there is one and only one upper triangular matrix R with positive diagonals such that $A = R^{\mathrm{T}}R$. This is the Cholesky factorization of A.
- A has an *LU* (or *LDU*) factorization with all pivots being positive. The *LDU* factorization is of the form $A = LDL^{\mathrm{T}} = R^{\mathrm{T}}R$, where $R = D^{1/2}L^{\mathrm{T}}$ is the Cholesky factor of A.
- The leading principal minors of A are positive.
- All principal minors of A are positive.

Any of the statements above can serve as the definition of a positive definite matrix.

A matrix A is *positive semidefinite* if for all $x \in \mathbb{R}^n$, $x^{\mathrm{T}}Ax \geq 0$. The following statements are equivalent and can serve as the definition of a positive semidefinite matrix:

- All eigenvalues of A are nonnegative.
- $A = B^{\mathrm{T}}B$ for some B with $\mathrm{rank}(B) = r$.
- All principal minors of A are nonnegative.

The *Cholesky method* for solving a symmetric positive definite system $Ax = b$ using the factorization $A = LL^T$ computes the elements of the lower triangular matrix L as follows (see Appendix B7). For $k = 1, \cdots, n$, compute:

$$l_{kk} = \sqrt{a_{kk} - \sum_{i=1}^{k-1} l_{ki}^2}; \quad l_{jk} = \frac{a_{jk} - \sum_{i=1}^{k-1} l_{ji} l_{ki}}{l_{kk}} \text{ for } j = k+1, \cdots, n.$$

This process breaks down at a stage k if the computation of l_{kk} involves the square root of a negative number. This is the case if A is not positive definite.

QR factorization For each nonsingular matrix $A \in \mathbb{R}^{n \times n}$, there is a unique orthogonal matrix Q and a unique upper triangular matrix R with positive diagonal entries, such that $A = QR$ (see Appendix B10).

Symmetric eigenvalue decomposition (spectral decomposition) Suppose $A \in \mathbb{R}^{n \times n}$ is a real symmetric matrix. Then A can be factored as $A = Q \Lambda Q^T$, where $Q \in \mathbb{R}^{n \times n}$ is orthogonal and $\Lambda = \text{diag}(\lambda_1, \ldots, \lambda_n)$, where λ_i are the eigenvalues of A.

Singular value decomposition Suppose $A \in \mathbb{R}^{m \times n}$ with $\text{rank}(A) = r$. Then A can be factored as $A = U \Sigma V^T$, where $U \in \mathbb{R}^{m \times r}$ satisfies $U^T U = I$, $V \in \mathbb{R}^{n \times r}$ satisfies $V^T V = I$, and $\Sigma = \text{diag}(\sigma_1, \ldots, \sigma_r)$ with $\sigma_1 \geq \sigma_2 \geq \cdots \geq \sigma_r > 0$. The columns of U are called *left singular vectors* of A, the columns of V are called *right singular vectors*, and the numbers σ_i are the *singular value*.

The *inertia* of a symmetric matrix A is defined to be the triple (p, n, z) in which p, n, and z are the respective number of positive, negative, and zero eigenvalues, counting algebraic multiplicities.

The matrices A and B are *congruent* if there is a nonsingular matrix C such that $B = C^T AC$. The *Sylvester's law of inertia* states that: the symmetric matrices A and B are congruent if and only if A and B have the same inertia.

For a vector $x \in \mathbb{R}^n$, the following *norms* can be defined: $\|x\|_1 = \sum_{i=1}^n |x_i|$, $\|x\|_2 = (x^T x)^{1/2}$, and $\|x\|_\infty = \max_{i=1, \ldots, n} |x_i|$. The norm $\|.\|_2$ is called the *Euclidian norm*. On the other hand, $\|.\|_1$ is referred to as the l_1 *norm*, and $\|.\|_\infty$ is referred to as l_∞ *norm*. All these norms measure the length of the vector in some sense. They are equivalent in the sense that each one is bounded above and below by a multiple of the other one.

For the Euclidian norm, the *Cauchy-Schwarz inequality* holds: $|x^T y| \leq \|x\| \|y\|$, with equality if and only if one of these vectors is a nonnegative multiple of the other one. The *Hölder inequality*, a generalization of the Cauchy-Schwarz inequality, states that for all $a_i > 0, b_i > 0, i = 1, \ldots, n$, $p, q > 0$ such that $1/p + 1/q = 1$:

$$\sum_{i=1}^n a_i b_i \leq \left(\sum_{i=1}^n a_i^p \right)^{1/p} \left(\sum_{i=1}^n b_i^q \right)^{1/q}.$$

In particular, $|x^T y| \leq \|x\|_\infty \|y\|_1$.

The *matrix norms* induced by the vector l_1 norm and the vector l_∞ norm are as follows:

$$\|A\|_1 = \max_{\|x\|_1=1} \|Ax\|_1 = \max_j \sum_i |a_{ij}| = \text{the largest absolute column sum.}$$
$$\|A\|_\infty = \max_{\|x\|_\infty=1} \|Ax\|_\infty = \max_i \sum_j |a_{ij}| = \text{the largest absolute row sum.}$$

The matrix norm induced by the Euclidian vector norm is $\|A\|_2 = \max_{\|x\|_2=1} \|Ax\|_2 = \sqrt{\lambda_{\max}}$, where $\lambda_{\max}$ is the largest eigenvalue of A^TA.

The *Frobenius norm* of $A \in \mathbb{R}^{m\times n}$ is defined as $\|A\|_F^2 = \sum_{i,j} |a_{ij}|^2 = tr(A^TA)$, where for the matrix $A^{n\times n} = (a_{ij})$, $tr(A) = a_{11} + \cdots + a_{nn}$, is the trace of A.

The condition number of a nonsingular matrix $A \in \mathbb{R}^{n\times n}$ denoted as $cond(A)$ or $\kappa(A)$ is defined as: $cond(A) = \|A\|_2 \|A^{-1}\|_2 = \sigma_{\max}(A)/\sigma_{\min}(A)$, where $\sigma_{\max}(A)$ and $\sigma_{\min}(A)$ are the largest and the smallest singular values of A, respectively.

Consider the block matrix $X = \begin{bmatrix} A & C \\ D & B \end{bmatrix}$, where A and B are nonsingular submatrices. The matrices $S = B - DA^{-1}C$ and $T = A - CB^{-1}D$ are called the *Schur complements* of A and B in X, respectively. The Schur complements arise in several contexts and appear to be an important ingredient in many theorems. For example, we have $\det X = \det A \det S$.

Sherman-Morrison formula. Let $A \in \mathbb{R}^{n\times n}$ be a nonsingular matrix and $a, b \in \mathbb{R}^n$ two vectors such that $1 + b^TA^{-1}a \neq 0$. Then, the inverse of the matrix $B = A + ab^T$ is:

$$B^{-1} = A^{-1} - \frac{A^{-1}ab^TA^{-1}}{1 + b^TA^{-1}a}.$$

If $1 + b^TA^{-1}a = 0$, then B is a singular matrix.

A2. Elements of Analysis

Let $\{x_k\}$ be a sequence of points from $\mathbb{R}^n$. A sequence $\{x_k\}$ *converges* to a point x, written as $\lim_{k\to\infty} x_k = x$, if for any $\varepsilon > 0$ there exists an index K such that $\|x_k - x\| \leq \varepsilon$ for all $k \geq K$. Given an index set $\bar{K} \subset \{1, 2, \ldots\}$, a *subsequence* of $\{x_k\}$corresponding to $\bar{K}$can be defined and denoted by $\{x_k\}_{k\in\bar{K}}$.

Theorem A2.1 (*Bolzano-Weierstrass theorem*) *Each bounded sequence in* $\mathbb{R}^n$ *has a convergent subsequence.*

The point $x^* \in \mathbb{R}^n$ is an *accumulation point* or a *limit point* for the sequence $\{x_k\}$ if there is an infinite set of indices $k_1, k_2, k_3, \ldots$ such that the subsequence $\{x_{k_i}\}_{i=1,2,3,\ldots}$ converges to x^*, i.e., $\lim_{i\to\infty} x_{k_i} = x^*$. A sequence is a *Cauchy sequence* if for any $\varepsilon > 0$ there exists an integer K such that $\|x_k - x_m\| \leq \varepsilon$ for all indices $k \geq K$ and $m \geq K$. A sequence converges if and only if it is a Cauchy sequence.

A function $f: \mathbb{R}^n \to \mathbb{R}^m$ is *continuous* at $x \in \mathbb{R}^n$ if for all $\varepsilon > 0$ there exists a $\delta > 0$ such that for any $y \in \mathbb{R}^n$, $\|y - x\|_2 \leq \delta \Rightarrow \|f(y) - f(x)\|_2 \leq \varepsilon$. The continuity can be described in terms of limits: whenever the sequence $\{x_k\}$ in $\mathbb{R}^n$ converges to a point $x \in \mathbb{R}^n$, the sequence $\{f(x_k)\}$ in $\mathbb{R}^m$ converges to $f(x)$, i.e., $\lim_{k\to\infty} f(x_k) = f(\lim_{k\to\infty} x_k)$. A function f is continuous if it is continuous at every point in $\mathbb{R}^n$.

Let $f: \mathbb{R} \to \mathbb{R}$ be a real-valued function of a real variable. The *first derivative* is defined by:

$$f'(x) \triangleq \lim_{\varepsilon\to 0} \frac{f(x+\varepsilon) - f(x)}{\varepsilon}.$$

The *second derivative* is defined by:

$$f''(x) \triangleq \lim_{\varepsilon\to 0} \frac{f'(x+\varepsilon) - f'(x)}{\varepsilon}.$$

The *directional derivative* of a function $f: \mathbb{R}^n \to \mathbb{R}$ in direction $p \in \mathbb{R}^n$ is given by:

$$D(f(x);p) \triangleq \lim_{\varepsilon\to 0} \frac{f(x+\varepsilon p) - f(x)}{\varepsilon}.$$

If f is continuously differentiable in a neighborhood of x, then:

$$D(f(x);p) = \nabla f(x)^{\mathrm{T}} p.$$

Let $f: \mathbb{R}^n \to \mathbb{R}$ be a continuously differentiable function. The conditions which characterize a minimum can be expressed in terms of the *gradient* $\nabla f(x)$ with the first partial derivatives defined as:

$$\nabla f(x) = \left[\frac{\partial f}{\partial x_1}, \ldots, \frac{\partial f}{\partial x_n}\right]^{\mathrm{T}}$$

and $n \times n$ *Hessian* matrix $\nabla^2 f(x)$ with the second partial derivatives whose (i,j)th element is.

$$(\nabla^2 f(x))_{ij} = \partial^2 f(x)/\partial x_i \partial x_j,\ i,j = 1, \ldots, n.$$

When f is twice continuously differentiable, the Hessian matrix is always symmetric.

As a simple example, let us consider the quadratic function $f: \mathbb{R}^n \to \mathbb{R}$, $f(x) = (1/2)x^T A x + b^T x + a$, where $A \in \mathbb{R}^{n\times n}$ is a symmetric matrix. Then $\nabla f(x) = Ax + b$. The Hessian of f is given by $\nabla^2 f(x) = A$, i.e., the second-order approximation of a quadratic function is itself.

When $f = [f_1, \ldots, f_m]^{\mathrm{T}}$ is a vector-valued function, that is, $f: \mathbb{R}^n \to \mathbb{R}^m$, then $\nabla f(x)$ is a $m \times n$ matrix called *Jacobian*, whose (i,j) element is $\partial f_i(x)/\partial x_j$, $i = 1, \ldots, m$ and $j = 1, \ldots, n$.

Consider the nonlinear optimization problem

$$\min\{f(x) : c(x) \leq 0, \quad h(x) = 0\}$$

where $f: \mathbb{R}^n \to \mathbb{R}$, $c: \mathbb{R}^n \to \mathbb{R}^m$, and $h: \mathbb{R}^n \to \mathbb{R}^p$ are twice continuously differentiable functions defined on $\mathbb{R}^n$, with components $c(x) = [c_1(c), \ldots, c_m(x)]^{\mathrm{T}}$ and $h(x) = [h_1(x), \ldots, h_p(x)]^{\mathrm{T}}$. In the theory of nonlinear optimization problems, the *Lagrange function* (or *Lagrangian*) is defined as:

$$\begin{aligned} L(x, \mu, \lambda) &\triangleq f(x) + \mu^{\mathrm{T}} c(x) + \lambda^{\mathrm{T}} h(x) \\ &= f(x) + \sum_{i=1}^{m} \mu_i c_i(x) + \sum_{j=1}^{p} \lambda_j h_j(x), \end{aligned}$$

where $\mu = [\mu_1, \ldots, \mu_m] \in \mathbb{R}^m$ and $\lambda = [\lambda_1, \ldots, \lambda_p] \in \mathbb{R}^p$ are known as the *Lagrange multipliers*. For this function we have

$$\begin{aligned} \nabla L(x, \mu, \lambda) &= \nabla f(x) + \nabla c(x)^{\mathrm{T}} \mu + \nabla h(x)^{\mathrm{T}} \lambda \\ &= \nabla f(x) + \sum_{i=1}^{m} \mu_i \nabla c_i(x) + \sum_{j=1}^{p} \lambda_j \nabla h_j(x), \end{aligned}$$

where $\nabla c(x)$ and $\nabla h(x)$ are the Jacobian matrices of $c(x)$ and $h(x)$, respectively.

$$\nabla^2 L(x, \mu, \lambda) = \nabla^2 f(x) + \sum_{i=1}^{m} \mu_i \nabla^2 c_i(x) + \sum_{j=1}^{p} \lambda_j \nabla^2 h_j(x),$$

where $\nabla^2 c_i(x)$ and $\nabla^2 h_j(x)$ are the Hessian matrices of $c_i(x)$ and $h_j(x)$,respectively.

Theorem A2.2 (*Mean value theorem*). *Given a continuously differentiable function $f: \mathbb{R} \to \mathbb{R}$ and two real numbers x_1 and x_2 that satisfy $x_2 > x_1$, then*

$$f(x_2) = f(x_1) + f'(\xi)(x_2 - x_1),$$

for some $\xi \in (x_1, x_2)$.

For a multivariate function $f: \mathbb{R}^n \to \mathbb{R}$, the mean value theorem says that for any vector $d \in \mathbb{R}^n$.

$$f(x + d) = f(x) + \nabla f(x + \alpha d)^{\mathrm{T}} d,$$

for some $\alpha \in (0, 1)$.

Theorem A2.3 (*Taylor's theorem*). *If f is continuously differentiable in a region containing the line segment $[x_1, x_2]$, then there is a θ, $0 \leq \theta \leq 1$, such that:*

$$f(x_2) = f(x_1) + \nabla f(\theta x_1 + (1 - \theta) x_2)^{\mathrm{T}} (x_2 - x_1).$$

Moreover, if f is twice continuously differentiable in a region containing the line segment $[x_1, x_2]$, *then there is a* θ, $0 \le \theta \le 1$, *such that:*

$$f(x_2) = f(x_1) + \nabla f(x_1)^T (x_2 - x_1) + \frac{1}{2}(x_2 - x_1)^T \nabla^2 f(\theta x_1 + (1-\theta)x_2)(x_2 - x_1).$$

Theorem A2.4 (*Implicit function theorem*). *Let* $h: \mathbb{R}^n \times \mathbb{R}^m \to \mathbb{R}^n$ *be a function such that:*

1. $h(z^*, 0) = 0$ *for some* $z^* \in \mathbb{R}^n$.
2. *The function* $h(.,.)$ *is continuously differentiable in some neighborhood of* $(z^*, 0)$.
3. $\nabla_z h(z,t)$ *is nonsingular at the point* $(z,t) = (z^*, 0)$.

Then, there are the open sets $N_z \subset \mathbb{R}^n$ *and* $N_t \subset \mathbb{R}^m$ *containing* z^* *and* 0, *respectively, and a continuous function* $z: N_t \to N_z$ *such that* $z^* = z(0)$ *and* $h(z(t), t) = 0$ *for all* $t \in N_t$. $z(t)$ *is uniquely defined. If h is q times continuously differentiable with respect to both its arguments for some* $q > 0$, *then* $z(t)$ *is also q times continuously differentiable with respect to t and*

$$\nabla z(t) = -\nabla_t h(z(t), t)\left[\nabla_z h(z(t), t)\right]^{-1},$$

for all $t \in N_t$.

The implicit function theorem is applied to parameterized systems of linear equations in which z is obtained as the solution of $M(t)z = g(t)$, where $M(.) \in \mathbb{R}^{n \times n}$ has $M(0)$ nonsingular and $g(t) \in \mathbb{R}^n$ (see the algebraic characterization of a tangent space). To apply the theorem, define $h(z,t) = M(t)z - g(t)$. If $M(.)$ and $g(.)$ are continuously differentiable in some neighborhood of 0, the theorem implies that $z(t) = M(t)^{-1} g(t)$ is a continuous function of t in some neighborhood of 0.

Convergence Let $\{x_k\}$ be a sequence from $\mathbb{R}^n$ that converges to x^*. This sequence converges *Q–linear* if there is a constant $r \in (0,1)$ such that

$$\frac{\|x_{k+1} - x^*\|}{\|x_k - x^*\|} \le r,$$

for all k sufficiently large. The convergence is *Q–superlinear* if

$$\lim_{k \to \infty} \frac{\|x_{k+1} - x^*\|}{\|x_k - x^*\|} = 0.$$

The convergence is *Q–quadratic* if

$$\frac{\|x_{k+1} - x^*\|}{\|x_k - x^*\|^2} \le M,$$

for all k sufficiently large where M is a positive constant, not necessarily smaller than 1.

Order notation The *order notation* is a concept used to see how the members of a sequence behave when we get far enough along in the sequence. Let us consider two nonnegative sequences of scalars $\{\eta_k\}$ and $\{\theta_k\}$. We write $\eta_k = o(\theta_k)$ if the sequence of ratios $\{\eta_k/\theta_k\}$ approaches zero, i.e., $\lim_{k\to\infty}\eta_k/\theta_k = 0$. We write $\eta_k = O(\theta_k)$ if there is a positive constant c such that $|\eta_k| \le c|\theta_k|$, for all k sufficiently large. If $\eta : \mathbb{R} \to \mathbb{R}$ is a function, then we write $\eta(t) = o(t)$ to specify that the ratio $\eta(t)/t$ approaches zero either as $t \to 0$ or $t \to \infty$. Similarly, we write $\eta(t) = O(t)$ if there is a constant c such that $|\eta(t)| \le c|t|$ for all $t \in \mathbb{R}$. A slight variant of the above definitions is as follows. We write $\eta_k = o(1)$ to specify that $\lim_{k\to\infty}\eta_k = 0$. Similarly, we write $\eta_k = O(1)$ to indicate that there is a constant c such that $|\eta_k| \le c$ for all k. Sometimes, in the above definitions, there are vectors or matrices quantities as arguments. In these cases the definitions apply to the norms of these quantities. For instance, if $f: \mathbb{R}^n \to \mathbb{R}^n$, then we write $f(x) = O(\|x\|)$ if there is a positive constant c such that $\|f(x)\| \le c\|x\|$ for all x in the domain of f.

A3. Elements of Topology in the Euclidian Space $\mathbb{R}^n$

The *open ball* of radius ε centered at x^* is defined as the set $B(x^*, \varepsilon) = \{x \in \mathbb{R}^n : \|x^* - x\| < \varepsilon\}$, in any norm.

A subset $D \subset \mathbb{R}^n$ is *open* if for every $x \in D$ there exists a positive number $\varepsilon > 0$ such that the ball of radius ε centered at x is contained in D, i.e., $\{y \in \mathbb{R}^n : \|y - x\| \le \varepsilon\} \subset D$. The intersection of finitely many open sets is open. Any union of open sets is open.

A point $x \in \mathbb{R}^n$ is an *interior point* of the set D if there is an open ball $B(x, \varepsilon)$ such that $B(x, \varepsilon) \subset D$. The *interior* of a set D, denoted by int$D$, is the set of the interior points of D. The interior of a set is the largest open set contained in D.

A point $x \in \mathbb{R}^n$ is an *exterior point* of D if it is an interior point of $\mathbb{R}^n \setminus D$. Notice that the set D is open if every point of D is an interior point of D. Obviously, if D is open then int $D = D$.

A point $\tilde{x}$ is said to be a *limit point* of the set D if every open ball $B(\tilde{x}, \varepsilon)$ contains a point $x \neq \tilde{x}$ such that $x \in D$. Note that $\tilde{x}$ does not necessarily have to be an element of D to be a limit point of D.

The set D is *closed* if for all possible sequences of points $\{x_k\}$ in D, all limit points of $\{x_k\}$ are elements of D. The union of finitely many closed sets is closed. Any intersection of closed sets is closed.

The set D is *bounded* if there is some real number $M > 0$ such that $\|x\| \le M$, for all $x \in D$.

The set D is *compact* if every sequence $\{x_k\}$ of points in D has at least one limit point and all such limit points are in D. A central result in topology is that in $\mathbb{R}^n$ the set D is compact if it is both closed and bounded.

The *closure* of the set D is the set $cl(D) \triangleq D \cup L$, where L denotes the set of all limit points of D.

For a given point $x \in \mathbb{R}^n$, a *neighborhood* of x is an open set containing x. A useful neighborhood is the open ball of radius ε centered at x.

A point $x \in \mathbb{R}^n$ is a *boundary point* of the set D if every neighborhood of x contains points both inside and outside of D. The set of boundary points of D is denoted by ∂D.

Let $f: \mathbb{R}^n \to \mathbb{R}^m$. Then f is *Lipschitz continuous* on an open set $D \subset \mathbb{R}^n$ if there is a constant $L > 0$ such that $\| f(x) - f(y) \| \leq L \|x - y\|$ for all $x, y \in D$. L is called the *Lipschitz constant.*

A4. Elements of Convexity: Convex Sets and Convex Functions

A set $C \subset \mathbb{R}^n$ is a *convex set* if for every points $x, y \in C$, the point $z \triangleq \lambda x + (1-\lambda)y$ is also in the set C for any $\lambda \in [0, 1]$, The intersection of any family of convex sets is a convex set.

Let $a \in \mathbb{R}^n$ be a vector and $c \in \mathbb{R}$ a scalar. Then, the set $H \triangleq \{x : a^T x = c\}$ is called a *hyperplane*. The set $H^+ \triangleq \{x : a^T x \geq c\}$ is called a *half-space*.

A nonempty set $C \subset \mathbb{R}^n$ is a *cone* if for every point $x \in C$, $\alpha x \in C$ for any $\alpha \geq 0$. If, besides this, C is convex, then C is a *convex cone*. The cone generated by the vectors $\{x_1, \ldots, x_m\}$is the set of all vectors x of the form $x = \sum_{i=1}^m \alpha_i x_i$, where $\alpha_i \geq 0$ for all $i = 1, \ldots, m$. Notice that all cones of this form are convex.

The following two results on the separation of the convex sets are used in specifying the optimality conditions for nonlinear programming problems. We present them here without proof (see (Bazaraa, Sherali, & Shetty 1993)).

Proposition A4.1 (*Separation of a convex set and a point*). *Let $C \subset \mathbb{R}^n$ be a nonempty and convex set. Consider a point $y \notin C$. Then, there exist a nonzero vector $a \in \mathbb{R}^n$ and a scalar $c \in \mathbb{R}$ such that $a^T y > c$ and $a^T x \leq c$ for any $x \in C$.* ♦

Proposition A4.2 (*Separation of two convex sets*). *Let C_1 and C_2 be two nonempty and convex sets in $\mathbb{R}^n$. Suppose that $C_1 \cap C_2 = \emptyset$. Then, there exists a hyperplane that separates C_1 and C_2,* i.e., *there is a nonzero vector $p \in \mathbb{R}^n$ such that $p^T x_1 \geq p^T x_2$ for any $x_1 \in cl(C_1)$ and for any $x_2 \in cl(C_2)$.* ♦

Theorem A4.1 (*Farkas' theorem*). *Let A be an $m \times n$ matrix and c an n vector. Then, just one of the following two statements holds:*

System 1. *There exists $x \in \mathbb{R}^n$ such that $Ax \leq 0$ and $c^T x > 0$.*

System 2. *There exists $y \in \mathbb{R}^m$ such that $A^T y = c$ and $y \geq 0$.*

Proof. Suppose that System 2 has a solution, i.e., there exists $y \geq 0$ such that $A^T y = c$. Let x be such that $Ax \leq 0$. Then, $c^T x = y^T Ax \leq 0$. Therefore, System 1 has no solution.

Now suppose that System 2 has no solution. Consider the set $X = \{x : x = A^T y, y \geq 0\}$. Observe that X is a closed and convex set. By Proposition A4.1, there exist a vector

$p \in \mathbb{R}^n$ and a scalar α such that $p^T c > \alpha$ and $p^T x \leq \alpha$ for all $x \in X$. Since $0 \in X$, then $\alpha \geq 0$ and so $p^T c > 0$. Also, $\alpha \geq p^T A^T y = y^T Ap$ for all $y \geq 0$. Since $y \geq 0$ can be made arbitrarily large, it follows that the last inequality implies that $Ap \leq 0$. Therefore, we have constructed a vector $p \in \mathbb{R}^n$ such that $Ap \leq 0$ and $c^T p > 0$, i.e., System 1, has a solution. ♦

Theorem A4.2 (*Gordan's theorem*). *Let A be an* $m \times n$ *matrix. Then just one of the following two statements holds:*

System 1. *There exists* $x \in \mathbb{R}^n$ *such that* $Ax < 0$.

System 2. *There exists* $y \in \mathbb{R}^m$, $y \neq 0$ *such that* $A^T y = 0$ *and* $y \geq 0$.

Proof. System 1 can be equivalently written as $Ax + es \leq 0$ for some $x \in \mathbb{R}^n$ and $s > 0$, $s \in \mathbb{R}$, where e is a vector of m ones. Now, rewriting this system in the form of System 1 of Theorem A4.1, we obtain $[A \ \ e]\begin{bmatrix} x \\ s \end{bmatrix} \leq 0$ and $(0, \ldots, 0, 1)\begin{pmatrix} x \\ s \end{pmatrix} > 0$ for some $\begin{pmatrix} x \\ s \end{pmatrix} \in \mathbb{R}^{n+1}$. By Theorem A4.1, the associated System 2 states that $\begin{bmatrix} A^T \\ e^T \end{bmatrix} y = (0, \ldots, 0, 1)^T$ and $y \geq 0$ for some $y \in \mathbb{R}^m$, that is, $A^T y = 0$, $e^T y = 1$, and $y \geq 0$ for some $y \in \mathbb{R}^m$. But, this is equivalent to System 2. ♦

Theorem A4.3 *Let P and Q be two symmetric matrices such that* $P \geq 0$ *and* $P > 0$ *on the null space of Q* (i.e., $y^T P y > 0$ *for any* $y \neq 0$ *with* $Qy = 0$). *Then, there exists* $\bar{c} > 0$ *such that* $P + cQ > 0$ *for any* $c > \bar{c}$.

Proof. Assume the contrary. Then, for any $k > 0$ there exists x^k, $\|x^k\| = 1$ such that $(x^k)^T P(x^k) + k(x^k)^T Q(x^k) \leq 0$. Consider a subsequence $\{x^k\}_K$ convergent to some $\bar{x}$ with $\|\bar{x}\| = 1$. Dividing the above inequality by k and taking the limit as $k \in K \to \infty$, we get $\bar{x}^T Q \bar{x} \leq 0$. On the other hand, Q being semipositive definite, we must have $\bar{x}^T Q \bar{x} \geq 0$, hence $\bar{x}^T Q \bar{x} = 0$. Therefore, using the hypothesis, it follows that $\bar{x}^T P \bar{x} \geq 0$. But this contradicts the fact that $\bar{x}^T P \bar{x} + \limsup_{k \to \infty, k \in K} k(x^k)^T Q(x^k) \leq 0$. ♦

A function $f: C \to \mathbb{R}^n$ defined on a convex set $C \subset \mathbb{R}^n$ is a *convex function* if $f(\lambda x + (1-\lambda)y) \leq \lambda f(x) + (1-\lambda) f(y)$, for every $x, y \in C$ and every $\lambda \in (0, 1)$. Moreover, f is said to be *strictly convex* if $f(\lambda x + (1-\lambda)y) < \lambda f(x) + (1-\lambda) f(y)$, for every $x, y \in C$ and every $\lambda \in (0, 1)$. A function that is not convex is said to be *nonconvex*.

Proposition A4.3 (*Convexity of level set*). *Let C be a convex set in* $\mathbb{R}^n$, *and let* $f: C \to \mathbb{R}$ *be a convex function. Then, the level set* $C_\alpha \triangleq \{x \in C : f(x) \leq \alpha\}$, *where* α *is a real number, is a convex set.*

Proof. Let $x_1, x_2 \in C$. Of course, $x_1, x_2 \in C_\alpha$. $f(x_1) \leq \alpha$ and $f(x_2) \leq \alpha$. Now, let $\lambda \in (0, 1)$ and consider $x = \lambda x_1 + (1-\lambda)x_2$. By convexity of C, it follows that $x \in C$. On the other hand, by convexity of f on C,

$$f(x) \leq \lambda f(x_1) + (1-\lambda) f(x_2) \leq \lambda\alpha + (1-\lambda)\alpha = \alpha,$$

i.e., $x \in C_\alpha$. ♦

Proposition A4.4 (*Convexity of a domain defined by a set of convex functions*). *Let C be a convex set in $\mathbb{R}^n$, and let $c_i : C \to \mathbb{R}$, $i = 1, \ldots, m$, be convex functions on C. Then, the set defined by $X \triangleq \{x \in C : c_i(x) \leq 0, i = 1, \ldots, m\}$ is convex.*

Proof. The result follows from Proposition A4.3 and from the property of the intersection of convex sets. ♦

The following two propositions give *differential criteria* of checking the convexity of a function.

Proposition A4.5 (*First-order condition of convexity*). *Let C be a convex set in $\mathbb{R}^n$ with a nonempty interior. Consider the function $f : C \to \mathbb{R}$ which is continuous on C and differentiable on* int(C). *Then, f is convex on* int(C) *if and only if $f(y) \geq f(x) + \nabla f(x)^T (y - x)$ for any points $x, y \in C$.* ♦

Proposition A4.6 (*Second-order condition for convexity*). *Let C be a convex set in $\mathbb{R}^n$ with a nonempty interior. Consider the function $f : C \to \mathbb{R}$ which is continuous on C and twice differentiable on* int(C). *Then, f is convex on* int(C) *if and only if the Hessian $\nabla^2 f(x)$ is positive semidefinite at each $x \in$* int(C). ♦

The convexity of the objective function and of the constraints is crucial in nonlinear optimization. The convex programs have very nice theoretical properties which can be used to design efficient optimization algorithms. Therefore, it is important to know how to detect convexity and the operations that preserve the convexity of functions.

Proposition A4.7 (*Linear combination with nonnegative coefficients*). *Let C be a convex set in $\mathbb{R}^n$. If $f : C \to \mathbb{R}^n$ and $g : C \to \mathbb{R}^n$ are convex functions on C, then their linear combination $\lambda f + \eta g$, where the coefficients λ and η are nonnegative, is also convex on C.* ♦

Proposition A4.8 (*Composition with affine mapping*). *Let C and D be convex sets in $\mathbb{R}^m$ and $\mathbb{R}^n$, respectively. If $g : C \to \mathbb{R}$ is a convex function on C and $h : D \to \mathbb{R}^m$ is an affine mapping,* i.e., *$h(x) = Ax + b$, with range(h) $\subset C$, then the composite function $f : D \to \mathbb{R}$ defined as $f(x) \triangleq g(h(x))$ is convex on D.* ♦

Notes and References

The material in this appendix is covered in (Dennis and Schnabel 1983; Bazaraa, Sherali, & Shetty 1993; Demmel 1997; Lay 1982; Meyer 2000).

Appendix B: Solving Linear Equation Systems

This appendix concentrates on methods for solving systems of linear equations

$$Ax = b,$$

where $A \in \mathbb{R}^{n \times n}$ and $b \in \mathbb{R}^n$. This is a basic problem that arises in many optimization algorithms and is crucial in the efficiency of the algorithms. Assume that A is nonsingular, so the solution is unique for all vectors b and is given by $x = A^{-1}b$. The matrix A is often called the *coefficient matrix*, and the vector b is called the *right-hand side* term. Firstly we present some cases for which $Ax = b$ can be easily solved. In these cases the coefficient matrix has some special structures. Further on we will focus on general systems where A has no structure (Meyer 2000; Golub and Van Loan 1996; Demmel 1997; Higham 1996; and Trefethen and Bau 1997).

B1. Systems with Diagonal Matrices

Suppose that $A \in \mathbb{R}^{n \times n}$ is a diagonal and nonsingular matrix, i.e., for all i, $a_{ii} \neq 0$. In this case the set of linear equations $Ax = b$ can be written as $a_{ii}x_i = b_i$, $i = 1, \ldots, n$. Therefore, the solution is simply given by $x_i = b_i/a_{ii}$, $i = 1, \ldots, n$.

B2. Systems with Upper Triangular Matrices (Back Substitution)

Consider the system $Ux = b$, where the coefficient matrix $U \in \mathbb{R}^{n \times n}$ is an *upper triangular matrix* in which $u_{ii} \neq 0$, $i = 1, \ldots, n$, i.e., there are no zero pivots.

N. Andrei, *Continuous Nonlinear Optimization for Engineering Applications in GAMS Technology*, Springer Optimization and Its Applications 121,
DOI 10.1007/978-3-319-58356-3

$$\begin{bmatrix} u_{11} & u_{12} & \cdots & u_{1n} \\ 0 & u_{22} & \cdots & u_{2n} \\ \vdots & \vdots & \ddots & \vdots \\ 0 & 0 & \cdots & u_{nn} \end{bmatrix} \begin{bmatrix} x_1 \\ x_2 \\ \vdots \\ x_n \end{bmatrix} = \begin{bmatrix} b_1 \\ b_2 \\ \vdots \\ b_n \end{bmatrix}.$$

For solving this system, the general *back substitution* is as follows:

1. Firstly, compute $x_n = b_n / u_{nn}$.
2. Determine x_i, $i = n-1, n-2, \ldots, 1$, recursively as:

$$x_i = \frac{1}{u_{ii}}(b_i - u_{i,i+1}x_{i+1} - u_{i,i+2}x_{i+2} - \cdots - u_{i,n}x_n) = \frac{1}{u_{ii}}\left(b_i - \sum_{k=i+1}^{n} u_{ik}x_k\right).$$

B3. Systems with Lower Triangular Matrices (Forward Substitution)

Consider the system $Lx = b$, where the coefficient matrix $L \in \mathbb{R}^{n \times n}$ is a *unit lower triangular matrix*, that is,

$$\begin{bmatrix} 1 & 0 & \cdots & 0 \\ l_{21} & 1 & \cdots & 0 \\ \vdots & \vdots & \ddots & \vdots \\ l_{n1} & l_{n2} & \cdots & 1 \end{bmatrix} \begin{bmatrix} x_1 \\ x_2 \\ \vdots \\ x_n \end{bmatrix} = \begin{bmatrix} b_1 \\ b_2 \\ \vdots \\ b_n \end{bmatrix}.$$

For solving this system, the general *forward substitution* is as follows:

1. Firstly, compute $x_1 = b_1$.
2. Determine x_i, $i = 2, 3, \ldots, n$ recursively as:

$$x_i = b_i - (l_{i,1}x_1 + l_{i,2}x_2 + \ldots + l_{i,i-1}x_{i-1}) = b_i - \sum_{k=1}^{i-1} l_{ik}x_k.$$

B4. Systems with Orthogonal Matrices

A matrix $A \in \mathbb{R}^{n \times n}$ is orthogonal if $A^T A = I$, i.e., $A^{-1} = A^T$. In this case the solution of the system $Ax = b$ can be computed by a simple matrix-vector product $x = A^T b$.

B5. Systems with Permutation Matrices

Let $\pi = (\pi_1, \pi_2, \ldots, \pi_n)$ be a permutation of $(1, 2, \ldots, n)$. A permutation matrix is a square matrix obtained from the identity matrix by a permutation of its rows. Every row and column of a permutation matrix has a single 1 with 0 s everywhere else. If A is a permutation matrix, then solving $Ax = b$ is very simple: x is obtained by permuting the entries of b by π^{-1}.

B6. Gaussian Elimination (LU Factorization)

Gaussian elimination (Gauss method) is a *direct method* for solving linear systems of equations $Ax = b$,where $A \in \mathbb{R}^{n \times n}$ is a real matrix. Let A be a nonsingular matrix, then the end result of applying the Gaussian elimination to A is an upper triangular matrix with nonzero elements on the main diagonal, i.e.:

$$A \xrightarrow{\text{Gaussian elimination}} \begin{bmatrix} * & * & \cdots & * \\ 0 & * & \cdots & * \\ \vdots & \vdots & \ddots & \vdots \\ 0 & 0 & \cdots & * \end{bmatrix}.$$

Gaussian Elimination for a General Small Example

Let us illustrate the Gaussian elimination by means of a general small example:

$$\begin{bmatrix} a_{11} & a_{12} & a_{13} \\ a_{21} & a_{22} & a_{23} \\ a_{31} & a_{32} & a_{33} \end{bmatrix} \begin{bmatrix} x_1 \\ x_2 \\ x_3 \end{bmatrix} = \begin{bmatrix} b_1 \\ b_2 \\ b_3 \end{bmatrix}.$$

The Gaussian elimination transforms this system to triangular form as follows. Suppose that $a_{11} \neq 0$. Multiplying the first row by a_{21}/a_{11} and subtracting it from the second row lead to the equivalent system

$$\begin{bmatrix} a_{11} & a_{12} & a_{13} \\ 0 & a_{22}^{(2)} & a_{23}^{(2)} \\ a_{31} & a_{32} & a_{33} \end{bmatrix} \begin{bmatrix} x_1 \\ x_2 \\ x_3 \end{bmatrix} = \begin{bmatrix} b_1 \\ b_2^{(2)} \\ b_3 \end{bmatrix},$$

where

$$a_{22}^{(2)} = a_{22} - (a_{21}/a_{11})a_{12},$$
$$a_{23}^{(2)} = a_{23} - (a_{21}/a_{11})a_{13},$$

and

$$b_2^{(2)} = b_2 - (a_{21}/a_{11})b_1.$$

Now, multiplying the first row by a_{31}/a_{11} and subtracting it from the third row lead to the equivalent system

$$\begin{bmatrix} a_{11} & a_{12} & a_{13} \\ 0 & a_{22}^{(2)} & a_{23}^{(2)} \\ 0 & a_{32}^{(2)} & a_{33}^{(2)} \end{bmatrix} \begin{bmatrix} x_1 \\ x_2 \\ x_3 \end{bmatrix} = \begin{bmatrix} b_1 \\ b_2^{(2)} \\ b_3^{(2)} \end{bmatrix},$$

where

$$a_{32}^{(2)} = a_{32} - (a_{31}/a_{11})a_{12},$$
$$a_{33}^{(2)} = a_{33} - (a_{31}/a_{11})a_{13},$$

and

$$b_3^{(2)} = b_3 - (a_{31}/a_{11})b_1.$$

Finally, assuming that $a_{22}^{(2)} \neq 0$, multiplying the new second row by $\left(a_{32}^{(2)}/a_{22}^{(2)}\right)$ and subtracting it from the third row lead to the system

$$\begin{bmatrix} a_{11} & a_{12} & a_{13} \\ 0 & a_{22}^{(2)} & a_{23}^{(2)} \\ 0 & 0 & a_{33}^{(3)} \end{bmatrix} \begin{bmatrix} x_1 \\ x_2 \\ x_3 \end{bmatrix} = \begin{bmatrix} b_1 \\ b_2^{(2)} \\ b_3^{(3)} \end{bmatrix},$$

where

$$a_{33}^{(3)} = a_{33}^{(2)} - \left(a_{32}^{(2)}/a_{22}^{(2)}\right)a_{23}^{(2)}$$

and

$$b_3^{(3)} = b_3^{(2)} - \left(a_{32}^{(2)}/a_{22}^{(2)}\right)b_2^{(2)}.$$

Observe that the system obtained at the end of this process has the upper triangular form $Ux = c$, where

$$U = \begin{bmatrix} a_{11} & a_{12} & a_{13} \\ 0 & a_{22}^{(2)} & a_{23}^{(2)} \\ 0 & 0 & a_{33}^{(3)} \end{bmatrix} \quad \text{and} \quad c = \begin{bmatrix} b_1 \\ b_2^{(2)} \\ b_3^{(3)} \end{bmatrix},$$

which can be solved by back substitution.

General Gaussian Elimination

The above process may be performed in general by creating zeros in the first column, then in the second one, and so forth. For $k = 1, 2, \ldots, n-1$ the Gaussian elimination is defined by the following formulae:

$$a_{ij}^{(k+1)} = a_{ij}^{(k)} - \left(a_{ik}^{(k)} / a_{kk}^{(k)}\right) a_{kj}^{(k)}, \quad i, j > k$$

and

$$b_i^{(k+1)} = b_i^{(k)} - \left(a_{ik}^{(k)} / a_{kk}^{(k)}\right) b_k^{(k)}, \quad i > k,$$

where $a_{ij}^{(1)} = a_{ij}$, $i, j = 1, 2, \ldots, n$.

To be well defined, the only assumption required is that $a_{kk}^{(k)} \neq 0$, $k = 1, 2, \ldots, n$, hold. In the Gaussian elimination, these entries are called *pivots*. Usually, the following notation is used:

$$A^{(k)} x = b^{(k)},$$

as the system obtained after $(k-1)$ stages, $k = 1, 2, \ldots, n$, where $A^{(1)} = A$ and $b^{(1)} = b$. The final matrix $A^{(n)}$ is upper triangular.

Zero Pivots: Row Interchanges

The above-described Gaussian process breaks down when a pivot is zero, say $a_{kk}^{(k)} = 0$. In this case, in order to continue the Gaussian process, row interchanges are needed. We illustrate zero pivots and row interchanges by using a small general example.

Suppose we have executed two stages of the Gaussian elimination on a system of order 5, and at the third stage the system is $A^{(3)} x = b^{(3)}$, in the following form:

$$\begin{bmatrix} a_{11}^{(1)} & a_{12}^{(1)} & a_{13}^{(1)} & a_{14}^{(1)} & a_{15}^{(1)} \\ 0 & a_{22}^{(2)} & a_{23}^{(2)} & a_{24}^{(2)} & a_{25}^{(2)} \\ 0 & 0 & 0 & a_{34}^{(3)} & a_{35}^{(3)} \\ 0 & 0 & a_{43}^{(3)} & a_{44}^{(3)} & a_{45}^{(3)} \\ 0 & 0 & a_{53}^{(3)} & a_{54}^{(3)} & a_{55}^{(3)} \end{bmatrix} \begin{bmatrix} x_1 \\ x_2 \\ x_3 \\ x_4 \\ x_5 \end{bmatrix} = \begin{bmatrix} b_1^{(1)} \\ b_2^{(2)} \\ b_3^{(3)} \\ b_4^{(3)} \\ b_5^{(3)} \end{bmatrix}.$$

In this case, if $a_{43}^{(3)} \neq 0$ or $a_{53}^{(3)} \neq 0$ holds, then the third row is interchanged with either the fourth or the fifth row, and we may continue the Gaussian process. This interchanging to obtain nonzero pivots is called *pivoting*.

On the other hand, the Gaussian elimination breaks down if $a_{33}^{(3)} = a_{43}^{(3)} = a_{53}^{(3)} = 0$. In this case the matrix is singular. i.e.:

$$\det A^{(3)} = a_{11}^{(1)} a_{22}^{(2)} \det \begin{bmatrix} 0 & a_{34}^{(3)} & a_{35}^{(3)} \\ 0 & a_{44}^{(3)} & a_{45}^{(3)} \\ 0 & a_{54}^{(3)} & a_{55}^{(3)} \end{bmatrix} = 0.$$

Relationship with LU Factorization

For solving the system $Ax = b$, where A is nonsingular, the Gaussian elimination consists of the following four steps (Demmel 1997):

1. Factorize the matrix A as $A = PLU$, where:
   P is a *permutation* matrix.
   L is a *unit lower triangular* matrix.
   U is a *nonsingular upper triangular* matrix.
2. Solve the system $PLUx = b$ subject to LUx by permuting the entries of b, i.e., $LUx = P^{-1}b = P^T b$.
3. Solve the system $LUx = P^{-1}b$ subject to Ux by *forward substitution*, i.e., $Ux = L^{-1}(P^{-1}b)$.
4. Solve the system $Ux = L^{-1}(P^{-1}b)$ subject to x by *backward substitution*, i.e., $x = U^{-1}(L^{-1}(P^{-1}b))$.

The following result is central in the Gaussian elimination:

The following two statements are equivalent:

1. *There exists a unique unit lower triangular matrix L and a nonsingular upper triangular matrix U such that A = LU. This is called LU factorization of A.*
2. *All leading principal submatrices of A are nonsingular.*

LU factorization without pivoting can fail on nonsingular matrices, and therefore we need to introduce permutations into the Gaussian elimination.

If A is a nonsingular matrix, then there exist permutation matrices P_1 and P_2, a unit lower triangular matrix L and a nonsingular upper triangular matrix U such that $P_1AP_2 = LU$. Observe that P_1A reorders the rows of A. AP_2 reorders the columns of A. P_1AP_2 reorders both the rows and the columns of A.

If A is nonsingular, then it has a nonzero entry. Therefore, we choose the permutations P'_1 and P'_2 so that the (1,1) entry of $P'_1AP'_2$ is nonzero. Now we write the factorization and solve for the unknown components:

$$P'_1AP'_2 = \begin{bmatrix} a_{11} & A_{12} \\ A_{21} & A_{22} \end{bmatrix} = \begin{bmatrix} 1 & 0 \\ L_{21} & I_{n-1} \end{bmatrix} \begin{bmatrix} u_{11} & U_{12} \\ 0 & \bar{A}_{22} \end{bmatrix} = \begin{bmatrix} u_{11} & U_{12} \\ L_{21}u_{11} & L_{21}U_{12} + \bar{A}_{22} \end{bmatrix},$$

where $A_{22}, \bar{A}_{22} \in \mathbb{R}^{(n-1)\times(n-1)}$ and $L_{21}, U_{12}^T \in \mathbb{R}^{n-1}$. Solving for the components of this 2×2 block factorization, we get:

$u_{11}=a_{11}\neq 0$, $U_{12}=A_{12}$, and $L_{21}u_{11}=A_{21}$. Since $u_{11}=a_{11}\neq 0$, we can solve it to get $L_{21}=\frac{A_{21}}{a_{11}}$. Finally, $L_{21}U_{12}+\bar{A}_{22}=A_{22}$ implies that $\bar{A}_{22}=A_{22}-L_{21}U_{12}$. Observe that

$$\det P'_1AP'_2 = \det\begin{bmatrix} 1 & 0 \\ L_{21} & I_{n-1} \end{bmatrix} \det\begin{bmatrix} u_{11} & U_{12} \\ 0 & \bar{A}_{22} \end{bmatrix} = 1(u_{11}\det\bar{A}_{22}).$$

Since $\det P'_1AP'_2 = \pm\det A \neq 0$, it follows that $\det\bar{A}_{22}$ must be nonzero. Therefore, the factorization process may continue.

Indeed, by induction there exist the permutation matrices $\bar{P}_1$ and $\bar{P}_2$ so that $\bar{P}_1\bar{A}_{22}\bar{P}_2=\bar{L}\bar{U}$, where $\bar{L}$ is a unit lower triangular matrix and $\bar{U}$ is an upper triangular and nonsingular matrix. Substituting this in the above 2×2 block factorization, we get:

$$\begin{aligned} P'_1AP'_2 &= \begin{bmatrix} 1 & 0 \\ L_{21} & I \end{bmatrix}\begin{bmatrix} u_{11} & U_{12} \\ 0 & \bar{P}_1^T\bar{L}\bar{U}\bar{P}_2^T \end{bmatrix} = \begin{bmatrix} 1 & 0 \\ L_{21} & I \end{bmatrix}\begin{bmatrix} 1 & 0 \\ 0 & \bar{P}_1^T\bar{L} \end{bmatrix}\begin{bmatrix} u_{11} & U_{12} \\ 0 & \bar{U}\bar{P}_2^T \end{bmatrix} \\ &= \begin{bmatrix} 1 & 0 \\ L_{21} & \bar{P}_1^T\bar{L} \end{bmatrix}\begin{bmatrix} u_{11} & U_{12}\bar{P}_2 \\ 0 & \bar{U} \end{bmatrix}\begin{bmatrix} 1 & 0 \\ 0 & \bar{P}_2^T \end{bmatrix} \\ &= \begin{bmatrix} 1 & 0 \\ 0 & \bar{P}_1^T \end{bmatrix}\begin{bmatrix} 1 & 0 \\ \bar{P}_1L_{21} & \bar{L} \end{bmatrix}\begin{bmatrix} u_{11} & U_{12}\bar{P}_2 \\ 0 & \bar{U} \end{bmatrix}\begin{bmatrix} 1 & 0 \\ 0 & \bar{P}_2^T \end{bmatrix}. \end{aligned}$$

Therefore, the desired factorization of A is:

$$P_1AP_2 = \left(\begin{bmatrix} 1 & 0 \\ 0 & \bar{P}_1 \end{bmatrix}P'_1\right)A\left(P'_2\begin{bmatrix} 1 & 0 \\ 0 & \bar{P}_2 \end{bmatrix}\right) = \begin{bmatrix} 1 & 0 \\ \bar{P}_1L_{21} & \bar{L} \end{bmatrix}\begin{bmatrix} u_{11} & U_{12}\bar{P}_2 \\ 0 & \bar{U} \end{bmatrix}.$$

The next two results state simple ways to choose the permutation matrices P_1 and P_2 to guarantee that the Gaussian elimination will run on nonsingular matrices.

Gaussian Elimination with Partial Pivoting

We can choose the permutation matrices $P'_2=I$ and P'_1 in such a way that a_{11} is the largest entry in absolute value in its column, which implies that $L_{21}=\frac{A_{21}}{a_{11}}$ has entries bounded by 1 in absolute value. More generally, at step i of the Gaussian elimination, where we are computing the ith column of L, we reorder rows i through n so that the largest entry in the column is on the diagonal. This is called "Gaussian elimination with partial pivoting" or GEPP for short. GEPP guarantees that all entries of L are bounded by 1 in absolute value.

Gaussian Elimination with Complete Pivoting

We can choose the permutation matrices P'_2 and P'_1 in such a way that a_{11} is the largest entry in absolute value in the whole matrix. More generally, at step i of the Gaussian elimination, where we are computing the ith column of L, we reorder rows and columns i through n so that the largest entry in this submatrix is on the *diagonal. This is called "Gaussian elimination with complete pivoting" or GECP for short.*

The following algorithm is an implementation of the results mentioned above by performing permutations, by computing the first column of L and the first row of U, and then by updating A_{22} to get $\bar{A}_{22} = A_{22} - L_{21}U_{12}$.

Algorithm GE (LU Factorization with Pivoting)

```
for i = 1 to n − 1
    apply permutations so that a_ii ≠ 0 (permute L and U too)
      /* for example, for GEPP, swap rows j and i of A and of L
         where |a_ji| is the largest entry in |A(i:n, i)|; for GECP, swap
         rows j and i of A and of L, and columns k and i of A and
         of U, where |a_jk| is the largest entry in |A(i:n, i:n)| */
    /* compute column i of L */
    for j = i + 1 to n
        l_ji = a_ji / a_ii
    end for
    /* compute row i of U */
    for j = i to n
        u_ij = a_ij
    end for
    /* update A_22 */
    for j = i + 1 to n
      for k = i + 1 to n
        a_jk = a_jk − l_ji u_ik
      end for
    end for
end for
```

Remark Once the column i of A has been used to compute the column i of L, it will never be used later in the algorithm GE. Similarly, row i of A is never used after computing row i of U. This property allows us to overwrite L and U on top of A as soon as they are computed. Therefore, there is no need for extra space to store these matrices. L can occupy the strict lower triangle of A (the ones on the diagonal of L are not stored explicitly). Similarly, U can occupy the upper triangle of A. Therefore, the algorithm can be simplified as:

Algorithm LU (LU Factorization with Pivoting, Overwriting L and U on A)

for $i = 1$ *to* $n-1$
    apply permutations so that $a_{ii} \neq 0$
    for $j = i+1$ *to* n
        $a_{ji} = a_{ji}/a_{ii}$
    end for
    for $j = i+1$ *to* n
        for $k = i+1$ *to* n
            $a_{jk} = a_{jk} - a_{ji}a_{ik}$
        end for
    end for
end for

B7. Cholesky Factorization

If $A \in \mathbb{R}^{n \times n}$ is symmetric and positive definite, then it can be factored as

$$A = LL^T,$$

where L is a lower triangular and nonsingular matrix with positive diagonal elements. This is called the *Cholesky factorization* of A and can be interpreted as a symmetric LU factorization with $L = U^T$. The matrix L, which is uniquely determined by A, is called the *Cholesky factor* of A. The algorithm is as follows:

Cholesky Factorization

for $j = 1$ *to* n
    $l_{jj} = \left(a_{jj} - \sum_{k=1}^{j-1} l_{jk}^2\right)^{1/2}$
    for $i = j+1$ *to* n
        $l_{ij} = \left(a_{ij} - \sum_{k=1}^{j-1} l_{ik}l_{jk}\right)/l_{jj}$
    end for
end for

If A is not positive definite, then the Cholesky factorization will fail by attempting to compute the square root of a negative number or by dividing by zero. This is the cheapest way to test if a symmetric matrix is positive definite.

The Cholesky factorization can be used to solve the system $Ax = b$ when A is symmetric and positive definite.

Solving linear systems by Cholesky factorization:

1. *Cholesky factorization*. Factor A as $A = LL^T$.
2. *Forward substitution*. Solve $Lz = b$.
3. *Back substitution*. Solve $L^T x = z$.

B8. The Factor-Solve Method

For solving the linear system $Ax = b$, the basic approach is based on expressing A as a product of nonsingular matrices:

$$A = A_1 A_2 \cdots A_r.$$

Therefore, the solution is given by:

$$x = A^{-1} b = A_r^{-1} A_{r-1}^{-1} \ldots A_1^{-1} b.$$

The solution x is computed working from right to left as:

$$\begin{aligned}
z_1 &= A_1^{-1} b, \\
z_2 &= A_2^{-1} z_1 = A_2^{-1} A_1^{-1} b, \\
&\vdots \\
z_{r-1} &= A_{r-1}^{-1} z_{r-2} = A_{r-1}^{-1} A_{r-2}^{-1} \cdots A_1^{-1} b, \\
x &= A_r^{-1} z_{r-1} = A_r^{-1} A_{r-1}^{-1} \cdots A_1^{-1} b.
\end{aligned}$$

We see that the ith step of this process requires computing $z_i = A_i^{-1} z_{i-1}$, i.e., solving the linear system $A_i z_i = z_{i-1}$. The step of expressing A in factored form is called the *factorization step*. On the other hand, the process of computing $x = A^{-1} b$ recursively by solving a sequence of systems of the form $A_i z_i = z_{i-1}$ is called the *solve step*. The idea of the factor-solve method is to determine the factors A_i, $i = 1, \ldots, r$, as simple as possible, i.e., diagonal, lower, or upper triangular, permutation, orthogonal, etc.

Factor-Solve Method and LU Factorization

Assume that in the general Gaussian elimination, $a_{kk}^{(k)} \neq 0$ hold for every $k = 1, \ldots, n$. Referring to the general Gaussian elimination, we see that $l_{ik} = a_{ik}^{(k)} / a_{kk}^{k}$ for $i > k$ is exactly what is used to multiply the kth row and subsequently subtract it from the ith row in building the new ith row. l_{ik} is called a *multiplier*.

Now, let $L^{(k)}$ be the unit lower triangular matrix which differs from the identity matrix only in the kth column below the main diagonal, where the negatives of the multipliers l_{ik} appear. These matrices are called *elementary lower triangular matrices*. With these matrices, the general Gaussian elimination can be expressed in matrix notation as:

$$A^{(k+1)} = L^{(k)}A^{(k)},$$

where $A^{(1)} = A$. Using these relations for all values of k, we get:

$$U = A^{(n)} = L^{(n-1)}L^{(n-2)}\cdots L^{(1)}A.$$

The inverse of $L^{(k)}$ is very easy to be computed, by changing the sign of the multipliers.

$$L^{(k)} = \begin{bmatrix} 1 & & & & & & \\ & \ddots & & & & & \\ & & 1 & & & & \\ & & & 1 & & & \\ & & & -l_{k+1,k} & 1 & & \\ & & & \vdots & & \ddots & \\ & & & -l_{n,k} & & & 1 \end{bmatrix},$$

$$\left(L^{(k)}\right)^{-1} = \begin{bmatrix} 1 & & & & & & \\ & \ddots & & & & & \\ & & 1 & & & & \\ & & & 1 & & & \\ & & & l_{k+1,k} & 1 & & \\ & & & \vdots & & \ddots & \\ & & & l_{n,k} & & & 1 \end{bmatrix}$$

Therefore, from the above relations, we get:

$$A = \left(L^{(1)}\right)^{-1}\left(L^{(2)}\right)^{-1}\cdots\left(L^{(n-1)}\right)^{-1}U.$$

The solution of the linear system $Ax = b$ is very easy to be computed by using the structure of the $L^{(k)}$ and U matrices.

B9. Solving Underdetermined Linear Systems

Let us consider the linear system of equations $Ax = b$, where $A \in \mathbb{R}^{m\times n}$, and $m < n$. Assume that $\mathrm{rank}(A) = m$, so there is at least one solution for all b. In many applications it is sufficient to know one particular solution $\bar{x}$. In other situations it is necessary to have a parameterization of all solutions as:

$$\{x : Ax = b\} = \{Zy + \bar{x} : y \in \mathbb{R}^{n-m}\},$$

where Z is a matrix whose columns form a basis for the null space of A.

The solution of the underdetermined system $Ax = b$ is very easy to be determined if a $m \times m$ nonsingular submatrix of A is known. Assume that the first m columns of A are linearly independent. The system can be written as:

$$Ax = \begin{bmatrix} A_1 & A_2 \end{bmatrix} \begin{bmatrix} x_1 \\ x_2 \end{bmatrix} = A_1 x_1 + A_2 x_2 = b,$$

where $A_1 \in \mathbb{R}^{m \times m}$ is nonsingular. Therefore, we can express x_1 as:

$$x_1 = A_1^{-1}(b - A_2 x_2).$$

A particular solution for the system $Ax = b$ is $\bar{x}_2 = 0$ and $\bar{x}_1 = A_1^{-1}b$. All solutions of $Ax = b$ can be parameterized using $x_2 \in \mathbb{R}^{n-m}$ as a free parameter. We can write:

$$x = \begin{bmatrix} x_1 \\ x_2 \end{bmatrix} = \begin{bmatrix} -A_1^{-1}A_2 \\ I \end{bmatrix} x_2 + \begin{bmatrix} A_1^{-1}b \\ 0 \end{bmatrix}.$$

This gives the following parameterization:

$$Z = \begin{bmatrix} -A_1^{-1}A_2 \\ I \end{bmatrix}, \bar{x} = \begin{bmatrix} A_1^{-1}b \\ 0 \end{bmatrix}.$$

B10. The QR Factorization

The matrix $A \in \mathbb{R}^{n \times m}$ with $m \leq n$ and $rankA = m$ can be factored as:

$$A = \begin{bmatrix} Q_1 & Q_2 \end{bmatrix} \begin{bmatrix} R \\ 0 \end{bmatrix},$$

where $Q_1 \in \mathbb{R}^{n \times m}$ and $Q_2 \in \mathbb{R}^{n \times (n-m)}$ satisfy.

$$Q_1^T Q_1 = I, \quad Q_2^T Q_2 = I, \quad Q_1^T Q_2 = 0,$$

and $R \in \mathbb{R}^{m \times m}$ is upper triangular with nonzero diagonal elements. This is called the *QR factorization* of A.

The QR factorization can be used for solving the underdetermined systems of linear equations $Ax = b$, where $A \in \mathbb{R}^{m \times n}$ with $m < n$. Consider that

$$A^T = [Q_1 \quad Q_2] \begin{bmatrix} R \\ 0 \end{bmatrix}$$

is the QR factorization of A^T. Therefore, $\bar{x} = Q_1 \left(R^T\right)^{-1} b$ satisfies the equations:

$$A\bar{x} = R^T Q_1^T Q_1 \left(R^T\right)^{-1} b = b.$$

The columns of Q_2 form a basis for the null-space of A. Therefore, the complete solution set of the above system can be parameterized as:

$$\{x = \bar{x} + Q_2 z : z \in \mathbb{R}^{n-m}\}.$$

Usually, the QR factorization is used for solving underdetermined systems of linear equations. The main drawback of this method is that it is difficult to exploit the sparsity of the matrix. Even if A is sparse, the factor Q is usually dense.

B11. LU Factorization of Rectangular Matrices

The matrix $A \in \mathbb{R}^{n \times m}$ with $m \leq n$ and $rankA = m$ can be factored as

$$A = PLU,$$

where $P \in \mathbb{R}^{n \times n}$ is a permutation matrix, $L \in \mathbb{R}^{n \times m}$ is unit lower triangular (i.e., $l_{ij} = 0$ for $i < j$ and $l_{ii} = 1$), and $U \in \mathbb{R}^{m \times m}$ is nonsingular and upper triangular. If the matrix A is sparse, then the LU factorization usually includes row and column permutation, i.e., A is factored as

$$A = P_1 LUP_2,$$

where $P_1 \in \mathbb{R}^{n \times n}$, $P_2 \in \mathbb{R}^{m \times m}$ are permutation matrices. The LU factorization of a sparse rectangular matrix can be calculated efficiently at a cost that is much lower than for dense matrices.

The LU factorization can be used for solving underdetermined systems of linear equations. Consider the system of linear equations $Ax = b$, where $A \in \mathbb{R}^{m \times n}$ with $m < n$. Suppose that the matrix A^T is LU factored as $A^T = PLU$ and L is partitioned as

$$L = \begin{bmatrix} L_1 \\ L_2 \end{bmatrix},$$

where $L_1 \in \mathbb{R}^{m \times m}$ and $L_2 \in \mathbb{R}^{(n-m) \times m}$. Then the solution set of the system can be parameterized as

$$\{x : Ax = b\} = \{Zz + \bar{x} : z \in \mathbb{R}^{n-m}\},$$

with

$$Z = P\begin{bmatrix} -\left(L_1^T\right)^{-1} L_2^T \\ I \end{bmatrix},$$

$$\bar{x} = P\begin{bmatrix} \left(L_1^T\right)^{-1} \left(U^T\right)^{-1} b \\ 0 \end{bmatrix}.$$

The LU factorization of rectangular matrices is used in MINOS and SNOPT packages (Saunders 2015a, b, c).

Notes and References

For dense numerical linear algebra, the standard references are (Golub and Van Loan 1996; Demmel 1997; Higham 1996; and Trefethen and Bau 1997). The direct methods for sparse matrices are covered in (Duff, Erismann, & Reid 1986; Schendel 1989; George and Liu 1981; and Osterby and Zlatev 1983). A recent addition is the book by Davis (2006), which deals with sparse direct solution methods and contains plenty of helpful details on sparse matrices. The iterative methods for sparse linear systems are discussed in (Saad 1996, 2011). The numerical linear algebra that focuses on problems arising in nonlinear optimization are presented in (Gill, Murray, & Wright 1981; Nocedal and Wright 2006; Boyd and Vandenberghe 2006; Luenberger 1984; and Luenberger and Ye 2008), to mention only some of them.

References

Abadie, J. (1978). Un nouvel algorithme pour la programmation non lineaire. *RAIRO/Recherche Operationnelle, 12*(2., Mai 1978), 233–238.
Abadie, J. (1979). Une modification de la methode GRG. *RAIRO/Recherche Operationnelle, 13* (3., Aout 1979), 323–326.
Abadie, J., & Carpentier, J. (1969). Generalization of the Wolfe reduced gradient method to the case of nonlinear constraints. In R. Fletcher (Ed.), *Optimization* (pp. 37–47). London, UK: Academic.
Abadie, J., & Guigou, J. (1970). Numerical experiments with GRG method. In J. Abadie (Ed.), *Integer and nonlinear programming* (pp. 529–536). Amsterdam, Europe: North-Holland.
Abadie, J., & Haggag, A. (1979). Performance du gradient reduit generalise avec une methode quasi Newtonienne pour la programmation non lineaire. *RAIRO/Recherche Operationnelle, 13* (2., Mai 1979), 209–216.
Andersen, E. D., & Andersen, K. D. (2000). The MOSEK interior point optimizer for linear programming: an implementation of the homogeneous algorithm. In T. T. H. Frenk, K. Roos, & S. Zhang (Eds.), *High performance optimization* (pp. 197–232). New York, NY, USA: Kluwer Academic Publishers.
Andrei, N. (1985). Application of sparse matrix techniques to the GRG algorithm for large-scale non-linear programming. *Revue Roumaine Sciences Techniques Electrotechn. et Energy, 30* (2), 175–186.
Andrei, N. (1987). Application of sparse matrix techniques in GRG algorithm for very large-scale non-linear programming. *Revue Roumaine Sciences Techniques Electrotechn et Energy, 32*(4), 457–464.
Andrei, N. (1995). *Computational experience with conjugate gradient algorithms for large-scale unconstrained optimization.* (Technical Report, Research Institute for Informatics-ICI, Bucharest, July 21, 1995).
Andrei, N. (1996a). *Computational experience with a modified penalty-barrier method for large-scale nonlinear constrained optimization.* (Working Paper No. AMOL-96-1, Research Institute for Informatics-ICI, Bucharest, February 6, 1996).
Andrei, N. (1996b). *Computational experience with a modified penalty-barrier method for large-scale nonlinear, equality and inequality constrained optimization.* (Technical Paper No. AMOL-96-2, Research Institute for Informatics-ICI, Bucharest, February 12, 1996).

N. Andrei, *Continuous Nonlinear Optimization for Engineering Applications in GAMS Technology*, Springer Optimization and Its Applications 121,
DOI 10.1007/978-3-319-58356-3

Andrei, N. (1996c). *Computational experience with "SPENBAR" a sparse variant of a modified penalty-barrier method for large-scale nonlinear, equality and inequality, constrained optimization.* (Working Paper No. AMOL-96-3, Research Institute for Informatics-ICI, Bucharest, March 10, 1996).

Andrei, N. (1996d). *Computational experience with "SPENBAR" a sparse variant of a modified penalty-barrier method for large-scale nonlinear, equality and inequality, constrained optimization.* (Technical Paper No. AMOL-96-4, Research Institute for Informatics-ICI, Bucharest, March 11, 1996).

Andrei, N. (1996e). *Numerical examples with "SPENBAR" for large-scale nonlinear, equality and inequality, constrained optimization with zero columns in Jacobian matrices.* (Technical Paper No. AMOL-96-5, Research Institute for Informatics-ICI, Bucharest, March 29, 1996).

Andrei, N. (1998a). Penalty-barrier algorithms for nonlinear optimization. Preliminary computational results. *Studies in Informatics and Control, 7*(1), 15–36.

Andrei, N. (1998b). Predictor-Corrector interior point methods for linear constrained optimization. *Studies in Informatics and Control, 7*(2), 155–177.

Andrei, N. (1998c). An interior point algorithm for nonlinear programming. *Studies in Informatics and Control, 7*(4), 365–395.

Andrei, N. (1999). Programarea matematică avansată. Teorie, Metode computaţionale, Aplicaţii. [Advanced Mathematical Programming. Theory, Computational Methods, Applications] Editura Tehnică, Bucureşti.

Andrei, N. (2001). *Numerical examples solved with SPENBAR – modified penalty barrier method for large-scale nonlinear programming problems.* (Technical Report No. 1/2001, Research Institute for Informatics ICI – Bucharest, February 2001).

Andrei, N. (2003). Modele, Probleme de Test şi Aplicaţii de Programare Matematică. [Models, Test Problems and Applications for Mathematical Programming] Editura Tehnică, Bucureşti.

Andrei, N. (2004). *Stationary temperature field computation in a rectangular area.* (Technical Report, Research Institute for Informatics-ICI, Bucharest, March 18, 2004).

Andrei, N. (2007a). *Large-scale unconstrained optimization MINPACK-2 applications solved with SCALCG.* (Technical Report, Research Institute for Informatics-ICI, Bucharest. September 4, 2007, pp.1–5).

Andrei, N. (2007b). Scaled conjugate gradient algorithms for unconstrained optimization. *Computational Optimization and Applications, 38*, 401–416.

Andrei, N. (2007c). *SCALCG: Scaled conjugate gradient algorithms for unconstrained optimization.* (Technical Report No. 17/2007. Research Institute for Informatics-ICI, Bucharest, March 30, 2007).

Andrei, N. (2007d). A scaled BFGS preconditioned conjugate gradient algorithm for unconstrained optimization. *Applied Mathematics Letters, 20*, 645–650.

Andrei, N. (2007e). Scaled memoryless BFGS preconditioned conjugate gradient algorithm for unconstrained optimization. *Optimization Methods and Software, 22*, 561–571.

Andrei, N. (2008a). A scaled nonlinear conjugate gradient algorithm for unconstrained optimization. *Optimization, 57*(4), 549–570.

Andrei, N. (2008b). Noether Theorem and fundamentals of mathematical modeling. *Revista Română de Informatică şi Automatică, 18*(4), 11–22.

Andrei, N. (2009). Critica Raţiunii Algoritmilor de Optimizare fără Restricţii. [Criticism of the Unconstrained Optimization Algorithms Reasoning]. Editura Academiei Române, Bucureşti.

Andrei, N. (2010a). Accelerated scaled memoryless BFGS preconditioned conjugate gradient algorithm for unconstrained optimization. *European Journal of Operational Research, 204*, 410–420.

Andrei, N. (2010b). *Performances of LBFGS-B for solving MINPACK-2 applications.* (Technical Report, Research Institute for Informatics-ICI, Bucharest, August 9, 2010).

Andrei, N. (2010c). *Performances of SPG for solving MINPACK-2 applications.* (Technical Report, Research Institute for Informatics ICI, Bucharest, September 14, 2010)

Andrei, N. (2011a) *Critica raţiunii algoritmilor de programare liniară. [Criticism of the Linear Programming Algorithms Reasoning]*. Editura Academiei Române, Bucureşti.

Andrei, N. (2011b). *CAON: A collection of nonlinear optimization applications in GAMS language*. (Technical Report, No.1/2011, Research Institute for Informatics-ICI, Bucharest, January 31, 2011. (105 pages and CD)).

Andrei, N. (2012). *Eseu despre fundamentele modelării matematice. [Essay on Fundamentals of Mathematical Modeling]* Editura Academiei Române, Bucureşti.

Andrei, N. (2013a). *Another collection of large-scale unconstrained optimization test function*. (Technical Report, Research Institute for Informatics-ICI, Bucharest, January 30, 2013).

Andrei, N. (2013b). *Nonlinear optimization applications using the GAMS technology*. New York, NY, USA: Springer Science + Business Media.

Andrei, N. (2013c). Another conjugate gradient algorithm with guaranteed descent and conjugacy conditions for large-scale unconstrained optimization. *Journal of Optimization Theory and Applications*, *159*, 159–182.

Andrei, N. (2014). An accelerated subspace minimization three-term conjugate gradient algorithm for unconstrained optimization. *Numerical Algorithms*, *65*, 859–874.

Andrei, N. (2015). *Critica Raţiunii Algoritmilor de Optimizare cu Restricţii. [Criticism of the Constrained Optimization Algorithms Reasoning]*. Bucureşti, Balkans: Editura Academiei Române

Andrei, N. (2016). An adaptive conjugate gradient algorithm for large-scale unconstrained optimization. *Journal of Computational and Applied Mathematics*, *292*, 83–91.

Aris, R. (1975). *The mathematical theory of diffusion and reaction in permeable catalysts*. Oxford.

Armijo, L. (1966). Minimization of functions having Lipschitz continuous first partial derivatives. *Pacific Journal of Mathematics*, *16*(1), 1–3.

Arrow, K. J., & Sollow, R. M. (1958). Gradient methods for constrained maxima, with weakened assumptions. In K. J. Arrow, L. Hurwicz, & H. Uzawa (Eds.), *Studies in linear and nonlinear programming* (pp. 166–176). Stanford, CA, USA: Stanford University Press.

Averick, B. M., Carter, R. G., & Moré, J. J. (1991). *The Minpack-2 test problem collection (Preliminary Version)*. (Technical Memorandum No.150, Mathematics and Computer Science Division, Argonne National Laboratory, 9700 South Cass Avenue, Argonne, Illinois, Technical Memorandum No. 150, May 1991).

Averick, B. M., Carter, R. G., Moré, J. J., & Xue, G. L. (1992). *The MINPACK-2 test problem collection*. (Mathematics and Computer Science Division, Argonne National Laboratory, 9700 South Cass Avenue, Argonne, Illinois, Preprint MCS-P153–6092, June 1992).

Ballard, D. H., Jelinek, C. A., & Schinzinger, R. (1974). An algorithm for the solution of constrained polynomial programming problems. *Computer Journal*, *17*, 261–266.

Banga, J. R., Alonso, A. A., & Singh, R. P. (1997). Stochastic dynamic optimization of batch and semicontinuous bioprocesses. *Biotechnology Progress*, *13*, 326–335.

Bartels, R. H. (1980). A penalty linear programming method using reduced-gradient basis-exchange techniques. *Linear Algebra and its Applications*, *29*, 17–32.

Bartels, R. H., & Golub, G. H. (1969). The simplex method of linear programming using LU decomposition. *Communications of the ACM*, *12*, 266–268.

Bartholomew-Biggs, M. C. (1976). *A numerical comparison between two approach to nonlinear programming problems*. (Technical Report, No. 77, Numerical Optimization Centre, Hatfield, England).

Bartholomew-Biggs, M. (2008). *Nonlinear optimization with engineering applications*. New York, NY, USA: Springer Science +Business Media.

Barzilai, J., & Borwein, J. M. (1988). Two point step size gradient methods. *IMA Journal of Numerical Analysis*, *8*, 141–148.

Bazaraa, M. S., Sherali, H. D., & Shetty, C. M. (1993). *Nonlinear programming theory and algorithms* (2nd ed.). New York, NY, USA: Wiley.

Beale, E. M. L. (1959). On quadratic programming. *Naval Research Logistics Quarterly*, *6*, 227–244.

Bebernes, J., & Eberly, D. (1989). Mathematical problems from combustion theory. In *Applied mathematical sciences* (vol. 83). Berlin, Europe: Springer.

Benson, H. Y., Shanno, D. F., & Vanderbei, R. J. (2002a). Interior-point methods for nonconvex nonlinear programming: Filter methods and merit functions. *Computational Optimization and Applications, 23*, 257–272.

Benson, H. Y., Shanno, D. F., & Vanderbei, R. J. (2002b). *A comparative study of large-scale nonlinear optimization algorithms*. (Technical Report ORFE-01-04, Operations Research and Financial Engineering, Princeton University, July 17, 2002).

Ben-Tal, A., & Zibulevsky, M. (1993). *Penalty-barrier multiplier methods for large-scale convex programming problems*. (Research Report 6/93. Optimization Laboratory, Faculty of Industrial Engineering and Management, Technion, Haifa, Israel).

Ben-Tal, A., Yuzefovich, I., & Zibulevsky, M. (1992). *Penalty/barrier multiplier methods for minimax and constrained smooth convex problems*. (Research Report 9/92, Optimization Laboratory, Faculty of Industrial Engineering and Management, Technion, Haifa, Israel).

Bergman, L. (1988). Energy policy modeling: a survey of general equilibrium approaches. *Journal of Policy Modeling, 10*(3), 377–399.

Bertsekas, D. P. (1976). On the Goldstein-Levitin-Polyak gradient projection method. *IEEE Transactions on Automatic Control, 21*, 174–184.

Bertsekas, D. P. (1982a). Projected Newton methods for optimization problems with simple constraints. *SIAM Journal on Control and Optimization, 20*, 221–246.

Bertsekas, D. P. (1982b). *Constrained optimization and lagrange multiplier methods*. New York, NY, USA: Academic.

Bertsekas, D. P. (1999). *Nonlinear programming* (2nd ed.). Belmont, CA, USA: Athena Scientific.

Biegler, L. T., Nocedal, J., & Schmid, C. (1995). A reduced Hessian method for large-scale constrained optimization. *SIAM Journal on Optimization, 5*, 314–347.

Biggs, M. C. (1972). Constrained minimization using recursive equality quadratic programming. In F. A. Lootsma (Ed.), *Numerical methods for nonlinear optimization* (pp. 411–428). London, UK: Academic.

Birgin, E. G., & Martínez, J. M. (2001). A spectral conjugate gradient method for unconstrained optimization. *Applied Mathematics and Optimization, 43*, 117–128.

Birgin, E. G., Martínez, J. M., & Raydan, M. (1999). *Nonmonotone spectral projected gradient methods on convex sets*. (Technical Paper, June 7, 1999).

Birgin, E. G., Martínez, J. M., & Raydan, M. (2000). Nonmonotone spectral projected gradient methods on convex sets. *SIAM Journal on Optimization, 10*, 1196–1211.

Birgin, E. G., Martínez, J. M., & Raydan, M. (2001). Algorithm 813: SPG - software for convex-constrained optimization. *ACM Transactions on Mathematical Software, 27*, 340–349.

Bisschop, J., & Meeraus, A. (1982). On the development of the general algebraic modeling systems in a strategic planning environment. *Mathematical Programming Study, 20*, 1–29.

Boggs, P. T., & Tolle, J. W. (1989). A strategy for global convergence in a sequential quadratic programming algorithm. *SIAM Journal on Numerical Analysis, 21*, 600–623.

Boggs, P. T., & Tolle, J. W. (1995). Sequential quadratic programming. *Acta Numerica, 4*, 1–51.

Bondarenko, A. S., Bortz, D. M., & Moré, J. J. (1999). *COPS: Large-scale nonlinearly constrained optimization problems*. (Technical Report ANL/MCS-TM-237, Argonne National Laboratory, Argonne, Illinois, September 1998, October 1999 (revision)).

Bonnans, J. F., Panier, E. R., Tits, A. L., & Zhou, J. L. (1992). Avoiding the Maratos effect by means of a nonmonotone line search. II. Inequality constrained problems – feasible iterates. *SIAM Journal on Numerical Analysis, 29*, 1187–1202.

Boyd, S., & Vandenberghe, L. (2006). *Convex optimization* (2nd ed.). Cambridge, UK: Cambridge University Press.

Breitfeld, M. G., & Shanno, D. F. (1994a). Preliminary computational experience with modified log-barrier functions for large-scale nonlinear programming. In W. W. Hager, D. W. Hearn, & P. M. Pardalos (Eds.), *Large scale optimization, state of the art* (pp. 45–67). Dordrecht/Boston/London: Kluwer Academic Publishers.

Breitfeld, M.G., & Shanno, D.F. (1994b). *Computational experience with penalty-barrier methods for nonlinear programming*. (RUTCOR Research Report, RRR 17–93, August 1993, Revised March 1994. Rutgers Center for Operations Research, Rutgers University, New Brunswick, New Jersey 08903, March 1994).

Breitfeld, M. G., & Shanno, D. F. (1994c). *A globally convergent penalty-barrier algorithm for nonlinear programming and its computational performance*. (RUTCOR Research Report, RRR 12–94, April 1994, Rutgers Center for Operations Research, Rutgers University, New Brunswick, New Jersey 08903, March 1994).

Brooke, A., Kendrick, D., Meeraus, A., Raman, R., & Rosenthal, R. E. (1998). *GAMS: A user's guide*. GAMS Development Corporation, December 1998.

Brooke, A., Kendrick, D., Meeraus, A., & Raman, R. (2005). *GAMS: A user guide*. GAMS Development Corporation, December 2005.

Brown, A. A., & Bartholomew-Biggs, M. C. (1987). *ODE vs SQP methods for constrained optimisation*. (Technical Report No. 179, Numerical Optimisation Centre, The Hatfield Polytechnic, Hatfield, June 1987).

Bunch, J. R., & Kaufman, L. (1977). *Indefinite quadratic programming*. (Computing Science Technical Report 61, Bell Labs., Murray Hill, NJ).

Bunch, J. R., & Parlett, B. N. (1971). Direct methods for solving symmetric indefinite systems of linear equations. *SIAM Journal on Numerical Analysis*, *8*, 639–655.

Byrd, R. H., Schnabel, R. B., & Shultz, G. A. (1987). A trust region algorithm for nonlinearly constrained optimization. *SIAM Journal on Numerical Analysis*, *24*, 1152–1170.

Byrd, R. H., Tapia, R. A., & Zhang, Y., (1990) *An SQP augmented Lagrangian BFGS algorithm for constrained optimization*. (Technical Report, University of Colorado at Boulder).

Byrd, R. H., Lu, P., Nocedal, J., & Zhu, C. (1994). *A limited memory algorithm for bound constrained optimization*. (Technical Report NAM-08, [Revised May 1994], Department of Electrical Engineering and Computer Science, Northwestern University,Evanston, Ilinois 60208).

Byrd, R. H., Nocedal, J., & Schnabel, R. B. (1994). Representation of quasi-Newton matrices and their use in limited memory methods. *Mathematical Programming*, *63*(2), 129–156.

Byrd, R. H., Lu, P., & Nocedal, J. (1995). A limited memory algorithm for bound constrained optimization. *SIAM Journal on Scientific and Statistical Computing*, *16*(5), 1190–1208.

Byrd, R. H., Lu, P., Nocedal, J., & Zhu, C. (1995). A limited memory algorithm for bound constrained optimization. *SIAM Journal on Scientific Computing*, *16*, 1190–1208.

Byrd, R. H., Liu, G., & Nocedal, J. (1997). On the local behavior of an interior point method for nonlinear programming. In D. F. Griffths & D. J. Higham (Eds.), *Numerical Analysis 1997* (pp. 37–56). Reading, UK: Addison-Wesley Longman.

Byrd, R. H., Hribar, M. E., & Nocedal, J. (1999). An interior point method for large scale nonlinear programming. *SIAM Journal on Optimization*, *9*, 877–900.

Byrd, R. H., Gilbert, J.-C., & Nocedal, J. (2000). A trust region method based on interior point techniques for nonlinear programming. *Mathematical Programming*, *89*, 149–185.

Byrd, R. H., Gould, N. I. M., Nocedal, J., & Waltz, R. A. (2002). *On the convergence of successive linear-quadratic programming algorithms*. (Technical Report OTC 2002/5, Optimization Technology Center, Northwestern University, Evanston, IL).

Byrd, R. H., Nocedal, J., & Waltz, R. A. (2003). Feasible interior methods using slacks for nonlinear optimization. *Computational Optimization and Applications*, *26*(1), 35–61.

Byrd, R. H., Gould, N. I. M., Nocedal, J., & Waltz, R. A. (2004). An algorithm for nonlinear optimization using linear programming and equality constrained subproblems. *Mathematical Programming, Series B*, *100*, 27–48.

Byrd, R. H., Marazzi, M., & Nocedal, J. (2004). On the convergence of Newton iterations to non-stationary points. *Mathematical Programming, Series A*, *99*, 127–148.

Byrd, R. H., Nocedal, J., & Waltz, R. A. (2006). KNITRO: An integrated package for nonlinear optimization. In G. Di Pillo & M. Roma (Eds.), *Large-scale nonlinear optimization* (pp. 35–59). Boston, MA, USA: Springer Science+Business Media.

Carolan, W. J., Hill, J. E., Kennington, J. L., Niemi, S., & Wichmann, S. J. (1990). An empirical evaluation of the KORBX algorithms for military airlift applications. *Operations Research, 38*, 240–248.
Castillo, E., Conejo, A. J., Pedregal, P., García, R., & Alguacil, N. (2001). *Building and solving mathematical programming models in engineering and science*. New York, NY, USA: Wiley.
Cesari, L. (1983). *Optimization – theory and applications*. Bonn, Europe: Springer.
Chachuat, B. C. (2007). *Nonlinear and dynamic optimization – from theory to Practice*. IC-31: Winter Semester 2006/2007. École Politechnique Fédérale de Lausanne.
Cheng, S. H. (1998). *Symmetric indefinite matrices: Linear system solvers and modified inertia problems*. Ph.D. Thesis. University of Mancester. Faculty of Science and Engineering. January 1998.
Cheng, S. H., & Higham, N. J. (1998). A modified Cholesky algorithm based on a symmetric indefinite factorization. *SIAM Journal on Matrix Analysis and Applications, 19*, 1097–1110.
Chin, C. M. (2002). *A global convergence theory of a filter line search method for nonlinear programming*. (Technical Report, Department of Electrical Engineering, University of Malaya, Kuala Lumpur, Malaysia, August 2002).
Chin, C. M., & Fletcher, R. (2003). On the local convergence of an SLP-filter algorithm that takes EQP steps. *Mathematical Programming (Series A), 96*, 161–177.
Cimatti, G., & Menchi, O. (1978). On the numerical solution of a variational inequality connected with the hydrodynamic lubrication of a complete journal bearing. *Calcolo, 15*, 249–258.
Coleman, T. F., & Conn, A. R. (1982a). Nonlinear programming via an exact penalty function: Asymptotic analysis. *Mathematical Programming, 24*, 123–136.
Coleman, T. F., & Conn, A. R. (1982b). Nonlinear programming via an exact penalty function: Global analysis. *Mathematical Programming, 24*, 137–161.
Coleman, T. F., & Hulbert, L. A. (1989). A direct active set algorithm for large sparse quadratic programs with simple bounds. *Mathematical Programming (Series A), 45*, 373–406.
Coleman, T. F., & Li, Y. (1994). On the convergence of interior-reflective Newton methods for nonlinear minimization subject to bounds. *Mathematical Programming, 67*, 189–224.
Coleman, T. F., & Li, Y. (1996). An interior trust region approach for nonlinear minimization subject to bounds. *SIAM Journal on Optimization, 6*, 418–445.
Coleman T. F., & Li, Y. (1997). *A trust region and affine scaling interior point method for nonconvex minimization with linear inequality constraints*. (Technical Report, Cornell University, Ithaca, NY).
Conejo, A. J., Castillo, E., Minguez, R., & Garcia-Bertrand, R. (2006). *Decomposition techniques in mathematical programming: engineering and science applications*. Heidelberg, Europe: Springer.
Conn, A. R. (1976). Linear programming via a nondifferentiable penalty function. *SIAM Journal on Numerical Analysis, 13*, 145–154.
Conn, A. R., Gould, N. I. M., & Toint, P. L. (1988a). Global convergence of a class of trust region algorithms for optimization with simple bounds. *SIAM Journal on Numerical Analysis, 25*(2), 433–460.
Conn, A. R., Gould, N. I. M., & Toint, P. L. (1988b). Testing a class of methods for solving minimization problems with simple bounds on the variables. *Mathematics of Computation, 50*, 399–430.
Conn, A. R., Gould, N. I. M., & Toint, P. L. (1991). A globally convergent augmented Lagrangian algorithm for optimization with general constraints and simple bounds. *SIAM Journal on Numerical Analysis, 28*, 545–572.
Conn, A. R., Gould, N. I. M., & Toint, Ph. L. (1992a). *A globally convergent Lagrangian barrier algorithm for optimization with general inequality constraints and simple bounds*. (Technical Report 92/07, Department of Mathematics, Faculté Universitaires de Namur, Namur, Belgium).

Conn, A. R., Gould, N. I. M., & Toint, P. L. (1992b). *LANCELOT - A fortran package for large-scale nonlinear optimization (Release A), Springer series in computational mathematics* (vol. 17). Berlin, Europe: Springer.

Conn, A. R., Gould, N. I. M., & Toint, P. L. (1996). Numerical experiments with the LANCELOT package (Release A) for large-scale nonlinear optimization. *Mathematical Programming (Series A), 73*(1), 73–110.

Conn, A. R., Gould, N. I. M., Sartenaer, A., & Toint, P. L. (1996). Convergence properties of an augmented Lagrangian algorithm for optimization with a combination of general equality and linear constraints. *SIAM Journal on Optimization, 6*(3), 674–703.

Conn, A. R., Gould, N. I. M., & Toint, P. L. (1997). A globally convergent lagrangian barrier algorithm for optimization with general inequality constraints and simple bounds. *Mathematics of Computation, 66*, 261–288.

Conn, A. R., Gould, N. I. M., & Toint, P. L. (2000). *Trust-region methods*. Philadelphia, PA, USA: MPS-SIAM Series on Optimization, SIAM.

Cottle, R. W. (2012). William Karush and the KKT Theorem. In: Grötschel, M. (Ed.) *Optimization stories* (pp. 255–269). Documenta Mathematica, Journal der Deutschen Methematiker-Vereinigung, Extra volume, 21st International Symposium on Mathematical Programming, Berlin, August 19–24, 2012.

Courant, R. (1943). Variational methods for the solution of problems of equilibrium and vibrations. *Bulletin of the American Mathematical Society, 49*, 1–23.

Cuthrell, J. E., & Biegler, L. T. (1989). Simultaneous optimization and solution methods for batch reactor control profiles. *Computers and Chemical Engineering, 13*, 49–62.

Daniel, J. W., Graggs, W. B., Kaufman, L., & Stewart, G. W. (1976). Reorthogonalization and stable algorithms for updating the Gram-Schmidt QR factorizations. *Mathematics of Computation, 30*, 772–795.

Dantzig, G. B. (1963). *Linear programming and extensions*. Princeton, NJ, USA: Princeton University Press.

Davis, T. A. (2006). *Direct methods for sparse systems*. Philadelphia, PA, USA: SIAM.

Debreu, G. (1952). Definite and semidefinite quadratic forms. *Econometrica, 20*, 295–300.

Dembo, R. S. (1976). A set of geometric programming test problems and their solutions. *Mathematical Programming, 10*, 192–213.

Dembo, R. S., & Steihaug, T. (1983). Truncated newton algorithms for large-scale unconstrained optimization. *Mathematical Programming, 26*, 190–212.

Dembo, R. S., & Tulowitzki, U. (1983). *On the minimization of quadratic functions subject to box constraints*. (Technical Report, School of Organization and Management, Yale University, New Haven, CT).

Demmel, J. W. (1997). *Applied numerical linear algebra*. Philadelphia, PA, USA: SIAM.

Deng, N. Y., Xiao, Y., & Zhou, F. J. (1993). Nonmonotonic trust-region algorithm. *Journal of Optimization Theory and Applications, 26*, 259–285.

Dennis, J. E., Jr., & Schnabel, R. B. (1983). *Numerical methods for unconstrained optimization and nonlinear equations*. Englewoods Cliffs, NJ, USA: Prentice-Hall.

Dennis, J. E., Heinkenschlos, M., & Vicente, L. N. (1998). Trust-region interior-point algorithms for a class of nonlinear programming problems. *SIAM Journal on Control and Optimization, 36*, 1750–1794.

Dikin, I. I. (1967). Iterative solution of problems of linear and quadratic programming. *Soviet Mathematical Doklady, 8*, 674–675.

Dikin, I. I. (1974). On the convergence of an iterative process. *Upravlyaemye Sistemi, 12*, 54–60.

Dolan, E. D., & Moré, J. J. (2002). Benchmarking optimization software with performance profiles. *Mathematical Programming, 91*, 201–213.

Dolan, E. D., Moré, J. J., & Munson, T. S. (2004). *Benchmarking optimization software with COPS 3.0*. (Preprint ANL/MCS-TM-273, Argonne National Laboratory, Argonne, Illinois, February 2004).

Drud, A. (1976). Application of sparse matrix techniques in large scale nonlinear programming. In: Prekopa, A., (Ed.) *Survey of Mathematical Programming*. vol.1, Part 3, *Nonlinear Programming*. (Proceedings of the 9th International Mathematical Programming Symposium, Budapest, August 23–27, 1976, pp.429–445.)

Drud, A. (1983). CONOPT: *A GRG code for large sparse dynamic nonlinear optimization problems*. (Report No. DRD59, Development Research Department, Economics and Research Staff, The World Bank, 1818 H Street, Washington, D.C. 20433, August 1893).

Drud, S. A. (1985). CONOPT: A GRG code for large sparse dynamic nonlinear optimization problems. *Mathematical Programming, 31*, 153–191.

Drud, S. A. (1994). CONOPT – A large-scale GRG code. *ORSA Journal on Computing, 6*, 207–216.

Drud, S. A. (1995). *CONOPT – A system for large-scale nonlinear optimization*. (Tutorial for CONOPT subroutine library, 16p. ARKI Consulting and Development A/S, Bagsvaerd, Denmark).

Drud, S. A. (1996). *CONOPT: A system for large-scale nonlinear optimization*. (Reference Manual for CONOPT subroutine library, 69p. ARKI Consulting and Development A/S, Bagsvaerd, Denmark).

Drud, S. A. (2005). *CONOPT*. In GAMS The Solver Manuals. GAMS Development Corporation, 1217 Potomac Street, N.W, Washington, DC 20007, USA, pp.39–82.

Drud, S. A. (2011). *CONOPT – A system for large scale nonlinear optimization. Tutorial for CONOPT dynamic link library. Version 3.15 (Fortran 90 examples)*. (ARKI Consulting and Development A/S, Bagsvaerd, Denmark, August 2011).

Duff, I. S. (1977). *MA28-A set of Fortran subroutines for sparse unsymmetric linear equation*. (Report AERE R8730, Atomic Energy Research Establishment, Harwell, England).

Duff, I. S., Erismann, A. M., & Reid, J. K. (1986). *Direct methods for sparse matrices*. Clarendon Press.

Duran, M., & Grossmann, I. E. (1986). An outer-approximation algorithm for a class of mixed-integer nonlinear programs. *Mathematical Programming, 36*, 307–339.

El-Bakry, A. S., Tapia, R. A., Tsuchiya, T., & Zhang, Y. (1996). On the formulation and theory of the Newton interior-point method for nonlinear programming. *Journal of Optimization Theory and Applications, 89*(3), 507–541.

Eldersveld, S. K. (1991). *Large-scale sequential quadratic programming algorithms*. (Ph.D. Thesis, Department of Operations Research, Stanford University, Stanfors, CA).

Facchinei, F., & Lucidi, S. (1992a). A class of penalty functions for optimization problems with bounds constraints. *Optimization, 26*, 239–259.

Facchinei, F., & Lucidi, S. (1992b). *A class of methods for optimization problems with simple bounds. Part 2: Algorithms and numerical results*. (Technical Report R.336, IASI-CNR, Roma, Italy).

Facchinei, F., Júdice, J., & Soares, J. (1998). An active set Newton's algorithm for large-scale nonlinear programs with box constraints. *SIAM Journal on Optimization, 8*, 158–186.

Facchinei, F., Lucidi, S., & Palagi, L. (2002). A truncated Newton algorithm for large-scale box constrained optimization. *SIAM Journal on Optimization, 4*, 1100–1125.

Fiacco, A. V., & McCormick, G. P. (1964). The sequential unconstrained minimization technique for nonlinear programming. *A primal-dual method. Management Science, 10*, 360–366.

Fiacco, A. V., & McCormick, G. P. (1966). Extensions of SUMT for nonlinear programming: equality constraints and extrapolation. *Management Science, 12*, 816–828.

Fiacco, A. V., & McCormick, G. P. (1968). *Nonlinear programming. Sequential unconstrained minimization techniques*. John Wiley & Sons, Inc., New York, 1968. [Reprinted as: Volume 4 of *SIAM Classics in Applied Mathematics*. SIAM Publications, Philadelphia, PA 19104–2688, 1990].

Fletcher, R. (1971). A general quadratic programming algorithm. *Journal of the Institute of Mathematics and its Applications, 7*, 76–91.

Fletcher, R. (1982). *Second order corrections for nondifferentiable optimization*. In: D. Griffith, (Ed.) Numerical analysis, Berlin, Europe: Springer, 85–114. Proceedings Dundee 1981.

Fletcher, R. (1987). *Practical methods of optimization* (2nd ed.). New York, NY, USA: Wiley.

Fletcher, R. (2000). Stable reduced Hessian updates for indefinite quadratic programming. *Mathematical Programming, 87*, 251–264.

Fletcher, R. (2011). *A package of subroutines for NLP and LCP*. Open Source Initiative OSI – Eclipse Public License 1.0 (ELP-1.0), Release 1.0.

Fletcher, R., & Leyffer, S. (1998). *User Manual for FilterSQP*. (Technical Report NA/181, Department of Mathematics, University of Dundee, Scotland, April 1998. [Updated, March 1999]).

Fletcher, R., & Leyffer, S. (1999). *A bundle filter method for nonsmooth nonlinear optimization*. (Numerical Analysis Report NA/195, Dundee University, April 1999).

Fletcher, R., & Leyffer, S. (2002). Nonlinear programming without a penalty function. *Mathematical Programming, Series A, 91*, 239–269.

Fletcher, R., & Leyffer, S. (2003). Filter-type algorithms for solving systems of algebraic equations and inequalities. In G. di Pillo & A. Murli (Eds.), *High performance algorithms and software for nonlinear optimization* (pp. 259–278). Norwell, MA, USA: Kluwer Academic Publishers.

Fletcher, R., & Sainz de la Maza, E. (1989). Nonlinear programming and nonsmooth optimization by successive linear programming. *Mathematical Programming, 43*, 235–256.

Fletcher, R., Leyffer, S., & Toint, Ph. (1999). *On the global convergence of an SLP-filter algorithm*. (Numerical Analysis Report NA/183, August 1998. (Revised October 1999)).

Fletcher, R., Gould, N. I. M., Leyffer, S., Toint, P. T., & Wächter, A. (2002). Global convergence of a trust-region SQP filter algorithm for general nonlinear programming. *SIAM Journal on Optimization, 13*, 635–659.

Fletcher, R., Leyffer, S., & Toint, P. (2002). On the global convergence of a filter-SQP algorithm. *SIAM Journal on Optimization, 13*, 44–59.

Fletcher, R., Leyffer, S., & Toint, Ph. (2006). *A brief history of filter methods*. (Argonne National Laboratory, Mathematics and Computer Science Division, Preprint ANL/MCS-P1372–0906, September 26, 2006, revised October 9, 2006.)

Fletcher, R., Leyffer, S., & Shen, C. (2009). *Nonmonotone filter method for nonlinear optimization*. (Argonne National Laboratory, Mathematics and Computer Science Division, Preprint ANL/MCS-P1679–0909, October 14, 2009).

Floudas, C. A., & Pardalos, P. M. (1990). *A collection of test problems for constrained global optimization algorithms, Lecture notes in computer science* (vol. 455). Berlin, Europe: Springer.

Forsgren, A., Gill, E., & Wright, M. H. (2002). Interior points for nonlinear optimization. *SIAM Review, 44*, 525–597.

Fourer, R. (1983). Modeling languages versus matrix generators for linear programming. *ACM Transactions on Mathematical Software, 9*, 143–183.

Fourer, R., Gay, M., & Kernighan, B. W. (2002). *AMPL: A modeling language for mathematical programming* (2nd ed.). Belmont, NY, USA: Duxbury Press.

Friedlander, M. P., & Saunders, M. (2005). A globally convergent linearly constrained Lagrangian method for nonlinear optimization. *SIAM Journal on Optimization, 15*, 863–897.

Friedlander, A., Martínez, J. M., & Santos, S. A. (1994). A new trust region algorithm for bound constrained minimization. *Applied Mathematics and Optimization, 30*, 235–266.

Frisch, K. R. (1955). *The logarithmic potential method for convex programming*. (Manuscript. Institute of Economics, University of Oslo, Oslo, May 1955.)

Fukushima, M. (1986). A successive quadratic programming algorithm with global and superlinear convergence properties. *Mathematical Programming, 35*, 253–264.

Gabay, D. (1982). Reduced quasi-Newton methods with feasibility improvement for nonlinearly constrained optimization. *Mathematical Programming Study, 16*, 18–44.

GAMS – A user guide. (2005). GAMS Development Corporation.

GAMS – The solver manuals. (2005). GAMS Development Corporation.

Garner, J., & Benedek, R. (1990). Solution of Ginzburg-Landau equations for inhomogeneous superconductors by nonlinear optimization. *Physical Review B*, *42*, 376–385.

Gay, D. M. (1985). Electronic mail distribution of linear programming test problems. *Mathematical Programming Society COAL Newsletter*.

Gay, D. M., Overton, M. L., & Wright, M. H. (1997). *A primal-dual interior method for nonconvex nonlinear programming*. (Technical Report 97–4-08, Bell Lab. Murray Hill, July 29, 1997).

George, A., & Liu, J. W. H. (1981). *Computer solution of large-sparse positive definite systems*. Englewood Cliffs, NJ, USA: Prentice-Hall.

Gill, P. E., & Leonard, M. W. (2003). Limited-memory reduced-Hessian methods for unconstrained optimization. *SIAM Journal on Optimization*, *14*, 380–401.

Gill, P. E., & Murray, W. (1974a). Methods for large-scale linearly constrained problems. In P. E. Gill & W. Murray (Eds.), *Numerical methods for constrained optimization* (pp. 93–147). London\New York\San Francisco: Academic.

Gill, P. E., & Murray, W. (1974b). Newton-type methods for unconstrained and linearly constrained optimization. *Mathematical Programming*, *7*, 311–350.

Gill, P. E., & Murray, W. (1978). Numerically stable methods for quadratic programming. *Mathematical Programming*, *14*, 349–372.

Gill, Ph. E., & Murray, W., (1979). *Conjugate gradient methods for large-scale nonlinear optimization*. (Report SOL 79–15. Department of Operations Research, Stanford University, Stanford).

Gill, P. E., Golub, G. H., Murray, W., & Saunders, M. A. (1974). Methods for modifying matrix factorizations. *Mathematics of Computation*, *28*, 505–535.

Gill, P. E., Murray, W., & Wright, M. H. (1981). *Practical optimization*. Academic.

Gill, P. E., Murray, W., Saunders, M. A., & Wright, M. H. (1984). *User's guide for SOL/QPSOL*. (Technical Report SOL84–6, Department of Operations Research, Stanford University, Stanford, California).

Gill, P. E., Murray, W., Saunders, M. A., & Wright, M. H. (1986). *User's guide for NPSOL (version 4.0): A Fortran package for nonlinear programming*. (Report SOL 86–2, Department of Operations Research, Stanford University, Stanford, CA, 1986).

Gill, P. E., Murray, W., Saunders, M. A., & Wright, M. H. (1987). Maintaining LU factors of a general sparse matrix. *Linear Algebra and its Applications*, *88/89*, 239–270.

Gill, P. E., Murray, W., Saunders, M. A., & Wright, M. H. (1992). Some theoretical properties of an augmented Lagrangean merit function. In P. M. Pardalos (Ed.), *Advances in optimization and parallel computing* (pp. 101–128). Amsterdam, Europe: North-Holland.

Gill, P. E., Murray, W., & Saunders, M. A. (1997). *User's guide for SNOPT (version 5.3): A Fortran package for large-scale nonlinear programming*. (Technical Report NA 97–4, Department of Mathematics, University of California, San Diego).

Gill, P. E., Murray, W., & Saunders, M. A. (2002). SNOPT: A SQP algorithm for large-scale constrained optimization. *SIAM Journal on Optimization*, *12*, 979–1006.

Gill, P. E., Murray, W., & Saunders, M. A. (2005). SNOPT: An SQP algorithm for large-scale constrained optimization. *SIAM Review*, *47*, 99–131.

Gill, P. E., Murray, W., & Saunders, M. A. (2006). *User's guide for SQOPT Version 7: Software for large-scale nonlinear programming*. (Report, Department of Mathematics, University of California, San Diego).

Glowinski, R. (1984). *Numerical methods for nonlinear variational problems*. Berlin, Europe: Springer.

Gödel, K. (1931). Über formal unentscheidbare Sätze der principia mathematica und verwandter Systeme I. *Monatshefte für Mathematik und Physik*, *38*, 173–199.

Goldfarb, D. (1975). Matrix factorizations in optimization of nonlinear functions subject to linear constraints. *Mathematical Programming*, *10*, 1–31.

Goldfarb, D. (1976). Using the steepest-edge simplex algorithm to solve sparse linear programs. In J. R. Bunch & D. Rose (Eds.), *Sparse matrix computations* (pp. 227–240). New York, NY, USA: Academic.

Goldfarb, D., & Idnani, A. (1983). A numerically stable dual method for solving strictly convex quadratic programs. *Mathematical Programming*, *27*, 1–33.

Goldfarb, D., & Reid, J. K. (1977). A practicable steepest-edge simplex algorithm. *Mathematical Programming*, *12*, 361–371.

Goldfarb, D., Liu, S., & Wang, S. (1991). A logarithmic barrier function algorithm for quadratically constrained convex quadratic programming. *SIAM Journal on Optimization*, *1*(2), 252–267.

Golub, G. H., & Van Loan, C. F. (1996). *Matrix computation* (3rd ed.). Baltimore, MD, USA: Johns Hopkins University Press.

Gonçalves, A. S. (1972). A primal-dual method for quadratic programming with bounded variables. In F. A. Lootsma (Ed.), *Numerical Methods for Nonlinear Optimization* (pp. 255–263). London, UK: Academic.

Goodman, J., Kohn, R., & Reyna, L. (1986). Numerical study of a relaxed variational problem from optimal design. *Computer Methods in Applied Mechanics and Engineering*, *57*, 107–127.

Gould, N. I. M. (1986). On the accurate determination of search directions for simple differentiable penalty functions. *IMA Journal of Numerical Analysis*, *6*, 357–372.

Gould, N. I. M. (1991). An algorithm for large scale quadratic programming. *I.M.A. Journal on Numerical Analysis*, *11*, 299–324.

Gould, N. I. M., & Robinson, D. P. (2008). *A second derivative SQP methods: Local convergence.* (Numerical Analysis Report 08/21, Oxford University Computing Laboratory. 2008).

Gould, N. I. M., & Robinson, D. P. (2010). A second derivative SQP methods: Global convergence. *SIAM Journal on Optimization*, *20*, 2023–2048.

Gould, N. I. M., & Toint, P. L. (2002a). Numerical methods for large-scale non-convex quadratic programming. In A. H. Siddiqi & M. Kocvara (Eds.), *Trends in industrial and applied mathematics* (pp. 149–179). Dordrecht, Europe: Kluwer Academic Publishers.

Gould, N. I. M., & Toint, P. L. (2002b). An iterative working-set method for large-scale nonconvex quadratic programming. *Applied Numerical Mathematics*, *43*, 109–128.

Gould, N. I. M., Toint, Ph. L. (2012). *A quadratic programming bibliography*, (RAL Numerical Analysis Group Internal Report 2000–1, March 28, 2012).

Gould, N. I. M., Hribar, M. E., & Nocedal, J. (2001). On the solution of equality constrained quadratic problems arising in optimization. *SIAM Journal on Scientific Computing*, *23*, 1375–1394.

Gould, N. I. M., Orban, D., & Toint, P. L. (2005). Numerical methods for large-scale nonlinear optimization. *Acta Numerica*, *14*, 299–361.

Gould, N. I. M., Sainvitu, C., & Toint, P. L. (2005). A filter-trust-region method for unconstrained optimization. *SIAM Journal on Optimization*, *16*, 341–357.

Griffith, R. E., & Stewart, R. A. (1961). A nonlinear programming technique for the optimization of continuous processing systems. *Management Science*, *7*(4), 379–392.

Grippo, L., Lampariello, F., & Lucidi, S. (1986). A nonmonotone line search technique for Newton's method. *SIAM Journal Numerical Analysis*, *23*, 707–716.

Grippo, L., Lampariello, F., & Lucidi, S. (1989). A truncated Newton method with nonmonotone line search for unconstrained optimization. *Journal of Optimization Theory and Applications*, *60*, 401–419.

Grippo, L., Lampariello, F., & Lucidi, S. (1991). A class of nonmonotone stabilization methods in unconstrained optimization. *Numerische Mathematik*, *59*, 779–805.

Grötschel, M. (Ed.) (2012). *Optimization stories*. Documenta Mathematica, Journal der Deutschen Mathematiker-Vereinigung, Berlin.

Guéret, C., Prins, C., & Sevaux, M. (2002). *Applications of optimization with Xpress-MP*. Dash Optimization Ltd., Blisworth House, UK.

Hager, W. W., & Zhang, H. (2005). A new conjugate gradient method with guaranteed descent and an efficient line search. *SIAM Journal on Optimization, 16*, 170–192.

Hager, W. W., & Zhang, H. (2006a). A new active set algorithm for box constrained optimization. *SIAM Journal on Optimization, 17*, 526–557.

Hager, W. W., & Zhang, H. (2006b). Recent advances in bound constrained optimization. In F. Ceragioli, A. Dontchev, H. Furuta, K. Marti, & L. Pandolfi (Eds.), *System Modeling and Optimization* (pp. 67–82). Berlin, Europe: Springer.

Han, S. P. (1976). Superlinearly convergent variable metric algorithms for general nonlinear programming problems. *Mathematical Programming, 11*, 263–282.

Han, S. P. (1977). A globally convergent method for nonlinear programming. *Journal of Optimization Theory and Applications, 22*, 297–309.

Han, S. P., & Mangasarian, O. L. (1979). Exact penalty functions in nonlinear programming. *Mathematical Programming, 17*, 251–269.

Heinkenschlos, M., Ulbrich, M., & Ulbrich, S. (1999). Superlinear and quadratic convergence of affine-scaling interior-point Newton methods for problems with simple bounds without strict complementarity assumption. *Mathematical Programming, 86*, 615–635.

Heinz, J., & Spellucci, P. (1994). A successful implementation of the Pantoja-Mayne SQP-method. *Optimization Methods and Software, 4*, 1–28.

Helgason, R. V., & Kennington, J. L. (1980). Spike swapping in basis reinversion. *Naval Research Logistics Quarterly, 27*(4), 697–701.

Helgason, R. V., & Kennington, J. L. (1982). A note on splitting the bump in an elimination factorization. *Naval Research Logistics Quarterly, 29*(1), 169–178.

Hellerman, E., & Rarick, D. (1971). Reinversion with the preassigned pivot procedure. *Mathematical Programming, 1*, 195–216.

Hellerman, E., & Rarick, D. (1972). The partitioned preassigned pivot procedure (P4). In D. J. Rose & R. A. Willoughby (Eds.), *Sparse Matrices and their Applications* (pp. 67–76). New York, NY, USA: Plenum Press.

Hestens, M. R. (1969). Multiplier and gradient methods. *Journal of Optimization Theory and Applications, 4*, 303–320.

Higham, N. (1988). Fortran codes for estimating the one-norm of a real or complex matrix, with applications to condition estimation. *ACM Transactions on Mathematical Software, 14*, 381–396.

Higham, N. J. (1996). *Accuracy and stability of numerical algorithms*. Philadelphia, PA, USA: SIAM.

Hock, W., & Schittkowski, K. (1981). *Test examples for nonlinear programming codes, Lecture notes in economics and mathematical systems* (vol. 187). Berlin, Europe: Springer.

Holmström, K. (1997). *Tomlab – A general purpose, Open MATLAB environment for research and teaching in optimization*, (Technical Report IMa-TOM-1997-3, Department of Mathematics and Physics, Mälardalen University, Sweden. [Presented at the 16th International Symposium on Mathematical Programming, August 24–29, 1997, Lausanne, Switzerland]).

Hürlimann, T. (1999). *Mathematical modeling and optimization: An essay for the design of computer-based modeling tools*. Berlin, Europe: Springer.

ILOG CPLEX 8.0. (2002). *User's manual*. France, Europe: ILOG SA Gentilly.

Irizarry, R. (2005). A generalized framework for solving dynamic optimization problems using the artificial chemical process paradigm: Applications to particulate processes and discrete dynamic systems. *Chemical Engineering Science, 60*, 5663–5681.

Jensen J. (2006). *Welcome to GAMS*. TECA TRAINING ApS, September 2006.

Jensen, D. L., & Polyak, R. (1992). *The convergence of a modified barrier method for convex programming*. (Research Report RC 18570, IBM Research Division, T.J. Watson Research Center, Yorktown Heights, New York, 1992).

Jittorntrum, K., & Osborne, M. (1980). A modified barrier function method with improved rate of convergence for degenerate problems. *Journal of the Australian Mathematical Society (Series B), 21*, 305–329.

John, F. (1948). Extremum problems with inequalities as subsidiary conditions. In *Studies and essays, presented to R. Courant on his 60th birthday* (pp. 187–204). New York, NY, USA: Interscience.

Kalan, J. E. (1971). Aspects of large-scale in-core linear programming. *Proceedings of the 1971 26th annual conference*. ACM, New York, pp. 304–313.

Kallrath, J., & Wilson, J. M. (1997). *Business optimization using mathematical programming*. Basingstoke, UK: MacMillan (PalGrave).

Kalvelagen, E. (2001). *Nonlinear programming models 1: a collection of simple models*. (GAMS Development Corp., August 2, 2001).

Kalvelagen, E. (2002). *Solving systems of linear equations with GAMS*. (GAMS Technical Report, November 12, 2002).

Kalvelagen, E. (2003). *An elementary Ramsey growth model*. (GAMS Technical Report, March 12, 2003).

Karmarkar, N. (1984). A new polynomial time algorithm for linear programming. *Combinatorica, 4*, 373–395.

Karush, W. (1939) *Minima of functions of several variables with inequalities as side constraints*. M.Sc. Dissertation. Deptartment of Mathematics, University of Chicago, Chicago, IL.

Keller, C., Gould, N. I. M., & Wathen, A. J. (2000). Constraint preconditioning for indefinite linear systems. *SIAM Journal on Matrix Analysis and Applications, 21*, 1300–1317.

Kelley, C. T. (1999). *Iterative methods for optimization*. Philadelphia, PA, USA: SIAM, Frontiers in Applied Mathematics.

Kortanek, K. O., Potra, F. A., & Ye, Y. (1991). On some efficient interior point methods for nonlinear convex programming. *Linear Algebra and its Applications, 152*, 169–189.

Kristjansson, B. (1993). *MPL user manual*. Iceland, Europe: Maximal Software Inc.

Kuhn, H. W., & Tucker, A. W. (1951). *Nonlinear programming*. Proceedings of 2nd Berkeley Symposium. Berkeley: University of California Press, pp. 481–492.

Lalee, M., Nocedal, J., & Plantenga, T. (1998). On the implementation of an algorithm for large-scale equality constrained optimization. *SIAM Journal on Optimization, 8*, 682–706.

Lapidus, L., & Luus, R. (1967). The control of nonlinear systems: Part II: convergence by combined first and second variations. *A.I.Ch. E. Journal, 13*, 108–113.

Larrosa, J.A.E. (2008). *New heuristics for global optimization of complex bioprocesses*. Ph.D. Thesis, Universidade de Vigo, Departamento de Enxeñería Química, Vigo.

Lasdon, L. S., & Waren, A. D. (1978). Generalized reduced gradient software for linearly and nonlinearly constrained problems. In H. J. Greenberg (Ed.), *Design and implementation of optimization software* (pp. 335–362). Holland, Europe: Sijthoff and Noordhoff.

Lasdon, L. S., & Waren, A. D. (1980). A survey of nonlinear programming applications. *Operations Research, 28*, 34–50.

Lasdon, L. S., & Waren, A. D. (1982). *GRG2 user's guide*. (Department of General Business, School of Business Administration, University of Texas, Austin, May 1982).

Lasdon, L. S., Fox, R. L., & Ratner, M. W. (1974). Nonlinear optimization using the generalized reduced gradient method. *RAIRO, 3*, 73–104.

Lasdon, L. S., Waren, A. D., Jain, A., & Ratner, M. (1978). Design and testing of a generalized reduced gradient code for nonlinear programming. *ACM Transactions on Mathematical Software, 4*, 34–50.

Lay, S. R. (1982). *Convex sets and their applications*. Whitehouse Station, NJ: Wiley.

Lemaréchal, C., Nemirovskii, A., & Nesterov, Y. (1995). New variants of bundle methods. *Mathematical Programming, 69*(1005), 111–147.

Lemke, C. E. (1962). A method of solution for quadratic programs. *Management Science, 8*, 442–453.

Lenstra, J. K., Rinnooy Kan, A. H. G., & Schrijver, A. (1991). *History of Mathematical Programming. A collection of personal reminiscences*. Amsterdam, the Netherlands: CWI Amsterdam and North-Holland.

Lescrenier, M. (1991). Convergence of trust region algorithms for optimization with bounds when strict complementarity does not hold. *SIAM Journal on Numerical Analysis*, *28*, 476–495.

Levitin, E. S., & Polyak, B. T. (1966). Constrained minimization problems. *USSR Computational Mathematics and Mathematical Physics*, *6*, 1–50.

Leyffer, S., & Mahajan, A. (2010). *Foundations of constrained optimization.* (Preprint ANL/MCS-P1767–0610. Argonne National Laboratory, Mathematics and Computer Science Division, June 17, 2010).

Lin, Y., & Cryer, C. W. (1985). An alternating direction implicit algorithm for the solution of linear complmentarity problems arising from free boundary problems. *Applied Mathematics and Optimization*, *13*, 1–7.

Lin, C.-J., & Moré, J. J. (1999). Newton's method for large bound-constrained optimization problems. *SIAM Journal on Optimization*, *9*, 1100–1127.

Liu, D. C., & Nocedal, J. (1989). On the limited memory BFGS method for large scale optimization. *Mathematical Programming*, *45*, 503–528.

Luenberger, D. G. (1973). *Introduction to linear and nonlinear programming*. Reading, UK: Addison-Wesley Publishing Company.

Luenberger, D. G. (1984). *Introduction to linear and nonlinear programming* (2nd ed.). Reading, UK: Addison-Wesley Publishing Company.

Luenberger, D. G., & Ye, Y. (2008). *Linear and nonlinear programming, International series in operations research & management science* (vol. 116, 3rd ed.). New York, NY, USA: Springer.

Lustig, I. J., Marsten, R. E., & Shanno, D. F. (1990). The primal-dual interior point method on the Cray supercomputer. In T. F. Coleman & Y. Li (Eds.), *Large-scale numerical optimization* (pp. 70–80). Philadelphia, PA, USA: SIAM.

Lustig, I. J., Marsten, R. E., & Shanno, D. F. (1991). Computational experience with a primal-dual interior point method for linear programming. *Linear Algebra and its Applications*, *152*, 191–222.

Lustig, I. J., Marsten, R. E., & Shanno, D. F. (1992). On implementing Mehrotra's predictor-corrector interior-point method for linear programming. *SIAM Journal on Optimization*, *2*(3), 435–449.

Lustig, I. J., Marsten, R. E., & Shanno, D. F. (1994). Computational experience with a globally convergent primal-dual predictor-corrector algorithm for linear programming. *Mathematical Programming*, *66*, 123–135.

Luus, R. (1993). Piecewise linear continuous optimal control by iterative dynamic programming. *Industrial and Engineering Chemistry Research*, *32*, 859–865.

Manne, A. S. (1977). ETA-MACRO: A model of energy-economy interactions. In C. J. Hitch (Ed.), *Modeling energy-economy interactions: Five approaches*. Washington, DC, USA: Resources for the Future.

Maratos, N. (1978). *Exact penalty function algorithms for finite-dimensional and control optimization problems.* Ph. D. Thesis, University of London.

Markowitz, H. M. (1952). Portfolio selection. *Journal of Finance*, *8*, 77–91.

Markowitz, H. M. (1957). The elimination form of the inverse and its application to linear programming. *Management Science*, *3*, 255–269.

Markusen, J. R. (2005). *Introduction to GAMS for equilibrium problems.* (University of Colorado Boulder, July 4, 2005).

Mayne, D. Q., & Polak, E. (1982). A superlinearly convergent algorithm for constrained optimization problems. *Mathematical Programming Studies*, *16*, 45–61.

McCarl, B. A. (1998). *Agricultural impact analysis using GAMS. Introduction to GAMS.* [http://www.gams.com/mccarl/aggamint.pdf].

McCarl, B. A., Meeraus, A., Van der Eijk, P. Bussieck, M. Dirkse, S., Steacy, P., & Nelissen, F. (2011). *McCarl GAMS user guide.* [http://www.gams.com/dd/docs/bigdocs/gams2002/mccarlgamsuserguide.pdf].

McKinney, D. C., & Savitsky, A. G. (2003). *Basic optimization models for water and energy management.* June 1999 (revision 6, February 2003).

Megiddo, N. (1989). Chapter 8. Pathways to the optimal set in linear programming. In N. Megiddo (Ed.), *Progress in mathematical programming: Interior-point and related methods* (pp. 131–158). New York, NY, USA: Springer.

Meyer, C. D. (2000). *Matrix analysis and applied linear algebra.* Philadelphia, PA, USA: SIAM.

Morales, J. L. (2002). A numerical study of limited memory BFGS methods. *Applied of Mathematical Letters, 15*, 481–487.

Moré, J. J., & Toraldo, G. (1989). Algorithms for bound constrained quadratic programming problems. *Numerische Mathematik, 55*, 377–400.

Moré, J. J., & Toraldo, G. (1991). On the solution of large quadratic programming problems with bound constraints. *SIAM Journal on Optimization, 1*, 93–113.

Murtagh, B. A., & Saunders, M. A. (1978). Large-scale linearly constrained optimization. *Mathematical Programming, 14*, 41–72.

Murtagh, B. A., & Saunders, M. A. (1980). *MINOS/AUGMENTED user's manual.* (Technical Report SOL 80–14, Systems Optimization Laboratory, Department of Operations Research, Stanford University, Stanford, California, CA 94305).

Murtagh, B. A., & Saunders, M. A. (1982). A projected lagrangian algorithm and its implementation for sparse nonlinear constraints. *Mathematical Programming Study, 16*, 84–117.

Murtagh, B. A., & Saunders, M. A. (1987). *MINOS 5.0. user's guide.* (Report SOL 83–20, Department of Operations Research, Stanford University, 1983. [Revised as *MINOS 5.1 User's Guide*, Report SOL 83-20R]).

Murtagh, B. A., & Saunders, M. A. (1995). *MINOS 5.4 user's guide.* (Technical Report SOL 83-20R, Systems Optimization Laboratory, Department of Operations Research, Stanford University, Stanford, California, CA 94305, February 1995).

Murty, K. G., & Kabadi, S. N. (1987). Some NP-complete problems in quadratic and nonlinear programming. *Mathematical Programming, 19*, 200–212.

Nash, S. G. (1984a). Newton type minimization via the Lanczos method. *SIAM Journal on Numerical Analysis, 21*, 770–788.

Nash, S. G. (1984b). *User's guide for TN/TNBC: Fortran routines for nonlinear optimization.* (Report 397, Baltimore, MD: Mathematical Sciences Department, The John Hopkins University).

Nash, S. G. (1985). Preconditioning of truncated-Newton methods. *SIAM Journal on Scientific and Statistical Computing, 6*, 599–616.

Nash, S. G. (1998). SUMT (Revisited). *Operations Research, 46*, 763–775.

Nash, S. G. (2000). A survey of truncated-newton methods. *Journal of Computational and Applied Mathematics, 124*, 45–59.

Nash, S. G., & Nocedal, J. (1991). A numerical study of the limited memory BFGS method and the truncated-Newton method for large scale optimization. *SIAM Journal of Optimization, 1*, 358–372.

Nash, S. G., Polyak, R., & Sofer, A. (1994). A numerical comparison of barrier and modified-barrier methods for large-scale bound-constrained optimization. In W. W. Hager, D. W. Hearn, & P. M. Pardlos (Eds.), *Large scale optimization: State of the art* (pp. 319–338). Dordrecht/Boston/London: Kluwer Academic Publishers.

Nemirovskii, A., & Todd, M. (2008). Interior point methods for optimization. *Acta Numerica, 17*, 181–234.

Nesterov, Y., & Nemirovskii, A. (1994). *Interior-point polynomial algorithms in convex programming*. Philadelphia, PA, USA: SIAM.

Ni, Q., & Yuan, Y. (1997). A subspace limited memory quasi-Newton algorithm for large-scale nonlinear bound constrained optimization. *Mathematics of Computation, 66*, 1509–1520.

Nishida, N., Liu, Y. A., Lapidus, L., & Hiratsuka, S. (1976). An effective computational algorithm for suboptimal singular and/or bang-bang control. *AICHE Journal, 22*(3), 505–513.

Nocedal, J. (1980). Updating quasi-Newton matrices with limited storage. *Mathematics of Computation, 35*(151), 773–782.

Nocedal, J., & Wright, S. J. (2006). *Numerical optimization, Springer series in operations research* (2nd ed.). New York, NY, USA: Springer Science+Business Media.

Nocedal, J., Wächter, A., & Waltz, R. A. (2005). *Adaptive barrier strategies for nonlinear interior methods*. (Technical Report RC 23563, IBM Watson Research Center, Yorktown Heights, NY, USA).

O'Sullivan, M. J., & Saunders, M. A. (2002). *Sparse rank-revealing LU factorization via threshold complete pivoting and threshold rook pivoting*. (Presented at Householder Symposium XV on Numerical Linear Algebra, Peebles, Scotland).

Omojokun, E. (1989). *Trust region algorithms for optimization with nonlinear equality and inequality constraints*. Ph.D. Disertation, Department of Computer Science, University of Colorado.

Osterby, O., & Zlatev, Z. (1983). *Direct methods for sparse matrices*. New York, NY, USA: Springer.

Paige, C. C., & Saunders, M. A. (1975). Solution of sparse indefinite systems of linear equations. *SIAM Journal on Numerical Analysis, 12*, 617–629.

Panier, E. R., & Tits, A. L. (1991). Avoiding the Maratos effect by means of a nonmonotone line search. I: General constrained problems. *SIAM Journal on Numerical Analysis, 28*, 1183–1195.

Pant, M., Thangaraj, R., & Singh, V. P. (2009). Particle swarm optimization with crossover operator and its engineering applications. *IAENG International Journal of Computer Science, 36*, 112–121.

Pantoja, J. F. A., & Mayne, D. Q. (1991). Exact penalty function algorithm with simple updating of the penalty parameter. *Journal of Optimization Theory and Applications, 69*, 441–467.

Peressini, A. L., Sullivan, F. E., & Uhl, J. J., Jr. (1988). *The mathematics of nonlinear programming*. New York, NY, USA: Springer.

Peyret, R., & Taylor, T. D. (1985). *Computational methods for fluid flow*. New York, NY, USA: Springer.

Polyak, B. T. (1969). The conjugate gradient method in extremal problems. *USSR Computational Mathematics and Mathematical Physics, 9*, 94–112.

Polyak, R. (1992). Modified barrier functions (theory and methods). *Mathematical Programming, 54*, 177–222.

Powell, M. J. D. (1969). A method for nonlinear constraints in optimization problems. In R. Fletcher (Ed.), *Optimization* (pp. 283–297). New York, NY, USA: Academic.

Powell, M. J. D. (1977). *A fast algorithm for nonlinearly constrained optimization calculations*. (Technical Report 77/NA2, Department of Applied Mathematics and Theoretical Physics, University of Cambridge).

Powell, M. J. D. (1978a). A fast algorithm for nonlinearly constrained optimization calculations. In G. A. Watson (Ed.), *Numerical analysis.*, Dundee 1977, *Lecture notes in mathematics* (vol. 630, pp. 144–157). Berlin, Europe: Springer.

Powell, M. J. D. (1978b). The convergence of variable metric methods of nonlinearly constrained optimization calculations. In O. L. Mangasarian, R. R. Meyer, & S. M. Robinson (Eds.), *Nonlinear programming 3* (pp. 27–63). New York, NY, USA: Academic.

Powell, M. J. D. (1983). *ZQPCVX: A Fortran subroutine for convex quadratic programming*. (Technical Report, Department of Applied Mathematics and Theoretical Physics, Cambridge University).

Powell, M. J. D. (1986). Convergence properties of algorithms for nonlinear optimization. *SIAM Review, 28*, 487–500.

Powell, M. J. D. (2009). *The BOBYQA algorithm for bound constrained optimization without derivatives*. (Technical Report, Cambridge, England, UK: Department of Applied Mathematics and Theoretical Physics, August, 2009).

Price, W. L. (1983). Global optimization by controlled random search. *Journal of Optimization Theory and Applications, 55*, 333–348.

Raydan, M. (1993). On the Barzilai and Borwein choice of steplength for the gradient method. *IMA Journal of Numerical Analysis, 13*, 321–326.

Raydan, M. (1997). The Barzilai and Borwein gradient method for the large scale unconstrained minimization problem. *SIAM Journal on Optimization, 7*, 26–33.

Reid, J. K. (1971). On the method of conjugate gradients for the solution of large sparse systems of linear equations. In J. K. Reid (Ed.), *Large sparse sets of linear equations* (pp. 231–254). New York, NY, USA: Academic.

Reid, J. K. (1975). Sparse in core linear programming. In G. Watson (Ed.), *Numerical analysis Dundee, Lecture notes in mathematics* (vol. 506, pp. 176–189).

Reid, J. K. (1976). *Fortran subroutines for handling sparse linear programming bases.* (Report AERE R-8269, Harwell, Oxon, UK: Computer Science and Systems Division, A.E.R.E. January 1976).

Reid, J. K. (1982). A sparsity-exploiting variant of the Bartels-Golub decomposition for linear programming bases. *Mathematical Programming, 24*, 55–69.

Rijckaert, M. J. (1973). Engineering applications of geometric programming. In M. Avriel, M. J. Rijckaert, & M. Wilde (Eds.), *Optimization in design*. Englewood Cliffs, NJ, USA: Prentice-Hall.

Robichaud, V. (2010). *An introduction to GAMS*. http://www.pep-net.org/fileadmin/ medias/pdf/ GAMSManual.pdf. July 2010.

Robinson, S. M. (1972). A quadratically convergent algorithm for general nonlinear programming problems. *Mathematical Programming, 3*, 145–156.

Robinson, S. M. (1974). Perturbed Kuhn-Tucker points and rates of convergence for a class of nonlinear programming algorithms. *Mathematical Programming, 7*, 1–16.

Rosen, J. B. (1960). The gradient projection method for nonlinear programming. Part I, Linear constraints. *SIAM Journal of Applied Mathematics, 8*, 181–217.

Rosen, J. B., & Kreuser, J. (1972). A gradient projection algorithm for nonlinear constraints. In F. Lootsma (Ed.), *Numerical methods for nonlinear optimization* (pp. 297–300). London, UK: Academic.

Rosenthal, R.E. (2011). *GAMS – A user's guide*. (GAMS Development Corporation, Washington, DC. July 2011).

Saad, Y. (1996). *Iterative methods for sparse linear systems*. Boston, MA, USA: PWS Publishing Company.

Saad, Y. (2011). *Numerical methods for large eigenvalue problems* (2nd ed.). Philadelphia, PA, USA: SIAM.

Sainvitu, C., & Toint, Ph. L. (2006). *A filter-trust-region method for simple-bound constrained optimization*. (Technical Report, Belgium, Europe: Department of Mathematics, University of Namur).

Sargent, E., & Murtagh, B. A. (1973). Projection methods for nonlinear programming. *Mathematical Programming, 4*, 245–268.

Saunders, M. (2015a). *LUSOL – a basis factorization package*. (Notes 6. Stanford University, Management Science & Engineering. Spring 2015).

Saunders, M. (2015b) *Augmented Lagrangian methods*. (Notes 9. Stanford University, Management Science & Engineering. Spring 2015).

Saunders, M. (2015c). *NPSOL and SNOPT – SQP methods*. (Notes 11. Stanford University, Management Science & Engineering. Spring 2015).

Schendel, U. (1989). *Sparse matrices – Numerical aspects with applications for scientists and engineers*. Chichester, UK: Ellis Horwood Limited.

Schittkowski, K. (1981). The nonlinear programming method of Wilson, Han and Powell with an augmented Lagrangean type line search function, Part I: Convergence analysis. *Numerische Mathematik, 38*, 83–114.

Schittkowski, K. (1983). On the convergence of a sequential quadratic programming method with an augmented Lagrange line search function. *Mathematical Operationsforschung und Statistik (Series Optimization), 14*(2), 197–216.

Schittkowski, K. (1985). NLPQL: A Fortran subroutine for solving constrained nonlinear programming problems. *Annals of Operations Research*, *5*(2), 485–500.

Schittkowski, K. (1986). NLPQL: A Fortran subroutine solving constrained nonlinear programming problems. *Annals of Operations Research*, *5*, 485–500.

Schittkowski, K. (1987). *More test examples for nonlinear programming codes, Lecture notes in economics and mathematical systems* (vol. 282). Berlin, Europe: Springer.

Schittkowski, K. (2002). *NLPQLP: A Fortran implementation of a sequential quadratic programming algorithm. User's guide*. (Report, Department of Mathematics, University of Bayreuth).

Schittkowski, K. (2005). *QL: A Fortran code for convex quadratic programming. User's guide, Version 2.11*. (Technical report, Department of Mathematics, University of Bayreuth, July 2005).

Schittkowski, K. (2009). *NLPQO: A Fortran implementation of a sequential quadratic programming algorithm with distributed and non-monotone line search. User's guide, version 3.0*. (Technical Report, Department of Computer Science, University of Bayreuth).

Schittkowski, K. (2010). *A robust implementation of a sequential quadratic programming algorithm with successive error restoration*. (Technical Report, Department of Computer Science, University of Bayreuth, 2010).

Schlick, T., & Fogelson, A. (1992a). TNPACK – A truncated newton minimization package for large scale problems: I. Algorithm and usage. *ACM Transactions on Mathematical Software*, *18*, 46–70.

Schlick, T., & Fogelson, A. (1992b). TNPACK – A truncated newton minimization package for large scale problems: II. Implementation examples. *ACM Transactions on Mathematical Software*, *18*, 71–111.

Schrage, L. (1997). *Optimization modeling with LINDO*. Belmont, CA, USA: Duxbury Press.

Shanno, D. F. (2012). Who invented the interior-point method? In: Grötschel, M. (Ed.) *Optimization stories* (pp. 55–64). Documenta Mathematica, Journal der Deutschen Methematiker-Vereinigung, Extra volume, 21st International Symposium on Mathematical Programming, Berlin, August 19–24, 2012.

Shanno, D. F., & Simantiraki, E. M. (1997). Interior-point methods for linear and nonlinear programming. In I. S. Duff & G. A. Watson (Eds.), *The state of the art in numerical analysis* (pp. 339–362). New York, NY, USA: Oxford University Press.

Shanno, D. F., Breitfeld, M. G., & Simantiraki, E. M. (1996). Implementing barrier methods for nonlinear programming. In T. Terlaky (Ed.), *Interior point methods of mathematical programming* (pp. 399–414). Dordrecht, Europe: Kluwer Academic Publishers.

Spellucci, P. (1981). Han's method without solving QP. In A. Auslender, W. Oettli, & J. Stoer (Eds.), *Optimization and optimal control, Lecture notes in control and information sciences* (vol. 30, pp. 123–141). Berlin, Europe: Springer.

Spellucci, P. (1985). Sequential quadratic programming: Theory, implementation, problems. In M. J. Beckmann, K. W. Gaede, K. Ritter, & H. Schneeweiss (Eds.), *Methods of operations research* (vol. 53, pp. 183–213).

Spellucci, P. (1993). *A new technique for inconsistent QP-problems in the SQP-method*. (Preprint 1561, 64829 Darmstadt, Germany: Technical University at Darmstadt, Department of Mathematics).

Spellucci, P. (1994). *A simplified SQP method with equality constrained subproblems*. (64829 Darmstadt, Germany: Technical University at Darmstadt, Department of Mathematics).

Spellucci, P. (1995). *DONLP2: Program and user's guide*. (Available from netlib/opt.)

Spellucci, P. (1998). An SQP method for general nonlinear programs using only equality constrained subproblems. *Mathematical Programming*, *3*, 413–448.

Strogatz, S. H. (1994). *Nonlinear dynamics and chaos: With applications to physics, biology, chemistry, and engineering*. Cambridge, MA, USA: Perseus Books.

Suhl, U. H., & Suhl, L. M. (1990). Computing sparse LU factorizations for large-scale linear programming bases. *ORSA Journal on Computing*, *2*, 325–335.

Suhl, L. M., & Suhl, U. H. (1991). *A fast LU-update for linear programming*. (Arbeitspapier des Instituts fur Wirtschaftsinformatik, Freie Universitaet – Berlin, August 1991).

Sun, W., & Yuan, Y. X. (2006). *Optimization theory and methods. Nonlinear programming*. New York, NY, USA: Springer Science + Business Media.

Surry, P. D., Radcliffe, N. J., & Boyd, I. D. (1995). A multi-objective approach to constrained optimization of gas supply networks: The COMOGA method. In T. C. Fogarty (Ed.), *Evolutionary computing: AISB Workshop, number 993, Lecture notes in computer science* (pp. 166–180). Berlin, Europe: Springer.

Tapia, R. A. (1974). A stable approach to Newton's method for general mathematical programming problems in *C. Journal of Optimization Theory and Applications, 14*, 453–476.

Tits, A. L., Wächter, A., Bakhtiari, S., Urban, T. J., & Lawrence, C. T. (2003). A primal-dual interior-point method for nonlinear programming with strong global and local convergence properties. *SIAM Journal on Optimization, 14*, 173–199.

Toint, P. L. (1996). An assessment of nonmonotone line search techniques for unconstrained optimization. *SIAM Journal on Scientific Computing, 17*, 725–739.

Toint, P. L. (1997). A nonmonotone trust-region algorithm for nonlinear optimization subject to convex constraints. *Mathematical Programming, 77*, 69–94.

Trefethen, L. N., & Bau, D., III. (1997). *Numerical linear algebra*. Philadelphia, PA, USA: SIAM.

Tröltzsch, A. (2007). *Benchmarking of bound-constrained optimization software*. (pp. 1–39) (CERFACS working note: WN/PA/07/143).

Ulbrich, S. (2004). On the superlinear local convergence of a filter-SQP method. *Mathematical Programming, 100*(1), 217–245.

Ulbrich, M., Ulbrich, S., & Heinkenschlos, M. (1999). Global convergence of affine-scaling interior-point Newton methods for infinite-dimensional nonlinear problems with pointwise bounds. *SIAM Journal on Control and Optimization, 37*, 731–764.

Ulbrich, M., Ulbrich, S., & Vicente, L. N. (2004). A globally convergent primal-dual interior-point filter method for nonconvex nonlinear programming. *Mathematical Programming, 100*, 379–410.

Van der Pol, B. (1927). Forced oscillations in a circuit with nonlinear resistance (receptance with reactive triode). *London, Edinburg, and Dublin Philosophical Magazine, 3*, 65–80.

Vanderbei, R. J. (1990). *ALPO: Another linear program optimizer*. (Technical Report AT&T Bell Laboratories).

Vanderbei, R. J. (1994). *An interior point code for quadratic programming*. (Technical Report SOR 94–15, Princeton University).

Vanderbei, R. J. (1995). *LOQO: An interior point code for quadratic programming*. (Technical Report SOR 94–15, Princeton University).

Vanderbei, R. J. (2001). *Linear programming: Foundations and extensions* (2nd ed.). New York, NY, USA: Springer.

Vanderbei, R. J., & Shanno, D. F. (1997). *An interior-point algorithm for nonconvex nonlinear programming*. (Technical Report SOR 97–21, Princeton University).

Vanderbei, R. J., & Shanno, D. F. (1999). An interior point algorithm for nonconvex nonlinear programming. *Computational Optimization and Applications, 13*, 231–252.

Vanderplaats, G. N. (1995). *DOT users manual. Version 4.20*. Colorado Springs, USA: Vanderplaats Research & Development, Inc.

Vardi, A. (1985). A trust region algorithm for equality constrained minimization: convergence properties and implementation. *SIAM Journal on Numerical Analysis, 22*, 575–591.

Vavasis, S. A. (1990). Quadratic programming is NP. *Information Processing Letters, 36*, 73–77.

Von Stryk, O. (1999). *User's guide for DIRCOL (Version 2.1): A direct collocation method for the numerical solution of optimal control problems*. (Technical Report, Technische Universität München, 1999).

Wächter, A. (2002). *An interior point algorithm for large scale nonlinear optimization with applications in process engineering*. [Ph.D. Thesis]. Pittsburg, PA: Carnegie Mellon University, January 2002.

Wächter, A., & Biegler, L. T. (2000). Failure of global convergence for a class of interior point methods for nonlinear programming. *Mathematical Programming, 88*, 565–574.

Wächter, A., & Biegler, L. T., (2001). *Line search filter methods for nonlinear programming: Motivation and global convergence*. (Technical Report RC 23036, Yorktown Heights, NY: IBM T.J. Watson Research Center, 2002; revised 2004).

Wächter, A., & Biegler, L. T. (2005a). Line search filter methods for nonlinear programming: Motivation and global convergence. *SIAM Journal on Optimization, 16*, 1–31.

Wächter, A., & Biegler, L. T. (2005b). Line search filter methods for nonlinear programming: Local convergence. *SIAM Journal on Optimization, 16*, 32–48.

Wächter, A., & Biegler, L. T. (2006). On the implementation of an interior-point filter line-search algorithm for large-scale nonlinear programming. *Mathematical Programming, 106*, 25–57.

Waltz, R. A. (2004). *KNITRO 4.0 user's manual*. (Evanston, IL: Technical Report, Ziena Optimization Inc., October 2004).

Waltz, R. A., Morales, J. L., Nocedal, J., & Orban, D. (2003). *An interior algorithm for nonlinear optimization that combines line search and trust region step*. Vol 107. (Technical Report 2003-6, Optimization Technology Center, Northwestern University, Evanston, IL, USA, June 2003. [Mathematical Programming, 2006:391–408.]).

Wilson, R. B. (1963). *A simplicial algorithm for concave programming*. Ph.D Thesis, Harvard University.

Wolfe, P. (1967). Methods of nonlinear programming. In J. Abadie (Ed.), *Nonlinear programming* (pp. 97–131). Amsterdam, Europe: North-Holland.

Wolfe, P. (1969). Convergence conditions for ascent methods. *SIAM Review, 11*, 226–235.

Wolfe, P. (1971). Convergence conditions for ascent methiods. II: Some corrections. *SIAM Review, 13*, 185–188.

Wright, M. H. (1976). *Numerical methods for nonlinearly constrained optimization*. (SLAC Report No.193, 1976, Stanford University, California, 1976. (Ph.D. Dissertation)).

Wright, S. J. (1990). Implementing proximal point methods for linear programming. *Journal of Optimization Theory and Applications, 65*, 531–554.

Wright, M. H. (1991). Interior methods for constrained optimization. *Acta Numerica, 1*, 341–407.

Wright, S. J. (1997). *Primal-dual interior-point methods*. Philadelphia, PA, USA: SIAM.

Yamashita, H. (1998). A globally convergent primal-dual interior-point method for constrained optimization. *Optimization Methods and Software, 10*, 443–469.

Yang, E. K., & Tolle, J. W. (1991). A class of methods for solving large convex quadratic programs subject to box constraints. *Mathematical Programming, 51*, 223–228.

Ye, Y. (1997). *Interior point algorithms: Theory and analysis*. New York, NY, USA: Wiley.

Zhang, Y. (2004). *Interior-point gradient methods with diagonal-scaling for simple-bound constrained optimization*. (Houston, TX: Technical Report TR04–06, Department of Computational and Applied Mathematics, Rice University).

Author Index

N. Andrei, *Continuous Nonlinear Optimization for Engineering Applications in GAMS Technology*, Springer Optimization and Its Applications 121,
DOI 10.1007/978-3-319-58356-3

Subject Index

N. Andrei, *Continuous Nonlinear Optimization for Engineering Applications in GAMS Technology*, Springer Optimization and Its Applications 121,
DOI 10.1007/978-3-319-58356-3

Printed by Printforce, the Netherlands